Developmental NeuroPsychobiology

BEHAVIORAL BIOLOGY

AN INTERNATIONAL SERIES

A list of books in this series is available from the publisher on request.

Developmental NeuroPsychobiology

Edited by

William T. Greenough
Department of Psychology, Department of Anatomical Sciences,
and Neural and Behavioral Biology Program
University of Illinois
Champaign, Illinois

Janice M. Juraska
Department of Psychology
Indiana University
Bloomington, Indiana

1986

ACADEMIC PRESS, INC.
Harcourt Brace Jovanovich, Publishers
Orlando San Diego New York Austin
London Montreal Sydney Tokyo Toronto

ACADEMIC PRESS, INC.
Orlando, Florida 32887

United Kingdom Edition published by
ACADEMIC PRESS INC. (LONDON) LTD.
24–28 Oval Road, London NW1 7DX

LIBRARY OF CONGRESS CATALOGING IN PUBLICATION DATA

Main entry under title:

Developmental neuropsychobiology.

Includes index.
1. Developmental neurology. 2. Developmental psychobiology. I. Greenough, William T. II. Juraska, Janice M. [DNLM: 1. Behavior—physiology. 2. Neurophysiology. 3. Psychophysiology. WL 103 D489]
QP363.5.D48 1985 599′.0188 85-6121
ISBN 0–12–300270–2 (hardcover) (alk. paper)
ISBN 0–12–300271–0 (paperback) (alk. paper)

PRINTED IN THE UNITED STATES OF AMERICA

86 87 88 89 9 8 7 6 5 4 3 2 1

In memory of Ryo Arai, editor for Academic Press, who instigated and nurtured this volume and affected our lives in other ways as well.

Contents

4 The Normal and Abnormal Development of the Mammalian Visual System

Raymond D. Lund and Fen-Lei F. Chang

5 Do Neurotransmitters, Neurohumors, and Hormones Specify Critical Periods?

Jean M. Lauder and Helmut Krebs

6 Sexual Differentiation of the Brain

C. Dominique Toran-Allerand

7 Behavioral Neuroembryology: Motor Perspectives

Robert R. Provine

8 Ontogeny of the Encephalization Process

Dennis J. Stelzner

9 Neuronal Activity as a Shaping Factor in Postnatal Development of Visual Cortex

Wolf Singer

10 Experience and Development in the Visual System: Anatomical Studies

Ronald G. Boothe, Elsi Vassdal, and Marilyn Schneck

11 Experience and Visual Development: Behavioral Evidence

Richard C. Tees

12 Neural Correlates of Development and Plasticity in the Auditory, Somatosensory, and Olfactory Systems

Ben M. Clopton

13 What's Special about Development? Thoughts on the Bases of Experience-Sensitive Synaptic Plasticity

William T. Greenough

14 Sex Differences in Developmental Plasticity of Behavior and the Brain

Janice M. Juraska

15 The Development of Olfactory Control over Behavior

Elliott M. Blass

16 New Views of Parent–Offspring Relationships

Jeffrey R. Alberts

Contributors

Numbers in parentheses indicate the pages on which the authors' contributions begin.

JEFFREY R. ALBERTS (449), Psychology Department, Indiana University, Bloomington, Indiana 47405

ELLIOTT M. BLASS (423), Department of Psychology, The Johns Hopkins University, Baltimore, Maryland 21218

RONALD G. BOOTHE[1] (295), Department of Psychology, and Department of Ophthalmology, University of Washington, Seattle, Washington 98195

FEN-LEI F. CHANG (95), Department of Anatomy and Cell Biology, Center for Neuroscience, University of Pittsburgh School of Medicine, Pittsburgh, Pennsylvania 15261

BEN M. CLOPTON (363), Kresge Hearing Research Institute, The University of Michigan School of Medicine, Ann Arbor, Michigan 48109

JOHN S. EDWARDS[2] (73), Max-Planck-Institut für Verhaltenphysiologie, Abteilung Huben, D8131 Seewiesen, Federal Republic of Germany

WILLIAM T. GREENOUGH (387), Department of Psychology, Department of Anatomical Sciences, and Neural and Behavioral Biology Program, University of Illinois, Champaign, Illinois 61820

LAWRENCE S. HONIG[3] (1), Laboratory for Developmental Biology, University of Southern California, Los Angeles, California 90089

[1]Present address: Yerkes Primate Center, Department of Psychology, and Department of Ophthalmology, Emory University, Atlanta, Georgia 30322.

[2]Present address: Department of Zoology, University of Washington, Seattle, Washington 98195.

[3]Present address: University of Miami, School of Medicine, Miami, Florida 33101.

JANICE M. JURASKA[4] (409), Department of Psychology, Indiana University, Bloomington, Indiana 47405

HELMUT KREBS (119), Laboratory of Developmental Neurobiology, University of North Carolina School of Medicine, Chapel Hill, North Carolina 27514

JEAN M. LAUDER (119), Laboratory of Developmental Neurobiology, University of North Carolina School of Medicine, Chapel Hill, North Carolina 27514

PAUL C. LETOURNEAU (33), Department of Anatomy, University of Minnesota, Minneapolis, Minnesota 55455

RAYMOND D. LUND (95), Department of Anatomy and Cell Biology, Center for Neuroscience, University of Pittsburgh School of Medicine, Pittsburgh, Pennsylvania 15261

ROBERT R. PROVINE (213), Department of Psychology, University of Maryland Baltimore County, Catonsville, Maryland 21228

MARILYN SCHNECK (295), Department of Psychology, and Department of Ophthalmology, University of Washington, Seattle, Washington 98195

WOLF SINGER (271), Max-Planck-Institut für Hirnforschung, 6000 Frankfurt am Main 71, Federal Republic of Germany

DENNIS J. STELZNER (241), Department of Anatomy, State University of New York Upstate Medical Center, Syracuse, New York 13210

DENNIS SUMMERBELL (1), The National Institute for Medical Research, The Ridgeway, Mill Hill, London NW7 1AA, England

RICHARD C. TEES (317), Department of Psychology, The University of British Columbia, Vancouver, British Columbia, Canada V6T 1W5

C. DOMINIQUE TORAN-ALLERAND (175), Center for Reproductive Sciences, and Department of Neurology, and Department of Anatomy and Cell Biology, Columbia University College of Physicians and Surgeons, New York, New York 10032

ELSI VASSDAL (295), Department of Psychology, and Department of Ophthalmology, University of Washington, Seattle, Washington 98195

[4]Present address: Department of Psychology, University of Illinois, Champaign, Illinois 61820.

Preface

A motivating force behind the organization of this book was our perception of the mutual isolation of the fields of developmental neuroscience and developmental psychobiology. Each has its separate societies, journals, and international meetings, and there is remarkably little overlap in membership and even less in attendance at meetings. The editors of this book have often found it difficult to classify themselves as either developmental psychobiologists or developmental neuroscientists (although we feel that those other members who identify themselves with one group tend to class us in the other group). Similarly, perhaps because we borrow paradigms and techniques from both, we have had difficulty in perceiving a natural break between the two disciplines.

The two developmental disciplines fall on a continuum, stretching from the molecular through the cellular and systems level to the behavioral level of analysis, and investigations at one level often have important implications for others. For example, issues of genetic constraints upon or determination of species-typical patterns have potential answers at all these levels, as do questions about the roles of hormones, neurotransmitters, and other neuromodulaters, and even about the roles of experience in behavioral development.

In deciding upon the chapters for this book, we attempted to pick representative examples of particularly fruitful approaches across the range of studies of neurobehavioral development. The selections emphasize fundamental issues rather than paradigms. The ordering imposed upon these selections may appear arbitrary, but it reflects one possible view of the developmental neuro/psychobiology continuum. The list is not meant to be comprehensive. Indeed, it is difficult to conceive of one's being comprehensive in an area with as much good research as this one has.

Although we do not feel that a formal organization of the book into subsections would be appropriate, in the sense that a definition of categorical boundaries would break up the continuity with which this field should be viewed, the topics we have chosen fall into logical groupings in terms of the research problems they address. The initial chapters deal largely with intrinsic mechanisms, specifying pattern formation in the developing nervous system. Summerbell and Honig begin by describing the coordinate systems that appear to govern peripheral patterns in limb development and the matching of spinal nerve ordering to the peripheral map. Letourneau discusses mechanisms of expression of pattern information at the cellular level, particularly as discerned from neurite–substrate interactions *in vitro.* Edwards then describes the generation of pattern *in vivo* in insects, where individual fibers can be reidentified across the developmental sequence, and their roles in pioneering pathways, as well as their responses to experimental and genetic manipulations, can be specifically assessed. In the final chapter on this topic, Lund and Chang consider the mechanisms whereby patterns are established in the vertebrate visual system, noting especially the nearly ubiquitous phenomenon of exuberant overprojection, followed by competitive restriction of terminal fields, as well as the evidence for the association of neuronal death with terminal loss in more peripheral projections.

The chapters of Provine and Stelzner also deal with pattern, but at a higher order, focusing upon the implications of the establishment of systems and how the sequences whereby these systems become established are manifested in the development of behavior. Working from a broad comparative perspective, Provine describes embryonic and postnatal behavioral manifestations of somatic neural maturation and their neural substrates. Stelzner then deals with encephalization, the process whereby supraspinal centers exert control, in a partly nonreversible manner, over spinal circuitry. In this context, Stelzner discusses a phenomenon often found in the developing mammalian nervous system, that of greater sparing or recovery of behavioral function when CNS damage is sustained earlier in life.

The chapter by Lauder and Krebs and that by Toran-Allerand discuss intrinsic and extrinsic modulatory influences, both upon the timing and sequencing of the genesis of neurons and synapses and upon the functional organization that results from these influences. Noting the mounting evidence for a morphogenetic role for neurotransmitters in embryonic development, Lauder and Krebs speculate upon the orchestrating roles of neurotransmitters, neurohumors, and hormones as differentiation and recognition signals from the time of neural tube development through postnatal ontogeny and suggest a broadened use of the term "critical period." The best understood example of hormonal interactions with the developing mammalian brain, that of the role of gonadal steroid hormones in sexual differentiation, may serve as a model for the less behaviorally specific effects of other hormones. Toran-Allerand reviews work indicating structural and functional sexual dimorphisms in the brain and describes *in vitro* approaches to determination of the mechanisms whereby steroid hormones alter brain organization. Further implications of possible hormonal influences at a more molar anatomical and behavioral level are discussed in the chapter by Juraska, who reviews sex differences in the developing brain's responses to environmental conditions, in the context of human data that suggest greater male vulnerability to insult.

The chapter by Singer could easily be grouped with the previous topic, since among its several foci are the forces controlling the periods of greatest sensitivity to experience. However, since it also concentrates upon the incorporation of information gained from experience—the organism's interactions with its environment—it seems better grouped with others discussing the roles of experience in the development of mammalian sensory systems. Singer reviews the visual and nonvisual experiential controls of the maturation of mammalian visual system physiology, as well as its central modulation, and proposes rules and mechanisms for Hebbian synapses that could underlie the effects of experience. Boothe, Vassdal, and Schneck consider the structural development of the visual system, with an emphasis upon synaptic changes underlying the physiological and behavioral effects of experience manipulations. Tees casts a critical eye upon those who argue that behavioral assessments of visual deprivation effects are too imprecise for comparison with physiological and anatomical measures, arguing that although much work *has* been deficient, such work need not be so. Clopton considers the degree to which observations on the visual system are paralleled by the much less frequent studies of other sensory systems. Greenough presents evidence that many of the consequences of early sensory manipulations upon brain structure are also seen after later manipulations of the complexity and learning opportunities the environment provides, but argues for a fundamental difference in the manner in which the changes originate in sensory development versus later life.

Finally, the chapters of Alberts and Blass represent a separate category in that they focus on the organism as part of a system within which development takes place. Blass describes the role of olfactory information from the pre- and postnatal environment in maintaining social orientation and organization during development. Alberts evaluates previous models of the forces driving parent–offspring relations and suggests a new model in which both infant–infant interactions and offspring–parent interactions result in mutually beneficial outcomes.

The papers are intended to be comfortably read by advanced followers of the biological and/or psychological sciences, whether or not they are familiar with work in the specific areas discussed. The chapters vary in the degree to which technical details are discussed, largely because the technical details are more critical to an understanding of theories and interpretations of experimental tests in some areas than in others. Where the degree of detailed knowledge necessary to understand an argument or explanation exceeds the knowledge expected of, say, a graduate student in biology or experimental-biological psychology, we have tried to ensure that the necessary information is included within the chapter (unless it is available within others to which the reader is referred). The book's greatest potential will be realized if readers not really familiar with any of the areas covered will gain a perspective on the coherence that actually exists across the broad range of approaches to brain–behavior development, an appreciation of the progress that has been made, and a feel for the excitement with which workers within the area view the prospects for major advances in the immediate future.

Finally, we thank the staff at Academic Press for consistent encouragement over the far-too-long time it took to bring this book together, and we thank for their patience the authors who got their chapters in on schedule.

WILLIAM T. GREENOUGH
JANICE M. JURASKA

Developmental
NeuroPsychobiology

1

Embryonic Mechanisms

DENNIS SUMMERBELL
The National Institute for Medical Research
The Ridgeway, Mill Hill
London, England

LAWRENCE S. HONIG[1]
Laboratory for Developmental Biology
University of Southern California
Los Angeles, California

I. Pattern Formation

During embryogenesis, a single fertilized egg cell develops into a complex organism consisting of many (~200) cell types arranged in a reproducible pattern. The development of spatial organization of the differentiated cells

[1] Present address: University of Miami, School of Medicine, Miami, Florida 33101.

is called pattern formation. Many different processes occur, including cell movements, changes in cell division, appearances of overt differences between cells and restriction of developmental potentialities (see Fig. 1). We will not be concerned with the molecular mechanism by which a cell adopts a particular terminal state of differentiation, but rather with how each cell "knows" its appropriate behavior within the entire organism. Nor will we restrict our discussion to the nervous system, for as it happens it has not been a particularly popular subject for the study of pattern formation. The research that has been done does, however, suggest that many of the principles and rules of development discovered in other systems will be equally applicable to the central nervous system (CNS).

The operational unit of pattern formation is the embryonic field (Huxley and de Beer, 1934), which is a region in which cells of identical or nearly so characteristics, autonomously develop into a pattern of different cell types. Developmentally, an early example of an embryonic field is the amphibian blastula. From a simple undifferentiated ball of cells it is transformed into an embryo in which there are three distinct layers: ectoderm, mesoderm, and endoderm, and in which the primary body plan of the embryo is evident. The embryo is marked off into a head and 30-plus segments anteroposteriorly, and into four main tissue types mediolaterally (Cooke, 1981). This original pattern is initially marked out in the mesoderm, forming notochord, somites, nephrogenic tissue, and lateral plate, with the somites becoming segmented. Meanwhile, the ectodermal tissue under the control of the underlying mesoderm, becomes *determined* (irreversibly committed to a particular developmental pathway) to form specific parts of the nervous system (Gordon and Jacobson, 1978). The primordial regions of the brain (i.e., prosencephalon, mesencephalon, metencephalon, myencephalon), the spinal cord, and the adjacent neural crest (which will form parts of the peripheral nervous system) are all present in a recognizable form from this very early stage. The central nervous system is therefore roughly mapped out as part of the primary body plan. The spinal nerves appear to form at gaps between the somites so that there is a one-to-one correspondence. The general properties of embryonic fields are best illustrated in a classical experiment concerning the reorganization of gastrulation. When a particular small group of cells is grafted from the dorsal lip of the blastopore of a donor blastula to a different position in a host embryo, then the host forms a second primary body plan and nervous system, giving rise to mirror-image twinned parallel body axes. The grafted cells do not themselves give rise to much of this second axis, rather they reorganize host cells to form two neural planes. Such regions are therefore called "organizers."

This first embryonic field also maps out the axial positions of subsequent secondary fields. It creates the necessary boundary conditions for these fields

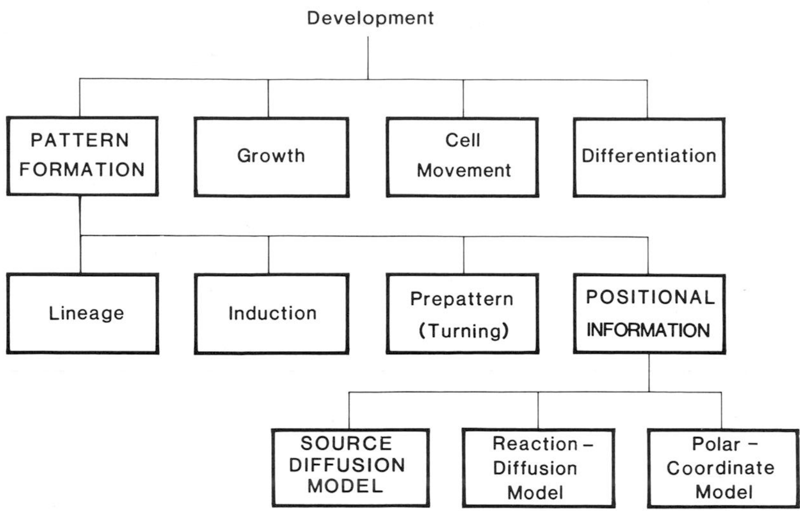

Fig. 1. A representation of the levels of relationship between different observations and ideas in developmental biology. Many more categories are known than are shown. More space has been devoted to the topics shown in capitals.

so that the embryo develops harmoniously. Much recent research has concentrated on these secondary fields such as (1) the eye rudiment, which develops into a laminar pattern of pigmented retinal epithelium, ganglion cells which must connect to the brain in a spatially ordered fashion, and other neural layers of retina, lens, and cornea (Jacobson, 1978); (2) the mandibular branchial arch, which becomes the lower jaw with specific pattern of cartilage, muscle, nerves, membranous bone, and teeth (Hall, 1978); and (3) the limb rudiment, which develops into a proximodistally segmented array of cartilage-model skeletal elements, muscles, tendons, blood vessels, and nerves (Hinchliffe and Johnson, 1980).

The further development of the nervous system is a special case. It usually involves the matching of two fields. The retina's neural ganglion cells send out axons that grow back to the brain and make connections on the optic tectum (the visual center in Amphibia) that form a topographic map of the surface of the retina and hence of the visual field (Gaze, 1978). Similarly, the spinal cord sends out axons to the limbs so that again there are topographic maps between the motor pools in the ventral horn and the appropriate muscles in the limb (Landmesser, 1980; Summerbell and Stirling, 1982). In another example, the sensory vibrissae on the muzzle of the mouse project accurately to the barrels in the appropriate part of the cortex despite the occurrence of two intervening synapses, suggesting the serial matching of several successive secondary fields (Van Der Loos, 1979). This successive matching is also found in the visual system of higher vertebrates. In other parts of the nervous system, topographic matching is not so obvious, but evidence still suggests that most nerve–nerve and nerve–target (sensory or motor) synapses make important use of information as to position in the primary formation of connections. In refining these connections, experience and electrical activity have a subsequent role (Lund, 1978), as the following chapters make clear.

A. Some Proposed Mechanisms

Development involves a number of cell phenomena (Fig. 1). In this chapter we concentrate on pattern formation although associated behavior such as cell growth (e.g., the extension of axons), cell proliferation, cell migration (e.g., the movement of neural crest cells), and overt differentiation (e.g., neuronal commitment to a neurotransmitter) are described in succeeding chapters. We will now introduce several mechanisms of pattern formation and the concept of positional information.

1. Lineage Restriction

In invertebrates there is a very limited ability for one cell to take over the function of another. Cell division is asymmetrical so that the two daughter cells differ in a defined and predictable way. The cells occupy or move to appropriate determined positions and divide again to produce unequal granddaughters. The process continues until the entire cell complement of the adult is present. Each cell is the product of its history of division and can be precisely defined by its family tree. The mechanism produces organisms of extreme invariability in which the numbers, origins, positions, and functions of each cell can be plotted accurately. Possible molecular mechanisms include successive positioning of cytoplasmic determinants, or quantal cell cycles. In certain simple organisms such as leeches (Muller *et al.*, 1981), or the nematode worm (Kimble *et al.*, 1979), the entire lineage for the nervous system has now been described and the work continues by mapping the connections of every cell. As a rule, each neuron is strictly determined by its lineage but limited regulation is possible; for example, following destruction of a neuroblast, the corresponding cell of the other side can in some cases perform an extra cell division to replace the missing cell.

Lineage is now also being studied in vertebrates, particularly in the amphibian nervous system (Jacobson, 1982). Because of the ability of primary and secondary developmental fields to regulate, it seems unlikely that this mechanism can have a very important role (Cooke, 1980). However, continuing improvements in techniques for tracing the progeny of embryonic cells should advance our knowledge.

2. *Induction*

Induction is often used in a very general sense to describe all types of developmental interactions between cells. We have already described its original use: the formation of neural structures (brain and spinal cord) from the ectoderm overlying the chordamesoderm. However, induction also has a more specific connotation that includes mechanisms in which information is transferred from one tissue to another tissue, usually of different cell type. Inductive interactions are divided into two categories, directive and permissive. Directive interactions cause the induced tissue to behave in one of several appropriate ways. The information exchanged must vary in different situations so there must be the facility to encode complicated instructions through intercellular signaling. Intermediary molecules such as messenger RNA have often been suggested, but no specific examples have been confirmed. Neural induction is apparently an example of directive (instructional) interaction where the type of epidermal or neuroepithelial (e.g., brain or spinal cord) structure formed seemingly depends on the character of the underlying chordamesoderm. Permissive interactions allow the induced tissue to express the phenotype to which it is already predisposed. Examples include the epithelial–mesenchymal interactions and formation of various glands (e.g., salivary, mammary, pancreas, lung), teeth, and integumentary structures. The subject has been extensively reviewed by Saxen (1979) and in a recent symposium (Sawer and Fallon, 1983).

3. *Prepatterns*

Prepattern models propose that the generation of each spatial pattern of a particular differentiation cell type is a reflection of an underlying distribution of some quality or morphogen that varies in value or concentration in a manner like that of the desired pattern. To obtain identical differentiation at two positions in one field, one must reproduce at each position an identical distribution of morphogen. Most commonly suggested mechanisms are based on reaction–diffusion equations as described by Turing (Meinhardt, 1982). For example, cells may each produce two substances: an activator (A) that diffuses very slowly from cell to cell across the field; and an inhibitor (I) that diffuses comparatively rapidly. The activator stimulates the production of both activator (autocatalysis) and inhibitor whereas the inhibitor reduces production of activator. A field may start with relatively homogeneous, low concentrations of activator and inhibitor. This is unstable; chance local fluctuations elevate activator concentration above a threshold (Fig. 2A) whereupon the positive feedback action of the activator amplifies its own production and the production of the inhibitor (Fig. 2B). Because the inhibitor is free to diffuse away from the site of production, it inhibits any rise in activator in the surrounding area, but the initial site of activation

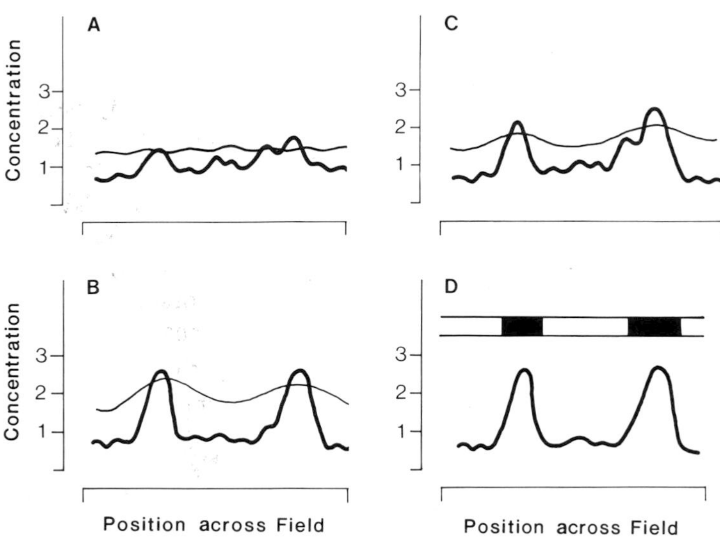

Fig. 2. Pattern formation using a prepattern reaction–diffusion mechanism. (A) Initial state, random local fluctuations raise activator above threshold to escape from inhibition. (B) Concentration of activator rises locally by autocatalysis, concentration of inhibitor also rises but diffuses away from site of production, suppressing formation of nearby secondary activator peaks. (C) Concentration reaches equilibrium state with areas of stable high and low concentrations of activator and inhibitor. (D) Above a particular threshold an appropriate set of genes is switched on, in this case to produce alternating black and white stripes across the field. If the activator > 1, then the black stripe appears.

escapes from inhibition. The situation reaches an equilibrium with areas of stable high concentration and areas of stable low concentration (Fig. 2C). The high concentration of activator switches on the appropriate set of genes (Fig. 2D). In some models there are several interacting substances to produce the prepattern. Complicated patterns need increasingly complicated patterns of interaction. The model is at its best when explaining simple two-color patterns on two-dimensional sheets: zebra stripes, leopard spots, and so on (Murray, 1981).

B. Positional Information

The theory of positional information proposed by Wolpert (1969) supposes that pattern formation involves two steps. First, each cell's position in a field must be specified with respect to reference points associated with the field. These reference points are likely to be at field boundaries and may well be "organizers"; as unique points they may have an active role in setting up the information that is recorded in the cell as positional value. This specification process involves the creation of a quantitatively varying characteristic or morphogen across the field. Superficially, this resembles the prepattern hypothesis. The distinction is that the form of the concentration profile need have no resemblance to the final pattern (Fig. 3). Indeed, it may be represented as a monotonic gradient with each cell (even those that will eventually differentiate identically) having a unique morphogen level: its positional value.

In the second phase, there is "interpretation" of the positional value by the cell. The cell must use this value, and its genome, to enact expression of the appropriate differentiated state. The relevant gene products must be synthesized. Interpretation is a difficult problem tied to control of gene expression but is a problem on which almost no experimental work has been done. Many ideas on gene-switching networks have been discussed (Meinhardt, 1982).

The important distinction between positional information and other mechanistic frameworks is that cells that will differentiate into the same histological type (e.g., muscle cells of the shoulder and of the hand) are nevertheless different. The presence of positional codings has been described in different ways: cells are nonequivalent, possess a hidden label or secret name, or show covert differentiation. The presence of cryptic labeling can be demonstrated in some cases by cell-sorting experiments and by experiments in regenerating systems. Nonequivalence cannot be accounted for with prepattern models, since all cells that are to develop in the same way are provided with identical information and are developmentally equivalent.

C. Models for Positional Specification

Specification of positional values has been explained using several different classes of mechanism: source–diffusion, reaction–diffusion, and local intercalation, for example. These models must account for the generation of positional values over an embryonic field consisting of a homogeneous cell population. They must also account for the regulatory ability of such a field to respond to accidents or experimental manipulations. It is often erroneously supposed that the models are dependent on particular molecular mechanisms. In fact, there is no requirement, and little evidence, favoring any one mechanism at a molecular level. However it is sensible to predicate simulations on parameters that are biologically plausible, the underlying equations should accurately model the processes, wherever and whatever the molecules involved.

1. Source–Diffusion

Source-diffusion models suppose that at one boundary there is a production source of morphogen, a hypothetical substance that is the positional signal (Wolpert, 1971). If there is continuous degradation of morphogen elsewhere in the field, then at steady state an exponential

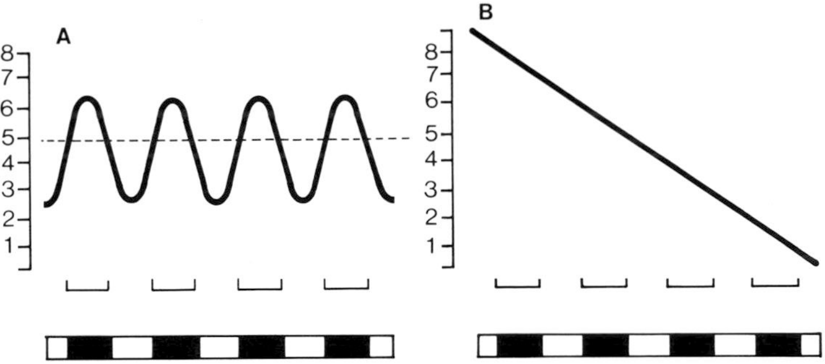

Fig. 3. Comparison of pattern formation by concentration profiles using prepattern or positional information-type mechanisms. (A) Prepattern sets up an equilibrium state with areas of stable high and low concentrations. Above a particular threshold an appropriate set of genes is switched on, in this case to produce alternating black and white stripes across the field. If concentration (C) > 3.0, then black. (B) Positional information produces a monotonic concentration profile. Particular concentration ranges switch on appropriate sets of genes, again producing black and white stripes. If 7 < C < 8, then black; if 5 < C < 6, then black; if 3 < C < 4, then black; if 1 < C < 2, then black.

concentration gradient of morphogen will result (Fig. 4A). This sort of model explains adequately the pattern formation in "semiinfinite" fields such as the vertebrate limb, as discussed below. However, this particular model does not provide the system with size invariance in cases in which the initial field is of finite size. For example, in many species, when only one of the blastomeres at the two-cell embryonic stage is allowed to develop, it nevertheless grows into a normally proportioned, albeit small, animal. There should be insufficient tissue for degradation to reduce the concentration of morphogen to its lowest normal value (Fig. 4B), but we observe that the animal is not distally truncated. The slope of the gradient has adjusted to the width of the field. One solution to this problem is the presence of a second important field boundary. If the other field boundary is an exclusive sink, maintaining morphogen at zero concentration (Fig. 4C), then size invariance obtains. All positional values will be found, regardless of the width of the field (Fig. 4D). In systems in which there is little evidence for size invariance, such as the chick limb, one special boundary region, whether source or sink, suffices. It is possible to devise field size sensing mechanisms such that size invariance will obtain even with only one boundary region, but such models are more complicated.

2. *Reaction–Diffusion*

The reaction–diffusion model can be used not only for generation of prepatterns but also to provide a mechanism for creating a single source

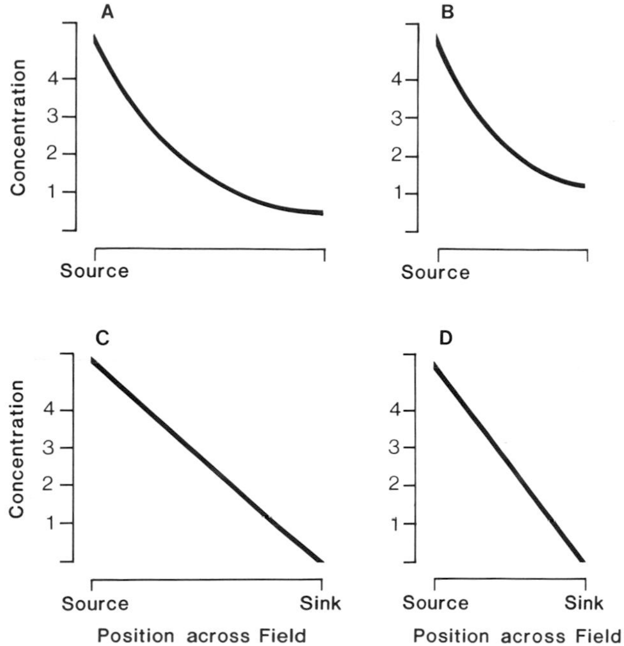

Fig. 4. Source–diffusion and size invariance. (A) Source plus continuous degradation gives an exponential monotonic gradient. (B) Truncation of field causes loss of some positional values and distal truncation of organism. (C) Source plus distal sink gives linear monotonic gradient. (D) Truncation of field produces size invariance, a small but well-proportioned embryo.

(Meinhardt, 1982). Such a model may use very similar equations and molecules to the periodic prepattern model (Fig. 2) except that the parameters are such that less than one chemical wavelength occupies the embryonic field. In such a case, the concentration profile resembles that found in source–diffusion models. The boundary at which the gradient forms will be determined by a small initial spatial bias in morphogen level, which is greatly magnified by autocatalysis and lateral inhibition. These properties also allow the single source of high morphogen to be self-maintaining. Decay of morphogen provides that an exponential concentration profile will result at steady state, as in the source diffusion model. Reaction–diffusion models are formally very similar. Short of understanding the underlying biochemical reactions, one would find it difficult to distinguish between them.

Reaction–diffusion kinetics have also been used to formulate models for gene activation and repression. This is an attractive idea because it would mean that the same molecular mechanisms could be used to control both pattern formation and differentiation.

3. *Intercalation*

Intercalation is a means for generating intermediate positional values during growth. Intercalation must presuppose some initial positional values. Then, local cell–cell averaging can generate new positional values by, for example, any new cell averaging the values of its two neighbor cells. Evidence for intercalation is strongest in regenerating systems (such as amphibian and insect limbs and the imaginal discs of higher insects) where a set of positional values are clearly present already. In these systems there are no obvious boundary regions and all positions in the field seem equally active during morphogenesis.

II. The Limb as a Model System

The limb is a well-studied model system for understanding the development of pattern. Historically, the three-dimensional structure of the limb has been anatomically partitioned into three orthogonal coordinate axes: dorsoventral, anteroposterior, and proximodistal (Fig. 5). These are not totally independent but do seem to represent three embryonic fields (Wolpert, 1978). Differentiation at any point in the limb depends on the positional value along all three axes. Much experimental work on the developing limb, including extensive studies on limb innervation, has been performed in the chick. It seems that the anatomical pattern of nerves within the limb is determined by axons growing along routes marked out within the connective tissue, and it is probable that the routes themselves are determined by some form of positional information. More controversial is the question as to whether any form of direct chemical specificity is involved in the actual formation of connections (see below).

A. Dorsoventral Axis of the Chick Limb

The dorsoventral axis of the limb is poorly understood, but it is fixed quite early, before the limb bud bulges from the flank (Fig. 5A). While the skeletal pattern of the limb (especially the chick wing) is rather symmetrical dorsoventrally, there are very significant differences in skin structures (feathers and scales) as well as in muscle and nerve patterns. When the limb ectoderm is rotated with respect to the mesoderm, the dorsoventral pattern of muscles (distally) is claimed to follow that of the ectoderm. Similarly, when disaggregated chick leg mesodermal cells are placed in an ectodermal jacket the dorsoventral polarity of the resulting recombinant outgrowth is reported to conform to that of the ectoderm. Unfortunately, adequate confirmation by examination of the limb muscle patterns is lacking. Work on amphibian limbs has shown the dangers of relying solely on external form. Thus the

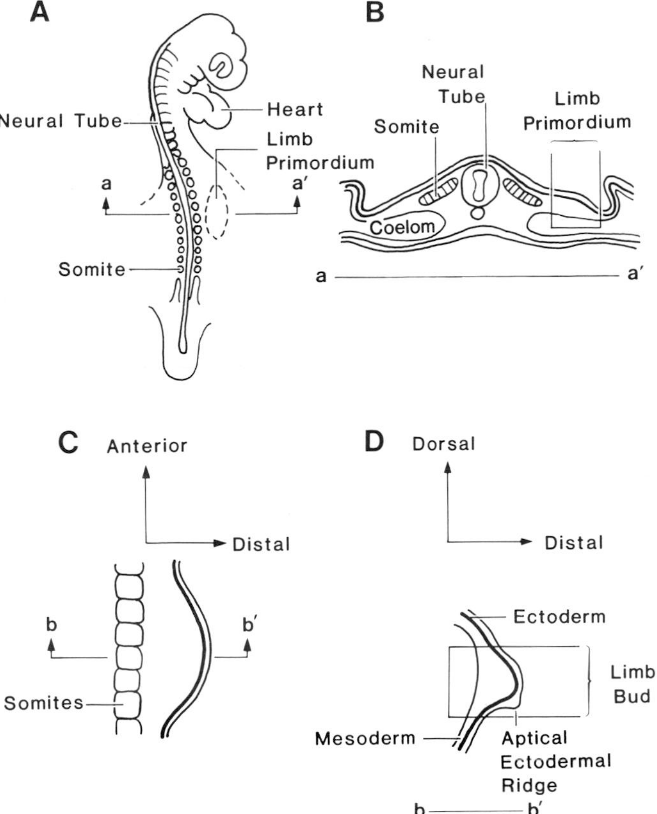

Fig. 5. The development of the limb bud and the major axes. (A) The embryo at day 2 seen from the dorsal surface. The general body plan is apparent and a disc of tissue in the lateral plate is already determined as the limb primordium. (B) Section through a to a′ of (A), showing limb primordium in relation to general body plan. (C) Forelimb bud of embryo seen at day 3.5 from dorsal surface, anterior–posterior and proximal–distal axes indicated above. (D) Section through b to b′ of (C) showing limb bud. The flank tissue, shown in (B), has folded ventrally so that the limb bud grows out laterally. The dorsal–ventral and proximal distal axes are indicated above.

role of ectoderm in dorsoventral polarity remains unclear. If the ectoderm is responsible, it is a unique mechanism since pattern phenomena generally seem to be governed by mesoderm.

B. Proximodistal Axis of the Limb

The proximodistal axis to the limb has been the subject of study for many years. Specification, differentiation, and innervation all proceed sequentially in a proximal-to-distal fashion. The apical ectodermal ridge (AER) is a thickened specialization of the distalmost ectoderm. It appears to be responsible for keeping distal mesenchyme in a labile, undetermined state. This region, which histologically is undifferentiated (up to ca. stage 28) for a distance of about 350 μm from the tip of the limb, has been called the progress zone, for reasons apparent below. All cells in the progress zone are dividing, and proliferation is more rapid there than in regions more proximal. During growth, those cells left behind the progress zone are found to be determined, and differentiate prior to those more distal. Removal of the apical ectodermal ridge both causes a drop in the rate of distal outgrowth and results in a truncated limb. The stage of AER removal determines the level at which limb structures are lost (Fig. 6). The later the operation is performed, the more complete the resulting limb until by stage 29, AER removal does not affect the final form of the limb. The proximodistal axis is relatively mosaic; there is some ability to regulate for excess or deficiency at early stages (before stage 21) where a substantial portion of the limb consists of progress zone, but at later stages (e.g., stage 22; nevertheless, well before determination is completed), different parts of the limb bud, including the progress zone, all

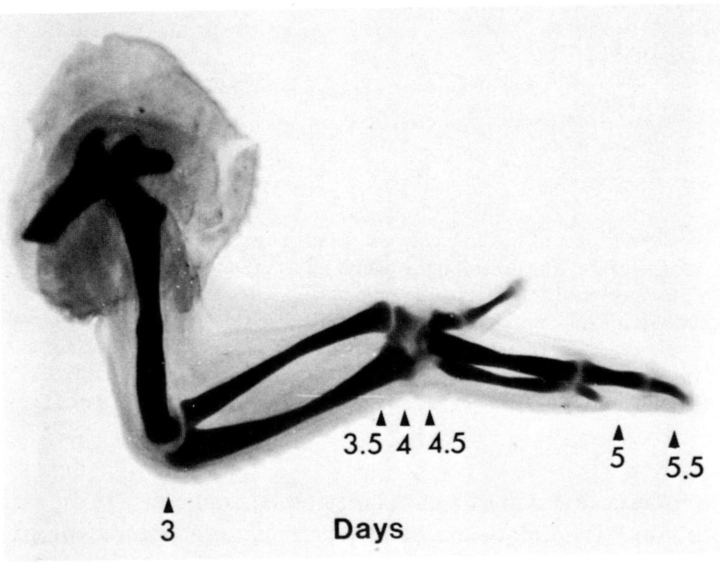

Fig. 6. Effect of apical ridge removal at different times during limb development. The arrows indicate the mean level of truncation following apical ridge removal at 12-h intervals. The time has been normalized so that day 3 is equivalent to stage 18 (Hamburger and Hamilton, 1951). The level of truncation is strongly time dependent.

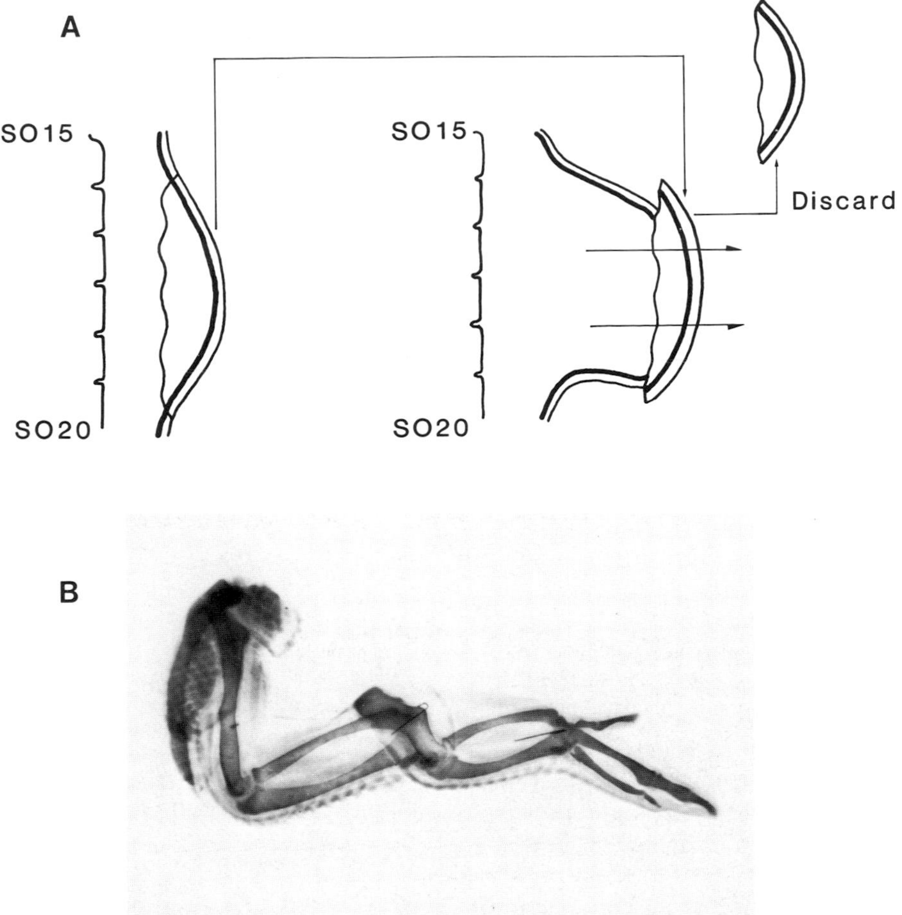

Fig. 7. Autonomous behavior of limb fragments and serial reduplication. (A) The tip of a stage-24 limb bud is removed and an entire stage 19 is grafted in its place. (B) The result at day 10 is serial duplication with stylopod (humerus) and zeugopod (radius/ulna) repeated. SO, Somite numbers.

behave autonomously. If an early stage limb tip is combined with a later stage stump, serial duplication of elements occurs (Fig. 7). If a late stage bud tip is grafted onto an early stump, a limb with deficiencies will result.

The progress zone model for the proximodistal limb field postulates that cells in the progress zone suffer a continual decrease in positional value and that this decrementation only ceases when the cells leave the tip zone. Thus, each cell's positional value ultimately depends on the length of its period of residence in the progress zone (Fig. 8). One way of measuring this is by

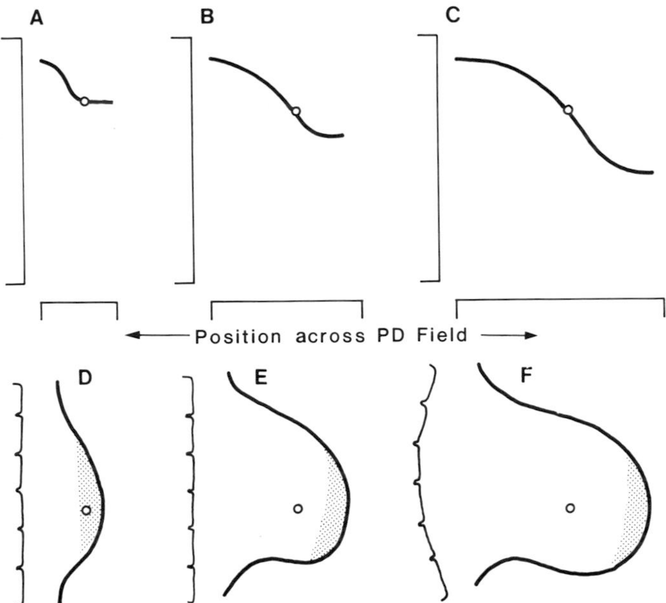

Fig. 8. The progress zone model. (A–C) Graphs of positional value against distance along the proximal–distal axis for three different stages at 12-h intervals. The progress zone is the horizontal part of the positional value profile. Within the progress zone, positional value changes autonomously toward a more distal character. Due to cell division within the progress zone, a cell (open circle) is eventually pushed out of the zone (B) and its positional value becomes fixed (C) at the level it had reached at the time that it left the progress zone. (D–F) Dorsal views of a limb bud at equivalent stages to (A–C). Open circle indicates the cell and hatching the progress zone.

counting the number of cell divisions. It has been suggested that the number of cell divisions during the cell determination process is about seven, and thus equal to the number of proximodistal limb segments (wrist counts as two), so that one cycle is equivalent to one segment.

It is possible that similar counting mechanisms may be important in determining positional values of neurons. Retinal ganglia are produced in a narrow peripheral zone at the outer margin of the expanding retina and seem to carry both circumferential and radial coordinates of positional information. It is possible that the radial coordinate is produced by a progress zone. Motor neurons in the spinal cord similarly go through their final cell division in a sequence that results in lateral motor neurons being "born" before medial motor neurons. This could again be used to provide positional information, as will be discussed later.

An informative deliberation over the relation of this lineage counting model to positional information may be found in Wolpert *et al.* (1975). While the progress zone accounts for the data, its seems that a diffusion model, in which the morphogen issues from the tip and is slow diffusing, would equally provide an acceptable fit as would a modified reaction–diffusion model in which a time-dependent feedback loop causes increased morphogen levels at the distal tip (Meinhardt, 1982).

C. Anteroposterior Axis of the Limb

There is an organizing region at the posterior margin of the limb bud, which, when transplanted anteriorly, causes outgrowth of additional tissue, resulting in the formation of a duplicate limb (Fig. 9). This region uniquely is able to signal in the limb bud causing production of extra digits. We will describe its action as that of a source of morphogen, creating an anteroposterior gradient of positional information (Fig. 10). The genesis of this limb organizer (the positional signaling region) is presumably a consequence of its specification at the time of the determination of the primary body axis (Fig. 5). An alternative explanation for its appearance could be that it is a result of a reaction–diffusion mechanism's magnifying some slight polarity bias initially present. The two are formally similar and have enormous predictive power (Summerbell and Honig, 1982).

The polarizing region arises quite early (Fig. 11), before the limb bud bulges from the flank. It stays active throughout the whole period of limb determination. Always located posteriorly, it moves distally so as always to lie near the distal tip (and indeed it seems that it can only act on cells in the progress zone). If the region is grafted to the anterior edge of the limb bud, time is then required for the signal to spread across the responding limb tissue and for new positional values to be established. If a graft is totally removed (with some surrounding tissue) before determination, an incomplete set of extra limb elements is obtained. Factors involved are loss of graft (the source), loss of some neighboring tissue, and continued decay of the signal. About 18 hours of graft–host contact are required to ensure production of at least some extra digits following graft removal. Even in normal development, removal of limb tissue from the influence of the limb polarizing region early enough results in digit deficiencies. This can be demonstrated by emplacing an impermeable barrier (tantalum foil, see Fig. 10D) between polarizing region and the limb field or by simply excising the polarizing region. At later stages, approaching the time of determination, positional values are remembered (Summerbell and Honig, 1982).

The gradient nature of the polarizing region signal can be seen in the apparent summation of signals from multiple sources in a single field. If a single region is grafted to a very anterior position, there results a limb with

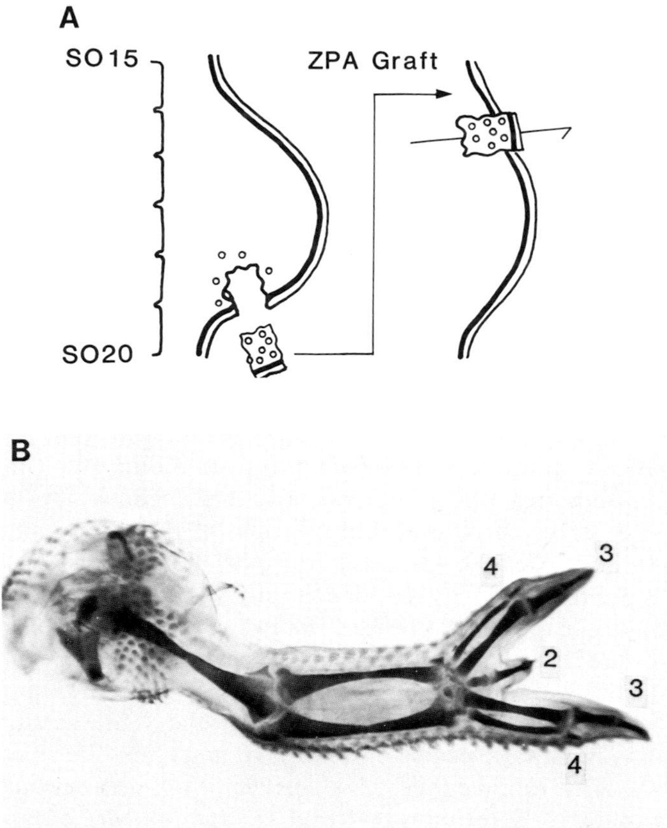

Fig. 9. Graft of organizing region from the posterior limb margin. (A) The posterior limb margin is grafted to a prepared site on the anterior border of a host limb. Organizer region shown stippled. (B) Result at 10 days is supernumerary limb in mirror image symmetry to the original host limb. SO, Somite numbers; ZPA, zone of polarizing activity.

two full complements of structures. The two sources are at a distance so large that the gradients do not interact. As the sources are moved closer together (by making the grafts less anterior), the gradients summate (see Fig. 10C), and the lowest positional values are lost. The observations and the predictions of the model are in general accord, but the limb widening that accompanies duplication must be taken into account in accurate modeling.

The limb bud is normally considered as growing only along the proximodistal axis and not substantially in an anteroposterior or dorsoventral fashion. However, the limb widening following a positional signaling region graft involves increased cell division. This increased cell division starts near

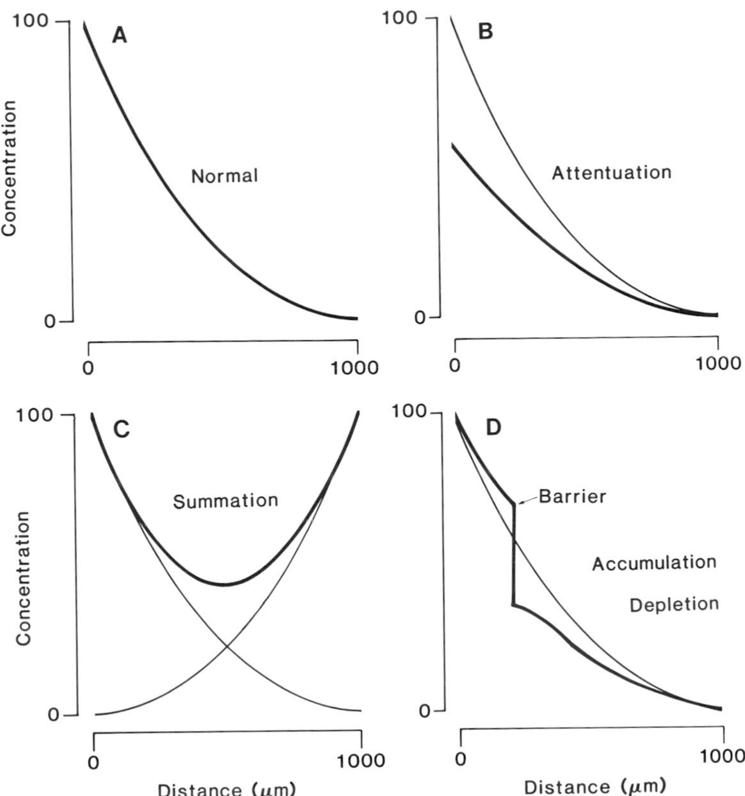

Fig. 10. Morphogen concentration profiles and the source-diffusion model. (A) Normal equilibrium profile with concentration of morphogen held constant at the organizer region (source = 100) and allowed to diffuse freely across the limb where it has been destroyed. (B) Attentuation—the effect of reducing the source concentration. (C) Summation—the effect of two identical sources interacting over the same developmental field. (D) Accumulation and depletion—a barrier inserted in the field leads to accumulation of morphogen on one side and depletion on the other; excision of tissues gives similar effects.

the graft but sweeps across the entire limb bud over an 18-h period. This does not result in an increased rate of proximodistal elongation but rather results in a distal widening of the limb bud. This widening is quantitated easily following an experiment in which a defined limb field is created by grafting two additional polarizing regions to the limb bud margin (Fig. 12). The field between them grows at a reproducible rapid rate and the intergraft distance accurately determines the number of digits obtained from the field. To obtain two full digit complements, one must have 600 μm of responding tissue, implying that the presumptive digit field for a normal limb is about

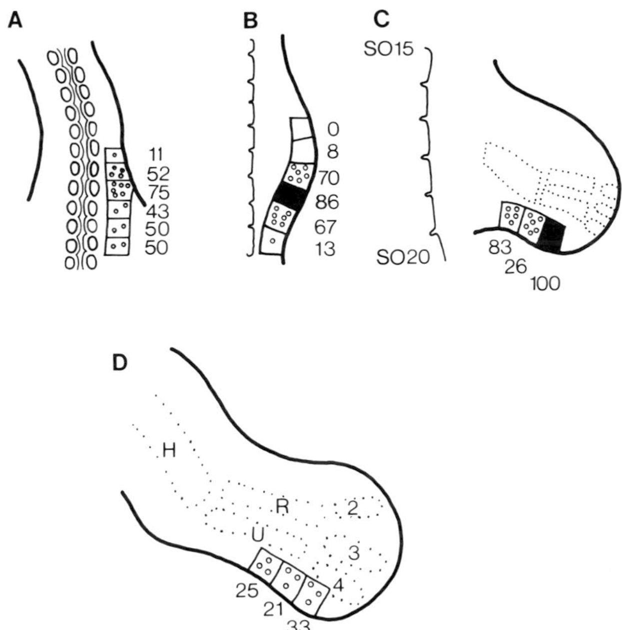

Fig. 11. Maps of polarizing activity of posterior limb organizer. The strength of activity is shown by density of stippling and by the figures giving percentage polarizing activity. (A) Stage 16. (B) Stage 18. (C) Stage 23, fate map superimposed. (D) Stage 28, skeletal pattern superimposed. SO, Somite numbers; H, humerus; U, ulna; R, radius; 2, digit 2; 3, digit 3; 4, digit 4.

300 μm wide. This estimate is in accord, after widening, with estimates from fate maps using marked (^{3}H-labeled thymidine) tissue (Summerbell and Honig, 1982). This shows that appearances are deceptive and that despite the much larger width (~ 1mm) of the limb bud, the width of the part generating digits is only about one-quarter as much. Hence, the notion of the semiinfinite field (see above) applies. At the posterior margin we can describe a boundary and an area governed by positional information. As we move more anterior, we eventually leave the area under the direct influence of the organizer; in this direction the field effectively extends ''infinitely.''

While summation causes the progressive loss of anterior (midline) positional values, it is possible to selectively lose posterior (high) positional values. Treatment of the polarizing region with various agents does not result in all-or-none behavior in which either full or no limb duplications occur. Rather, the signal of the region is attenuated in a graded manner. When the donor is subjected to increasing doses of γ irradiation, ultraviolet radiation, or various biochemical inhibitors, the ability to specify extra digits 4, 3, and

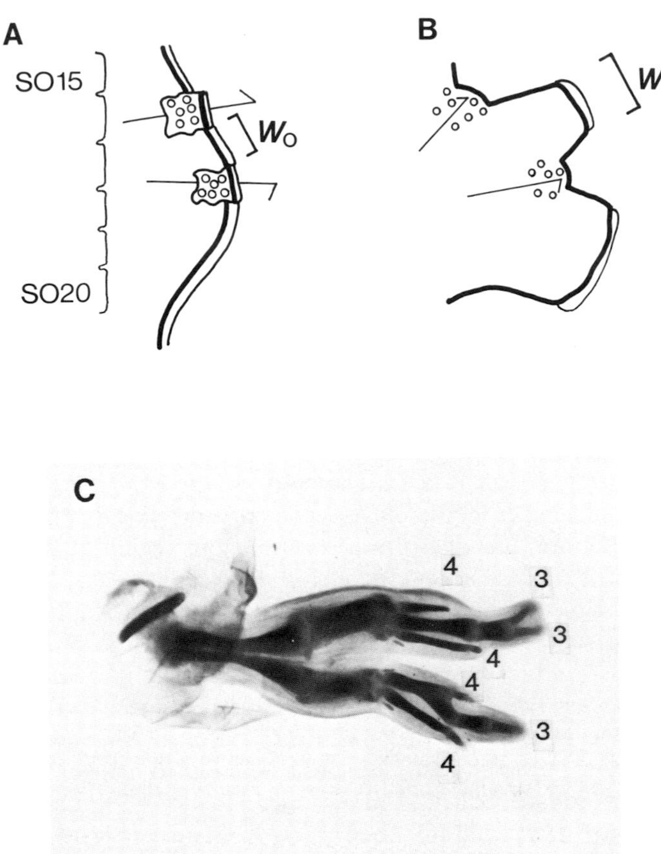

Fig. 12. The width of responding tissue in the field. The experiment illustrates widening of the responding field under the influence of the two grafts ($W_t > W_o$), and provides an estimate of the amount of tissue required to produce various combinations of digits (in this case 434). (A) Diagram of operation, two grafts share a responding field between them. (B) The limb outline 24 h later. (C) The result at day 10 with digit formula G434G4334.

2 is lost progressively (Fig. 11B). Such attenuation may be the result of the output of each and every cell being reduced, or of their ability to communicate being disturbed, or of the number of cells capable of signaling simply being reduced. For example, there is direct evidence that decreasing the number of functional cells in a graft also causes progressive attenuation of signaling activity (Tickle, 1980). Polarizing regions may be disaggregated into signal cell suspension, reaggregated, and grafted without any loss of activity. If the polarizing region cells are diluted with an equal number of anterior cells,

the ability to specify a digit 4 is lost; further dilution eliminates the specification of digits 3 and 2.

Despite all the evidence compatible with the diffusible morphogen model, there is still no direct demonstration of the nature of the morphogen. As we have pointed out earlier there may be no morphogen, only some communicable cellular activity that we can model using the diffusion equations. However the evidence is good enough to make a strong case that some of our scientific effort should be spent on a direct search for morphogens. Fortunately, at least one research group has profitably tackled this formidable problem. They have at least one substance that has been isolated from polarizing regions that has a morphogenic effect. So far, this substance has been assayed by its ability to prevent cell death and apical ridge flattening in pieces of test apical tissue *in vitro* but has not been conclusively related to pattern formation (MacCabe and Richardson, 1982).

The polarizing region is a unique part of the limb bud with special properties. The same properties have been confirmed in avian fore- and hind limbs. Furthermore, tissues from the homologous region of the limb buds of a wide variety of amniote species, including mice, hamsters, humans, pigs, alligators, and turtles, are all capable of inducing chick limb duplications when grafted to the anterior margin of the chick limb bud. Still more impressive, Hornbruch and Wolpert have reported that the only other nonlimb organizer, that of the primary body axis (Hensen's node in the chick), is also capable of causing chick limb duplications. More recently, local application of retinoic acid has been shown to cause similar limb reduplications (Summerbell and Harvey, 1983; Tickle, 1983). One of the main predictions of the original theory of positional information was "universality," the principle that different systems might use the same positional signals.

Ideally we could now introduce some similar example in the nervous system. Unfortunately there is none. Positional signaling is demonstrated uniquely well in the limb bud, and there are no comparable examples for the nervous system. It remains possible, however, that the nervous system may use many of the same processes if not the same morphogens.

D. Polar-Coordinate Description

The polar-coordinate model (Bryant *et al.*, 1976; Bryant *et al.*, 1981) originated as a means of explaining the results of tissue manipulation experiments in experiments on regenerating insect and amphibian limbs. The distinctive features in these limb systems are that (1) there does not appear to be any unique signaling region; (2) all parts of the field are capable of interaction; and (3) tissue interactions apparently occur over a very short

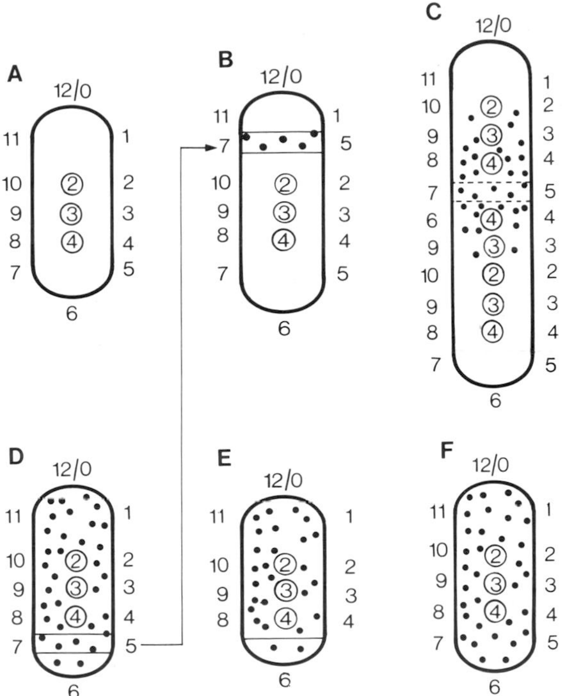

Fig. 13. A polar coordinate interpretation of grafts of posterior limb organizer to anterior margin. (A) Host limb. (B) Host limb plus graft. (C) Full intercalation giving digit formula 234G43234. (D) Donor limb. (E) Donor limb minus graft. (F) Full intercalation and deficiency.

range, that is, between adjacent cells. The polar coordinate model considers the three-dimensional limb as being collapsed into a two-dimensional field that can be described using polar coordinates. Radial values correspond to the proximodistal limb axis. Circumferential coordinates run around each cross section of the limb and are labeled as on the face of a clock, with numerals 1 through 12 (Fig. 13). There is no significance to the 12/0 boundary, which is merely a result of the numbering scheme. The model is based on two rules. One rule concerns distal transformation. The model originally proposed that a complete circle of values was required for distal outgrowth, but it has now been modified to suggest that the amount of outgrowth is simply related to the proportion of a complete circle that is present. The second key rule is that local intercalation (along the shortest route) occurs whenever tissues of disparate positional value are opposed. While the model is rather successful in accounting for regeneration phenomena (where there already exist complete sets of positional labels), it does not seem to apply to chick limb development.

An important set of experiments was initiated by Iten (1982) to test the applicability of the polar coordinate model to the chick limb. Transplants of anterior limb bud tissue were placed in posterior host positions and were found to give extra limb elements (Fig. 14). These might have represented intercalated values, but it has been shown that they actually represent self-differentiation of the grafted anterior, responding tissue under the influence of the hosts's polarizing region. The extra structures are nearly entirely graft-derived as can be seen by using marked (quail tissue) grafts. This is in contrast to the extra digits resulting from a polarizing region graft (of posterior tissue anteriorly) in which the graft tissue frequently makes no contribution to the resulting reduplication. Not only can this fact be demonstrated by use of quail grafts with their distinctive nucleolar cell marker but it can also be shown through use of irradiated chick or quail grafts in which the cells ultimately die and thus cannot contribute at all to the definitive limb. An additional test for intercalation in the chick limb involves the use of two polarizing region grafts as shown in Fig. 12. The resulting limbs usually have digit patterns (between the grafts) of the type 434 or 4334 but only in rare cases do digits 2 result. In an unmodified intercalation model, no positional values should be lost, and many more digits (three or four full hands of either 432 or 234 pattern) should result. Instead, the results are as expected from a gradient model with summation of morphogen concentration profiles causing loss of low threshold digits. These experiments together imply that the positional signaling region organizes field of about 300 μm width.

There is also direct evidence that the signal from the positional signaling region is not local but extends over a distance of 20 to 30 cell diameters (Summerbell and Honig, 1982). A polarizing region is grafted to an anterior host limb site with a piece of marked (leg or quail) tissue immediately adjacent to it posteriorly (Fig. 15); the polarizing region is separated from the host wing field by this marked "barrier" tissue. The distance over which the polarizing region signal is transmitted can then be measured by examination of the width of the barrier through which the influence of the polarizing region can still be demonstrated in the posterior–distal host wing tissue. If interaction were purely local, one would expect the marked tissue only to contribute to duplicated limb elements. However, "barriers" up to 300 μm wide can still allow the formation of duplicated host elements on the other side of the barrier. Thus the signal can travel over about 30 cell diameters. This distance is also about the anteroposterior extent of the digit field at these stages, as estimated from various fate-map studies.

Although the polar coordinate description does not fit the development of the chick limb bud, the work that has arisen out of the model has proved valuable in suggesting further lines of research.

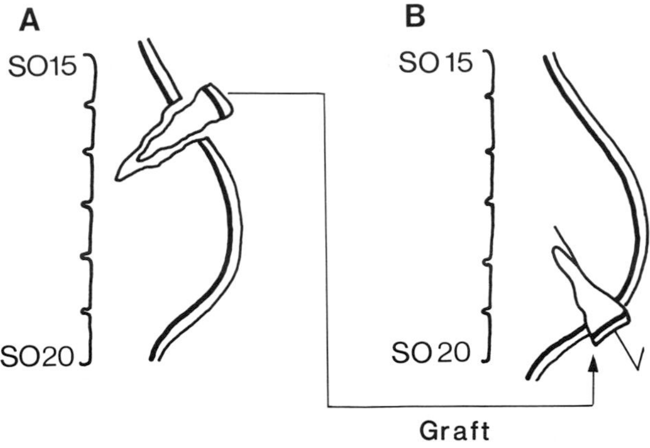

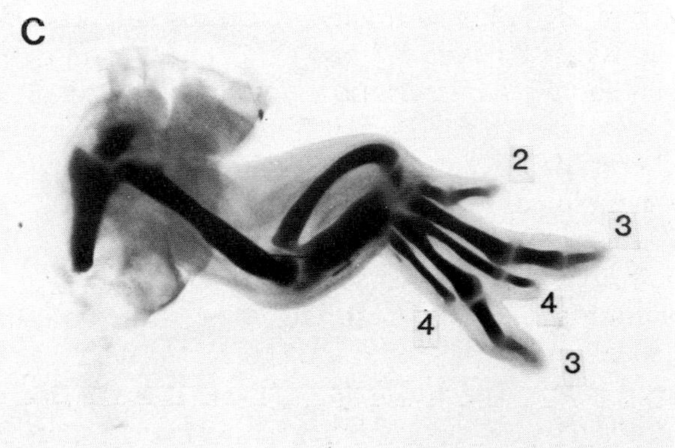

Fig. 14. Self-differentiation of an anterior graft under the influence of the host organizer. (A) Anterior donor tissue is grafted to a posterior position in a host embryo. (B) Result at day 10, digit formula 23443. Analysis of the additional digits using the quail marker demonstrates that they are derived from the graft and that there has been respecification rather than intercalation.

E. Interactions between the Axes

There is no evidence for circumferential values in the developing chick limb, but there are interactions between the axes. The apical ectodermal ridge seems to keep the distal tip of the limb unspecified with respect to any axis. If the limb ectodermal jacket is rotated 180° around the limb mesoderm, dorsoventral reversal is observed only distally. For the anteroposterior axis,

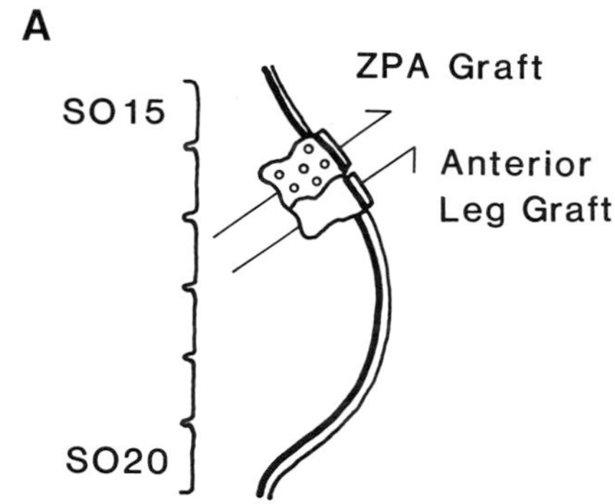

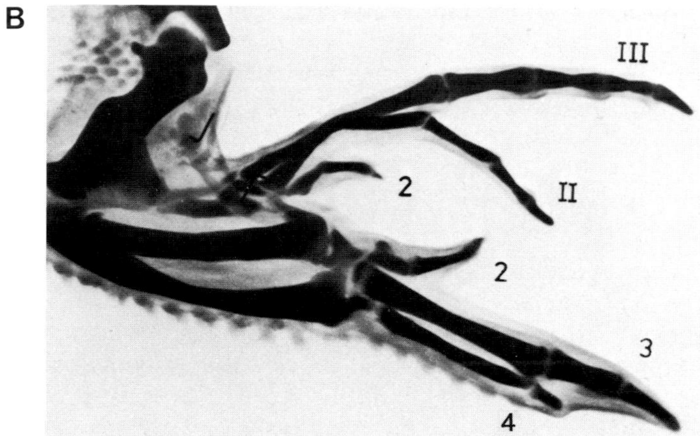

Fig. 15. Range of the positioning signals. (A) An organizer is grafted to the anterior border of a host limb and a piece of anterior leg tissue is grafted adjacent to it. (B) Result at day 10 with digit formula III II 2234. The organizer has caused the formation of extra leg digits II and III from the anterior margin graft that has also signaled through this tissue to more distal host tissue so that it forms an extra digit 2. ZPA, Zone of polarizing activity.

grafts of polarizing regions result in duplications distally, and the most proximal duplicated point is found more distally the later the stage of the host limb. At stages 16 to 18, reduplication may start at levels as proximal as the humerus; while at stages 23 to 24, only the phalanges will show duplications. In addition, partial removal of the apical ectodermal ridge results in loss of limb structures fated to develop from the subjacent tissue.

F. Role of Other Embryonic Processes

We have considered the limb bud as a homogeneous mass of mesenchymal cells surrounded by the ectodermal jacket. However, there are various migrations into the limb buds of cell populations with their own separate lineages. Neural-crest-derived cells migrate into the early limb bud and give rise to the melanocytes that are responsible for skin pigmentation. At very early stages (14–16), while the limb primordium is still simply a part of the flank, there is migration of cells from the somites. These somitic cells are not found in the distal tip of the wing bud as it grows. Grafting operations between chicks and quail demonstrate that the somitic cells only form muscle and that all or nearly all myoblasts are derived from this separate somitic cell lineage. For both melanoblasts and myoblasts it seems that these originally migratory cells respond to the positional information present in their host limb bud. At later stages, while a well-formed limb bud is growing (stages 24 onward), there is invasion of the chick limb bud by spinal cord neurons of the dorsal root ganglia and the ventral horns. The nerve cell axons also seem to respond to limb positional information, as discussed in Section VI.

III. Peripheral Innervation

Limb innervation is a special case of pattern formation involving the matching of the spinal cord nerve cells with the pattern of muscle tissues in the limb so as to produce a topographic map. Sensory axons originate from cell bodies in the dorsal root ganglia and motor axons from cells in the ventral (motor) horns of the cord (Fig. 16). There is not yet any known anatomical pattern within the dorsal root ganglia, except that axons from anterior ganglia innervate relatively anterior targets, and those from posterior ganglia innervate relatively posterior targets.

There is a similar arrangement of the ventral (motor) horns along the anterior/posterior dimension, but also axons from cell bodies lying medial and dorsal in the horn innervate ventral targets in the limb while axons from lateral and ventral cell bodies innervate dorsal targets. Innervation is by active invasion. The axons grow into the limb after the basic pattern of the tissues has been established. They form mixed (motor and sensory fibers) nerves that branch at predictable points such as the wrist to form a repeatable and relatively constant pattern (Fig. 17). Each muscle has its own nerve branch.

During normal development, the path of individual axons within the nerve is simple and direct with axons maintaining their neighbor–neighbor relationship from their relative position in the cord, through the nerve, to their final destination (Fig. 18). The farther the axons travel from the center,

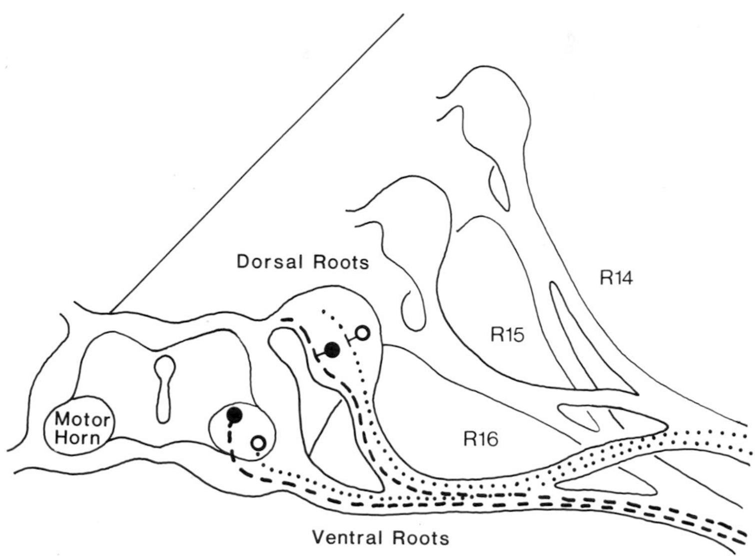

Fig. 16. Diagram of the spinal cord and right plexus. Axons from medial motor neurons (solid circle, dashed line) travel ventrally to pass down ventral (medioulnar) nerve to reach flexor muscles. Axons from lateral motor neurons (open circle, dotted line) travel dorsally to pass down dorsal (radial) nerve to reach extensor muscles. The arrangement of neurons in the dorsal root ganglion is conjectural, but note that some sensory axons must cross motor axons in the plexus so as to reach ventral targets.

the farther this ordering deteriorates but nevertheless axons from neighboring neurons (i.e., a particular part of the cord) tend to reach the same general target area (therefore showing apparent specificity), having traversed the same route. There is a vigorous debate as to how much of the perceived order is due to this passive deployment or whether it is necessary to propose active mechanisms that can increase the proportion of fibers that find their way to appropriate targets (Landmesser, 1980; Summerbell and Stirling, 1982).

A. Neuronal Specificity

Neuronal specificity supposes that each cell in the cord knows its appropriate target in the limb and that there are active mechanisms to ensure that it makes a functional connection to its target (Gaze, 1978). This implies that the motor neurons and the targets each carry some label or positional value that uniquely identifies them. We have already discussed how medial motor neurons are formed earlier than lateral motor neurons. This would

Fig. 17. A day-10 right limb stained to show the nerve pattern. (A) Viewed from dorsal side. (B) Viewed from ventral side.

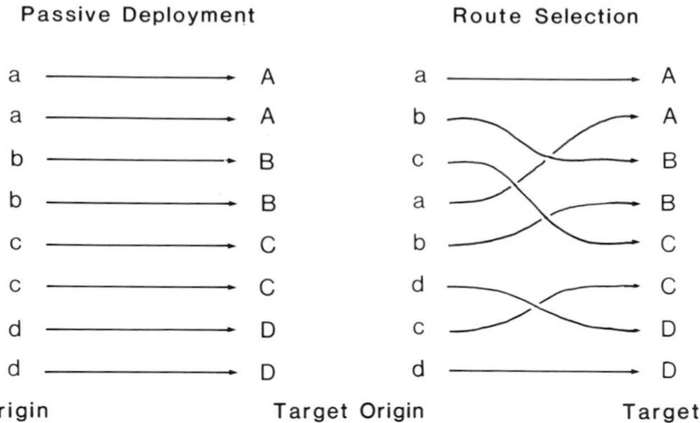

Fig. 18. Passive deployment and route selection hypothesis. (A) The cord and limb (origin and target) are so arranged that if the axons grow straight out they will encounter their appropriate target. (B) Axons actively seek out their appropriate target.

provide a mechanism (analogous to the progress zone) for ascribing one part of the necessary label. We must further suppose that this information in some way matches positional values carried by the muscles in the limb, so that medial corresponds to flexor (or ventral) and lateral corresponds to extensor (or dorsal).

Two main classes of mechanism have been proposed to increase the proportion of neurons that connect to appropriate targets above that which would be obtained by passive deployment. The first class of mechanism is based on the observation that during normal development about half of the motor neurons die during the period that innervation is established. It has been suggested that there is initial redundancy and that only those cells that have succeeded in connecting to their appropriate target are rescued from dying, so that cell death is an active mechanism for tidying mistakes and so improving selectivity (Lamb, 1983).

The second class of mechanism assumes that in principle each axon ''seeks out'' its appropriate target as it grows into the limb. Each time than an axon encounters a choice between possible routes, it favors the direction appropriate to its correct target. Though there may be mistakes, most axons make appropriate initial connections. To test this we make the axons enter the limb by an abnormal route by changing the spatial relationship between limb and spinal cord (Stirling and Summerbell, 1983; Landmesser *et al.*, 1983). Provided that the change is not too great so that the axons pass nearby their normal route, then they are sometimes able to correct and find their normal target. However, if the operation carries the axons well off their normal position or into entirely new territories, then they are unable to make their correct connections and seem to obey the passive deployment rules (Fig. 19).

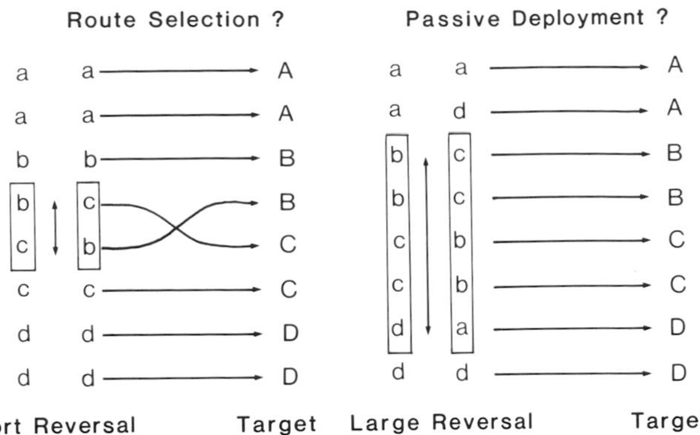

Fig. 19. Experimental tests of route selection and passive deployment. (A) If small changes are made in either origin or target, then the axons seem to find their way to appropriate muscles, implying route selection. (B) If large changes are made in either origin or target, then the axons innervate the nearest target and seem to obey passive deployment rules.

There is no simple hypothesis that will account for all of the experimental results. It is possible that selective cell death, route selection, and passive deployment all have a part to play. The evidence favors the presence of a short-range interaction or selective signal guiding axons, but which of these is unknown (Lewis *et al.*, 1983).

B. Positional Information of the Target Determines the Nerve Pattern

It is well established that the developmental control of the pattern of nerves resides in the limb tissue itself and is not a function of the innervating axons. Thus, a wing bud grafted in place of a leg bud receives axons from the lumbosacral plexus but the pattern of nerves in the graft is appropriate to wing and not to leg and vice versa (Summerbell and Stirling, 1982). It seems probable that the presumptive track of the nerve is marked out in some hidden way in the limb tissue before the axons arrive. The axons invade the limb, following the pattern of preferred tracks. At branches in the track the axons are able to choose the route that goes to their normal target, but if that particular branch is not present because they are in the wrong track, then they simply take the branch that is nearest them (Fig. 20). If a barrier is inserted prior to innervation so as to block the track, then the axons encountering the barrier are deflected off their normal course and forced to divert around the barrier. If the barrier is large, then the axons may find themselves near to a foreign branch, which they will enter, eventually innervating abnormal targets. If the barrier is smaller, then the axons, having passed the barrier, will set off "across country" back to their normal track,

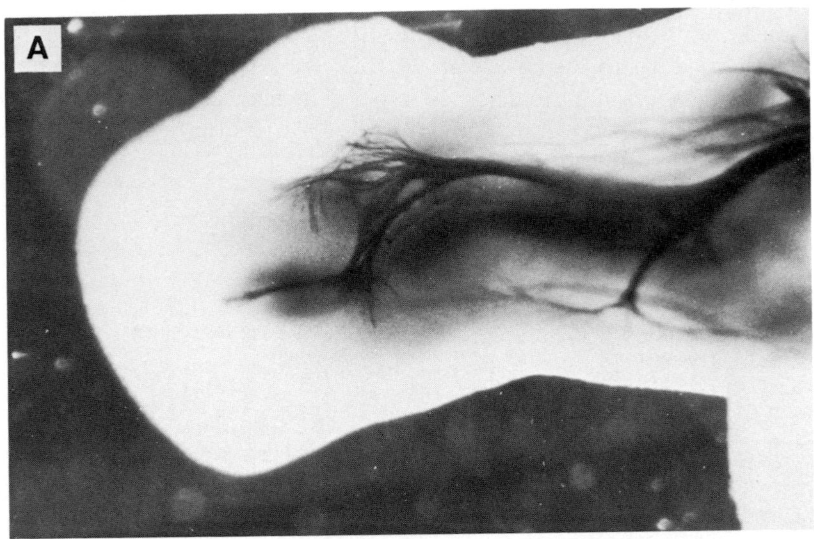

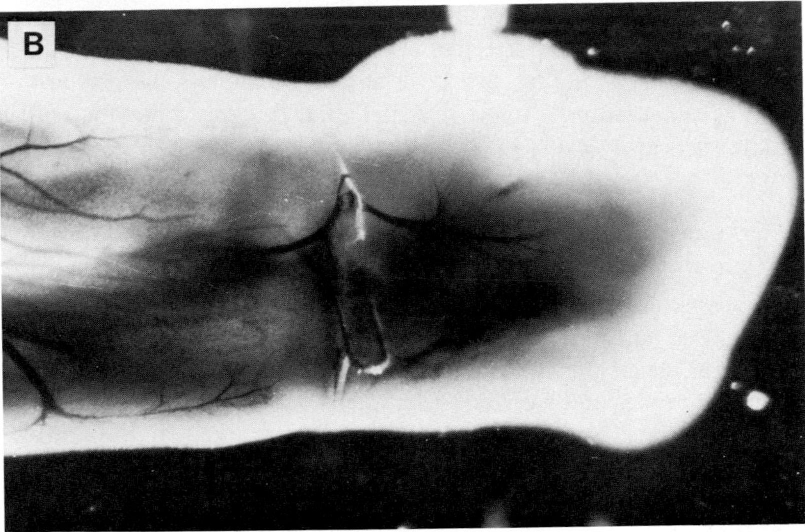

Fig. 20. Track recognition. (A) Advancing wavefront of growing tips at branch point in wrist. (B) Nuclepore barrier placed so as to block same branch point. The nerve diverts around the anterior edge of the barrier but then is able to return to its normal path. Some axons are able to find their way through holes in the barrier, seeking the shortest route. Axons "find" their way into near branches but are unable to enter more distant posterior branch.

apparently being able to recognize its presence at a distance. Many suggestions have been made as to the nature of the tracks and the succeeding chapters consider this aspect in more detail. Whatever the mechanism, the tracks are probably an expression of the same processes of pattern formation that determined the structure of rest of the limbs.

References

Bryant, P. J., Bryant, S. V., and French, V. (1976). *Sci. Am.* **237**, 66–81.

Bryant, S. V., French, V., and Bryant, P. J. (1981). *Science* **212**, 993–1002.

Cooke, J. C. (1980). *Curr. Top. Dev. Biol.* **15**, 373–407.

Cooke, J. C. (1981). *Philos. Trans. R. Soc. Lond., Ser. B.* **295**, 509–524.

Gaze, R. M. (1978). *In* "Specificity of Embryological Interactions" (D. R. Garrod, ed.), Vol. 4, pp. 53–93. Chapman & Hall, London.

Gordon, R., and Jacobson, A. G. (1978). *Sci. Am.* **238**, 106–113.

Hall, B. K. (1978). "Developmental and Cellular Skeletal Biology." Academic Press, New York.

Hamburger, V., and Hamilton, H. L. (1951). *J. Morph.* **88**, 49–92.

Hinchliffe, J. R., and Johnson, D. R. (1980). "The Development of the Vertebrate Limb." Oxford Univ. Press (Clarendon), London and New York.

Huxley, J. S., and de Beer, G. R. (1934). "The Elements of Experimental Embryology." Cambridge Univ. Press. Reprinted (1963), Hafner, New York.

Iten, L. E. (1982). *Am. Zool.* **22**, 117–129.

Jacobson, M. (1978). "Developmental Neurobiology." Plenum, New York.

Jacobson, M. (1982). *In* "Neuronal Development" (N. Spitzer, ed.), pp. 45–99. Plenum, New York.

Kimble, J. E., Sulston, J. E., and White, J. G. (1979). In "Stem cells, Cell Lineages, and Cell Determination" (N. le Douarin and A. Monsory, eds.), pp. 59–68. Elsevier, Amsterdam.

Lamb, A. H. (1983). *Prog. Clin. Biol. Res.* **110**, 227–236.

Landmesser, L. T. (1980). *Annu. Rev. Neurosci.* **3**, 279–302.

Landmesser, L. T., O'Donovan, M. J., and Honig, M. (1983). *Prog. Clin. Biol. Res.* **110**, 207–216.

Lewis, J., Al-Ghaith, L., Swanson, G., and Akbar Khan. (1983). *Prog. Clin. Biol. Res.* **110**, 195–205.

Lund, R. D. (1978). "Development and Plasticity of the Brain." Oxford Univ. Press, London and New York.

MacCabe, J. A., and Richardson, K. E. Y. (1982). *J. Embryol. Exp. Morph.* **67**, 1–12.

Meinhardt, H. (1982). "Models of Biological Pattern Formation." Academic Press, New York.

Muller, K. J., Nicholls, J. G., and Stent, G. S. (1981). "The Neurobiology of the Leech." Cold Springs Harbor Laboratory, Cold Spring Harbor, New York.

Murray, J. D. (1981). *Philos. Trans. R. Soc. Lond. Ser. B.* **295**, 473–496.

Sawer, R. H., and Fallon, J. F. (1983). "Epithelial–Mesenchymal Interactions in Development." Praeger, New York.

Saxen, L. (1979). *In* "Cell Interactions in Differentiation" (M. Karkinen-Jaaskelainen, L. Saxen, and L. Weiss, eds.), pp. 145–151. Academic Press, New York.

Stirling, R. V., and Summerbell, D. (1983). *Prog. Clin. Biol. Res.* **110**, 217–226.

Summerbell, D., and Harvey, F. (1983). *Prog. Clin. Biol. Res.* **110**, 109–118.

Summerbell, D., and Honig, L. S. (1982). *Am. Zool.* **22**, 105–116.

Summerbell, D., and Stirling, R. V. (1982). *Am. Zool.* **20**, 173–184.
Tickle, C. (1980). *In* "Development in Mammals" (M. H. Johnson, ed.), Vol. 4, pp. 101–136. Elsevier, Amsterdam.
Tickle, C. (1983). *Prog. Clin. Biol. Res.* **110**, 89–98.
Van Der Loos, H. (1979). *In* "Neural Growth and Differentiation" (A. Meisami and M. A. B. Brazier, eds.), pp. 331–336. Raven, New York.
Wolpert, L. (1969). *J. Theor. Biol.* **25**, 1–47.
Wolpert, L. (1971). *Curr. Top. Dev. Biol.* **6**, 183–224.
Wolpert, L. (1978). *Sci. Am.* **239**, 154–164.
Wolpert, L., Lewis, J., and Summerbell, D. (1975). *Ciba Found. Symp.* **29**.

2

Regulation of Nerve Fiber Elongation during Embryogenesis

PAUL C. LETOURNEAU
Department of Anatomy
University of Minnesota
Minneapolis, Minnesota

I. Introduction

Specificity in neural function is largely determined by the organization of neuronal connections. How is nerve fiber elongation controlled during morphogenesis of these pathways? I focus here on *in vitro* examination of

Developmental
NeuroPsychobiology

this question. My principal thesis will that regulation of nerve fiber elongation occurs mainly by interactions of the extending nerve tip with environmental cues.

A central theme in studying the development of neuronal connectivity for over 20 years has been neuronal specificity; that is, before nerve fibers are initiated, neurons acquire individual affinities that are used to establish synapses (Jacobson, 1978; Sperry, 1963). Experimental manipulations reveal that axons can be made to take abnormal pathways and will bypass available, but inappropriate, targets en route to their normal partners (Arora and Sperry, 1962; Attardi and Sperry, 1963; Lance-Jones and Landmesser, 1978, 1980). Yet, it is also known that environmental features can influence the pathways of axonal growth and connection (Constantine-Paton, 1978; Katz and Lasek, 1980; Summerbell and Stirling, 1981). What are the interactions that regulate axonal growth and to what extent do axons actively seek particular targets versus being directed to targets by environmental features? What level of selectivity exists in such environmental pathways or influences and how are pathways defined? As a counterpoint to the idea of neuronal specificity, recent proposals suggest that more general spatial and temporal factors during axonal growth determine much of synaptic topography without the need for individual specific axonal labels (Bodick and Levinthal, 1980; Bonhoeffer and Gierer, 1984; Horder and Martin, 1978; Rager, 1980).

In vitro approaches to these problems involve examination of the intrinsic mechanisms of nerve fiber growth and investigation of environmental influences on this process. With an *in vitro* approach, one can clearly observe growing nerve fibers, and high resolution morphological and cytochemical studies of growing nerve fibers are practical. In addition, potential modulators of nerve fiber growth can be thoroughly evaluated *in vitro*. In this chapter, we will describe a model for nerve fiber elongation that emphasizes the motile nerve tip. The *in vitro* evidence for regulation of nerve fiber elongation by four factors—adhesive contacts with the substratum, interactions among nerve fibers, chemotactic responses, and electrical currents in the environment—are discussed in relation to the activity of the nerve fiber tip. At the end, we return to *in vivo* work and exercise conclusions from *in vitro* work to help us understand how nerve fibers extend from neuronal perikarya to synaptic partners. A more extensive treatment of this subject is made by Rogers and Letourneau (1986).

II. Neurite Formation *in Vitro*

A. What Is the Behavior of a Growing Neurite?

In tissue culture embryonal neurons form cytoplasmic extensions that cannot be distinguished as axons or dendrites without additional electrophysiological or structural evidence. This may not matter if the growth of immature axons and dendrites share many features. In any case, the general terms neurite or nerve fiber will be used for these structures.

1. Protrusion and Regression

The most striking feature of neurite elongation is motility of the neurite tip, called the growth cone because of its shape and presumed function (Cajal, 1890). There is a continuous advance and retreat of the growth cone margin, like the ocean's lapping on a sandy shore. The cell margin is protruded as filopodia and lamellipodia, which extend, wave, and contact other surfaces to explore their environment (Godina, 1963; Harrison, 1910; Hughes, 1953; Luduena and Wessells, 1973; Nakai and Kawasaki, 1959). Most protrusions are short-lived and eventually regress into the growth cone, but those which adhere to other surfaces persist and can play a significant role in neurite elongation.

The regression of filopodia and lamellipodia typifies the backward movements that are characteristic of motility at the nerve tip. Frequently, particles are conveyed back on top of a protrusion and stop where the growth cone joins the neurite (Bray, 1970). Smaller complexes of plant lectins with their polysaccharide surface receptors also move back along filopodia and lamellipodia and form aggregates or caps on the growth cone (Carbonetto and Argon, 1980; Letourneau, 1979a). All these backward movements are inhibited by cytochalasin B, which disrupts microfilaments, suggesting that a common mechanism exerts force to move structures within the growth cone. Rearward movements may retrieve nonadherent protrusions and may collect surface-bound substances (including active molecules, like nerve growth factor) for internalization. However, we will discuss a more central role for these forces in neurite growth.

The vigorous motility of the growth cone contrasts with a low level of protrusion along the neurite itself. Few filopodia are extended along a neurite, and particles attached to the neurite membrane do not move forward or backward (Bray, 1970). Thus, the nerve tip expresses locomotory properties involving the cell membrane and cytoskeleton that are not common elsewhere in the neuron. A good source for further reading on nerve growth cones is Kater and Letourneau (1985).

2. Adhesion

As important to neurite elongation as protrusion are the adhesive interactions of a growth cone. Cellular adhesions are noncovalent bonds between cell surface molecules and components of other cells or surfaces (Bell, 1978; Curtis, 1973; Edwards, 1977; Frazier and Glaser, 1979; Trinkaus, 1984; Weiss, 1977). These bonds may involve molecular recognition, as between enzymes and substrates or antibodies and antigens. Neurite extension requires

Table I
Percentage of Neurons with an Axon(s) at 24 h *in Vitro*

Age of embryo[a] (days)	Substratum	Neurons with axon(s) (% ± SD)
8	PORN or PLYS	54 ± 8.7 (n = 739)
8	Tissue culture	25 ± 2.6 (n = 642)
6	PORN or PLYS	31 (n = 165)
6	Tissue culture	7 (n = 123)
4	PORN or PLYS	21 ± 3.9 (n = 439)
4	Tissue culture	11 ± 1.6 (n = 626)

[a]Data for 8-day embryonic neurons are the combined results from five experiments: 6-day neurons from one experiment, and 4-day neurons from three experiments. Neurons were identified as spherical, refractile cells with an approximate diameter of 15 μm. Axons were identified as thin processes with a terminal growth cone, which had visible microspikes. n, Number of neurons counted. From Letourneau (1975a), with permission of Academic Press, Inc.

a surface to which they can adhere (Harrison, 1914), whether a thread of spider web, a matrix of agar (Strassman *et al.*, 1973) or collagen (Ebendal, 1976), or a tissue culture dish. When the growth of embryonal sensory neurons was compared on two substrata differing in a coating of a polycationic molecule, polyornithine (PORN), growth cones adhered avidly to the coated surface, and the initiation, elongation, and branching of nerve fibers were much greater on PORN (Letourneau, 1975a, Tables I and II; Fig. 1). In addition, growth cones were larger, longer, and more filopodia were protruded, and nerve fibers were crooked (Figs. 2, 3, 4), not straight and taut as on the untreated substratum (Letourneau, 1975a, 1975b; Luduena, 1973b). We propose that the enhancement of neurite initiation and growth on polyornithine is due to the stabilizing force of adhesive bonds.

Cell adhesion to glass coverslips can be seen with interference reflection microscopy, which visualizes the contours of the lower cell surface, particularly adhesive sites separated from the substratum by 10–30 nm (Izzard and Lochner, 1976). Neurons on untreated glass have a few adhesive contacts beneath the growth cone and the cell soma (Letourneau, 1979b). Filopodia lack adhesive contacts for the most part or adhere for only a short duration. Just as surface movements are infrequent along a neurite, few adhesions occur beneath neurites, again contrasting the locomotory properties of the nerve tip versus the nerve fiber proper.

On a polyornithine-treated surface, adhesion is much greater. Expanded growth cones have large contacts with the substratum, but more specifically

Table II
Average Axon Lengths after 24 h *in Vitro*

Expt[a]	Substratum	Average lengths: μm/axon	Average lengths: μm of axon/neuron
1	PORN	291	626
	Tissue culture	68	106
2	PORN	245	524
	Tissue culture	115	189
3	PORN	285	651
	Tissue culture	105	168
Total	PORN	275	600
Total	Tissue culture	96	154

[a] The lengths of the axons of 20 randomly selected neurons were measured on each experimental substratum. In each of the three experiments the statistical significance of the differences in μm of axon/neuron between neurons on PORN and on tissue culture plastic was $p < 0.0002$, by use of the Mann-Whitney U Test. From Letourneau (1975a), with permission of Academic Press, Inc.

related to motility, filopodia and lamellipodia make linear adhesions which often extend from their distal tip inward beneath the growth cone (Figs. 5, 6). Only filopodia and lamellipodia make these linear adhesions. Although the proximal portions of growth cones, neurites, and neuronal somata make extensive adhesions to PORN, these contacts are not discrete linear adhesions.

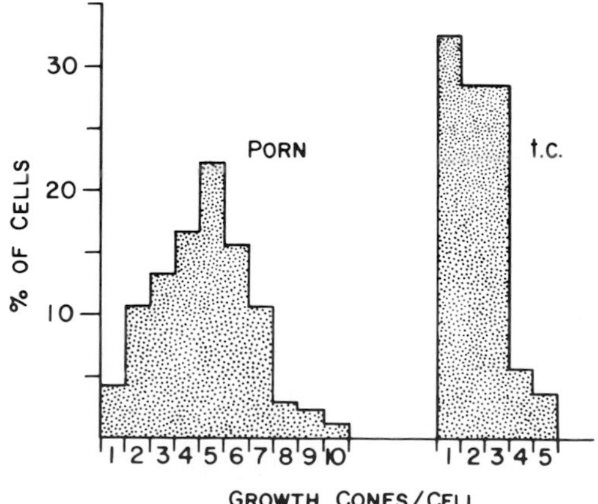

Fig. 1 Frequency distribution of the number of growth cones per neuron on PORN-treated plastic and on untreated tissue culture (t.c.) plastic, after h *in vitro*. From Letourneau (1975a).

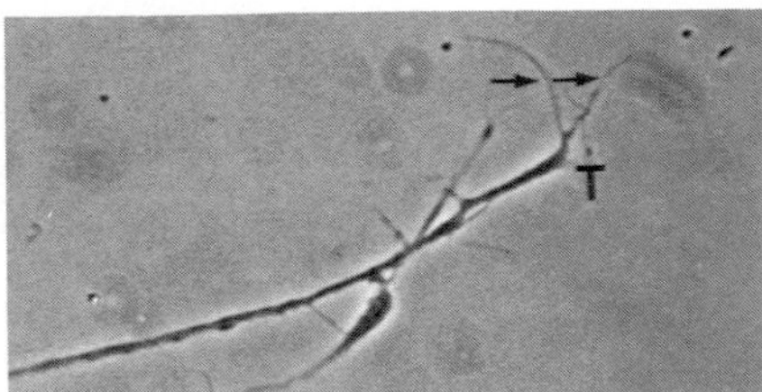

Fig. 2. Neurite tip of a chick embryo sensory neuron cultured on untreated glass. Filopodia (arrows) protrude from the neurite, but the tip itself (T), is only slightly expanded. From Letourneau (1979b). ×1050.

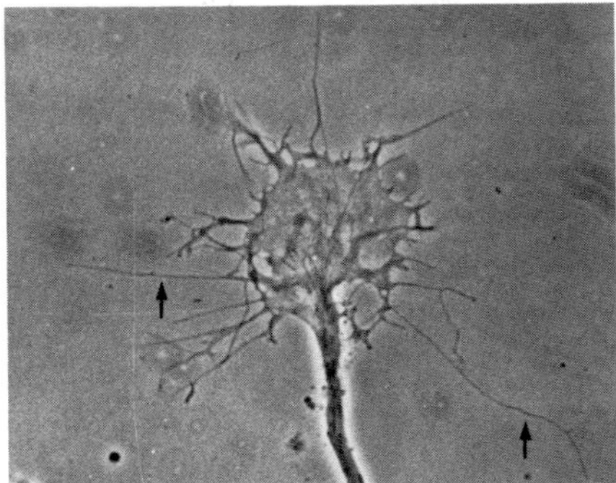

Fig. 3. Neurite tip of a sensory neuron cultured on polyornithine-treated glass. Spreading of the neurite tip is clearly evident, as is the extreme length of some filopodia (arrows). From Letourneau (1979b). ×1125.

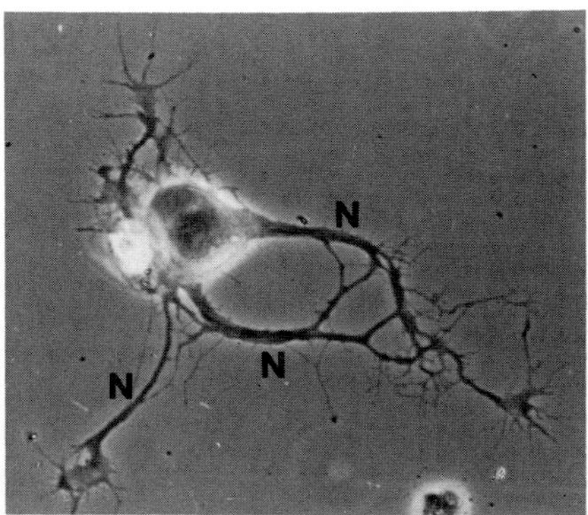

Fig. 4. A sensory neuron cultured 20 h on polyornithine-treated glass. The curved neurites (N) indicate high adhesion to the substratum. From Letourneau (1975a). ×550.

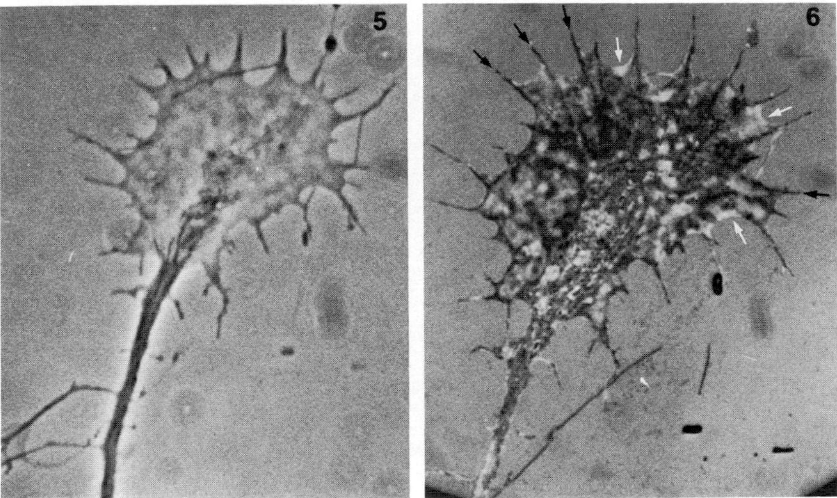

Fig. 5. Phase contrast photo of a well-spread growth cone on polyornithine-treated glass. From Letourneau (1979b). ×1500.

Fig. 6. Interference reflection micrograph revealing the adhesive contacts of this growth cone. Black-appearing adhesive contacts underlie filopodia and continue inward beneath the growth cone margin (black arrows). White arrows indicate concave portions of cell margin widely separated from substratum. From Letourneau (1979b). ×1800.

3. *Summary*

The behavior of a growing nerve fiber can be summarized as continuous protrusion and regression of cytoplasmic processes from the nerve tip. These protrusions can establish adhesive contacts that counteract the backward forces that otherwise retract the protrusions. In the following, we shall relate the motile behavior of the nerve tip to nerve fiber elongation.

B. Structure of a Growing Neurite

The differences in surface movements and adhesive contacts between growth cones and neurites are based upon structural differences. Before the mechanism of nerve fiber elongation is considered, the structure of growth cones and nerve fibers will be compared.

1. *The Neurite*

At the core of a neurite is the cytoskeleton. Ten-nm neurofilaments and 25-nm microtubules are oriented along the neurite axis (Bunge, 1973; Palay and Chan-Palay, 1977; Lasek and Hoffman, 1976; Tennyson, 1970; Yamada

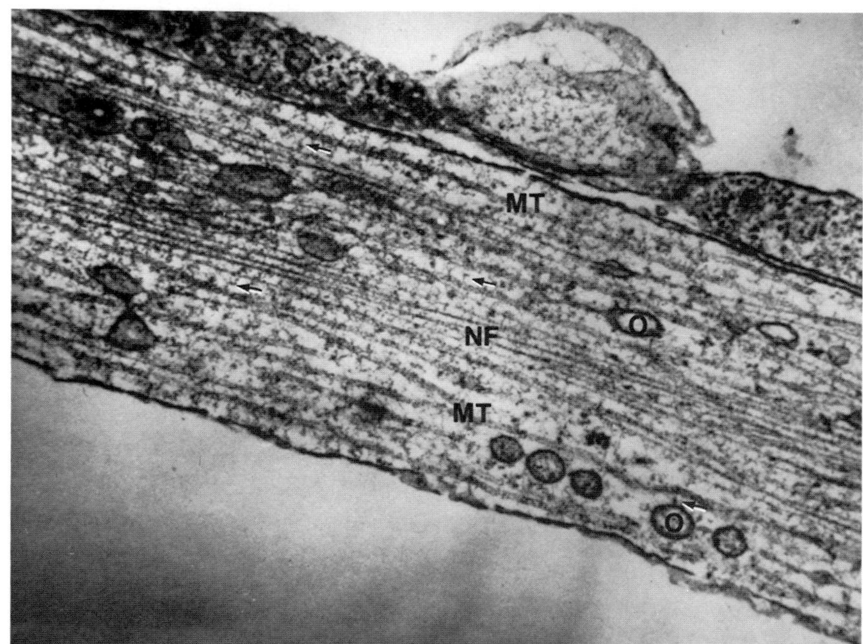

Fig. 7. Longitudinal section along a neurite of a sensory neuron. Neurofilaments (NF) and microtubules (MT) are conspicuously long and straight. Microfilaments form an indistinct network beneath the plasmalemma and centrally among the other cytoskeletal fibers. Short microfilament segments (arrows) seem to contact neurofilaments and microtubules, as well as membranous organelles (O) that may be in transit within the neurite, From Yamada *et al.* (1971), with permission of the authors and Rockefeller University Press. ×31,000.

et al., 1971). These fibers extend into the base of the growth cone, though it is not clear how long the individual fibers are (Bray and Bunge, 1981). The remarkably straight course of these fibrous structures suits them for support and for acting in axonal transport (Fig. 7). Pharmacological studies implicate microtubules in these functions (Daniels, 1973, 1975), but no strong experimental evidence supports a role for neurofilaments in particular activities of neurite elongation. A third cytoskeletal component, microfilaments, do not form long parallel fibers. Rather, microfilaments, as well as other cytoskeletal-associated proteins, form a three-dimensional network, or matrix, composed of crisscrossed fibers, short segments which meet at many points, or articulated, sharply bent fibers (Bunge, 1973; Hueser and Kirschner, 1980; Tsui *et al.*, 1983; Yamada *et al.*, 1971). This network lacks orientation, fills the cell cortex, attaches to the plasma membrane at many points, and penetrates among the neurofilaments and microtubules apparently to form short bridges between the longer fibers and other neuritic components (Ellisman and Porter, 1980; Hirokawa, 1982; Schnapp and Reese, 1982; Yamada *et al.*, 1971). Because microfilaments contain actin, the network may

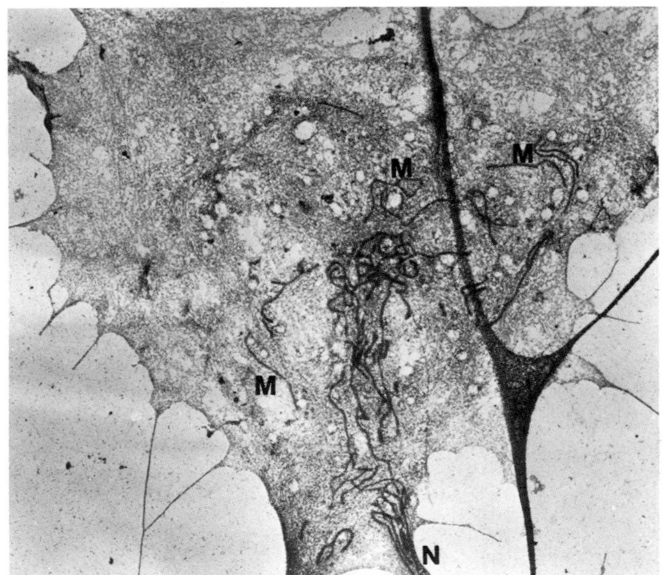

Fig. 8. A whole growth cone cultured on a Formvar-coated gold grid. Mitochondria (M) swirl around within the base of the growth cone. The neurite (N) extends from below the photograph. ×3300.

be involved in such movements as axonal transport and the infrequent protrusion of filopodia along a neurite (Bray *et al.*, 1978; Chang and Goldman, 1973; Ellisman and Porter, 1980; Fernandez and Samson, 1973; Wessells *et al.*, 1978).

Besides the plasma membrane, membranous structures within the neurite include mitochondria, lysosomes, smooth sacs, and vesicles (Bunge, 1973; Droz, *et al.*, 1975; Teichberg and Holtzman, 1973). Presumably, many are in transit between the cell soma and nerve tip in both directions. Many of the sacs and vesicles rapidly incorporate exogenous tracers (Birks *et al.*, 1972; Bunge, 1977; Wessells *et al.*, 1974), suggesting endocytic origin, but other membranous structures may be outward-bound precursors of the plasma membranes of the neurite and growth cone or may contain secretory products, such as neurotransmitters and trophic substances.

2. *The Growth Cone*

At the neurite tip, neurofilaments and microtubules splay out into the base of the growth cone, entangled with mitochondria and other vesicular elements (Fig. 8). The motile periphery of the growth cone contains a network of microfilaments, contacting the plasmalemma at many points (Bunge, 1973; Yamada *et al.*, 1971). Much of this network is not polarized, but in filopodia

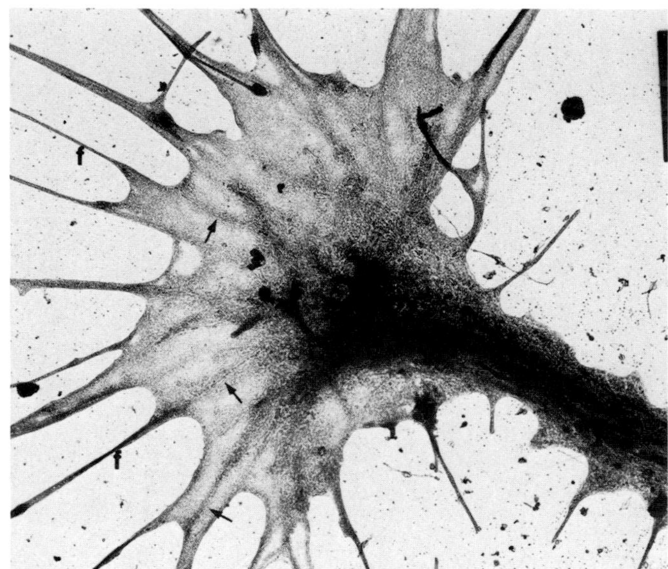

Fig. 9. A whole growth cone cultured on a Formvar-coated gold grid. Note the linear cytoplasmic densities (arrows) aligned on the same axes as filopodia (f) extended from the cell margins (arrows). The distribution of these denser cytoplasmic areas resembles the linear adhesive contacts seen with interference reflection microscopy. From Letourneau *et al.* (1979b). ×3700.

and lamellipodia, microfilaments align into bundles, which often extend centrally and have interesting associations with other organelles (Letourneau, 1979b; Luduena and Wessells, 1973; Figs. 9, 10, 11), particularly microtubules that extend forward from the neurite. Actin and myosin are concentrated in the growth cone into linear arrays that correspond to these microfilament bundles (Kuczmarski and Rosenbaum, 1979a,b; Letourneau, 1981; Figs. 12, 13). In addition, these linear concentrations of actomycin are often situated where the growth cone margin makes linear adhesions beneath filopodia and lamellipodia (Letourneau, 1979b, 1981; Figs. 14, 15). Thus, the margins of growth cones contain actomyosin complexes that attach to adhesive sites and may act like a simple sarcomere to exert tension on other surfaces or intracellular structures.

C. Mechanism of Nerve Fiber Elongation

This section presents a model for neurite elongation to be used in discussing the means by which axons find their targets. The model proposes that the motility of the nerve tip, a common cellular activity, promotes the assembly of a unique cellular structure, the neurite. In the model, expansion of the neurite membrane and elongation of the neurite cytoskeleton will be defined

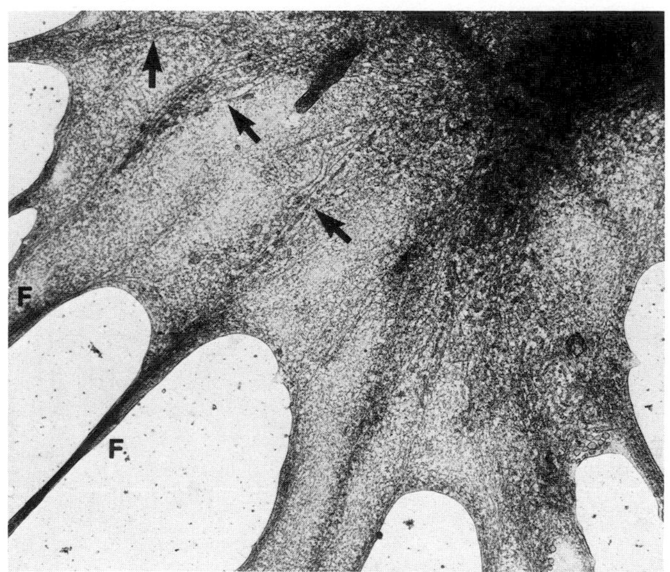

Fig. 10. Higher magnified view of same growth cone as seen in Fig. 10, showing that microtubules and small agranular vesicles (arrows) are aligned on the axis of several filopodia (F). From Letourneau (1979b). ×10,500.

in tcrms of basic motile behaviors: protrusion, adhesion, and the generation of force.

1. Expansion of the Neurite Membrane

The growth cone margin expands when filopodia and lamellipodia extend, but regresses, with endocytotic uptake of membrane, when protrusions withdraw. If these activities are equal, the net membrane area does not change. However, adhesion will stabilize protrusions against retraction so that the growth cone margin becomes greatly expanded on an adhesive surface (Letourneau, 1975a, 1979b). In this case much surface membrane is available to be incorporated into the elongating neurite. Since protrusive activity is largely restricted to the nerve tip, the neurite membrane may be expanded principally at the nerve tip. This conclusion is supported by indirect evidence that the neurite membrane does not move proximal to the nerve tip as the neurite grows and can even be immobilized by high adhesion to a surface without slowing neurite growth (Letourneau, 1975a, 1979b; Ludueña, 1973b).

How the membrane expands at the nerve tip is controversial. Preformed membrane vesicles may migrate distally in the neurite and fuse with the surface of the nerve tip as during exocytosis (Bray, 1973b; Pfenninger and Maylie-Pfenninger, 1981; Pfenninger and Johnson, 1983). Tracer studies indicate

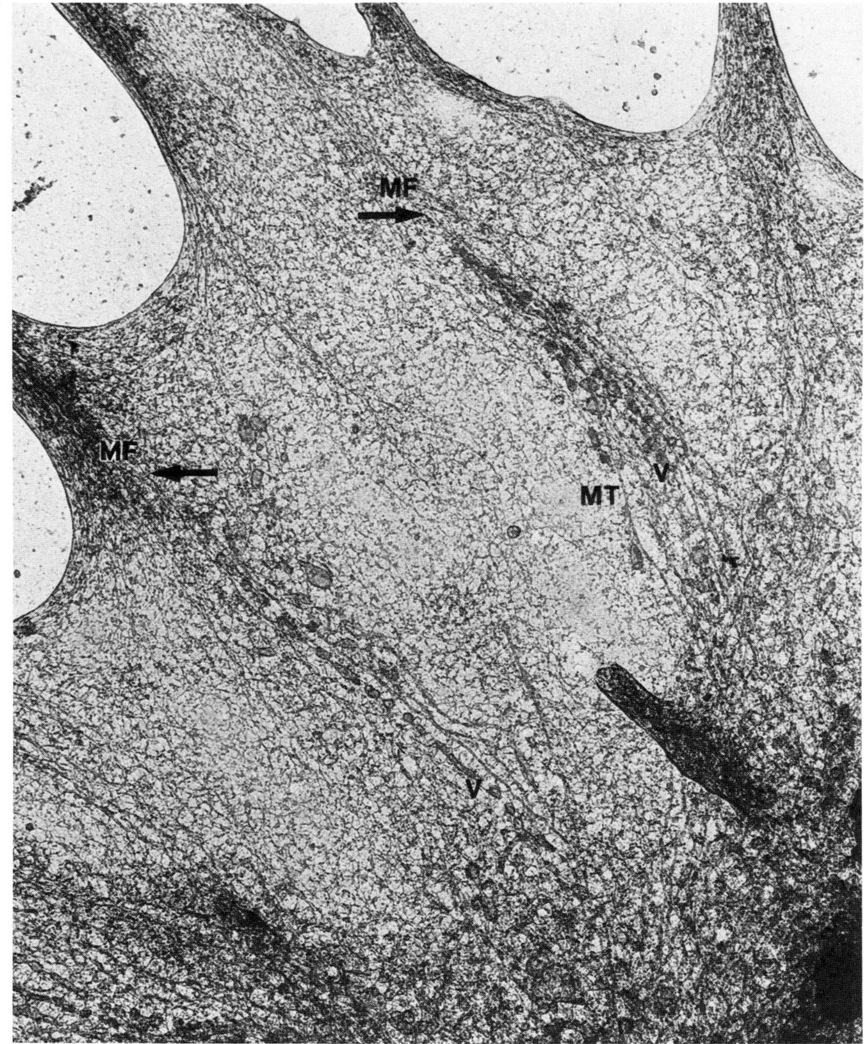

Fig. 11. Same growth cone as seen in Figs. 9 and 10. Contacts of the aligned vesicles (V) and microtubules (MT) with bundles of microfilaments (MF) can be seen (arrows). This interaction may be responsible for the positions of these vesicles and microtubules in the growth cone margin. From Letourneau (1979b). ×24,500.

that some vesicles in the growth cone are endocytic and may be recycled (Steinmann *et al.*, 1983), but others may be new plasma membrane precursors. As an alternative, single molecules or small aggregates may be inserted into the plasma membrane from the cytoplasm, and perhaps also migrate in the

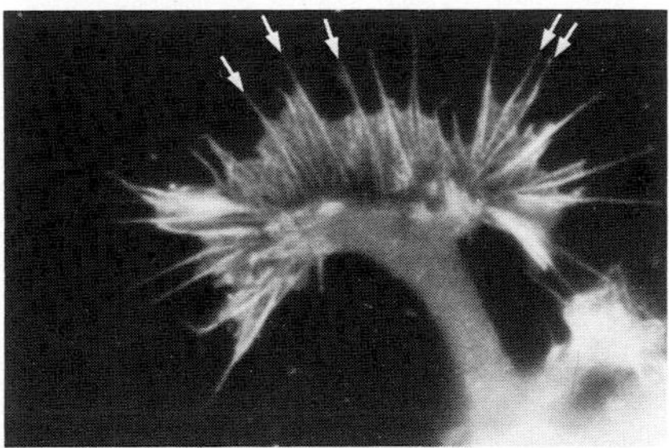

Fig. 12. Immunocytochemical localization of actin in a growth cone of a sensory neuron. Fluorescent antibodies are concentrated into linear arrays (arrows) in the growth cone margin, which often is aligned with filopodia. These may coincide with microfilament bundles. From Letourneau (1981). ×1075.

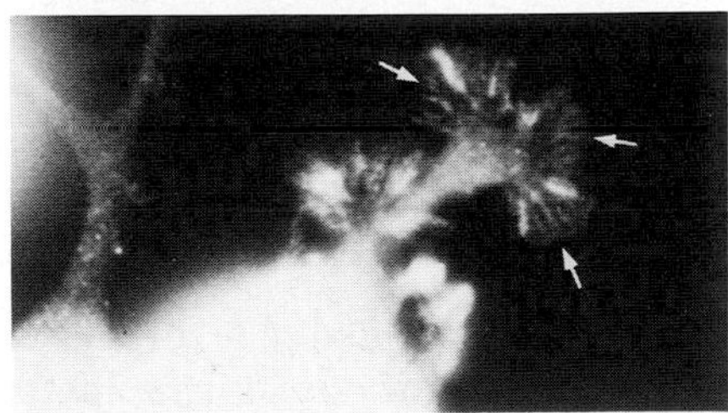

Fig. 13. Immunocytochemical localization of myosin in a growth cone of an NGF-treated PC12 cell. Linear arrays of fluorescence (arrows) are seen with the same type of distribution as actin, indicating that actomyosin complexes may exist in the growth cone margin. From Letourneau (1981). ×1075.

plane of the membrane (Carbonetto and Fambrough, 1979; Small and Pfenninger, 1984; Small *et al.*, 1984). These forms of membrane expansion are not mutually exclusive, but both may occur, depending on the nature of particular membrane molecules (Rothman and Lenard, 1977).

2. *Growth of the Cytoskeleton*

Microtubules and neurofilaments are present all along the neurite, and many terminate in the growth cone. Microtubules always extend farther forward than neurofilaments and occasionally reach the bases of filopodia (Letourneau, 1979, 1983). This is significant because these cytoskeletal fibers grow by addition of subunits to an end and not by intercalation within a fiber (Kirschner, 1980; Margolis and Wilson, 1978). Thus, the ends of these fibers, particularly microtubules, may determine where the cytoskeleton is assembled (Heidemann *et al.*, 1981). Because far more fibers terminate in the growth cone than elsewhere in the neurite, it maybe the major site for cytoskeletal

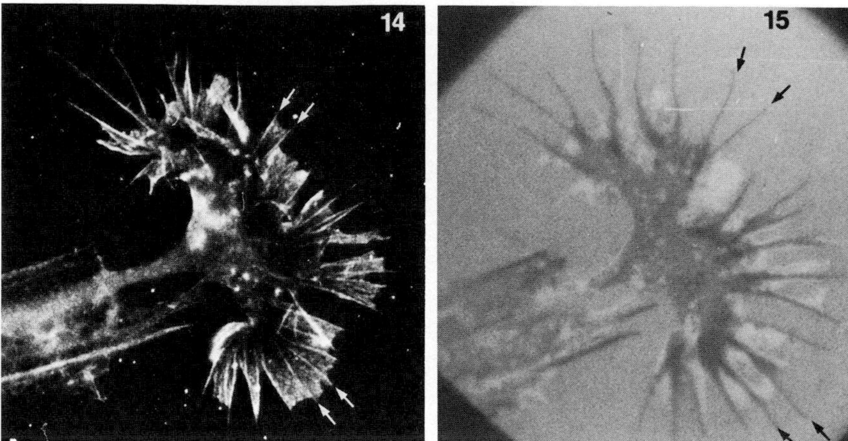

Figs. 14, 15. Immunocytochemical localization of actin in a growth cone and interference reflection image of growth cone–substratum adhesion of a sensory neuron. Many linear actin arrays (white arrows) coincide with linear adhesions (black arrows) of the growth cone margin to the substratum. From Letourneau (1981). ×1075.

growth in the elongating neurite, provided that sufficient tubulin monomer is present to drive polymerization at the distal microtubule ends (Letourneau, 1982).

3. Microfilaments and Actomyosin

We have noted that actin and myosin are colocalized into bundles of microfilaments in the growth cone margin (Kuczmarski and Rosenbaum, 1979a,b; Letourneau, 1981, 1983; Roisen *et al.*, 1978; Figs. 14, 15). This establishes a potential for a localized generation of actomyosin-like mechanical forces. In the contracting sarcomere of striated muscle, actin filaments slide past immobile myosin filaments, exerting force on the attachments of actin filaments to the Z-line. In the nonmuscle cell, contractile filaments must also be associated with other structures to transmit mechanical energy, but the degree to which actin and myosin filaments slide may vary, depending both on the masses of structures linked to the actin and myosin filaments and on other forces impinging on these structures. Actin filaments in nonmuscle cells are linked to the plasma membrane with the same polarity as actin filaments are linked to the Z-line (Begg *et al.*, 1979; Kuczmarski and Rosenbaum, 1979b, Mooseker, 1976; Small *et al.*, 1978). The associations of myosin in nonmuscle cells are not clear, but may include the plasma membrane, membranous organelles, and cytoskeletal fibers (Herman and Pollard, 1981; Mooseker *et al.*, 1978; Shizuta *et al.*, 1976). Equally important to the presence of actin and myosin is the distribution of actin-associated proteins, which regulate several aspects of actin organization, as well as the interaction of actin with other cytoskeletal components and membranous structures (Stossel, 1984; Vallee *et al.*, 1984).

4. *Action of Actomyosin*

How do actomyosin complexes work in nerve fiber elongation? The motility of the nerve tip includes protrusion of the growth cone margin, formation of adhesive contacts, and frequent withdrawal or retraction of the extended protrusions. Retraction may occur when tension is exerted on the filament-to-membrane attachments of protrusions that are only weakly adherent. If, however, a protrusion has strong adhesions, actin filaments may not slide, but will, instead, remain anchored to the membrane at the contact site and draw the myosin filaments and associated structures forward, Thus, precursors of the plasmalemma, neurofilaments, and microtubules will be advanced in the growth cone if they are linked to the actomysin complex (Pollard *et al.*, 1984). This may promote neurite growth by drawing new membrane in the form of vesicles to the front of the growth cone, as well as by pulling neurofilaments and microtubules forward.

Electron microscopy of growth cones presents a picture that, although static, supports our hypothesis of how actomyosin acts in the growth cone. Microtubules, neurofilaments, membranous sacs, and vesicles are generally absent from the growth cone margin. However, where filopodia protrude, small bundles of microfilaments often project inward within the cell margin to meet microtubules, neurofilaments, and membranous structures, emerging from the neurite and aligned with the axis of the filopodia (Letourneau, 1979b). It seems as though these organelles are drawn out from the base of the growth cone by virtue of their associations with microfilament bundles (Figs. 9, 10, 11).

Until recently, the protrusion of filopodia or lamellipodia from growth cones was viewed as necessary for neurite elongation. In a previous study, neurite growth was rapidly inhibited by addition of the drug cytochalasin B, which disrupts the actin filament network at the neurite tip (Yamada *et al.*, 1971). However, we found that on an adhesive substratum neurites elongate in the presence of cytochalasin B without exhibiting filopodial or lamellipodial motility (Marsh and Letourneau, 1984). Actin is seen in these neurites as small densities from which short actin filaments project, and what functions actin may have in the presence of cytochalasin are unknown. Yet, these results illustrate that the typical cyclic protrusive behavior and the tensions produced within filopodia and lamellipodia are not a required part of neurite elongation. Rather, we suggest that the elongation of a nerve fiber involves most strictly the transportation and organization of the cytoskeletal and membranous components of neurite structure (Lasek, 1982; Lasek *et al.*, 1984; Wujek and Lasek, 1983). These activities can be influenced and promoted by the tensions and other motile behavior at the growth cone, but they are not dependent upon such behavior at the neurite tip. Environmental cues to direct neurite elongation may operate by modulating the motile behavior of the neurite tip and the way it influences the elongation process.

III. Regulation of the Route of Neurite Growth *in Vitro*

Normally, elongating axons reach synaptic targets along characteristic routes. Yet, axons can also reach their usual targets when forced to enter a foreign environment (Arora and Sperry, 1962; Attardi and Sperry, 1963; Hibbard, 1965; Lance-Jones and Landmesser, 1978, 1980, 1981b). In addition, axons respond to some manipulations in their environment with repeatable new growth patterns (Constantine-Paton and Capranica, 1975; Katz and Lasek, 1979; Stirling and Summerbell, 1978). These observations indicate that both intrinsic cellular properties and extrinsic environmental features determine axonal pathways.

The importance of the nerve tip makes it the focus of our examination of the regulation of nerve fiber growth. Because we believe that the motility of the nerve tip promotes neurite elongation, potential modulators of neurite growth will be examined in terms of their effects on growth cone motility.

A. Cell–Substratum Adhesion

1. Effects of Adhesion of Growth Cone Activity

When growth cone–substratum adhesion is strong, neurons sprout many neurites that grow faster and are more branched (Letourneau, 1975a) than they are on a less adhesive surface. These data can be explained by the influence of adhesive contacts on the application of mechanical forces in the growth cone. Actomyosin complexes linked to plasmalemmal areas with strong adhesions are more able to pull the neurite and its organelles forward than are complexes in weakly adherent areas. Thus, adhesive substrata increase the effectiveness of growth cone motility in advancing the neurite, leading to faster growth.

The role of adhesion in neurite initiation is illustrated by Collins's finding that embryonal ciliary neurons sprout neurites shortly after exposure to a substratum that has absorbed components of heart conditioned medium (Collins, 1978a). Just before neurites emerge, filopodia, which previously protruded but adhered infrequently in the absence of conditioned medium, attach to the substratum, which now contains components of the conditioned medium (Collins, 1978b). Perhaps actin filaments and their membrane attachments become well anchored and no longer withdraw the cell margin during actomyosin contractions. Instead, mechanical force is exerted in new

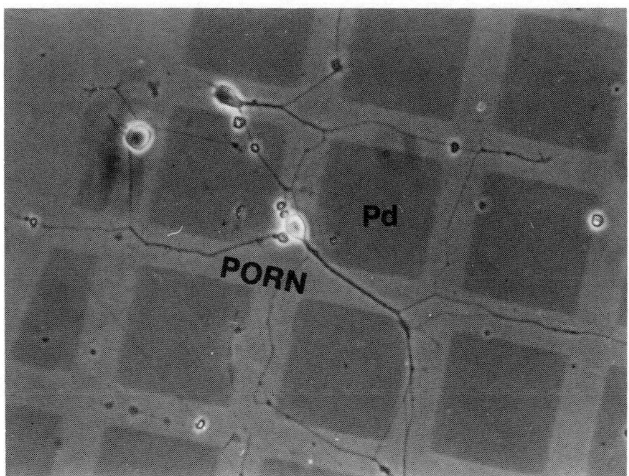

Fig. 16. Preferential extension of neurites on the more adhesive portions of a patterned substratum. Growth cones branch, turn, or continue straight upon the PORN-treated plastic (PORN) and rarely cross onto the palladium (Pd)-coated squares, which is a less adhesive surface. From Letourneau (1975b). ×140.

directions in the cell margin to initiate the organization and extension of a neurite from adjacent cytoskeletal components in the cell soma (Wessells, 1982).

Because adhesions stabilize actomyosin complexes in a local manner, neurite organelles are drawn forward only in the vicinity of adhesive contacts. Consequently, if the adhesive contacts of protrusions are stronger or more frequent at one side of the growth cone than elsewhere, neurite organelles will advance to the adherent side, and the neurite will grown in that direction. This has been demonstrated using substrata with patterned variations or adhesiveness (Letourneau, 1975b; Fig. 16). In addition to moving organelles, mechanical tensions may be transmitted via networks of cytoskeletal associations from regions of strong cell–substratum adhesion to weakly adherent areas, causing the less adherent portions to become detached (Bray, 1979, 1982). This action will also direct neurite growth toward stronger cell–substratum adhesion.

The local effects of adhesion on force generation may also explain increased neurite branching on adhesive substrata (Fig. 1). Neurite branches arise by the division of a growth cone into smaller growth cones (Bray, 1973a). On a PORN-treated substratum, neurite branching can be induced by detaching the middle of a broad growth cone with a microneedle (Wessells and Nuttall, 1978). This manipulation may isolate the outwardly directed tensions exerted in the remaining sides of the growth cone and prompt a division of the growth cone and its contents. On an adhesive substratum, neuritic precursors may

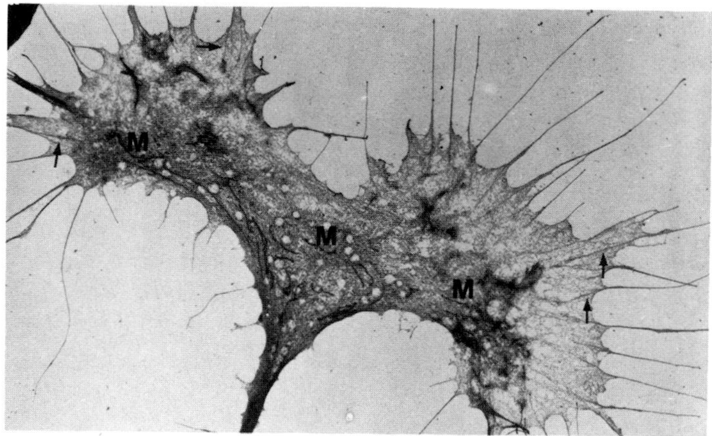

Fig. 17. A growth cone in the process of dividing into two neurite branches. Mitochondria (M) have moved into the two diverging sides of the growth cone, and bundles of microfilaments (arrows) are present in the growth cone margin at the two sides. Compare distribution of these bundles to the actin fibers seen in Fig. 14. ×4000.

often divide into several streams when they are pulled forward by actomyosin complexes firmly anchored at the periphery of a spread growth cone (Fig. 17).

2. *Contact Guidance of Neurite Growth*

These conclusions lead to the proposal that variations in neuron–substratum and growth cone–substratum adhesion can determine the time and place of neurite initiation, axonal branching patterns, and axonal routes. These variations may arise from the distribution of adhesion-mediating surface molecules in gradients and other spatial patterns, but even the mere shape of the substratum may influence axonal growth. Weiss (1934) showed that neurites follow tension lines in a stretched plasma clot, and named the phenomenon *contact guidance* (Weiss, 1941, 1961). Reevaluation of the motility of fibroblasts suggested that fibroblastic cell movement may be restricted along ridges or in grooves because microfilament bundles cannot assemble and operate in association with adhesive sites in the bent states required to protrude and spread the cell margin over sharp ridges or in steep grooves (Dunn, 1982; Dunn and Heath, 1976). As a corollary statement, protrusion and cell movement is favored along a surface whose shape allows the organization in motile regions of more or larger actin filament bundles with plasmalemmal insertions. The guidance of neurites along oriented collagen matrices may reflect the larger and greater number of contacts, with consequent influences on actomyosin activity, that filopodia can make when extended along the axis of collagen fibrils rather than by contacting fibrils from the side (Dunn and Ebendal, 1978; Ebendal, 1976). In addition, mechanical forces within filopodia will be effectively used to advance the nerve tip when exerted along the axis of a collagen fibril that has great tensile strength, but force will be lost for neurite growth if filopodia pull on fibrils from the side and move them sideways in the loose matrix.

3. *Spatial Variations in Adhesive Ligands*

There is *in vitro* evidence that the surfces of neurons contain variation in adhesive ligands that are related to cell position and may be important in directing nerve process growth. The adhesive affinities of embryonal retinal cells for explanted optic tecta mimic the pattern of retinotectal synapses, suggesting that neuronal specificity is expressed in the topographic distribution of cell surface adhesive components (Barbera, 1975; Gottlieb *et al.*, 1976; Gottlieb and Glaser, 1980; Marchase, 1977). However, to be more relevant, the adhesive preferences that retinal cell bodies show in adhesion assays must be expressed by the growth cones of retinal ganglion cells, and the tectal surfaces with which retinal cells interact in adhesion assays must resemble the surfaces encountered by the ganglion cells' growth cones *in vitro.* An *in vitro* system being developed that allows one to manipulate the growth and interactions of retinal axons with tectal surfaces is an exciting prospect for additional information (Bonhoeffer and Huf, 1980; Halfter and Deiss, 1984; Halfter *et al.*, 1983).

The elucidation of adhesive interactions that guide neurite elongation has primarily involved the use of *in vitro* surfaces coated with materials derived from cells or extracellular materials (Carbonetto, 1984). Two purified glycoproteins that have received increasing attention with respect to neurite elongation are fibronectin, a component of plasma, basement membranes, and many connective tissues, and laminin, demonstrated by immunohistochemistry to be in many basal laminae (Akers *et al.*, 1981; Baron-VanEvercooren *et al.*, 1982; Carbonetto *et al.*, 1983). Neuronal culture on treated culture surfaces indicates that there are two distinct regions of the fibronectin molecule that neurons may bind to and gain anchorage for outgrowth. Interestingly, central versus peripherally derived embryonic neurons interact differently with these two domains of fibronectin, perhaps due to differential expression of cell surface receptors or binding sites for distinct sequences of the fibronectin molecule (Rogers *et al.*, 1983, 1985). Treatment of *in vitro* substrata with laminin promotes neurite outgrowth by both central and peripheral neurons, and cellular interaction with laminin may also involve specific interaction of the extracellular molecule with cell surface laminin receptors (Lestot *et al.*, 1983; Rao *et al.*, 1983).

Several groups have described neurite-outgrowth promoting factors that are released by several cell types into culture media or by tissue extraction. These are acidic molecules that bind to cationic surfaces and induce neurite elongation by mediating neurite–substratum adhesion, but the nature and diversity of these molecules have not yet been well defined (Adler *et al.*, 1981; Collins, 1978, 1980). The neurite-promoting factor may be or contain a proteoglycan and some preparations contain fibronectin or laminin (Lander *et al.*, 1982, 1984), though antilaminin fails to remove the neurite-promoting activity of the factor. Perhaps the factor is a complex of several active molecules, arranged and bound to the *in vitro* surface with some resemblance to their *in vivo* configurations.

A number of *in vitro* studies has shown that spatial variation in these surface-bound molecules can limit neurite elongation to pathways defined by higher adhesivity (Adler and Varon, 1981; Collins and Lee, 1984; Hammarback *et al.*, 1985; Smallheiser *et al.*, 1984). It is not known whether neurites *in vivo* are guided by similar pathways of extracellular materials or other surface-bound molecules. In early chick embryos laminin is concentrated along the pathway of the ventral roots and may promote elongation of motor axons toward the base of the limb bud (Rogers *et al.*, 1984).

B. Regulation of Neurite Growth by Contact with Other Cells

Because elongating axons, neuroblasts, and supportive cells often migrate in close association with axons or glial fibers, it is thought that many cell movements in the nervous system are guided by other cell surfaces (Rakic, 1971, 1981; Sidman and Wessells, 1975). For example, pioneering fibers may become preferred substrata for later axons by virtue of strong adhesive bonds between the following growth cones and the pioneer neurites (Goodman *et al.*, 1984). We will propose a means by which nerve tips become associated with pioneering fibers and also suggest that the total adhesive environment must be considered in interpreting cell–cell interaction.

Protrusions from nerve tips *in vitro* can contact and pull on nearby neurites with enough force to distort the neurite (Nakai, 1960; Nakai and Kawasaki, 1959; Nakajima, 1965; Wessells *et al.*, 1980; Figs. 18–21). If the adhesion is weak, the contact may be withdrawn; but if the contact resists tension, the nerve tip may establish more contacts and come to elongate along the neurite. Because new material is added predominantly at the nerve tip, the neurite proximal to the extending tip need not release its adhesions with adjacent neurites as the growth cone advances.

Whether a growth cone follows a pioneer axon may be determined by the resistance of these initial filopodium–neurite contacts to tension applied within the filopodium. Thus, adhesive affinities between interacting neurites and nerve tips might regulate which fasicle a neurite joins and, hence, what target area is entered (Bray *et al.*, 1980; Goodman *et al.*, 1984; McKay *et al.*, 1983). Even within a bundle, neurite–neurite relationships may change as growth

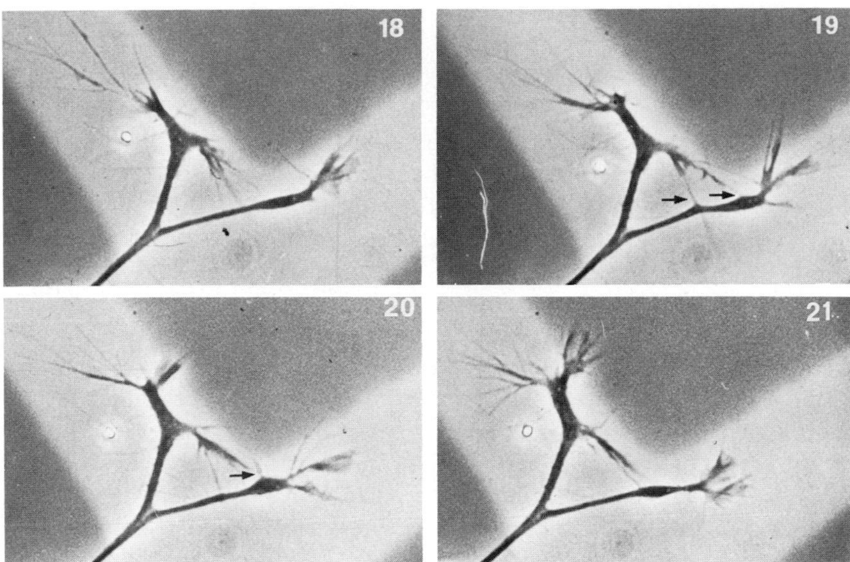

Figs. 18–21. Frames from a time lapse movie, showing the exertion of force by filopodia upon contact points with an adjacent neurite branch. The neurite branch is transiently distorted (arrows) at the point of attachment with a filopodium. The elapsed time between successive frames is 18–19, 230 sec; 19–20, 113 sec; 20–21, 524 sec. ×650.

cones maximize their adhesive contacts. This may organize a topographic distribution of fibers within a tract (Easter *et al.*, 1984; Rusoff and Easter, 1979), provided that fibers of neighboring cells adhere more to each other than to fibers of distant origin.

From study of cultures of sensory ganglia in plasma clots, Dunn (1971) described a repulsion between neurites. He proposed that this repulsion inhibits fasciculation and direct outgrowth radially. This reaction stemmed from growth cone withdrawal from sites of filopodial contact with adjacent neurites. Yet, growth cones from neurons cultured on adhesive flat surfaces, such as PORN, did not withdraw from cell–cell contacts; instead, filopodial contact was followed by movement closer to the other neurite or cell (Wessells *et al.*, 1980). We believe that the initial contacts were similar in both cases, but the different adhesive environments favored different results. In a clot, the contacts of growth cones and filopodia with thin, flexible fibers of the matrix are weak, so that when a filopodium pulls on another cell, few other adhesions help the filopodium resist the tension, and the filopodium is frequently retracted. Thus, actual repulsion is not needed to explain this withdrawal of neurites from one another. On the other hand, when the nerve tip gains traction from contact with an adhesive substratum, contractile forces in a protrusion are less likely to withdraw it from contact with another neurite, even if an initial cell–cell contact is broken.

On entering a target area, individual or small groups of axons leave a nerve bundle for particular zones of innervation. *In vitro* studies indicate that this exit may be prompted by decreases in fiber–fiber adhesion or by the availability of new, more adhesive substrata. When sensory ganglia are cultured on untreated substrata, the extending neurites form fascicles. In the presence of antisera to the nerve adhesion molecule (N-CAM), neurite–neurite adhesion is inhibited but not neurite–substratum adhesion, and fascicles do not form; instead neurites leave the explant as individual fibers (Edelman, 1983; Rutishauser, 1985; Rutishauser and Edelman, 1980; Rutishauser *et al.*, 1978). The same result occurs without addition of anti-CAM, when ganglia are cultured on substrata to which growth cones adhere very tightly (P. C. Letourneau, unpublished data). In this case, single fibers are extended, as growth cones prefer the adhesive substratum over the surfaces of other neurites. A recent model proposes that hierarchical adhesive interactions among optic fiber growth cones, other optic fibers, and tectal cells are expressed in sequence to determine retinotectal connectivity (Fraser, 1980). Optic fiber growth cones may follow other fibers and surfaces in the optic stalk until they reach the tectum, where the fibers spread across the newly available tectal surfaces of greater adhesivity. This is supported by recent evidence that retinal growth cones choose to extend on a layer of tectal cells rather than on retinal cells (Bonhoeffer and Huf, 1980).

How many adhesion molecules influence neurite growth in addition to N-CAM? Another surface protein, Ng-CAM, is present on neurites and binds a receptor on glia (Grumet *et al.*, 1984) and several other neuronal surface glycoproteins may also have adhesive functions (Schubert and LaCorbiere, 1982). N-CAM is heavily sialylated and the degree of sialylation modulates the self-association of N-CAM from cell-to-cell (Edelman, 1983). The precisely characteristic fasciculation that builds the developing central nervous system (CNS) of insects may depend on the distribution of adhesive ligands that give growth cones affinities for specific fascicles (Goodman *et al.*, 1984). In like manner, the spatial and temporal regulation of a rather limited repertoire of cell-surface ligands and receptors for molecules on other cells and extracellular matrices may determine the characteristic patterns of neuronal outgrowth without resorting to the need for a large mulitplicity of individual adhesive affinities.

C. Regulation of Neurite Growth by Chemotaxis

The notion that morphogenetic cell movements exhibit chemotaxis, that is, orientation relative to a gradient of soluble molecules, is old, though only recently has the chemotactic potential of vertebrate cells been examined in detail (Postlethwaite *et al.*, 1978; Ramsey, 1972; Trinkhaus, 1984; Schiffman, 1982; Zigmond, 1978). Cajal's initial proposal of neurotropism seems to be a proposal of chemotaxis, and since his time, both positive and negative evidence (Weiss and Taylor, 1944) has been reported. Explants of neural tissues will extend neurites preferentially toward a piece of tissue that is assumed to release a chemoattractant (Chamley *et al.*, 1973, Charlwood *et al.*, 1972; Coughlin, 1975; Ebendal and Jacobson, 1977). These results, however, are open to other interpretations, such as preferential survival or adhesion of those neurites closest to the tissue source. In addition, differences in the interactions of neurites with each other and with nonneuronal cells may produce the observed directed elongation. These possibilities are interesting and relevant to regulation of neurite growth, but they are *not* chemotaxis. To prove chemotaxis one should demonstrate that motility of the nerve tip is oriented by a chemical gradient.

The powerful action of nerve growth factor (NGF) on the metabolism and differentiation of sympathetic and sensory neurons is well known (Harper and Thoenen, 1980), but whether NGF is also a chemoattractant is a more controversial idea. *In vitro* studies show that neurites extend from receptive explants toward a source of NGF (Charlwood *et al.*, 1972), and a dramatic *in vivo* finding is that NGF, injected into the brains of young rats, induces abnormal growth of axons from peripheral sympathetic neurons into the spinal cord and up to the site of NGF injection (Levi-Montalcini, 1976). Because NGF is chemically well characterized, it is the focus of this discussion of the chemotactic response of growth cones.

When sensory neurons were cultured in agar matrices containing an NGF gradient, neurite growth displayed a partial, yet repeatable, orientation up the gradient (Letourneau, 1978). This showed a long-term response, but did not elucidate the short-term effects of a gradient on growth cones. However, when NGF was released from a pipette placed near nerve tips, the tips turned and extended toward the pipette within 20 min (Gunderson and Barrett, 1979, 1980). Thus, growth cone motility can be modulated by a chemical gradient.

The sensory responses of bacteria and leukocytes have been studied by rapidly changing the concentration of a chemoattractant (McNab and Koshland, 1972; Zigmond and Sullivan, 1979). When this method was applied to sensory neurons, we found, unexpectedly, that neurites rapidly retracted when NGF concentrations were elevated in a range corresponding to a large increase in occupancy of a high-affinity surface receptor (Griffin and Letourneau, 1980). Because retraction can be elicited from neurites severed from their somas, local NGF receptors on neurites are involved. We also found that cytochalasin B, a microfilament disrupter, inhibits the retraction. Though paradoxical, this retraction demonstrates rapid effects of changes in NGF concentration on growth cone motility, and may reflect a supernormal response of the motile apparatus to large increases in NGF receptor occupancy (Seeley and Greene, 1983).

The elements of a chemotactic response are (1) sensation of the attractant, (2) determination of the gradient, and (3) modulation of locomotion (Adler, 1976). Gradients may be sensed by spatial differences in occupancy of NGF receptors along a neurite or nerve tip (Carbonetto and Stack, 1982; Rohrer and Barde, 1982; Sutter *et al.*, 1979). As with leukocyte behavior, changes in NGF levels may rapidly alter ion fluxes, calcium binding, or even adhesivity (Schubert *et al.*, 1978). These responses may influence growth cone motility by stimulating protrusion, increased contractility, or increased ability of the cell surface to adhere at the point of greatest NGF binding. Any of these changes would direct neurite elongation up the NGF gradient. As previously postulated for regulation by adhesion, the key to a chemotactic response is local changes in growth cone motility.

D. Electrical Regulation of Neurite Growth

Years ago, theories of neurobiotaxis and electrodynamic control of neuronal development proposed that axonal growth is oriented by electrical potential differences (Jacobson, 1978). But lack of convincing evidence brought these theories into disfavor and, for years, the idea of electrical effects on nerve fiber growth was viewed as unlikely despite such positive results as those of Marsh and Beams (1946). Jaffe and co-workers (Jaffe and Poo, 1979) undertook careful studies that provided evidence that neurites that extend from explanted ganglia grow at asymmetric rates in a steady electrical field. Many neurites in fields of 70 mV/mm or more grew significantly faster toward the cathode than the anode.

As shown for surface receptors on other cells (Jaffe, 1977; Poo, 1981), it was proposed that the steady current electrophoreses plasmalemmal NGF receptors toward the cathodal side of neurons and neurites. Presumably, the interactions of NGF with the unequally distributed NGF receptors speeds neurite elongation toward the cathode but not in the direction of the anode. The means by which this occurs is unclear, but just as in a chemotactic response, asymmetric influences on protrusion, adhesion, or contraction may enhance neurite growth toward a cathode. In chemotaxis, asymmetric distribution of environmental NGF may produce local differences in receptor occupancy, while electrical fields may also generate asymmetries by inducing movements of membrane molecules. Electrophoresis of other membrane components, for example, adhesive ligands, could also influence neurite growth, but the possibility that a common cellular mechanism acts in chemotaxis and electrical effects on neurite growth is interesting.

In vitro neurite growth in electrical fields is being further investigated. Dispersed cell cultures can better show how growth cones and their surface components respond to electrical fields (Hinkle *et al.*, 1981; Patel and Poo, 1982; Patel *et al.*, 1985). Although most cell surfaces are negatively charged and carry cells toward an anode, growth cones are sufficiently specialized that they may carry a different charge which drags them toward a cathode. This directed response should be investigated in several situations (e.g., in plasma clots, agar matrices, and in serum-free or other media) to show more convincingly that electrical currents act on neurites and not on environmental features. Jaffe's thesis can be tested by visualization of NGF receptors (Rohrer and Barde, 1982) on neurites growing in electrical fields.

E. Intrinsic Regulation of Neurite Growth

The above data indicate that extrinsic agents modulate neurite growth *in vitro*, yet that neurite formation occurs at all *in vitro* reveals the stability of an intrinsic regulation of cytoskeleton and membranous and other organelles to general neuronal morphology. Do intracellular factors regulate finer details of neuronal morphogenesis, such as the distinctively shaped pyramidal (Van der Loos, 1965) and Purkinge's neurons or the remarkably characteristic shapes of invertebrate neurons (Goodman, 1974, 1978)?

When different vertebrate neurons are cultured, they do not assume a single generalized shape. Rather, *in vitro* morphologies can be related to *in vivo* identities (Banker and Cowan, 1978; Bray, 1973a), although rigorous comparisons of *in vitro* and *in vivo* neuronal shapes, as done for *in vivo* dendritic form (Berry *et al.*, 1980), have not been reported. Characteristic morphologies *in vitro* might reflect merely the regeneration of dendritic or axonal stumps by differentiated neurons. Alternatively, characteristic shapes may reveal intrinsic determinants of the initial morphologies of axons or dendrites (Solomon, 1979).

Several organelles in particular may influence neuronal morphogenesis. Positions of the Golgi complex and microtubular organizing centers may determine the site of axonal sprouting (Olmstead *et al.*, 1984; Spiegelman *et al.*, 1979). Local aggregation of molecules that mediate cell adhesion or generation of mechanical force would produce surface regions especially effective in protrusion of the cell margin and exertion of tension on organelles. These are potential sites for axon initiation.

Neurite branching may be intrinsically regulated. Each new branch must be furnished with microtubules and neurofilaments. This might occur through initiation of new cytoskeletal fibers at the nascent branch point or by the diversion into a branch of already assembled fibrils from the parent neurite. In neurites of cultured sensory neurons, microtubules are very long (Bray and Bunge, 1981), so ends available for diversion into a branch are uncommon, except at the neurite tip (Letourneau, 1982). This explains Bray's (1973b) finding that nearly all branches *in vitro* arise by bifurcation of the neurite tip. However, other nerve types may be capable of increased amounts of collateral sprouting if microtubule initiation occurred within neurites or if microtubules were short and had many ends along the neurites.

Intracellular asymmetries that influence neuronal morphology could be established by prior interactions with other cells and environmental features. An extracellular cue for polarizing neuronal organelles comes early as neuroblasts contact the basal lamina of the ventricular layer of the developing CNS (Jacobson, 1978). The orderly movements of glial and neuronal precursors during these early proliferative phases may preserve this initial cellular polarity (Letourneau *et al.*, 1980), thereby defining the apical–basal axis of mature cells.

Certainly the most intriguing property attributed to axons is neuronal specificity, derived from positional information acquired before axonal growth occurs (Fraser and Hunt, 1980; Gaze and Keating, 1972; Hunt and Jacobson, 1972; Jacobson, 1978). Axes and boundaries of positional information for a neuron are determined by extrinsic factors, but at some point the information becomes intrinsic and cannot be easily changed. It is a mystery how positional values are stored or expressed, although topographic variation of adhesive ligands on the surfaces of interacting axons and target cells is a way for positional information to influence axonal growth (Barbera, 1975; Fraser, 1980; Sperry, 1963).

IV. Regulation of the Route of Nerve Fiber Elongation *in Vivo*

Having discussed the regulation of neurite growth *in vitro* by extrinsic influences on growth cone behavior, we turn to axonal growth *in vivo*. Conclusions obtained from *in vitro* studies cannot be directly applied to the embryo, because critical parameters (e.g., cell–cell and cell–substratum adhesion, concentrations of chemoattractants. local electrical currents) are so difficult to measure *in vivo*. However, *in vitro* studies provide a lens to examine *in vivo* situations for clues as to what regulates axonal growth and for inspiration to future studies in which these parameters can be measured *in vivo*.

A. Influences on Nerve Fiber Growth *in Vivo*

Axonal contacts with other cells and nerve fibers may be the principal guiding influences in the embryonal CNS (Anderson *et al.*, 1980; Lopresti *et al.*, 1973; Sidman and Wessells, 1975). In the early retina and spinal cord, intercellular channels form in characteristically oriented arrays between neuroepithelial cells (Krayanek and Goldberg, 1981; Nordlander and Singer, 1982a,b; Silver and Sidman, 1980), and they are then invaded by the first axons to penetrate the area. Subsequent fibers follow to form fascicles that eventually fill the channels. At all times the advancing growth cones are situated at the periphery of a fascicle, where filopodia encounter the end-feet of neuroepithelia, the basal lamina, and adjacent nerve fibers (Bodick and Levinthal, 1980; Krayanek and Goldberg, 1981; Nordlander and Singer, 1982b). More general interactions such as these are included in alternative hypotheses to the specific chemoaffinity hypothesis for the development of topographic projections (Sperry, 1963). Instead of individual surface markers on axons, these models propose that spatial and temporal sequences of axon initiation, fasciculation, contact guidance, and related events act by relatively unspecific means to deliver a topographically ordered array of axons to target areas (Bodick and Levinthal, 1980; Grant and Rubin, 1980; Horder and Martin, 1978; Nornes *et al.*, 1980; Rager, 1980). Recent demonstrations that the cell adhesion molecule N-CAM is concentrated at the end-feet of neuroepithelial cells suggest that pathways may be marked by a neuron-specific ligand but one without great specificity (Silver and Rutishauser, 1984).

Supporting evidence for extrinsic guidance in the development of peripheral connections comes from studies of the innervation of limbs subject to distal rotation (Stirling and Summerbell, 1978; Summerbell and Stirling, 1981). The axonal pathways reflect the rotated polarity of the distal limb structures, so that inappropriate connections are made. However, motor neurons can be displaced by partial spinal cord reversals, and they can make normal connections by acquiring a proper heading in proximal limb areas (Lance-Jones and Landmesser, 1981a,b). Some other axons can also reach their normal muscle targets but do so by wildly aberrent pathways (Hollyday, 1981; Lance-Jones and Landmesser, 1981b). These data suggest that axons can respond selectively to environmental cues in certain localities. In contrast to the CNS, potential guidance pathways similar to the retinal channels have not been identified in the loose mesenchyme and connective tissues of developing limbs (Bennett *et al.*, 1980). Yet axons do not grow randomly through the developing limb to be followed by cell death and axon retraction to correct widespread random growth and achieve the normal connectivity (Lance-Jones and Landmesser, 1978, 1980; 1981a). In addition, the cord reversals show that axons need not follow fixed topographic paths to reach correct targets, although guidance along pioneer fibers does occur in peripheral tissues (Anderson *et al.*, 1980; Speidel, 1933).

Do growing axons exercise any intrinsic selection of their pathways? Neurons transplanted into abnormal CNS sites project axons along routes that are characteristically followed by the host axons in that region (Constantine-Paton, 1978; Katz and Lasek, 1980). Further studies involving more neuronal types, developmental periods, and transplantation sites may show differences in the formation of axonal pathways that will allow one to evaluate the selectivity of growing axons. Perhaps cell adhesion gradients are best demonstrated *in vivo* by inversion grafts in the epithelium of developing butterfly wings (Nardi, 1982; Nardi and Kafatos, 1976). It would be interesting to observe the growth cones of sensory neurons as they cross the epithelium and encounter grafts of differing orientations.

B. Relationship to *in Vitro* Studies

Can *in vitro* studies help answer this question? Axonal tracking along intercellular channels may simply be contact guidance along an oriented surface, yet cell movements *in vivo* are not always oriented by an oriented substratum (Nelson and Revel, 1975). Growth cones may have adhesive affinity for the end-feet of neuroepithelial cells or for basal laminae that is more crucial in drawing neurites to the periphery of a developing pathway than is the simple presence of a channel. N-CAM and laminin are two molecules with adhesive affinity for growth cones that are present at these sites (McLoon, 1984; Silver and Rutishauser, 1984). One can analyze *in vitro* the interactions of neurites with glia or with the molecular constituents of basal laminae (Hatten and Liem, 1981; Hatten *et al.*, 1984; Akers *et al.*, 1981; Rogers *et al.*, 1983, 1984, 1985), and such studies are beginning to reveal selective affinities of neurites for particular surfaces. The development of topographic order in fascicles may depend on selective adhesion of growth cones to the neurites of adjacent neurons or to immature, recently formed neurites. These affinities can also be probed *in vitro* (Bray *et al.*, 1980), although seeming contradictory findings suggest a greater adhesivity between dorsal and ventral retinal cells than the adhesion measured in dorsal–dorsal or ventral–ventral combinations (Gottlieb *et al.*, 1976). *In vitro* studies can be puzzling when an *in vivo* situation is poorly replicated or when novel circumstances, such as the juxtaposition of dorsal and ventral retinal cells, are constructed *in vitro*. This is both a weakness and strength of the *in vitro* approach.

C. The Growth Cone *in Vivo*

We have emphasized the nerve tip and its motility and role in neurite growth. The term *growth cone* arose from Cajal's histological studies (1890), and in many fixed tissues expanded nerve tips with filopodial protrusions are seen at the ends of presumably growing nerve givers (Hinds and Hinds,

1972; Meinertzhagen, 1973; Muller and Scott, 1980; Murphy and Kater, 1980). In addition, the ultrastructure of nerve tips *in vivo* resembles what is, *in vitro*, the active growth cone (Skoff and Hamburger, 1974; Tennyson, 1970; Vaughn *et al.*, 1974). Because growth cones are so labile and difficult to fix, even with glutaraldehyde (Bunge, 1977; Nuttall and Wessells, 1979), distortion, retraction, and disintegration may obscure some details of growth cone morphology in fixed tissues.

In vitro studies have shown that growth cone shape is related to the shape and adhesivity of the *in vitro* substratum (Letourneau, 1979b). Within intact tissues, growth cones show a variety of shapes that suggests high adhesion to planar surfaces like basal laminae, or more limited adhesive contacts, when growth cones are bulbous (Bodick and Levinthal, 1980; Meinertzhagen, 1973; Nordlander and Singer, 1982b). These variations in growth cone shape may be used to deduce the adhesiveness of growth cone interactions. A more detailed interpretation of growth cone contacts *in vivo* is difficult to make because there are no reliable morphological markers for the transient adhesions of filopodia or growth cones, although specialized contacts of filopodia have been described. Hypertonic fixatives induce cell shrinkage that highlights some adhesive points (Krayanek and Goldberg, 1981), but one does not know whether other important contacts have been ruptured.

Growth cone behavior *in vivo* would be better understood if filopodia and other protrusions could be better visualized. The exploratory behavior of filopodia and their primary role in directing neurite growth is well documented *in vitro* (Letourneau, 1975b; Nakai and Kawasaki, 1959; Wessells *et al.*, 1980). Injection of fluorescent dyes into invertebrate neurons within whole tissues reveals filopodia >50 μm long (Goodman and Bate, 1981; Goodman *et al.*, 1984; Murphy and Kater, 1980). Such great lengths occur *in vitro* when adhesion is high (Letourneau, 1979b). These filopodia may allow a growing neurite to sample large portions of the environment and initiate contacts or detect chemicals that direct axonal growth. Perhaps, vertebrate motor neurons or Rohon–Beard neurons might be injected and orthogradely filled by skillful manipulation of early embryos, thereby providing extremely useful information about "real" growth cone behavior in the CNS and peripheral tissues.

D. Regulation by Chemotaxis

The next most likely extrinsic influence on axonal growth is chemotactic growth toward the source of a soluble attractant. Many target organs of NGF-dependent sympathetic neurons synthesize substances *in vivo* with NGF-like immunoreactivity (Johnson *et al.*, 1971). A dramatic result is that injection of NGF into the brains of young rats induces abnormal growth of axons from peripheral sympathetic neurons into the spinal cord to the site of injection (Levi-Montalcini, 1976). An interesting observation was that the invading sympathetic axons followed characteristic routes of elongation in the spinal cord that are otherwise followed by the normally resident fibers. Perhaps the chemotactic response acts together with the influence of contact interactions with substrata to direct growth cone activity (Katz and Lasek, 1980).

In spite of accumulating *in vitro* evidence for chemotactic growth of neurites, the *in vivo* evidence is lacking except for the above study. It is controversial over what distances chemical gradients can be established in an embryo (Crick, 1970). Chemotaxis may act only when the distance between the source and neurite tip is short, such as during initial axonal growth when direction along a pathway is chosen, or it may act later in the local vicinity of the target organ. The degree of connection specificity that could be determined by a chemoattractant in unclear, since NGF is synthesized by many organs and could not alone direct axons to a particular target.

The sprouting factors produced by denervated tissues may be related to chemoattractants (Brown and Ironton, 1978; Edds, 1953). They may act on a nonmotile axon to induce motile activity and axonal sprouts in a similar manner to that of a chemoattractant's actions to turn a growth cone. This is supported by recent findings concerning NGF. Several adult organs with sympathetic innervation usually make no NGF but are induced to synthesize NGF by denervation *in vivo* (Ebendal *et al.*, 1980) or by placement into tissue culture (Harper *et al.*, 1980a,b). When the denervated iris is reinnervated, however, NGF synthesis stops (Ebendal *et al.*, 1980). Thus, NGF can be viewed as a target-derived sprouting factor that induces and directs growth of sympathetic fibers back to denervated tissue.

E. Regulation by Electrical Currents

Evidence for significance of electrical effects in directing neurites has begun to accumulate. Steady currents have been demonstrated in algal zygotes, chick blastoderms, and regenerating newt limbs (Borgens *et al.*, 1977, 1984; Jaffe and Nuccitelli, 1977; Jaffe and Stern, 1979). The biological actions of such currents are best understood in the egg of the alga *Fucus* (Jaffe and Nuccitelli, 1977; Quatrano *et al.*, 1979). More relevant to axonal growth, imposed currents induce partial regeneration, including nerve growth, from the normally nonregenerative frog limb, and large, steady currents enter the transected lamprey spinal cord (Borgens *et al.*, 1980). Elucidation of a direct effect of electrical currents on axonal growth will more probably come from *in vitro* studies, though understanding of the source and regulation of these currents requires *in vivo* study.

V. Is Regulation of Nerve Fiber Growth Related to Synaptogenesis?

Is regulation of axonal pathways related to synaptogenesis? These events may be unrelated mechanistically, for example, if axonal growth were controlled locally in the tissues that axons transiently cross. On the other hand, future postsynaptic cells may participate in regulation of axonal morphogenesis. For example, target cells may synthesize chemoattractants to guide the growing axons to them. Alternatively, target cells may constitute the top of adhesion gradients that guide axonal growth. However, the model for neurite growth presented earlier does not predict that neurite growth will necessarily stop at the top of an adhesion gradient. The model must be amended if synaptogenesis is to be included.

To do this, I propose that when growing axons encounter potential target cells, something new occurs in the interaction of nerve tips with other cells. A cascade of events is initiated in the interacting cells that leads to cessation of axonal growth and the formation of a synapse. Synaptogenesis may begin [e.g., clustering of acetylcholine receptors (Bloch and Geiger, 1980)] from filopodial contacts, so that axon growth and synaptogeneses become competing events at the axonal tip. One can propose growth cone–target cell interactions that would terminate axonal elongation as synaptogenesis begins. The application of serotonin to growth cones of certain identified neurons cultured from the snail *Helisoma* induces rapid withdrawal of filopodia and cessation of axonal elongation (Haydon *et al.*, 1984). This is a localized response and if serotonin is removed, filopodia reappear as elongation recommences. In such a manner a target cell may release substances to stop filopodia activity by responsive growth cones in the target area and promote synaptogenesis. Another hypothesis involves a potent Ca^{2+}-activated protease within nerve terminals that degrades cytoskeletal structures, particularly neurofilaments (Day, 1980; Fulton, 1985). Interaction of a growth cone with a target cell may prompt an inward Ca^{2+} flux or a release of Ca^{2+} from intracellular stores to activate the protease, degrading cytoskeletal fibers that project into the growth cone. This would also stop neurite elongation.

Both of these hypotheses for stopping neurite growth depend on interactions of a target-derived signal with particular components of the growth cone surface. Just as in the previous discussion of the determination of axonal pathways, the degree of specificity in such growth cone–target interactions is unknown. However, the significant point is that arrival in a target area brings exposure to new signals, modifying the dynamic nature of growth cones and promoting the transformation of the terminal to a maturing synapse. Synapses are generally regarded as fairly stable structures; however, neuronal plasticity, memory, and regeneration may require structural changes in neural connectivity that involve a return in the dynamic motility and other activities characteristic of the nerve growth cones.

VI. Summary

Nerve fiber growth is cell growth and cell movement. Most components are synthesized in the perikaryon, though this synthetic activity is neither necessary (over short intervals) nor sufficient for neurite growth to occur. Growth cone activity includes protrusion of cytoplasmic processes that explore the local environment and contact other cells and extracellular surfaces. Mechanical force exerted by these protrusions either breaks the contacts or pulls the nerve tip and its contents forward toward the associated adhesive points. Intracellular events that accompany this behavior include expansion of the surface membrane, the association of microfilament bundles with the adhesive contacts of protrusions, and the forward movements of cytoskeletal filaments and membranous vesicles via association with actomyosinlike forces exerted within the protrusions.

Interactions of the growth cone with extrinsic features of the *in vitro* environment regulate nerve fiber growth. Strong adhesion of growth cone protrusions to the substratum increases the transmission of contractile forces to advance intracellular organelles, rather than to retract the protrusion. If strong adhesions are limited to one side of the growth cone, then the neurite either turns toward that side or may branch. The association of growth cones with nonneuronal cells or with neurites depends on the adhesion of active nerve tips to particular surfaces, compared with other available adhesive interactions.

Growing neurites *in vitro* seem to exhibit chemotaxis toward explants of target organs or to gradients of NGF. The response to NGF involves turning of the nerve tip toward an NGF source. NGF receptors on the growing neurite may sense NGF, and formation of NGF–receptor complexes may initiate local signals to modify growth cone motility.

Neurites tend to grow faster toward the cathode in a steady current. How this occurs is unclear, but a current may induce electrophoresis of molecules in the neurite membrane, so that growth is faster in the direction that places the electrophoresed molecules forward.

These extrinsic factors are based on very different solid, liquid, and electrical properties of the embryonal environment. Yet, we propose that they act similarly to regulate neurite growth in the sense that they produce local changes in protrusion or adhesion, or the generation of mechanical tensions by the active nerve tip.

The *in vivo* pathways of axonal growth are influenced by extrinsic factors. The dominant influence is that of contacts of the nerve fiber with cells or extracellular materials, though the regulatory nature of the cell–cell contacts—that is, morphological and relatively nonspecific, or involving adhesive interactions which discriminate between cells or directions—is unknown.

Chemotactic responses and electrical currents may also direct axonal growth. Several extrinsic factors may act simultaneously or at different times.

In vivo investigations are limited by the difficulty of seeing growth cones and demonstrating responses to extrinsic factors. In most cases, axonal growth and interactions are deduced from what is preserved by fixation, which can distort growth cones. Recent methods of vital staining may allow observation of growth cones in their normal environment, although small protrusions may still not be seen.

Increasing numbers of neuronal types are being cultured *in vitro*. *In vivo* interactions that suggest the operation of selective affinities can now be further studied. The well-characterized neurons of invertebrates are particularly attractive candidates for *in vitro* study (Ready and Nicholls, 1979; Wong *et al.*, 1981), because they may reveal intrinsic cell-specific features that determine neuronal shape and connectivity. Molecules produced by glia or other nonneuronal cells that act as adhesive ligands or chemoattractants are being chemically and biologically examined with tissue culture (Fallon, 1983; Grumet *et al.*, 1984; Hatten *et al.*, 1984). These and other systems illustrate that *in vivo* and *in vitro* approaches can be combined to allow us to achieve understanding of the cellular and molecular bases for the morphogenesis of the extraordinary order of neuronal connections.

Acknowledgments

I thank Stanley Kater for comments on the manuscript. Helpful discussions on nerve fiber growth *in vivo* were also had with Corey Goodman, Stanley Kater, Ray Lasek, Lynn Landmesser, Ruth Nordlander, and Jerry Silver. Alice Ressler and Walter Gutzmer provided expert technical and photographic assistance, and Roberta Andrich and Jennifer Steinert skillfully typed the manuscript. Preparation of this chapter was supported by the National Science Foundation, National Institutes of Health, University of Minnesota Graduate School, and the American Cancer Society Institutional Grant to the University of Minnesota.

References

Adler, J. (1976). *J. Supramol. Struct.* **4**, 305–317.
Adler, R., Manthorpe, M., Skaper, S. D., and Varon, S. (1981). *Brain Res.* **206**, 129–144.
Adler, R., and Varon, S. (1981). *Dev. Biol.* **81**, 1–11.
Akers, R. M., Mosher, D. F., and Lilien, J. F. (1981). *Dev. Biol.* **86**, 179–188.
Anderson, H., Edwards, J. S., and Palka, J. (1980). *Annu. Rev. Neurosci.* **3**, 97–140.
Arora, H. L., and Sperry, R. W. (1962). *Am. Zool.* **2**, 389.
Attardi, D. G., and Sperry, R. W. (1963). *Exp. Neurol.* **7**, 46–64.
Banker, G. A., and Cowan, W. M. (1978). *J. Comp. Neurol.* **187**, 469–494.
Barbera, A. (1975). *Dev. Biol.* **46**, 167–191.

Baron-VanEvercooren, A., Kleinman, H. K., Ohno, S., Marangos, P., Schwartz, J. P., and Dubois-Dalcq, M. (1982). *J. Neurosci. Res.* **8**, 179–193.
Begg, D. A., Rodewald, R., and Rebhum, L. I. (1979). *J. Cell. Biol.* **79**, 846–852.
Bell, G. I. (1978). *Science* **200**, 618–627.
Bennett, M. R., Davey, D. F., and Uebel, K. E. (1980). *J. Comp. Neurol.* **189**, 335–357.
Berry, M., McConnell, P., and Sievers, J. (1980). *Curr. Top. Dev. Biol.* **15**, 67–101.
Birks, R. I., Mackey, M. C., and Weldon, P. R. (1972). *J. Neurocytol.* **1**, 311–430.
Bloch, R. J., and Geiger, B. (1980). *Cell* **21**, 25–35.
Bodick, N., and Levinthal, C. (1980). *Proc. Natl. Acad. Sci. U. S. A.* **77**, 4374–4378.
Bonhoeffer, F., and Gierer, A. (1984). *Trends in Neurosci.* **7**, 378–381.
Bonhoeffer, F., and Huf, J. (1980). *Nature (London)* **288**, 162–164.
Borgens, R. B., Vanable, J. W., and Jaffe, L. F. (1977). *Proc. Natl. Acad. Sci. U. S. A.* **74**, 4528–4532.
Borgens, R. B., Jaffe, L. F., and Cohen, M. J. (1980). *Proc. Natl. Acad. Sci. U. S. A.* **77**, 1209–1213.
Borgens, R. B., McGinnis, M. E., Vanable, J. W., Jr., and Miles, E. S. (1984). *J. Exp. Zool.* **231**, 249–256.
Bray, D. (1970). *Proc. Natl. Acad. Sci. U. S. A.* **65**, 905–910.
Bray, D. (1973a). *J. Cell. Biol.* **56**, 702–712.
Bray, D. (1973b). *Nature (London)* **244**, 93–96.
Bray, D. (1979). *J. Cell. Sci.* **37**, 391–410..
Bray, D. (1982). *In* "Cell Behavior" (R. Bellair, A. Curtis, and G. Dunn, eds.), pp. 299–318. Cambridge Univ. Press, Cambridge.
Bray, D., and Bunge, M. B. (1981). *J. Neurocytol.* **10**, 589–605.
Bray, D., Thomas, C., and Shaw, G. (1978). *Proc. Natl. Acad. Sci. U. S. A.* **75**, 5226–5229.
Bray, D., Wood. P., and Bunge, R. P. (1980). *Exp. Cell. Res.* **130**, 241–250.
Brown, M. C., and Ironton, R. (1978) *J. Physiol. (London)* **278**, 325–348.
Bunge, M. B. (1973). *J. Cell Biol.* **56**, 713–725.
Bunge, M. B. (1977). *J. Neurocytol.* **6**, 407–439.
Cajal, S. R. (1890). *Anat. Anz.* **5**, 609–613, 631–639.
Cajal, S. R. (1928). "Degeneration and Regeneration of the Nervous System" (R. M. May, trans.). Hafner, New York, 1959.
Carbonetto, S. (1984). *Trends in Neurosci.* **7**, 382–387.
Carbonetto, S., and Argon, Y. (1980). *Dev. Biol.* **80**, 364–378.
Carbonetto, S., and Fambrough, D. M. (1979). *J. Cell Biol.* **81**, 555–569.
Carbonetto, S., and Stach, R. W. (1982). *Dev. Brain Res.* **3**, 463–473.
Carbonetto, S., Gruver, M. M., and Turner, D. C. (1983). *J. Neurosci.* **3**, 2324–2335.
Chamley, J. H., Goller, I., and Burnstock, G. (1973). *Dev. Biol.* **31**, 362–379.
Chang, C.-M., and Goldman, R. D. (1973). *J. Cell Biol.* **57**, 867–874.
Charlwood, K. A., Lamont, D. M., and Banks, B. E. C. (1972). *In* "Nerve Growth Factor and its Antiserum" (E. Zaimis and J. Knight, eds.), pp. 102–107. Oxford Univ. Press (Athlone), London and New York.
Chen, L. B., Murray, A., Segal, R. A., Bushnell, A., and Walsh, M. L. (1978). *Cell* **14**, 377–385.
Collins, F. (1978a). *Proc. Natl. Acad. Sci. U. S. A.* **75**, 5210–5213.
Collins, F. (1978b). *Dev. Biol.* **65**, 50–57.
Collins, F. (1980). *Dev. Biol.* **79**, 247–252.
Collins, F., and Garrett, J. E. (1980). *Proc. Natl. Acad. Sci. U. S. A.* **77**, 6226–6228.
Collins, F., and Lee, M. R. (1984). *J. Neurosci.* **4**, 2823–2829.
Constantine-Paton, M. (1979). *BioScience* **29**, 526–532.
Constantine-Paton, M., and Capranica, R. R. (1975). *Science* **189**, 480–482.

Coughlin, M. D. (1975). *Dev. Biol.* **43**, 140–158.
Crick, F. (1970). *Nature (London)* **225**, 420–422.
Curtis, A. S. G. (1973). *Prog. Biophys. Mol. Biol.* **27**, 315–385.
Daniels, M. P. (1973). *J. Cell Biol.* **58**, 463–470.
Daniels, M. (1975). *Ann. N. Y. Acad. Sci.* **253**, 535–544.
Das, G. D., Lammert, G. L., and McAllister, J. P. (1974). *Brain Res.* **69**, 13–29.
Day, W. A. (1980). *J. Ultrastruct. Res.* **70**, 1–7.
Droz, B., Rambourg, A., and Koenig, H. L. (1975). *Brain Res.* **93**, 1–13.
Dunn, G. A. (1971). *J. Comp. Neurol.* **143**, 491–508.
Dunn, G. A. (1982). *In* "Cell Behavior" (R. Bellairs, A. Curtis, and G. Dunn, eds.), pp. 247–280. Cambridge Univ. Press, London.
Dunn, G. A., and Ebendal, T. (1978). *Zoon* **6**, 65–68.
Dunn, G. A., and Heath, J. P. (1976). *Exp. Cell Res.* **101**, 1–14.
Easter, S. S., Jr., Bratton, B., and Scherer, S. S. (1984). *J. Neurosci.* **4**, 2173–2190.
Ebendal, T. (1976). *Exp. Cell Res.* **98**, 159–169.
Ebendal, T., and Jacobson, C. O. (1977). *Exp. Cell Res.* **105**, 379–387.
Ebendal, T., Olson, L., Seiger, A., and Hedlund, K. O. (1980). *Nature (London)* **286**, 25–28.
Edds, M. V. (1953). *Q. Rev. Biol.* **28**, 260–276.
Edelman, G. M. (1983). *Quart. Rev. Biol.* **28**, 260–276.
Edwards, J. G. (1977). *In* "Mammalian Cell Membranes: Membranes and Cellular Functions" (G. A. Jamieson and D. M. Robinson, eds.), Vol 4, pp. 32–56. Butterworth, London.
Ellisman, M. H., and Porter, K. R. (1980). *J. Cell Biol.* **87**, 464–479.
Fallon, J. R. (1983). *Soc. Neurosci. Abstr.* **9**, 207.
Fernandez, H. L., and Samson, F. E. (1973). *J. Neurobiol.* **4**, 201–206.
Fraser, S. E. (1980). *Dev. Biol.* **79**, 453–464.
Fraser, S. E., and Hunt, R. K. (1980). *Annu. Rev. Neurosci.* **3**, 319–352.
Frazier, W., and Glaser, L. (1979). *Annu. Rev. Biochem.* **48**, 491–523.
Fulton, A. B. (1985). "The Cytoskeleton." Chapman and Hall, New York.
Gaze, R. M., and Keating, M. J. (1972). *Nature (London)* **237**, 375–378.
Godina, G. (1963). *In* "Cinemicrography in Cell Biology" (G. G. Rose, ed.), pp. 313–338. Academic Press, New York.
Goodman, C. S. (1974). *J. Comp. Physiol.* **95**, 185–201.
Goodman, C. S. (1978). *J. Comp. Neurol.* **182**, 681–706.
Goodman, C. S., and Bate, M. (1981). *Trends Neurosci.* **4**, 163–169.
Goodman, C. S., Bastiani, M. J., Doe, C. Q., Dulac, S., Helfand, S. L., Kuwada, J. Y., and Thomas, J. B. (1984). *Science* **225**, 1271–1279.
Gottlieb, D. I., and Glaser, L. (1980). *Annu. Rev. Neurosco.* **3**, 303–318.
Gottlieb, D. I., Rock, K., and Glaser, L. (1976). *Proc. Natl. Acad. Sci. U. S. A.* **73**, 410–414.
Grant, P., and Rubin, E. (1980). *J. Comp. Neurol.* **189**, 671–698.
Griffin, C. G., and Letourneau, P. C. (1980). *J. Cell Biol.* **86**, 156–161.
Grinnell, F. (1978). *Int. Rev. Cytol.* **29**, 65–144.
Grinnell, F., and Minter, D. (1978). *Proc. Natl. Acad. Sci. U. S. A.* **75**, 4408–4412.
Grumet, M., Hoffman, S., and Edelman, G. M. (1984). *Proc. Natl. Acad. Sci. U. S. A.* **81**, 267–271.
Gunderson, R. W., and Barrett, J. N. (1979). *Science* **206**, 1079–1080.
Gunderson, R. W., and Barrett, J. N. (1980). *J. Cell Biol.* **87**, 546–555.
Halfter, W., and Deiss, S. (1984). *Dev. Biol.* **102**, 344–355.
Halfter, W., Newgreen, D. F., Sauter, J., and Schwarz, U. (1983). *Dev. Biol.* **95**, 56–64.
Hammarback, J., Palm, S. L., Furcht, L. T., and Letourneau, P. C. (1983). *J. Neurosci. Res.* **13**, 213–222.

Harper, G. P., and Thoenen, H. (1980). *J. Neurochem.* **34**, 5-16.
Harper, G. P., Al-Saffer, A. M., Pearce, F. L., and Vernon, C. A. (1980a). *Dev. Biol.* **77**, 379-390.
Harper, G. P., Pearce, F. L., and Vernon, C. A. (1980b). *Dev. Biol.* **77**, 391-402.
Harrison, R. G. (1910). *J. Exp. Zool.* **9**, 787-848.
Harrison, R. G. (1914). *J. Exp. Zool.* **17**, 521-544.
Hatten, M. E., and Liem, R. K. (1981). *J. Cell. Biol.* **90**, 622-630.
Hatten, M. E., Liem, R. K., and Mason, C. A. (1984). *J. Cell Biol.* **98**, 193-204.
Hawrot, E. (1980). *Dev. Biol.* **74**, 136-151.
Haydon, P. G., McCobb, D. P., and Kater, S. B. (1984). *Science* **226**, 561-564.
Heidemann, S. R., Landers, J. M., and Hamborg, M. A. (1981). *J. Cell Biol.* **91**, 661-665.
Herman, I. M., and Pollard, T. D. (1981). *J. Cell Biol.* **88**, 346-351.
Hibbard, E. (1965). *Exp. Neurol.* **13**, 289-301.
Hinds, J. W., and Hinds, P. L. (1972). *J. Neurocytol.* **1**, 169-187.
Hinkle, L., McCaig, C. D., and Robinson, K. R. (1981). *J. Physiol.* **314**, 121-135.
Hirokawa, N. (1982). *J. Cell. Biol.* **94**, 129-142.
Hollyday, M. (1981). *J. Comp. Neurol.* **202**, 439-465.
Horder, T. J., and Martin, K. A. C. (1978). *Symp. Soc. Exp. Biol.* **32**, 275-358.
Hueser, J. E., and Kirschner, M. W. (1980). *J. Cell Biol.* **86**, 212-234.
Hughes, A. (1953). *J. Anat.* **87**, 150-162.
Hunt, R. K., and Jacobson, M. (1972). *Proc. Natl. Acad. Sci. U. S. A.* **69**, 780-783.
Izzard, C. S., and Lochner, L. R. (1976). *J. Cell. Sci.* **21**, 129-159.
Jacobson, M. (1978). "Developmental Neurobiology." Plenum, New York.
Jaffe, L. (1977). *Nature (London)* **265**, 600-602.
Jaffe, L. F., and Nuccitelli, R. (1977). *Annu. Rev. Biophys. Bioeng.* **6**, 476-495.
Jaffe, L. F., and Poo, M. M. (1979). *J. Exp. Zool.* **209**, 115-128.
Jaffe, L. F., and Stern, C. D. (1979). *Science* **206**, 79-81.
Johnson, D. G., Gordon, P., and Kopin, I. J. (1971). *J. Neurochem.* **18**, 2355-2362.
Kater, S. B., and Letourneau, P. C., eds. (1985). "Biology of the Nerve Growth Cone." Liss, New York.
Katz, M. J., and Lasek, R. J. (1979). *J. Comp. Neurol.* **183**, 817-832.
Katz, M. J., and Lasek, R. J. (1980). *Cell Motil.* **1**, 141-158.
Kirschner, M. (1980). *J. Cell Biol.* **86**, 330-334.
Krayanek, S., and Goldberg, S. (1981). *Dev. Biol.* **84**, 41-50.
Kuczmarski, E. R., and Rosenbaum, J. L. (1979a). *J. Cell Biol.* **80**, 341-355.
Kuczmarski, E. R., and Rosenbaum, J. L. (1979b). *J. Cell Biol.* **80**, 356-371.
Lance-Jones, E., and Landmesser, L. T. (1978). *Soc. Neurosci. Abstr.* **4**, 118.
Lance-Jones, E., and Landmesser, L. T. (1980). *J. Physiol. (London)* **302**, 559-580, 581-602.
Lance-Jones, C., and Landmesser, L. (1981a). *Proc. R. Soc. London, Ser. B* **214**, 1-18.
Lance-Jones, C., and Landmesser, L. (1981b). *Proc. R. Soc. London, Ser. B* **214**, 19-52.
Lander, A. D., Fujii, D. K., Gospodarowicz, D., and Reichardt, L. F. (1982). *J. Cell Biol.* **94**, 574-585.
Lander, A. D., Fujii, D. K., Gospodarowicz, D., and Reichardt, L. F. (1984). *Soc. Neurosci. Abstr.* **10**, 40.
Lasek, R. J. (1982). *Phil. Trans. R. Soc. London B* **299**, 313-327.
Lasek, R. J., and Hoffman, P. N. (1976). *In* "Cell Motility" (R. Goldman, T. Pollard, and J. Rosenbaum, eds.), pp. 1021-1049.. Cold Spring Harbor Laboratory, Cold Spring Harbor, New York.
Lasek, R. J., Garner, J. A., and Brady, S. T. (1984). *J. Cell Biol.* **99**, 212s-221s.
Lesot, H., Kuhl, U., and Vonder Mark, K. (1983). *EMBO J.* **2**, 861-865.

Letourneau, P. C. (1975a). *Dev. Biol.* **44**, 77–91.
Letourneau, P. C. (1975b). *Dev. Biol.* **44**, 92–101.
Letourneau, P. C. (1978). *Dev. Biol.* **66**, 183–196.
Letourneau, P. C. (1979a). *J. Cell Biol.* **80**, 128–140.
Letourneau, P. C. (1979b). *Exp. Cell Res.* **124**, 127–138,
Letourneau, P. C. (1981). *Dev. Biol.* **85**, 113–122.
Letourneau, P. C. (1982). *J. Neurosci.* **2**, 806–814.
Letourneau, P. C. (1983). *J. Cell Biol.* **11**, 963–973.
Letourneau, P. C., Ray, P. N., and Bernfield, M. R. (1980). *In* "Biological Regulation and Development" (R. F. Goldberger, ed.), pp. 339–376. Plenum, New York.
Levi-Montalcini, R. (1976). *Prog. Brain Res.* **45**, 235–258.
Linder, E., Vaheri, A., Ruoslahti, E., and Wartiovaara, J. (1975). *J. Exp. Med.* **142**, 41–49.
Lopresti, V., Macagno, E. R., and Levinthal, C. (1973). *Proc. Natl. Acad. Sci. U. S. A.* **70**, 433–437.
Ludueña, M. A. (1973a). *Dev. Biol.* **33**, 268–284.
Ludueña, M. A. (1973b). *Dev. Biol.* **33**, 470–476.
Ludueña, M. A., and Wessells, N. K. (1973). *Dev. Biol.* **30**, 427–440.
McKay, R. D. G., Hockfield, S., Johansen, J., Thompson, J., and Frederiksen, K. (1983). *Science* **222**, 788–794.
McLoon, S. C. (1984). *Soc. Neurosci. Abstr.* **10**, 466.
McNab, R. M., and Koshland, D. E. (1972). *Proc. Natl. Acad. Sci. U. S. A.* **69**, 2509–2512.
Marchase, R. B. (1977). *J. Cell Biol.* **75**, 237–257.
Margolis, R. L., and Wilson, L. (1978). *Cell* **13**, 1–8.
Marsh, L., and Letourneau, P. C. (1984). *J. Cell Biol.* **99**, 2041–2047.
Meinertzhagen, I. A. (1973). *In* "Developmental Neurobiology of Arthropods" (D. Young, ed.), pp. 51–104. Cambridge Univ. Press, London and New York.
Mooseker, M. S. (1976). *in* "Cell Motility" (R. Goldman, T. Pollard, and J. Rosenbaum, eds.), pp. 631–650. Cold Spring Harbor Laboratory, Cold Spring Harbor, New York.
Mooseker, M. S., Pollard, T. D., and Fujiwara, K. (1978). *J. Cell Biol.* **79**, 444–453.
Muller, K. J., and Scott, S. A. (1980). *Nature (London)*, **283**, 89–90.
Murphy, A. D., and Kater, S. B. (1980). *Brain Res.* **186**, 251–272.
Nakai, J. (1960). *Z. Zellforsch. Mikrosk. Anat.* **52**, 427–449.
Nakai, J., and Kawasaki, Y. (1959). *Z. Zellforsch. Mikrosk. Anat.* **51**, 108–122.
Nakajima, S. (1965). *J. Comp. Neurol.* **125**, 193–204.
Nardi, J. B. (1983). *Dev. Biol.* **95**, 163–174.
Nardi, J., and Kafatos, F. (1976). *J. Embryol. Exp. Morphol,* **36**, 489–512.
Nelson, G., and Revel, J. P. (1975). *Dev. Biol.* **42**, 315–333.
Nordlander, R. H., and Singer, M. (1978). *J. Comp. Neurol.* **180**, 349–374.
Nordlander, R. H., and Singer, M. (1982a). *Exp. Neurol.* **75**, 221–228.
Nordlander, R. H., and Singer, M. (1982b). *Dev. Brain Res.* **4**, 181–193.
Nornes, H. O., Hart, H., and Carry, M. (1980). *J. Comp. Neurol.* **92**, 119–132.
Nuttall, R. P., and Wessells, N. K. (1979). *Exp. Cell Res.* **119**, 163–174.
Olmsted, J. B., Cox, J. V., Asnes, C. F., Parysek, L. M., and Lyon, H. D. (1984). *J. Cell Biol.* **99**, 28s–32s.
Palay, S. L., and Chan-Palay, V. (1977). *In* "Handbook of Physiology I: The Nervous System" (E. R. Kandel, ed.), pp. 5–37. Waverly, Baltimore.
Patel, N., and Poo, M.-M. (1982). *J. Neurosci.* **2**, 483–496.
Patel, N. B., Xie, Z.-P., Young, S. H., and Poo, M.-M. (1985). *J. Neurosci. Res.* **13**, 245–256.
Pfenninger, K. H., and Johnson, M. P. (1983). *J. Cell Biol.* **97**, 1038–1042.
Pfenninger, K. H., and Maylie-Pfenninger, M. F. (1981). *J. Cell Biol.* **89**, 536–546, 547–559.

Pollard, T. D., Selden, S. C., and Maupin, P. (1984). *J. Cell Biol.* **99**, 33s–37s.
Poo, M. M. (1981). *Annu. Rev. Biophys.* **10**, 245–276.
Postlethwaite, A. E., Seyer, J. M., and Kang, A. H. (1978). *Proc. Natl. Acad. Sci. U. S. A.* **75**, 871–875.
Quatrano, R. S., Brawley, S. H., and Hogsett, W. E. (1979). *Symp. Soc. Dev. Biol.* **37**, 77–96.
Rager, G. (1980). *Trends Neurosci.* **3**, 43–44.
Rakic, P. (1971). *J. Comp. Neurol.* **141**, 283–312.
Rakic, P. (1981). *Trends Neurosci.* **4**, 181–185.
Ramsey, W. S. (1972). *Exp. Cell Res.* **70**, 129–139.
Rao, N. C., Barsky, S. H., Teranova, V. P., and Liotta, L. A. (1983). *Biochem. Biophys. Res. Comm.* **111**, 804–808.
Ready, D. F., and Nicholls, J. (1979). *Nature (Lcndon)* **281**, 67–69.
Rogers, S. L., and Letourneau, P. C. (1985). *In* "Handbook of Physiology Volume on Developmental Neurobiology" (W. M. Cowan, ed.). Amer. Physiol. Soc., Bethesda, Maryland, (in press).
Rogers, S. L., Letourneau, P. C., Palm, S. L., McCarthy, J. B., and Furcht, L. T. (1983). *Dev. Biol.* **98**, 212–220.
Rogers, S. L., McLoon, S. C., and Letourneau, P. C. (1984). *Soc. Neurosci. Abstr.* **10**, 39.
Rogers, S. R., McCarthy, J. B., Palm, S. L., Furcht, L. T., and Letourneau, P. C. (1985). *J. Neurosci.* **5**, 369–378.
Rohrer, H., and Barde, Y. A. (1982). *Dev. Biol.* **84**, 309–315.
Roisen, F., Inczedy-Marcsek, M., Hsu, L., and Yorke, W. (1978). *Science* **199**, 1445–1448.
Rothman, J. E., and Lenard, J. (1977). *Science* **195**, 743–753.
Rusoff, A. C., and Easter, S. S. (1979). *Science* **208**, 31–32.
Rutishauser, U. (1985). *J. Neurosci. Res.* **13**, 123–134.
Rutishauser, U., and Edelman, G. M. (1980). *J. Cell Biol.* **87**, 370–378.
Rutishauser, U., Gall, W. E., and Edelman, G. M. (1978). *J. Cell Biol.* **79**, 382–393.
Schiffmann, E. (1982). *Annu. Rev. Physiol.* **44**, 553–568.
Schnapp, B. J., and Reese, T. S. (1982). *J. Cell Biol.* **94**, 667–679.
Schubert, D., and LaCorbiere, M. (1982). *J. Neurosci.* **2**, 82–89.
Schubert, D., LaCorbiere, M., Whitlock, C., and Stallcup, W. (1978). *Nature (London)* **273**, 718–723.
Seeley, P. J., and Greene, L. A. (1983). *Proc. Natl. Acad. Sci. U. S. A.* **80**, 2789–2793.
Shizuta, Y., Davies, P., Olden, K., and Pastan, I. (1976). *Nature (London)* **261**, 414–415.
Sidman, R. L., and Wessells, N. K. (1975). *Exp. Neurol.* **48**, 237–251.
Silver, J., and Rutishauser, U. (1984). *Dev. Biol.* **106**, 485–499.
Silver, J., and Sidman, R. L. (1980). *J. Comp. Neurol.* **189**, 101–111.
Skoff, R. P., and Hamburger, V. (1974). *J. Comp. Neurol.* **153**, 107–148.
Smalheiser, N., Crain, S. M., and Reid, L. M. (1984). *Dev. Brain Res.* **12**, 136–140.
Small, R. K., and Pfenninger, K. H. (1984). *J. Cell Biol.* **98**, 1422–1433.
Small, J. V., Isenberg, G., and Celis, J. E. (1978). *Nature (London)* **272**, 638–639.
Small, R. K., Blank, J., Ghez, R., and Pfenninger, K. H. (1984). *J. Cell Biol.* **98**, 1434–1443.
Solomon, F. (1979). *Cell* **16**, 165–169.
Speidel, C. C. (1933). *Am. J. Anat.* **52**, 1–79.
Sperry, R. W. (1963). *Proc. Natl. Acad. Sci. U. S. A.* **50**, 703–710.
Spiegelman, B. M., Lopata, M. A., and Kirschner, M. W. (1979). *Cell* **16**, 253–263.
Steinman, R. M., Mellman, I. S., Muller, W. A., and Cohn, Z. A. (1983). *J. Cell Biol.* **96**, 1–28.
Stirling, R. V., and Summerbell, D. (1978). *J. Embryol. Exp. Morphol.* **41**, 189–207.
Stossel, T. P. (1984). *J. Cell Biol.* **99**, 15s–21s.
Strassman, R. J., Letourneau, P. C., and Wessells, N. K. (1973). *Exp. Cell Res.* **81**, 482–487.

Summerbell, D., and Stirling, R. V. (1981). *J.Embryol. Exp. Morphol.* **61**, 233–247.
Sutter, A., Ropelle, R. J., Harris-Warrick, R. M., and Shooter, E. M. (1979). *J. Biol. Chem.* **254**, 5972–5982.
Teichberg, S., and Holtzman, E. (1973). *J. Cell Biol.* **57**, 88–108.
Tennyson, V. M. (1970). *J. Cell Biol.* **44**, 62–79.
Trinkhaus, J. P. (1969). *In* "Cells into Organs. The Forces That Shape the Embryo." Prentice-Hall, Englewood Cliffs, New Jersey.
Trinkhaus, J. P. (1984). "Cells into Organs: The Forces That Shape the Embryo." Prentice-Hall, Englewood Cliffs, New Jersey.
Tsui, H.-C. T., Ris, H., and Klein, W. L. (1983). *Proc. Natl. Acad. Sci. U. S. A.* **80**, 5779–5783.
Vallee, R. B., and Bloom, G. S., and Theurkay, W. E. (1984). *J. Cell Biol.* **99**, 38s–44s.
Van der Loos, H. (1965). *Bull. Johns Hopkins Hosp.* **117**, 228–250.
Vaughn, J. E., Hendrikson, C. K., and Grieshaber, J. A. (1974). *J. Cell Biol.* **60**, 664–672.
Weiss, L. (1977). *In* "Cell Interactions in Differentiation" (M. Kartinen-Jaaskelainen, L. Saxen, and L. Weiss, eds.), pp. 279–289. Academic Press, New York.
Weiss, P. (1934). *J. Exp. Zool.* **68**, 393–448.
Weiss, P. (1941). *Growth (Suppl.)* **5**, 163–203.
Weiss, P. (1961). *Exp. Cell Res. Suppl.* **8**, 260–281.
Weiss, P., and Taylor, A. C. (1944). *J. Exp. Zool.* **95**, 233–257.
Wessells, N. K. (1982). *In* "Cell Behavior" (R. Bellairs, A. Curtis, and G. Dunn, eds.), pp. 225–245. Cambridge Univ. Press, Cambridge.
Wessells, N. K., and Nuttall, R. P. (1978). *Exp. Cell Res.* **115**, 111–122.
Wessells, N. K., Ludueña, M. A., Letourneau, P. C., Wrenn, J. T., and Spooner, B. S. (1974). *Tissue Cell* **6**, 757–776.
Wessells, N. K., Johnson, S. R., and Nuttall, R. P. (1978). *Exp. Cell Res.* **117**, 335–346.
Wessells, N. K., Letourneau, P. C., Nuttall, R. P., Ludueña-Anderson, M., and Geiduschek, J. M. (1980). *J. Neurocytol.* **9**, 647–664.
Wong, R. G., Hadley, R. D., Kater, S. B., and Hauser, G. C. (1981). *J. Neurosci.* **1**, 1008–1021.
Wujek, J. R., and Lasek, R. J. (1983). *J. Neurosci.* **3**, 243–251.
Yamada, K. M., Spooner, B. S., and Wessells, N. K. (1971). *J. Cell Biol.* **49**, 614–635.

3

Pathways and Changing Connections in the Developing Insect Nervous System

JOHN S. EDWARDS[1]
Max-Planck-Institut für Verhaltensphysiologie
Abteilung Huber
Seewiesen, Federal Republic of Germany

Insects may be tolerated in neurouiology, but they are rarely admired. More often they are disdained by students of higher organisms for their lack of relatedness to mammals. [Greenspan, 1981. Copyright 1981 by the American Association for the Advancement of Science.]

[1]Present address: Department of Zoology, University of Washington, Seattle, Washington 98195.

I. Introduction

A. Are Insects Relevant?

My viewpoint in presenting some aspects of insect neural development to readers more familiar with vertebrates is that I should at the outset justify the claim that the analysis of insect development is, far from being a pursuit in itself, highly relevant to and perhaps necessary for a full appreciation of the essentials of neurogenesis. A prevalent view that insects are merely lower and more rudimentary forms of life is based on a rather simplistic linear view of phylogeny. This linear ''higher–lower'' dichotomy may combine with a culturally instilled antagonism toward insects and perhaps a familiarity only with noxious examples to elicit in the researcher a prejudice against accepting them as valid models for the study of nervous systems *in general*. This despite the fact that the first Nobel laureates from the field of behavior, von Frisch and Tinbergen, worked with insects, and that key aspects of modern neurobiology, for example, neurosecretion and receptor mechanisms, had their origins in studies of invertebrates.

The relevance of invertebrate studies to general neurobiology is sometimes disputed on the grounds that the stereotyped behavior of invertebrates and the relative paucity of central neurons give them little in common with vertebrates, especially mammals; but many insects, for example, bees, locusts, cockroaches, and *Drosophila*, learn well. They have proven well suited to studies of learning mechanisms (e.g., Erber, 1975a,b; Hoyle, 1980) and analysis of motor flexibility in response to sensory stimuli (e.g., Kien, 1979).

B. An Evolutionary Perspective

A more informative perspective of our relationship, as animals, to the insects is to regard the later evolutionary history of the Animal Kingdom as comprising two great divergent branches, one encompassing the so-called annulate superphylum exemplified by the octopus, the crayfish, and the honeybee, while the other great branch, the chordate superphylum, includes among its further branches the vertebrates. By the time these two basic body plans diverged, such things as sense cells, neurons, synapses, myoneural junctions, hormones—indeed, most components of complex nervous systems—were already refined. During their divergence along contrasting evolutionary pathways, the two great groups have made different uses of these elements but the fundamental mechanisms are held in common.

C. Some Special Features of Arthropod Nervous Systems

Some useful features of the insect nervous system that can be exploited in studies of development are summarized below.

1. Sensory cells originate in the epidermis. The same cell transduces stimuli and projects to the central nervous system. Their position and number is often rigorously specified. They can regenerate from epidermal cells.
2. Interneurons and motoneurons occupy positions in the central nervous system such that they can often be identified from animal to animal and many have been characterized anatomically and physiologically.
3. The central nervous system develops from segmentally repeated populations of neuroblasts that yield a series of segmental ganglia in which serial homology can often be unequivocally established.
4. The parceling of neuropil through a series of segmental ganglia facilitates the localization of function, although the basic segmental plan becomes more obscure in the higher insects where segments are increasingly fused.
5. Higher insects, with larvae and adults separated by the pupal stage, provide the neurobiologist with examples of major neural reorganization during postembryonic development.
6. With *Drosophila*, an immense background of genetic data allows unique manipulation of the pathways of neural development (e.g., Hall *et al.*, 1982).
7. Technical advances in physiological and anatomical techniques using intracellular electrodes and dye filling bring neural development to the study of single cells.

I will consider four principal topics: (1) embryogenesis, the making of first contacts; (2) postembryonic development, the growth of arborizations and metamorphosis, the redesign of the nervous system; (3) some experimental approaches to neural plasticity and finally (4) genetic approaches to neural organization. Each of these areas has obvious relevance to comparable studies with vertebrates. I have arbitrarily emphasized sensory development at the expense of motor innervation in order to meet space limits.

The neurobiology of invertebrates, long considered a curiosity shop, now abundantly shows that the general principles of neural function and of development—the differentiation and basic properties of neurons and their interactions—are common throughout the Animal Kingdom. I have chosen to concentrate on the insects, but it should be emphasized that other invertebrate groups are making profoundly important contributions to our concepts of neural development. Nematodes, with their extreme precision and parsimony of neural organization provide material for a synapse-by-synapse analysis of a whole nervous system and its development (White, 1985).

Leeches (Nicholls *et al.*, 1977) and mollusks (Kandel, 1976; Kriegstein, 1977) also provide ideal preparations with identified neurons; indeed, the mollusks may well hold the key to cellular mechanisms of many psychological phenomena. Some recent advances in the developmental neurobiology of invertebrates are reviewed by Anderson *et al.* (1980) and by Goodman (1985).

II. Embryogenesis: Foundations for Neurogenesis

The intricate and complex connections of an adult nervous system appear too elaborate to have been generated by mechanisms that are accessible to the methods of cell biology, but by following the sequence of events from the origin of the first neuroblasts, one can obtain an appreciation of the process.

Insect neuroembryology, having lain fallow since the turn of the century, is in the full flush of a new spring, propelled by general questions concerning cellular relationships in morphogenesis, in neurogenesis, and supported by technical advances in cell labeling, recording, and microscopy. A special merit of the insect nervous system in this context is the small number of neurons, stemming from a very uniform population of segmentally repeated neuroblasts in the central nervous system (Bate, 1976a), and the relatively simple appendage structure, allowing the detailed tracing of earliest peripheral neurons. In both center and periphery, key findings concern the role of the first fibers as morphogenetic templates.

A. Development of the Sensory System

In the early development of the sensory system of insect appendages (Fig. 1) (e.g., legs and antennae of locusts, Bate, 1976b; Keshishian, 1980; and the cerci of crickets and locusts, Edwards, 1969; Edwards and Chen, 1979; Shankland, 1981a,b) Shankland and Bentley (1983), a small set, sometimes a pair, of cells situated at the apex of the rudimentary embryonic appendage send processes to the central nervous system long before sense organs differentiate, and usually before the appendages begin to elongate. These cell processes, which are axonlike in ultrastructure, reach the central nervous system via highly specific pathways. After appendages have elongated, these tracts of pioneer fibers provide the focal pathways along which functional sensilla dispatch their axons to the central nervous system later in embryogenesis. Evidence from locusts (e.g., Goodman *et al.*, 1982) clearly shows that distalmost cells do not necessarily pioneer the entire pathway to

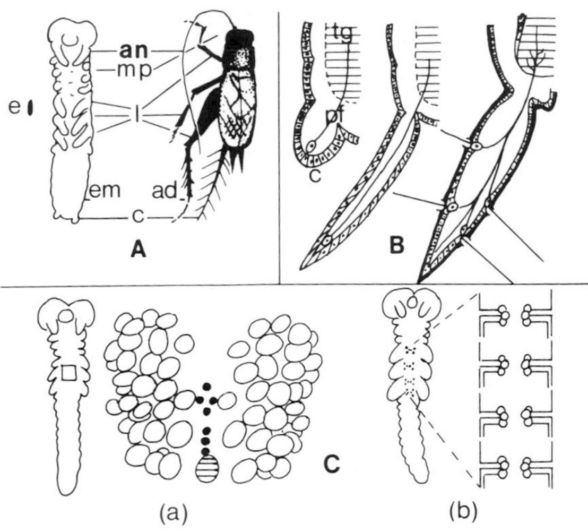

Fig. 1. Embryonic development of nervous system. (A) Comparison of embryo and adult. An egg (e) is shown at left in proportion to adult (ad) at right. Center, an embryo (em) is shown scaled to adult (right) size. an, Antenna; mp, mouthparts; 1, legs; c, abdominal cercus. (B) Embryonic development of sensory neurons in the abdominal cercus. Left, pioneer fibers (pf) with their cell bodies at the apex of the cercus (c) grow to the terminal ganglion (tg) before the cercal rudiment elongates. Center, embryonic cercus fully elongated. Pioneer fibers now traverse the full length of the cercus. Right, axons from cell bodies of functional sensilla follow pathway set up by pioneer fibers to the terminal ganglion. (C) Central nervous system development. (a) Map of neuronal precursor cells from the ventral aspect in a grasshopper embryo. The open circles are neuroblasts that will generate the central neurons. The hatched cell, a midline neuroblast generates a set of midline precursor cells (black). (b) From specific cells with the midline precursor group, a set of cell trios forms longitudinal tracts that function as the pioneer longitudinal tracts of the central ganglionic chain. Redrawn by permission from Bate and Grünewald (1981).

the center; a relay system may prove to be general. The first contacts between periphery and center are thus established over short distances, usually less than 1–200 μm.

In the relatively large embryo of the grasshopper *Schistocerca nitens*, where the process of pioneer fiber formation has been observed *in vivo* in cultured embryo preparations (Keshishian, 1980) a cell pair arises from an apical epidermal cell in a leg bud. Axonlike processes from these cells and others more proximally located, reach the central ganglia early in development, and their tract forms the pathway for subsequent leg afferents.

Shankland's (1981a,b) detailed study of the embryonic development of projections of cercal pioneer fibers in the grasshopper *Schistocerca nitens* shows that the pioneer fibers turn sharply as they enter the ganglion rudiment, to join axons of the primary longitudinal tract (see Section II,B), which is the central equivalent of the pioneers. They then grow several hundred micrometers along this tract without branching. These pioneers are followed to the ganglion from the cercus by the functional cercal afferents later in embryogenesis. Within the ganglion they project to their particular terminations without forming temporary colaterals; it seems that specific cues, rather than trial and error, govern the passage of the sensory fibers (Shankland, 1981b). It is clear that embryonic afferent innervation shapes the dendritic branching pattern of identified giant interneurons (Shankland *et al.*, 1982).

A demonstration that the pioneer fibers do play a necessary role in organizing the afferent nerves of the abdominal cerci comes from laser lesions of pioneer fiber cells at various stages in development. When the tip of the cercal rudiment was ablated in the region of the presumptive pioneer fiber cell body, so that the normal pioneer tract was not generated, then the afferent axons that arose from sensilla, and which, in the normal course of events, would have followed the pioneer fibers to the central nervous system, were unable to form a normal nerve and appeared to wander blindly within the cercus (Edwards *et al.*, 1981).

B. Development of the Central Nervous System

In the central nervous system too, the foundations on which the neuropil build are pioneer cell processes. They form longitudinal tracts and transverse "rungs" on the ladder that will serve as the template for subsequent neuropil formation. Bate and Grunewald (1981) have injected dye into the precursor cells of these central pioneers and have thus been able to follow their life history. Neuroblasts give rise to cell trios, located on each side of the midline in each segment. Specific members of the trio send processes to the anterior and posterior margins of each segment, where they cross the segmental border and join with their homologs of the neighboring segments. In this way, primary longitudinal tracts are established that serve as organizing foci for subsequent neuronal differentiation (Fig. 1). Aspects of pathfinding by neuronal growth cones in grasshopper embryos have been reviewed by Goodman *et al.* (1982), and Bastiani *et al.* (1985).

Together, the several studies of early neural development illustrate some basic general features of neurogenesis. The emergence of a basic pattern occurs when the scale is small, of the order of 1–200 μm. The cercal and limb rudiments, for example, are less than 100 μm from the central destinations of the pioneer peripheral fibers. Centrally, the earliest pathways are laid down on a segmental pattern in which each segment is about 150 μm in length. Second, the cell processes move over a basal lamina in establishing the first pathways. Once the pioneer fibers have made their connections, the embryo can grow, appendages can elongate, the pioneer fibers can elongate by passive stretching as noted by Harrison (1935), and the pathways can become complex due to differential growth.

These observations on the behavior of individual insect cells take to a higher resolution the long-standing observations on vertebrate neurogenesis dating back to the pioneers, His, Ramon y Cajal, Harrison, and others, at the turn of the century. The importance of surfaces as guides to axonal growth, a recurrent theme in vertebrate neural development (Hamburger, 1962; Singer *et al.*, 1979; Katz *et al.*, 1980), is open to study at the individual cell level in insect embryos and imaginal disks (Blair and Palka, 1985).

On the ladder template of the central pioneer fibers, embryonic neurons derived from the population of neuroblasts organize the process that will generate the functional neuropil. Some of these cells are large enough to be impaled with microelectrodes, and their differentiation has been followed from birth to maturation (Goodman and Spitzer 1979; Goodman *et al.*, 1979, 1980).

Neuroblasts are electrically coupled to other cell types initially, and are progressively uncoupled during development, first from nonneural cells and then from other neuroblasts. Neurites are first formed after a cell ceases to be dye-coupled with other cell bodies from the same neuroblasts, but while it is yet electrically coupled, implying that the cessation of interchange or larger molecules is a necessary part of differentiation. Many temporary supernumerary branches in the neuropil and in functionally incorrect segmental nerves are formed during the development of the neuron. The response to transmitter substances coincides with the limitation of neurite growth, and the onset of electrical activity occurs when the axon reaches its target, at which time it becomes electrically uncoupled from its siblings. Thus a sequence of events leads to the establishment of the final identity of the neuron, each of which further specifies its identity.

The process of forming multiple temporary branches by motor neurons is in contrast to the behavior of sensory axons within the neuropil, which seem to have no doubt about where they should go (Shankland, 1981b).

III. Postembryonic Development: Remodeling en Route

A. Direct Development: The Growth and Plasticity of Aborizations

Insects with direct development, and which resemble miniature adults when they hatch, do so with their full complement of motoneurons and with most of their interneurons. The cells increase in volume through the series of instars (stages) that lead to the adult. With each new instar a series of sensory axons arrive from added sense cells in the integument. They must find a place to synapse on their central targets. They reach the central nervous system, following existing bundles, but within the ganglion they must traverse the neuropil to reach their destination. The paths of such neurons can be reconstructed from cobalt-filled axons, and their target cells can similarly be mapped. It is therefore possible to follow the process of normal and variously modified growth throughout the postembryonic development of these insects.

1. The Growth and Variability of Dendritic Aborizations

Comparison of the record of growth patterns reflected in the form of mature identified interneurons and motor neurons show a large range of variability. Earliest studies using the cobalt filling technique emphasized the lack of variability in the arborization patterns of motor neurons (e.g., Burrows 1973; Pitman *et al.*, 1973), but later studies in which improved techniques were used have revealed significant variation in the form of identified cells. For example, an identified large mechanoreceptor, a wing stretch receptor axon of a locust, can enter the neuropil by two different routes, but project to the same regions (Altman and Tyrer, 1977). Comparison of superimposed projection patterns (Fig. 2) gives an indication of the amount of variation. Burrows (1975) compared the form of the right and left partners of pairs of an identified neuron and found as much variability within one animal as among individuals, and Pearson's and Goodman's (1979) work with the much-studied DCMD (descending contralateral movement detector), a large identified interneuron in a grasshopper, showed that few animals have all possible observed branches; indeed, in some animals, as many as half of the main branches are absent.

An identified population of brain interneurons associated with the ocelli or simple eyes has been surveyed by Goodman (1978), whose detailed comparison of the form of these neurons in a large number of adult grasshoppers revealed considerable variability, but the variants could be categorized by several groups. Neurons were sometimes absent, supernumerary cells were sometimes present, and a set of anomalous dendrite morphologies were recognized. One category of anomalous arborization, for example, terminated in the contralateral homolog of the normal site. It may be that some of this variability is a consequence of relaxed selection in laboratory cultures, but crayfish from wild populations were also found to have occasional major developmental anomalies of their identifiable neurons (Ballenger and Bittner, 1978). The lesson of these examples is that "errors" are prevalent and evidently tolerated even within relatively small neuronal pools in animals that are commonly considered to show extreme rigidity of behavior. But then perhaps what we consider rigid behavior will prove to have a comparable variance when its analysis has the requisite level of resolution.

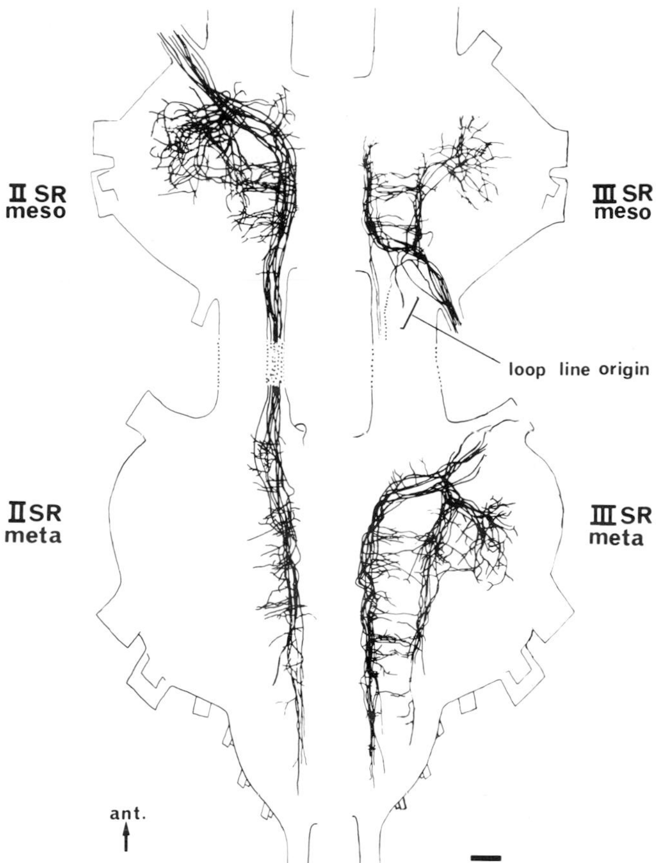

Fig. 2. Variation in arborization pattern of the central projection in the meso- and metathoracic ganglia of a locust. A particular identified single receptor cell was filled by cobalt impregnation in a series of animals and their arborizations are shown superimposed and normalized. A forewing stretch receptor (SR) is shown on the left. By permission from Altman and Tyrer (1977).

2. Impact of Experimental Manipulation on Arborization: Can Insect Dendrities Sprout?

Motor neuron structure and function seem to be very robust in arthropods, at least in decapod Crustacea, for removal of excitatory inputs to identified motor neurons in crayfish did not detectably alter their morphology (Wine, 1973), and removal of both target muscles as well as afferent input to lobster swimmeret motor neurons had no effect on their pattern of motor output (Davis and Davis, 1973). Similarly, locust motor neurons developed their normal mature pattern when deprived of the normal input from wing mechanoreceptors (Kutsch, 1974).

While the examples given above imply that the motor neurons have little or no capacity for growth responses to altered milieu, evidence has accumulated that indicates that at least some insect motor neurons do respond by sprouting. Thoracic motor neurons in a cricket survive for long periods when severed from the insects' cell bodies (Clark, 1976a,b), and make new outgrowths during a period of 50 days after operation, provided the cut is made within the neuropil; isolated distal segments of the same neuron degenerate. Vigorous sprouting of the metathoracic fast coxal depressor motor neuron in the cockroach *Periplaneta americana* has been reported by Pitman and Rand (1982). This identified neuron, which is well known in terms of function and arborization pattern, puts out extensive sprouts both after its axon has been cut peripherally and after other peripheral nerves and connectives associated with the ganglion have been cut without inflicting direct damage to the neuron in question. Sprouts usually arise from existing dendrites and they tend to follow tracts containing degenerating neurons but they also extend into areas of neuropil not normally traversed by that neuron. So far, the functional relationships of these sprouts to remaining neurons in the neuropil is unknown, but the future promises a detailed analysis of changed synaptic relationships in a much studied motoneuron. Motoneurons can also sprout distally in response to local denervation (Donaldson and Josephson, 1981).

The responses of identified interneurons to an altered milieu is best known from the giant interneurons in the cricket, both in the embryo (Shankland *et al.*, 1982) and during postembryonic development. The LGI (lateral giant interneuron) and MGI (medial giant interneuron) receive input from the abdominal cerci, where are large sensory appendages at the posterior end of the body (Fig. 1). Removal of the cerci deprives the giant fibers of their major source of input (Edwards and Palka, 1974, 1976; Palka *et al.*, 1977; Matsumoto and Murphey, 1977a). While the dendritic arborization is reduced in proportion to the reduction in neuropil volume (Murphey *et al.*, 1975), neither the general form nor the density of dendritic spines is changed significantly (Murphey and Levine, 1980) (Fig. 3). The effects are restricted

Fig. 3. Modification of properties of giant interneuron in the cricket *Acheta domesticus*. (A) Effects of deafferentation on the development of an identified giant neuron. The abdominal cercus supplying the right cercal sensory nerve was removed throughout postembryonic development, while the left cercal nerve (LCN) was allowed to develop normally. The same branch from each of the pair of giant interneurons are compared on their respective sides of the ganglion. The deprived dendrite is smaller and the spines shorter, but the density of spines per unit length is about the same on both. No sprouting was observed. From Murphey and Levine (1980), with permission. (B) Effect of sensory deprivation on responsiveness of an identified giant interneuron in the cricket. The response of giant interneurons to standard sound tone stimulation of mechanoreceptor hairs on the abdominal cerci. In treated animals (left) the cercal hairs of one cercus were immobilized throughout development but allowed to develop without restraint. Intracellular recordings obtained from the control and treated pair of giant interneurons show that action potentials were elicited only on the control side. The higher gain used on the treated side provides evidence for some excitatory input, but no action potentials were recorded. From Matsumoto and Murphey (1977b) by permission.

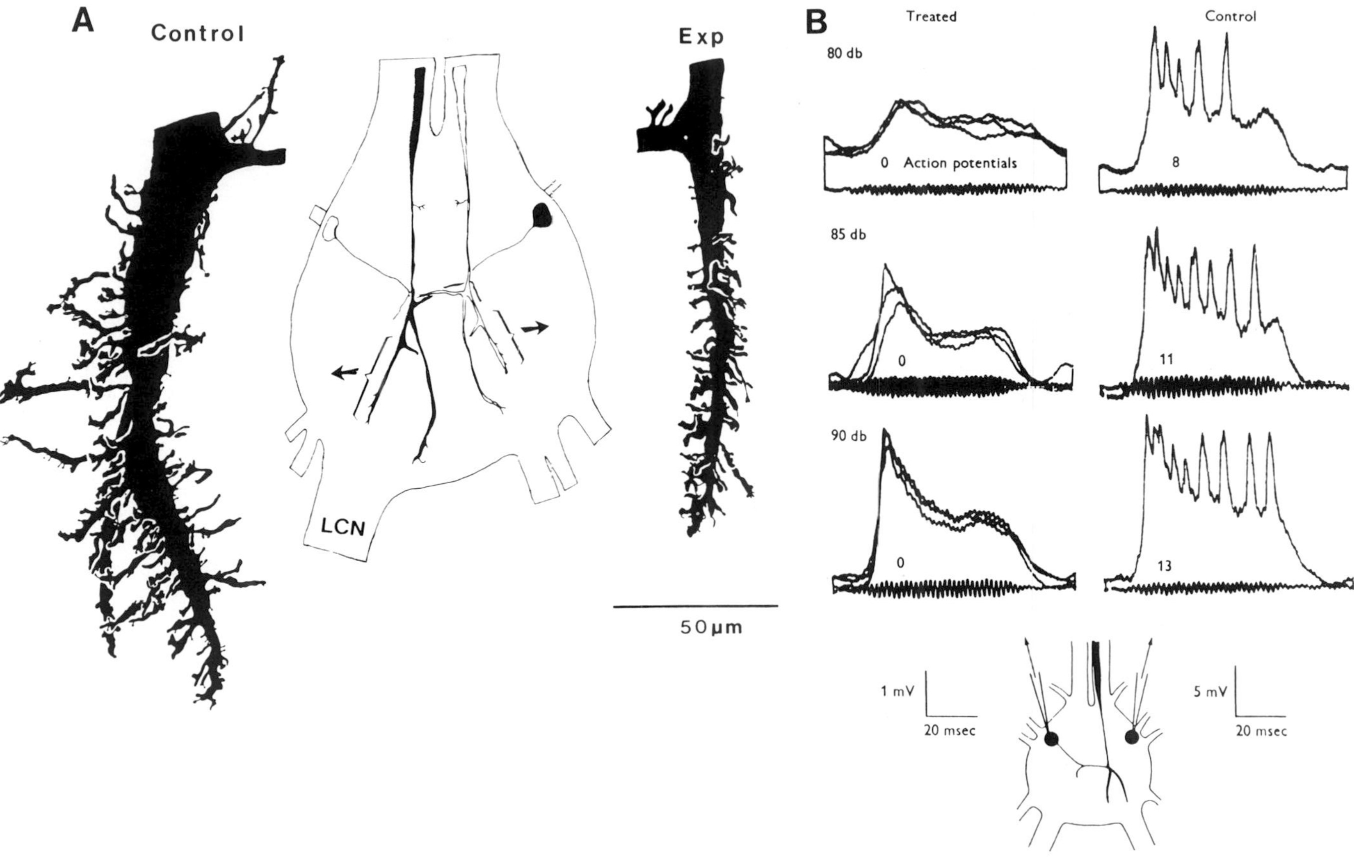
A
Control
LCN
Exp
50 µm
B
Treated
Control
80 db
0 Action potentials
8
85 db
0
11
90 db
0
13
1 mV
20 msec
5 mV
20 msec

to the deprived side of the ganglion within the neuropil, which lacks cercal sensory input; one interneuron can thus show localized responses, in this case, restriction of arborization on one side of the ganglion only. The arrested development is compensated when cerci regenerate in the course of postembryonic development (Murphey *et al.*, 1976). Growth of cerci is accelerated during regeneration until they reach normal length (J. S. Edwards and C. Daniels, unpublished), and comparable compensatory growth occurs in the interneurons. When sensory input from intact cerci was silenced by preventing the mechanosensory hairs from moving, the form and size of the giant interneuron arborization was normal (Murphey *et al.*, 1984) although mutant crickets that lack the cercal mechanoreceptor hairs and whose cercal nerves are thus intact but effectively silent showed minor reduction in dendrite growth (Bentley, 1975). The evidence from the cercus–giant interneuron system does not point to a major influence of cercal input on target cell arborization.

The anatomical effects of deafferentation, sensory deprivation, and regeneration are, however, much less striking than the physiological changes reflected in the evoked activity of the giant fibers, as monitored in the ventral nerve cord. For example, unilateral deafferentation (i.e., removal of one abdominal cercus) during postembryonic development causes increased contralateral excitation of the giant fibers on the deprived side. Contralateral sprouting was postulated as the mechanism of changed response since extra contralaterally terminating fibers were detected in the ganglia of asymmetric animals (Palka and Edwards, 1974) but the effect proved to be more significantly a modification of the balance between excitatory and inhibitory inputs to the giant fibers (Murphey *et al.*, 1984).

Plasticity of the escape response in *Periplaneta americana* after cercal ablation, involving correction of inadaptive escape responses over a period of 30 days, is correlated with changes in the response properties of their giant interneurons (Volman *et al.*, 1980). Adjustments in coordination of stridulation behavior following limb loss show a similar time course (Elsner and Hirth, 1978). Unilateral sensory deprivation of cercal sensory input by immobilizing the mechanosensory hairs of the cercus enhances contralateral inhibition and reduces the effect of contralateral excitation (Matsumoto and Murphey, 1977b), but these effects are only evident when the perturbation occurs during a sensitive period early in postembryonic development and is compensated for when sensory input is restored (Matsumoto and Murphey, 1978).

Another example of the effect of deafferentation in the cricket CNS, in this case the prothoracic ganglion, comes from Hoy *et al.*, (1978) who report that when a specific identified auditory interneuron was deprived of auditory input by amputation of the foreleg at hatching, before the ear developed on the tibia, that is, before the interneuron received any sensory input, the arborization of the lateral part of the dendrite was reduced, while the more medial dendrites, which normally remain strictly ipsilateral, crossed the midline and ramified in the equivalent contralateral neuropil.

These examples suffice to show that the insect neuropil is not the rigidly "hard-wired" system it was once thought to be, and with the techniques now available to characterize individual neurons, insects should prove to be well suited to the analysis of mechanisms of sprouting and changing synaptic relationships at the single cell level.

B. Metamorphosis: Massive Postembryonic Neural Reorganization

Insects with indirect development hatch from the egg as a larva, a relatively sedentary organism, a "feeding machine" with rudimentary sense organs, and later undergo a dramatic metamorphosis to become active adults in a far more complex world. This radical change of habitat and activity is reflected in the reorganization of the nervous system during metamorphosis. During the pupal phase that intervenes between larva and adult, a radical dismantling and restructuring of the body takes place, and this is reflected in the contrasting nervous systems of the larval and adult stages (Fig. 4). Much of the adult is built from groups of cells, the imaginal disks, which are formed and set aside in the embryo and carried through the larval stages to provide the source of adult structures during metamorphosis.

The implications of this radical change for the nervous system are profound: an almost blind caterpillar becomes a highly visual butterfly; a legless maggot becomes a skillfully maneuvering fly. Aspects of the means by which the profound changes in the nervous system are achieved during metamorphosis are covered in several reviews (Nuesch, 1968; Edwards, 1969; Bate, 1978; Pipa, 1978), and only two aspects that have a bearing on neural development in general will be examined here.

The first of these is the importance of continuity in neural development. The structures of the adult nervous system that are produced during metamorphosis from neuroblasts and imaginal disks depend for their connections to the central nervous system on pathways laid down in the embryo, either as part of the functional larval nervous system or as neural connections between imaginal disks and the CNS. In the development of the compound eye of the adult butterfly, for example, it is the slender nerve from the rudimentary larval eye that provides the bridge for axons from the large adult eye. The larval nerve splits up to form a scaffolding between the developing compound eye and the optic lobe of the brain (Nordlander and Edwards, 1969). The generation of highly structured neuropil in the optic lobes is an extension of this principle: neuronal outgrowth between neuropil

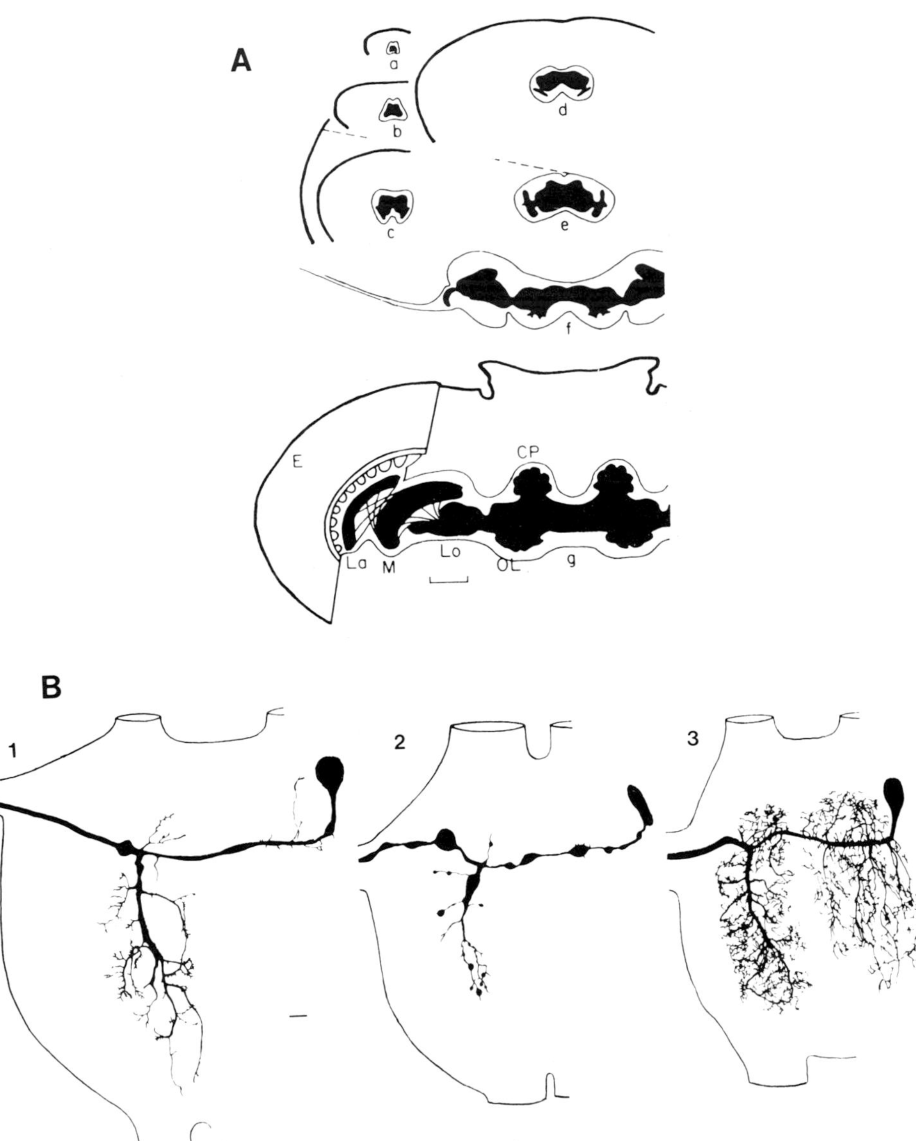

Fig. 4. Postembryonic neural charges associated with metamorphosis. (A) Postembryonic brain development in the monarch butterfly *Danaida plexippus*. Diagrammatic frontal sections through comparable regions of brain and head capsule at successive stages. (a) Newly hatched first instar larva; (b) early second instar; (c) early third instar; (d) early fourth instar; (f) early pupa; and (g) adult. Heavy line denotes head capsule—at far left for (e), not shown for (f). Thin line delimits brain. Black area denotes neuropil. Lateral extremities in larval brains are the optic lobe anlagen,

centers occurs in a precise sequence, coordinated in time and space, that ensures the formation of an ultimately highly complex neuropil (Edwards, 1977; Meinertzhagen, 1975; Palka, 1979a; Bate, 1978). This, of course, does not describe the mechanism but rather provides the set of rules from which one can search for mechanisms.

Cellular interrelationships in the developing compound eye of insects have been minutely analyzed (for reviews see Meinertzhagen, 1975; Bate, 1978; Palka 1979a). Analyses of retinotopic projections from the retina to the lamina neuropil (the most superficial neuropil of the optic lobe) in a range of arthropods (e.g., Anderson, 1978a,b; Macagno, 1978; Meinertzhagen, 1973, 1975) all point to the importance of temporal sequence, together with the guidance of axons along earlier formed pathways. The differentiation of the laminar neurons is triggered by the arrival of retinal cells in *Daphnia* where eye development has been followed in great detail (LoPresti *et al.*, 1973, 1974); destruction of retinal elements causes comparable reduction in lamina elements (Macagno, 1979). Cell death is a conspicuous feature of the normal developing insect optic lobe (Nordlander and Edwards, 1968; Anderson, 1978a), but the numbers of dying cells can be reduced if the number of receptor cells is experimentally increased (Mouze, 1978). There is evidently an oversupply of cells during development; those that fail to receive peripheral contacts die.

Antennal sensory neurons in debrained pupae of the moth *Manduca sexta* develop normally in the absence of their central targets. They have normal morphological, physiological, and biochemical characteristics (Sanes *et al.*, 1976). Their lack of dependence on central contacts is typical of insect sensillar development and stands in contrast to the response of vertebrate sensory neurons to the absence of a central target.

The motor system must also undergo radical changes in the transition from larva to adult. While sensory structures such as compound eyes arise from imaginal disks, the motor neurons, with their cell bodies surrounding the central neuropil, are already present in the larva and in many cases serve both larva and adult.

In the motor reorganization of an abdominal ganglion in the tobacco hornworm *Manduca sexta*, 89% of larval motor neurons are retained in the

---

which are prominent in the pupa and form the lamina (La), the medulla (M), and lobula (Lo) of the adult optic lobe. Calyces of the corpora pedunculata (CP) are prominent dorsally, as are the olfactory lobes (OL) ventrally. (E), Compound eye. Scale line = 20 μm. Adapted from Nordlander and Edwards (1969). (B) Dendritic reorganization of an identified motor neuron in the tobacco hornworm moth *Manduca sexta*, based on cobalt chloride impregnations of motor neuron MNI in the fourth abdominal ganglion in (1) the larval; (2) pupa; (3) adult. The cell body lies in the anterior right quadrant of the ganglion. Scale bar = 20 μm. From Truman and Reiss (1976). Copyright 1976 by the American Association for the Advancement of Science.

adult where they may serve a very different musculature generated during metamorphosis (Taylor and Truman, 1974) and they can undergo major restructuring of their arborizations (Truman and Reiss, 1976). Adult arborizations are derived in part from outgrowth from primary and secondary branches of the larval neuron, but new branches, unique to the adult, are also generated (Fig. 4).

Interneurons almost certainly undergo similar reorganization but they have not been followed, principally because cobalt filling of the interneurons does not permit a precise recognition of individual cells as does filling of motor neurons. The motor neurons, including those that undergo radical dendritic organization, remain electrically excitable throughout metamorphosis. They receive input, for example, from abdominal stretch receptors. While many pathways persist, some sources of innervation function only for a specific period of postembryonic development; for example the "gin traps" of the pupa require new elements to function only during the pupal phase (Levine and Truman, 1982).

A further example of central reorganization comes from a spectacularly asymmetrical crustacean, the snapping shrimp, in which loss of a snapping claw leads to transformation of the pincer claw of the opposite side into a snapper. This regenerative change is accompanied by major changes in position and size of the erstwhile pincer neurons when the limb becomes a snapper (Mellon, 1981).

IV. Ectopic Sensilla: An Experimental Approach to Pathfinding in the Nervous System

A. Transplant and Graft Experiments

The insect integument lends itself to experiments in the establishment of sensory connections, because the cell bodies of sensilla lie in the epidermis, the same cell transducing a stimulus and sending information to the central nervous system. Further, insect integument, that is the cuticle and its underlying epidermis, can be readily transplanted and grafted. Results of such experiments and the morphogenetic models generated by them are reviewed by Palka (1979a) in a neural context. This account is intended only to give some examples of conclusions that can be drawn about the behavior of sensory axons in establishing connections.

Transplanted appendages will grow at sites distant from their normal origin, and axons from such transplants can make functional contact with the central nervous system (Edwards and Sahota, 1967). When these experiments were done, the neuroanatomy of the giant interneurons that are the target cells of the cercal afferents was only partially known, and it was inferred that the ectopic afferent fibers formed synapses with the anterior projections of interneurons with which they normally connect in the posteriormost ganglion. With the advent of cobalt filling, we now know that the giant interneurons have extensive arborizations in the thoracic ganglia, and it transpires that the originally observed ectopic connections were not with the supposed interneurons (Murphey *et al.*, 1983). It remains an intriguing possibility, however, that the connections they do make are with the more anterior segmental homologs of those interneurons.

A more precise approach to the behavior of transplanted sensilla is possible with a population of wind-sensitive hairs on the heads of locusts. These hairs occur in groups or fields, each with a distinct central projection (Tyrer *et al.*, 1979). It is possible to excise and implant pieces of head integument from these fields, and by the use of the cobalt filling technique to ascertain the central projection of sensilla that developed on the implant. Axons from the grafts find the central nervous system and form arborizations in appropriate positions and with laterality consistent with their site of origin rather than with the site of implant (Anderson and Bacon, 1979). These and other results with the same system imply that the site of origin imparts a rigid specificity upon the sensilla. This conclusion is also supported by a reconstruction of the postembryonic growth patterns of a small population of sensilla on the cerci of crickets, where it was shown that the specific site of termination in the ganglion is determined by the site on the cercus and by its time of origin (Murphey *et al.*, 1980).

B. Genetic Analysis of Neural Connections

Among the great repertoire of mutations in *Drosophila*, some that affect the nervous system (Siddiqi *et al.*, 1980; Hall *et al.*, 1982; Campos Ortega, 1985; Blair and Palka, 1985) provide a new front for an attack on long-standing questions of neural development. In homeotic mutants for example, appendages are transformed so that an antenna becomes a leg, or a haltere (a balancing sense organ derived from the hind wing) becomes a second wing. The analysis of central projections of such transformed appendages offers an approach to the mechanisms that determine neural connectivity.

In *antennapedia* mutants, where the antenna takes the form of a leg, the central projections of the sensilla terminate in the olfactory lobe of the brain, as if they were antennal, and do not proceed to the more posterior terminations within the thorax that would be appropriate to a leg (Stocker *et al.*, 1976). Other genetic manipulations with such mutants as *bithorax*, where two pairs of wings, instead of wings and halteres, project to the thoracic ganglion (Ghysen, 1978, 1980; Palka *et al.*, 1979), are open to differences in interpretation; but the genetic approach, with further refinement of receptor identity and knowledge of gene action in receptor determination, promises the deepest analysis of mechanisms leading to the formation of neural connections.

V. Conclusions

Developmental processes in insect nervous systems are coming rapidly into sharper focus. Much of the detail reviewed above will doubtless soon be obsolete but some general features with broad relevance to other animal groups are clear. They are summarized below.

1. First connections between periphery and center and between successive segments of an embryo are established when the scale is on the order of tens of cell diameters. Paths taken by pioneering fibers are usually simple and direct. They follow defined corridors as their growth cones move over cells or basal laminae. Pioneer pathways provide routes for subsequent neuronal growth.
2. The later differentiation of central neurons builds upon the pathways established by pioneer fibers. The sequence of events involves the successive uncoupling of neuronal progenitors, and then neurons, from interchange with their neighbors. The acquisition of specific neuronal identity in some cases requires interactions with target cells. For example, a motor neuron may make multiple branches, then retract all but the appropriate process, and only then will the central arborization differentiate.
3. Detailed mapping of the arborizations of many identified neurons reveals the patterns of development throughout a rather extensive period of postembryonic growth and shows that a surprising degree of variability is tolerated in what was traditionally considered a highly specified network. The demonstration of sprouting and modification of synaptic function in many arthropod systems further opens up a picture of a much more flexible system than had hitherto been visualized.
4. In metamorphosis neuronal plasticity reaches a pinnacle. Individual identified neurons can serve both larva and adult. They can undergo radical reorganization and yet retain their capacity to function as integrating units throughout the process. In metamorphosis, as in embryogenesis, preexisting neuronal pathways serving as guides for adult neuronal development seem to be an invariable necessity.
5. Manipulations of neural organization by means of grafts emphasizes

the capacity of transplanted sensory cells to locate appropriate central connections as if imparted with an identity by their site of origin. Limitations on the accuracy of their projections are imposed by the major tracts encountered in an ectopic entry to the central nervous system.

With homeotic mutants too, manipulations of appendage form gives rise to responses of afferent fibers within the neuropil that imply a specific identity of sensilla, and interactions with central guide paths.

The question of the fundamental nature of metameric segmentation comes into critical focus in the interpretation of neural connections of homeotic mutants. Answers may hasten an understanding of developmental mechanisms for the evolution of complexity on a substrate of metameric segmentation in both the arthropods and the chordates.

Acknowledgments

I am grateful for support from an Alexander von Humboldt Award while preparing this work as a guest of Prof. Frana Huber, Max-Planck-Institut für Verhaltensphysiologie, Seewiesen, Federal Republic of Germany, and to Dr. Hilary Anderson for her critique of the manuscript. My work described above was supported by National Institutes of Health grant NB 07778. I thank Dr. O. Edwards, Ms. G. Stuart, and Frau I. von Casimir for help in preparing the manuscript.

References

Altman, J. S., and Tyrer, N. M. (1977). *J. Comp. Neurol.* **172**, 431-440.

Anderson, H. (1978a). *J. Embryol. Exp. Morphol.* **45**, 55-83.

Anderson, H. (1978b). *J. Embryol. Exp. Morphol.* **46**, 147-170.

Anderson, H., and Bacon, J. (1979). *Dev. Biol.* **72**, 364-373.

Anderson, H., Edwards, J. S., and Palka, J. (1980). *Annu. Rev. Neurosci.* **3**, 97-139.

Ballenger, M. A., and Bittner, G. D. (1978). *J. Neurobiol.* **9**, 301-307.

Bastiani, M. J., Doe, C. Q., Helfand, S. L., and Goodman, C. (1985). *Trends in Neurosci.* **8**, 257-266.

Bate, C. M. (1976a). *J. Embryol. Ex. Morphol.* **35**, 107-123.

Bate, C. M. (1976b). *Nature (London)* **260**, 54-56.

Bate, C. M. (1978). *In* "Handbook of Sensory Physiology" (M. Jacobson, ed.), Vol. 9, pp. 1-53. Springer-Verlag, Berlin and New York.

Bate, C. M., and Grünewald, E. G. (1981). *J. Embryol. Exp. Morphol.* **61**, 317-330.

Bentley, D. (1975). *Science* **187**, 760-764.

Blair, S., and Palka, J. (1985). *Trends in Neurosci.* **8**, 285–288.
Burrows, M. (1973). *J. Comp. Physiol.* **83**, 165–178.
Burrows, M. (1975). *J. Exp. Biol.* **62**, 189–219.
Campos, Ortega, J. A. (1985). *Trends in Neurosci.* **8**, 245–250.
Clark, R. (1976a). *J. Comp. Neurol.* **170**, 253–266.
Clark, R. (1976b). *J. Comp. Neurol.* **170**, 267–278.
Davis, W. J., and Davis, D. B. (1973). *Am. Zool.* **13**, 409–425.
Donaldson, P. L., and Josephson, R. K. (1981). *J. Comp. Neurol.* **196**, 317-327.
Edwards, J. S. (1969). *Adv. Insect Physiol.* **6**, 97–137.
Edwards, J. S. (1977). *In* "Identified Neurons and Behavior of Arthropods" (G. Hoyle, ed.), pp. 484–493. Plenum, New York.
Edwards, J. S., and Chen, S. W. (1979). *Wilhelm Roux Arch. Dev. Biol.* **186**, 151–178.
Edwards, J. S., and Palka, J. (1974). *Proc. R. Soc. London Ser. B* **185**, 83, 103.
Edwards, J. S., and Palka, J. (1976). *In* "Simpler Networks and Behavior" (J. C. Fentress, ed.), pp. 167–185. Sinauer, Sunderland, Massachusetts.
Edwards, J. S., and Sahota, T. (1967). *J. Exp. Zool.* **166**, 387–396.
Edwards, J. S., Chen, S. W., and Berns, M. W. (1981). *J. Neurosci.* **1**, 250–258.
Elsner, N., and Hirth, C. (1978). *Naturwissenschaften* **65**, 160.
Erber, J. (1975a). *J. Comp. Physiol.* **99**, 231–242.
Erber, J. (1975b). *J. Comp. Physiol.* **99**, 243–255.
Ghysen, A. (1978). *Nature (London)* **274**, 869–872.
Ghysen, A. (1980). *Dev. Biol.* **78**, 521–541.
Goodman, C. S. (1978). *J. Comp. Neurol.* **182**, 681–705.
Goodman, C. S., and Spitzer, N. C. (1979). *Nature (London)* **280**, 208–214.
Goodman, C. S. (1985). *Trends in Neurosci.* **8**, 229–230.
Goodman, C. S., O'Shea, M., McCaman, R., and Spitzer, N. C. (1979). *Science* **204**, 1219–1222.
Goodman, C. S., Pearson, K. G., and Spitzer, N. C. (1980). *Proc. Natl. Acad. Sci. U.S.A.* **77**, 1676–1680.
Goodman, C. S., Raper, J. A., Ho, R. K., and Chang, S. (1982). 40th *Symp. Soc. Devel. Biol.* 275–316.
Greenspan, R. (1981). *Science* **211**, 698–699.
Hall, J. C., Greenspan, R., and Harris, W. A. (1982). "Genetic Neurobiology." MIT Press, Cambridge, Massachusetts.
Hamburger, V. (1962). *J. Cell. Comp. Physiol.* **60**, Suppl. 1, 81–92.
Harrison, R. G. (1935). *Proc. R. Soc. London Ser. B* **118**, 155–196.
Hoy, R. R., Casaday, G. B., and Rollins, S. (1978). *Soc. Neurosci. Abstr.* **4**, 115.
Hoyle, G. (1980). *J. Neurobiol.* **11**, 323–354.
Kandel, E. R. (1976). "Cellular Basis of Behavior: An Introduction to Behavioral Neurobiology." Freeman, San Francisco.
Katz, M. J., Lasek, R. J., and Nauta, H. J. W. (1980). *Neuroscience.* **5**, 821–833.
Keshishian, H. (1980). *Dev. Biol.* **80**, 388–397.
Kien, J. (1979). *J. Comp. Physiol.* **134**, 55–68.
Kriegstein, A. R. (1977). *Proc. Natl. Acad. Sci. U.S.A.* **74**, 375–378.
Kutsch, W. (1974). *J. Comp. Physiol.* **88**, 413–424.
Levine, R. B., and Truman, J. W. (1982). *Nature (London)* **299**, 250–252.
LoPresti, V., Macagno, E. R., Levinthal, C. (1973). *Proc. Natl. Acad. Sci. U.S.A.* **70**, 433–437.
LoPresti, V., Macagno, E. R., Levinthal, C. (1974). *Proc. Natl. Acad. Sci. U.S.A.* **71**, 1098–1102.
Macagno, E. R. (1978). *Nature (London)* **275**, 318–320.
Macagno, E. R. (1979). *Dev. Biol.* **73**, 206–238.
Matsumoto, S. G., and Murphey, R. K. (1977a). *J. Comp. Physiol.* **119**, 319–330.
Matsumoto, S. G., and Murphey, R. K. (1977b). *J. Physiol.* **268**, 533–548.

Matsumoto, S. G., and Murphey, R. K. (1978). *Soc. Neurosci. Abstr.* **4**, 201.
Meinertzhagen, I. A. (1973). *In* "Developmental Neurobiology of Arthropods" (D. Young, ed.), pp. 55–104. Cambridge Univ. Press, London and New York.
Meinertzhagen, I. A. (1975). *Ciba Found. Symp.* **29**, 265–288.
Mellon, D. (1981). *Trends in Neurosci.* **4**, 245–248.
Mouze, M. (1978). *Wilhelm Roux Arch. Dev. Biol.* **184**, 325–349.
Murphey, R. K. (1978). *Soc. Neurosci. Abstr.* **4**, 202.
Murphey, R. K., and Levine, R. B. (1980). *J. Neurophysiol.* **43**, 367–382.
Murphey, R. K., Mendenhall, B., Palka, J., and Edwards, J. S. (1975). *J. Comp. Neurol.* **159**, 407–418.
Murphey, R. K., Matsumoto, S. G., and Mendenhall, B. (1976). *J. Comp. Neurol.* **169**, 335–346.
Murphey, R. K., Walthall, W. W., and Jacobs, G. A. (1984). *J. Exp. Biol.* **112**, 7–25.
Murphey, R. K., Jacklet, A., and Schuster, L. (1980). *J. Comp. Neurol.* **191**, 53–64.
Murphey, R. K., Bacon, J. P., Sakaguchi, D. S., and Johnson, S. E. (1983). *J. Neurosci.* **3**, 659–672.
Nicholls, J. G., Wallace, B., and Adal, M. (1977). *In* "Synapses" (G. A. Cottrell and P. N. R. Usherwood, eds.), pp. 249–263. Academic Press, New York.
Nordlander, R. H., and Edwards, J. S. (1968). *Nature (London)* **218**, 780.
Nordlander, R. H., and Edwards, J. S. (1969). *Wilhelm Roux Arch. Dev. Biol.* **163**, 197–220.
Nuesch, H. (1968). *Annu. Rev. Entomol.* **13**, 27–44.
Palka, J. (1979a). *Adv. Insect Physiol.* **14**, 256–349.
Palka, J. (1979b). *Soc. Neurosci. Symp.* **4**, 209–227.
Palka, J., and Edwards, J. S. (1974). *Proc. R. Soc. London Ser. B.* **185**, 105–121.
Palka, J., Levine, R., and Schubiger, M. (1977). *J. Comp. Physiol.* **119**, 267–283.
Palka, J., Lawrence, P. A., and Hart, H. S. (1979). *Dev. Biol.* **69**, 549–575.
Pearson, K. G., and Goodman, C. S. (1979). *J. Comp. Neurol.* **184**, 141–165.
Pipa, R. L. (1978). *Int. Rev. Cytol. Suppl.* **7**, 403–438.
Pitman, R. M., and Rand, K. A. (1982). *J. Exp. Biol.* **96**, 125–130.
Pitman, R. M., Tweedle, C. D., and Cohen, M. J. (1973). *In* "Intercellular Staining in Neurobiology" (S. B. Kater and G. Nicholson, eds.), pp. 83–97. Springer Publ., New York.
Sanes, J. R., Hildebrand, J. G., Prescott, D. J. (1976). *Dev. Biol.* **52**, 121–127.
Shankland, M. (1981a). *J. Embryol. Exp. Morphol.* **64**, 169–185.
Shankland, M. (1981b). *J. Embryol. Exp. Morphol.* **64**, 187–209.
Shankland, M., and Bentley, D. (1983). *Dev. Biol.* **97**, 468–482.
Shankland, M., Bentley, D., and Goodman, C. S. (1982). *Dev. Biol.* **92**, 507–520.
Siddiqi, O., Babu, P., Hall, L. M., and Hall, J. C. (1980). "Development and Neurobiology of *Drosophila*." Plenum, New York.
Singer, M., Nordlander, R. H., and Egar, M. (1979). *J. Comp. Neurol.* **185**, 1–22.
Stocker, R. F., Edwards, J. S., Palka, J., and Schubiger, G. (1976). *Dev. Biol.* **52**, 210–220.
Taylor, H. M., and Truman, J. W. (1974). *J. Comp. Physiol.* **90**, 367–388.
Truman, J. W., and Reiss, S. E. (1976). *Science* **192**, 447–479.
Tyrer, N. M., Bacon, J. P., and Davies, C. A. (1979). *Cell Tissue Res.* **203**, 79–92.
Volman, S. F., Camhi, J. M., and Vardi, N. (1980). *Soc. Neurosci. Abstr.* **6**, 679.
White, J. G. (1985). *Trends in Neurosci.* **8**, 277–283.
Wine, J. J. (1973). *Exp. Neurol.* **38**, 157–169.

4

The Normal and Abnormal Development of the Mammalian Visual System

RAYMOND D. LUND and FEN-LEI F. CHANG
Department of Anatomy and Cell Biology
Center for Neuroscience
University of Pittsburgh School of Medicine
Pittsburgh, Pennsylvania

I. Introduction

The central nervous system is made up of more than 10^{12} neurons, each of which makes on average between 5,000 and 10,000 connections with other neurons (Cragg, 1972). With the ever increasing sophistication in the methods of examining the brain, one is impressed that, despite the tremendous complexity, there is considerable order and predictability in patterns of connections. The literature since 1965 has emphasized, however, that certain perturbations introduced during development can modify the predicted patterns of organization substantially (Lund, 1978). The events leading to such changes result from genetic variants, specific injury, and from functional and environmental manipulation. In some cases at least, abnormal behaviors may be correlated with the anomalous connections (Schneider, 1973). Studies of the patterns of variation are in themselves extremely important, however, in elucidating the process of normal development and how it is regulated, but unless these variations are examined in the context of normal development, they run the risk of being viewed as interesting trivia.

The principle behind all work on normal and abnormal development of the nervous system is that the developmental process is guided by a program encoded in the genome. At present, very little is known of the instructions which underlie the details of neural organization. Furthermore, while it is clear that many events can modify the pattern of development, it is not at all clear in most cases at what level or by what mechanism the modification occurs. Some events may reflect controls of transcription or translation. Others may be unrelated to protein synthetic mechanisms but, rather, reflect variations in the environment in which the neuron or neuronal process is growing. Finally, it is apparent that the activity of neurons and the patterned functional inputs to which they respond during development are also important in determining detailed adult patterns.

While all parts of the nervous system are likely to respond similarly to developmental influences, the visual system has proven to be a particularly useful system in which to examine developmental plasticity for a variety of reasons. These include a relative simplicity of organization and the opportunity to control inputs and correlate structure and function more easily than in many systems. Equally important is the fact that pioneers in the field, most notably Roger Sperry (1963), showed how well this system could be exploited to address issues of general biological significance. A major focus of the earlier work in developmental plasticity was to show that abnormalities of organization could indeed be induced by altering various parameters of development. In recent years, due in large part to improved techniques for examining neural pathways in young animals, more attention has been given to understanding the basic mechanisms whereby normal development of the nervous system is achieved; and developmental plasticity has become viewed as a particularly important approach for gaining insight into the determinants of normal development. As a result, it is necessary to consider plasticity in the context of normal development.

In this review, we will summarize some of the more important organizational features of the visual system, how these features develop

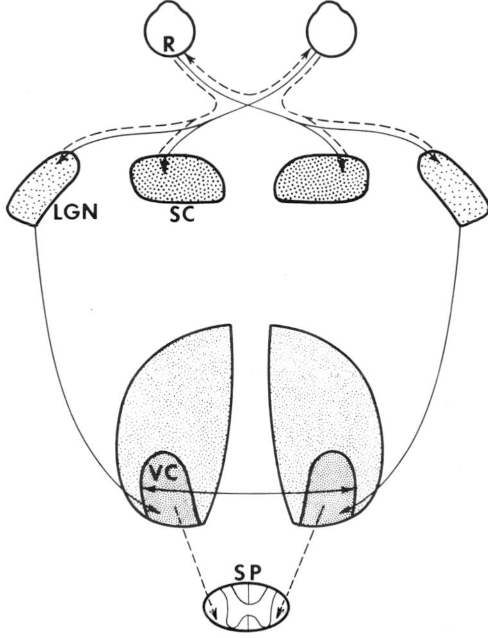

Fig. 1. Major interconnections in the mammalian visual system. Solid lines indicate permanent projections, while dotted lines indicate transient projections, or projections that are greatly diminished, for example, the ipsilateral retinogeniculate and ipsilateral retinotectal projections. LGN, Lateral geniculate nucleus; R, retina; SC, superior colliculus; SP, spinal cord; VC, visual cortex.

normally and what changes occurring after experimental intervention may signify with respect to the control mechanisms that operate in normal ontogeny.

II. Patterns of Organization

A summary diagram of major interconnections in the visual system is shown in Fig. 1. A number of smaller projections are omitted not because they are unimportant to visual function but rather because discussion of them will not add greatly to the principles reviewed here. Several principles of organization are not evident, however, from the diagram.

1. Each part of the visual system develops a certain characteristic set of features that are recognizable from one animal to another of a species and can often be seen among animals of widely different phylogeny. These include the laminar distribution of cells, the presence of certain cell types, the variety

and patterns of connections, and the transmitters used by different pathways. Thus, the layers and major cell classes identified in the retina are common to all vertebrates. The types of synapses are very similar from one vertebrate to another although the proportion of synapses made by a particular cell type vary from one species to another (e.g., Dowling and Boycott, 1966). Similarly, although there is some variability in transmitters used by a particular cell type in one species as compared with another, transmitter specificity is highly conserved within a single species. While there is somewhat more variability in the organization of central visual structures among animals of different species, individuals of a single species do not normally show substantial variability.

2. A logical map of the contralateral visual field or of the retina is projected upon the primary visual centers. This map is relayed to secondary regions such as area 17, the primary visual cortex, with reasonably good fidelity; but, with more distant projections, there is a tendency for a loss of retinal topography and sometimes for a representation of only part of the visual field (Tusa *et al.*, 1979, 1981).

3. There is a tendency to correlate inputs responding to comparable points in the visual field. This is seen in ordering of projections from each eye in the lateral geniculate nucleus (Bishop *et al.*, 1962), in the patterning of projections connecting two maps (e.g., from visual cortex to superior colliculus; Lund, 1966) and in the localized pattern of callosal projections to the 17/18 border of many mammals (Cusick and Lund, 1981; Ivy and Killackey, 1981).

4. Despite the correlation of inputs from the two eyes with respect to topography, there is, nevertheless, a tendency to segregate them, often to a higher level of organization. Segregation is seen in the laminae of the lateral geniculate nucleus and the ocular dominance stripes of the cerebral cortex (LeVay *et al.*, 1975).

5. The retina does not simply relay to the brain as a faithful and simple photographic image of the field of view, but rather transmits a series of parallel and somewhat independent parameters of vision, which are represented with different levels of resolution (Enroth-Cugell and Robson, 1966; Rodieck and Brening, 1983).

6. There is a degree of interdependence among individual parts of the visual system.

Each of these features is subject to variation under particular experimental conditions and they will be discussed separately after a brief review of normal development.

Table I

Visual System Developmental Timetable of Rat, Cat, and Monkey

Event	Rat[a] (22 days' gestation)	Cat[a] (65 days' gestation)	Monkey[a] (*Macaca*) (165 days' gestation)
Cell generation in retina	E12–18	E21–36	E30
Fibers out from retina	E14	—	—
Fibers crossing chiasm	E15	E30	—
Cell generation in superior colliculus	E12–18	—	E30–56
Fibers reaching superior colliculus	E16	E37	E64
Cell generation in lateral geniculate nucleus	E12–14	E22–32	E36–70
Fibers reaching lateral geniculate nucleus	E16	E37	—
Segregation in lateral geniculate nucleus	P5–10	E47–P0	E64–110
Cell generation in cortex	E14–21	—	E40–100
Projections (lateral geniculate nucleus) reaching cortex	E18–P0	E48–P14	E78
Segregation in cortex	—	P21–42	E144–P21
Corpus callosum projection	P0–12	P0–P60	—
Cortical projections to spinal cord	P0–14	—	—

[a]E, Embryonic day; P, postnatal day.

III. Normal Development

The course of development of the optic pathways follows a generally similar sequence in those mammals that have been studied in detail, although the relative timing of particular components of development varies considerably from one animal to another. This is exemplified in Table I where similar events are compared in rat, cat, and monkey.

Studies on morphology of the early developing retina suggest that the leading process of the earliest generated ganglion cells continues as the axon (Hinds and Hinds, 1974). The first axons to enter the optic nerve of mice are identified 1 day after the first ganglion cells become postmitotic (Sidman, 1961; Silver, 1984); in rats, a time difference of 2 days has been reported (Kuwabara and Weidman, 1974; R. D. Lund, unpublished results). The axons run in the ventral wall of the optic stalk in fascicles. There is a tendency for axons from different areas of the retina to occupy spatially different regions of the stalk with those arising from the ventral retina lying dorsolaterally and from the dorsal retina, more medially (Silver, 1984). Invasion of the more dorsal regions of the stalk appears to be prevented by a group of pigment-containing cells. Absence of pigment in albino animals results in optic axons occupying more dorsal positions (Oberdorfer *et al.*, 1981; Silver and Sapiro, 1981). As the stalk approaches the brain, the fascicles of optic fibers form a cylindrical structure. At the chiasm, axons from more nasal and ventral retinal positions tend to lie medially and anteriorly, while axons originating in the temporal retina are more lateral and posterior in the nerve.

Axons arriving at the optic chiasm project to the contralateral or ipsilateral optic tracts or to the opposite retina (Bunt *et al.*, 1983). In early development, very few axons appear to project to more than one location and this also appears to be true in adult rodents, both normal and those in which eye lesions were made during development (Jeffery *et al.*, 1981; Hsiao *et al.*, 1984; Hsiao, 1984). In rats the projection to the opposite retina arises predominantly from cells located in more ventral and nasal retina but fascicles of axons are seen running over the surface of the opposite retina. These fibers become reduced at late fetal times and disappear completely within 1 or 2 days postnatal (Bunt and Lund, 1981). A similar transient pathway has been seen in the developing chick (McLoon and Lund, 1982).

The first ipsilaterally directed axons in the optic tract are identified somewhat later than the first contralateral fibers (Bunt *et al.*, 1983; Shatz, 1983). Once they distribute in the primary optic centers, the uncrossed axons initially occupy much of the area invaded by contralateral axons, and at least some of these axons form synapses (Jeffery *et al.*, 1984). The cells of origin of the early ipsilateral pathway are not exclusively restricted to the temporal retina, there being in the rat, at least, a substantial projection arising from other retina areas (Bunt *et al.*, 1983). In the lateral geniculate nucleus, the overlapping projections from each eye gradually segregate to form the nonoverlapping projections characteristic of the adult nucleus (So *et al.*, 1978; Maxwell and Land, 1981). In the superior colliculus of rodents, the uncrossed pathway disappears from all but the most anterior part of the nucleus (Land and Lund, 1979). This period of development coincides in rodents with a period during which about 40–50% of the cells in the retina die (Potts *et al.*, 1982; Sengelaub and Finlay, 1982), and it is apparent that most of the ipsilaterally projecting cells located more nasally in the retina are lost during this stage (Bunt *et al.*, 1983; Jeffery and Perry, 1982). While it would seem likely that the disappearance of the exuberant ipsilateral projection is accompanied by loss of the parent cell bodies of these axons, the possibility cannot be excluded that some abnormally projecting cells may survive and redirect their terminal field. Indeed there is evidence to support this suggestion in the lateral geniculate nucleus of kittens (Sretavan and Shatz, 1984).

The cells of the lateral geniculate nucleus of rats become postmitotic between embryonic days 12 and 14, and axons from those cells first arrive below the visual cortex on embryonic day 18 (Lund and Mustari, 1977). As the cortex differentiates, the axons penetrate the deeper layers, ramifying in layer IV by postnatal day 4 and layer I by day 6. A similar sequence has been described for the monkey, where the axons wait in the subplate zone for approximately 46 days (Rakic, 1977).

Corticofugal pathways seem to develop relatively late. In rat for example the corticotectal pathway does not reach the superior colliculus until postnatal day 2 (P. W. Land and A. R. Harvey, unpublished results), even though the cells of origin in layer V are generated 7 days earlier on E15. Studies on the corticogeniculate pathway in monkeys show that this projection first forms a cap around the nucleus before invading it 14 days later. The initial projections seem to be at least grossly topographically appropriate (Shatz and Rakic, 1981).

The projections from one cerebral cortex to the other via the corpus callosum originate initially from cells located throughout the occipital area, but as development proceeds they become focused to the area of representation of the vertical meridian at the 17/18 border (Innocenti *et al.*, 1977; Ivy *et al.*, 1979). The exuberant origin of the callosal pathway has been identified in rodents and cats. Further studies on cats show that many of the cells with transient callosal projections survive, although it is not absolutely clear where they now project (Innocenti, 1981). Studies on the axonal distribution of the callosal pathway early in development indicate that like thalamic axons the fibers wait in the deeper layers for some days before invading the cortex (Wise and Jones, 1976). The growth of the axons into the cortex follows that of the thalamic fibers and is marked by an absence of significant growth into that part of area 17 in which are represented more peripheral parts of the visual field (Innocenti and Clarke, 1984; Lund *et al.*, 1984; Rhoades *et al.*, 1984).

To summarize, the main developmental features are the same from one species to another, although the relative timing may be different. A recurring theme is that some pathways are less precisely ordered early in development than at maturity. The mechanisms involved in the limitation of those exuberant projections may differ, being in some cases related to cell death and in others to reordering of projections.

IV. Variants of Normal Development

A. Capacity for Self-Organization

After a neuron has been generated, the next developmental event is for it to migrate to its final destination. It is natural to speculate that the path taken by the neuron during migration, and/or its final positioning may determine its future identity, including projection paths. During migration, cells often leave behind a trailing process that later develops into a projection pathway. Alternatively, the final location of the cell may permit access to certain neural or nonneuronal substrates, which allow the projection to reach certain specific targets.

To investigate the various roles of the migratory pathway and the cell location in establishment of specific connections, studies have taken advantage of genetic mutants, in which cell migration patterns are disrupted, as well as of experimental perturbations such as occur when tissue is transplanted to certain ectopic positions or when ionizing irradiation is used to disrupt early development.

While studying the mutant reeler mouse, Caviness and Sidman (1973) found that one of its prominent features is that the lamination in the cerebral cortex is reversed, with deeper cell layers located more superficially and vice versa. Subsequent studies have shown that even with the disruption of the radial organization and the abnormal projection paths of the thalamic afferents which go to layer I first, the tangential topography of the cerebral cortex is remarkably normal. The maps of the thalamic inputs from different nuclei to different cortical areas are similar in normals and mutants (Caviness and Frost, 1983). The callosal projection from the contralateral cerebral hemisphere is also comparable in both normals and mutants (Caviness and Yorke, 1976). On a finer scale, it is found that the synaptic organization is normal (Dräger, 1981). Furthermore, Lemmon and Pearlman (1981) found that even though the reeler corticotectal cells were not located in a single lamina (layer V) as normal, but rather spread throughout the cortical depth, the receptive field properties were largely normal. These studies showed that there is a robust tendency to form a constant connection pattern irrespective of the location and developmental history of the neuron.

The independence of patterns of connections from early developmental events may also be demonstrated using the transplant techniques. For example, Lund and his colleagues (Lund *et al.*, 1983) demonstrated that retinal, tectal, or cortical transplants placed into the host tectal region make specific connections with the host nervous system. These connections duplicate in some detail many of the connections made by the same regions in the intact brain. Thus, tectal transplants placed adjacent to the host tectal area receive a large projection from the visual cortical region, while retinal transplants placed at the same location receive no such projection (McLoon and Lund, 1980b). Retinal transplants located in the tectal region send abundant efferent projections to host tectum, while a retinal transplant located in the host cortex sends no or very limited projections to the host cortical areas (McLoon and Lund, 1984). These results strongly suggest that different brain tissue may have different "markers" that can either promote or retard the extension of projecting fibers.

To further investigate the role of the cell location on the final projection pattern, Jensen and Killackey (1984) used ionizing irradiation to disrupt cortical cell migration patterns. Irradiation on or before embryonic day 17 caused layer V corticospinal neurons to be mislocated to beneath the cortical white matter, but even in this ectopic position these cells still project into the spinal cord.

In summary, from the studies of mutants and cases in which transplantation and ionizing irradiation were used, it is clear that there is a specific pattern of connections that may be remarkably normal, even when the neurons involved have gone through an abnormal migratory process and occupy an abnormal location. This capacity to compensate for the early abnormalities represents one facet of developmental plasticity that allows for the formation of adequate connections under various conditions.

B. Representation of Visual Space in the Brain

Studies on early development of optic projections in frog and chick indicate that there is a period during which the topographic map from the retina to optic tectum is less precise than in the adult (McLoon, 1982) and the same appears to be true for early regenerating optic axons (Stelzner *et al.*, 1981). In the chick, the time during which the map becomes more precise coincides with a period during which half the ganglion cells die (McLoon and Lund, 1982). Recently, a similar sequence of events has been described in mammals (O'Leary *et al.*, 1984; Schneider and Jhaveri, 1984). Two possibilities are suggested. First, individual axons may have broad fields initially and these fields may be restricted as development proceeds. This may result in an individual neuron making too few synapses to sustain itself and lead to death. Alternatively, individual cells may make precise connections, some of which are nevertheless topographically inappropriate, and the wrong connections may subsequently be eliminated. Evidence for the first option comes from axon-filling studies either using horseradish peroxidase (HRP) or Golgi techniques (e.g., Lazar, 1973). The possible criticism that may be directed at these experiments is that it is frequently unclear whether, in the case of afferents to the tectum, these are indeed optic axons and if they are, whether they belong to the same classes of ganglion cells. This problem is being addressed by Sachs and Schneider (1984). In the lateral geniculate nucleus of young kitten embryos, it appears, however, that the early optic axons project in a highly restricted fashion with respect to the topographic map (Sretavan and Shatz, 1984). This would suggest that in this system early projections are precise and some are topographically appropriate and persist.

Others, however, are topographically inappropriate, and although their fields are small may be subsequently eliminated.

Studies on anophthalamic mutants show that a map from the lateral geniculate nucleus to the visual cortex is still present even in the absence of an optic input at any point during development (Kaiserman-Abramof *et al.*, 1975, 1980). It is apparent, however, that the map may not be as precise as normal. This has also been found in animals in which one eye was removed at birth (Jeffery, 1984), as well as in albino animals in which the primary optic pathway is abnormally distributed (J. G. Steedman and P. W. Land, unpublished results).

Other experiments in which the map may be modified during development follow the pattern of amphibian and fish regeneration studies, in that if the target structure is reduced in size, the axons will tend to form a logical map within the remaining region. Whether this reflects a different pattern of restriction of an initially broadly distributed projection or a redirection of axons to new target area is not known.

An important issue is the degree to which the order of axons in the primary optic pathway determines their orderly distribution on central structures. Many studies in nonmammalian vertebrates (e.g., Bunt *et al.*, 1979; Bunt, 1982; Stuermer and Easter, 1984), have shown that in some circumstances there is a logical relationship between the position of cells in the retina and the optic nerve. This also may be related to the position of termination on the tectum (Stuermer, 1984). However, the modifications of projection patterns that occur during development in some vertebrates together with the dissociation of fiber order and terminal patterns caused by intraocular administration of antibodies to neural cell adhesion molecules (N-CAM) (Rutishauser, 1984) suggest that while fiber order in the optic pathway may facilitate the orderly array of axons across the tectum, it is not an absolute determinant of retinotectal connectivity patterns.

C. Correlation of Maps

The fact the topographic maps of the periphery represented on different brain regions tend to correlate one with another in a logical fashion leads to an interesting question. If the periphery is modified or if one representation in the brain is altered differentially from another, would the maps still correlate in a logical fashion by adapting to the imbalance?

The first studies to address this issue directly concerned a visual input from the eye to the ipsilateral optic tectum of frogs by way of the nucleus isthmi. By carrying information from the eye opposite to the one projecting monosynaptically to that tectum, the pathway serves as a substrate for binocular vision. The question addressed by Keating and colleagues (Gaze *et al.*, 1970; Keating, 1977) was how this projection is able to map such that axons project to points on the tectum that respond to the same part of the visual field. They found that by rotating one eye sufficiently early in development, the pathway accommodated to ensure that axons carrying information from the same points in the visual field drove cells in the same area of the tectum. Subsequent studies have shown that axons in fact grow to the correct spot, assuming the eye to be in a normal orientation, and then turn to the new visuotopically appropriate spot (Udin and Keating, 1981; Udin, 1983). Studies on the projection from the mammalian homolog of the nucleus isthmi, the parabigeminal nucleus, show that in rats after neonatal eye removal there is an expanded projection to the opposite tectum that coincides with an enlarged ipsilateral projection derived from the remaining eye (Stevenson and Lund, 1982). This suggests that a similar mechanism may underlie the correlation of mapping of the homologous systems in rodents. In both frog and rat, if the eye manipulation is performed after a critical stage in development, no accommodation of the secondary pathway results.

A somewhat comparable situation is presented by the projection between the cerebral cortical visual areas of each hemisphere. As mentioned earlier, there is a substantial callosal projection that interconnects the border zone between V I (area 17) and V II (area 18 or 18a), where the vertical meridian of the visual field is represented. The upper cortical layers of that part of area 17 in which is represented the more peripheral part of the visual field are normally devoid of a callosal projection. Thus, in the same way that the isthmotectal projection plays a part in channeling inputs from the same part of the visual field by way of each eye to the same region of one tectum, the callosal projection interconnects points on the visual cortex receiving input from the same part of the visual field through both eyes. This parallel led to an expectation that misaligning the visual input through each eye might result in anomalous connections. Studies on cats with strabismus induced as kittens showed that the callosal projection spread more broadly within area 17 and arose from cells located more distant from the 17/18 border than normal (Lund *et al.*, 1978; Innocenti and Frost, 1979; Berman and Payne, 1983; Elberger *et al.*, 1983). There is indication from one study that there is an asymmetry in the enlarged projection, depending on the orientation of the squint (Lund and Mitchell, 1979b). This led to the proposal that regions and perhaps individual cells in area 17 tend to connect with cells responding to the same part of the visual field. Consistent with this was the further observation that the callosal projection was diminished by dark rearing or visual deprivation (Lund and Mitchell, 1979a; Innocenti and Frost, 1980).

Experiments performed on rodents failed to show the same dependence on vision as in cats in that callosal connections were not modified by dark rearing (Cusick and Lund, 1982). However, a number of studies on rats and hamsters have shown that the integrity of the input from the eye through the lateral geniculate nucleus is extremely important in defining the pattern of callosal projection (Cusick and Lund, 1979, 1982; Rhoades and Dellacroce, 1980; Rothblat and Hayes, 1982; Olavarria *et al.*, 1983). After neonatal eye removal, the projection arises from much more of area 17 than is normal and terminates more broadly also. There is a suggestion of a focus of projection in an intermediate position within area 17 as well as an enlargement of the projection from the 17/18 border. Optic tract section results in a reduced projection at the 17/18 border, while thalamic lesions result in an expanded projection within the major laminae of thalamic termination (Cusick and Lund, 1982). The callosal projection is also modified by infant tectal lesions (Mooney *et al.*, 1984).

In order to relate these findings to the pattern of development of the callosal pathway, a number of studies have been undertaken on the normal and abnormal development of the corpus callosum of rats. Axons first cross the midline shortly before birth or at early postnatal ages; cells over the whole cortex, including all of area 17, send axons to the opposite hemisphere. In area 17 at 4 days postnatal there is a gradual loss of callosally projecting cells, beginning medially and extending laterally towards the 17/18 border (Lund *et al.*, 1984). Studies of axon distribution indicate that in areas from which the callosally projecting cells disappear, the callosal axons invade only the deepest cortical layers (Innocenti and Clarke, 1984; Lund *et al.*, 1984; Rhoades *et al.*, 1984). If an eye removal is performed at birth, the cell loss from the medial part of area 17 still occurs but does not proceed as far as in normals. In normal animals axons begin to grow into the superficial cortical layers at the 17/18 border region at postnatal day 5. A similar ingrowth occurs in other parts of area 17 only if there has been prior eye removal. Thus at 4 days postnatal, the callosal projection of normal and eye-enucleated rats appears similar but by 10–12 days postnatal there are substantial differences even when the eye removal occurs as late as 4 days postnatal (Lund *et al.*, 1984).

Studies on the early development of the callosum in kittens show that about two-thirds of the axons are lost in a 3-month period (Koppel and Innocenti, 1983). Experiments performed on young kittens show that cells throughout area 17 contribute callosal axons in early development and that these normally lose their callosal projection. It is apparent that only cells in 17/18 border send axons into the contralateral cortex, although at this early postnatal period, cells located in a broad area of cortex contribute to the white matter underlying contralateral area 17 (Innocenti and Clarke, 1984). Double labeling studies have added the important information that cell death does not necessarily accompany the lost projection (O'Leary *et al.*, 1981). Whether the loss of callosal projection is accompanied by establishment of new connections elsewhere has yet to be determined. It appears, therefore, that rodents and kittens share many of the same developmental processes in the maturation of the visual callosal projection.

From these results, an overall pattern is beginning to emerge. In early development, cells over the whole cortical area send axons through the callosum, but the axons are unable normally to invade the regions of area 17 associated with the representation of the peripheral visual field. If the primary optic pathway has been disturbed, axons may innervate the upper cortical layers and, associated with this, cells in more medial regions of the contralateral area 17 maintain their callosal projection. While these studies, like those involving the isthmotectal pathway, show that the activity of the primary optic pathway plays a part in determining the distribution pattern of a secondary pathway, they still do not address directly the mechanism involved. There a number of problems as well.

The first callosal projection is not only dependent on functional correlative influences for its distribution, but there is some additional cue which ensures that even under the most extreme experimental conditions, there is always a projection to the 17/18 border zone (Cusick and Lund, 1982). A more serious concern is that, although the majority of the experimental results after manipulations of the primary optic pathway may be interpreted by proposing that axons attempt to connect with cells responding to similar peripheral stimulation, the timing of events in rodents requires that the period during which the system may be manipulated precedes the inception of visual function.

D. Segregation

The principle of segregation of crossed and uncrossed optic pathways is reflected in two ways in visual centers. First, the axons of the two pathways may terminate in alternating laminae and, second, the uncrossed pathway may be reduced to a small part of the area occupied by the crossed projection or, in extreme cases, may be totally absent.

The patterning of crossed and uncrossed projections may be varied in a variety of ways. First, since the patterning serves as a substrate for binocular vision, the extent of the uncrossed pathway relates directly to the degree of binocular overlap of the visual fields of each eye. Second, the uncrossed pathway is reduced in cases in which there is hypopigmentation of the pigment epithelium of the retina (Lund, 1965). Third, the uncrossed pathway from one eye is enlarged after damage to the opposite eye early in development (Lund and Lund, 1971; Lund *et al.*, 1973). Fourth, in certain groups of animals such as birds, the primary uncrossed pathway is absent and other decussations are used as a substrate for binocular vision (Pettigrew and Konishi, 1976). Fifth, while the segregation patterns of the primary optic pathways in mammals are not generally affected by visual function, those occurring at higher levels, such as primary visual cortex, may be distributed by restriction or modification of visual function during development (Hubel *et al.*, 1977; Shatz and Stryker, 1978).

What makes this principle particularly interesting is, as described in Section III, that early in development inputs from the two eyes have partially overlapping terminal distribution and that segregated patterns emerge from the overlapping projection. It is likely that some of the exuberant projections arise from inappropiate parts of the retina (Lund *et al.*, 1980). Segregation or restriction of the projections from each eye occurs during a period of cell death in the retina (Hughes and McLoon, 1979; McLoon and Lund, 1982). A substantial part of the uncrossed pathway to the superior colliculus in rodents is lost by the death of the cells providing the source of axons. This may also be true for the lateral geniculate nucleus but the alternative option cannot be ruled out, namely that the axonal arbor of single cells becomes progressively restricted, without death of the parent cell. Studies on the segregation process of geniculocortical axon terminals in area 17 of cats suggest again a progressive restriction of terminal arbors, not necessarily associated with cell death (LeVay and Stryker, 1979).

The mechanisms whereby the segregation process occurs are not clear at present. However, the progress of the event has been well worked out and the relationship of the process to the variations in normal patterns is beginning to be understood. In all animals so far studied, the overlapping of crossed and uncrossed projections is evident from the earliest time at which uncrossed axons enter the visual centers. At least some of the transient projections make functional synapses in the lateral geniculate nucleus and superior colliculus before disappearing. We do not yet know, however, how many synapses are made by transient projections compared with permanent axons. The processes of retraction of the uncrossed pathway and of segregation occur before visual function is possible. In rat, for example, they begin before ganglion cells receive synaptic inputs and are complete before an electroretinogram may be recorded from a retina (Weidman and Kuwabara, 1968). In monkey, it is complete by E110, 55 days before birth (Rakic, 1977). If an eye is removed at any stage before the process is complete, the distribution of axons from the remaining eye is retained at the appropriate embryonic state. This implies that the interaction between the two eyes plays a significant part in the segregation process. However, what the nature of the interaction may be is unclear. Comparable experiments in nonmammalian vertebrates in which additional eyes have been added can achieve segregated projections to one tectum from normal and supernumerary eyes (Constantine-Paton and Law, 1978; Law and Constantine-Paton, 1981). Such nonoverlapping stripes of the two projections can be inhibited by blocking electrical activity with the sodium channel blocker tetrodotoxin (TTX) (Meyer, 1981; Constantine-Paton and Reh, 1983). No study has been able to alter patterns of segregation at the lateral geniculate nucleus by modifying the visual environment and this is not surprising since the process is complete before function is possible, and no study has been able to manipulate the mature pattern after it has been achieved.

The events in the cortex are somewhat different. In such animals as the cat and monkey in which there is clear segregation of crossed and uncrossed visual inputs into ocular dominance columns, the projections arising from each eye remain overlapped until after the inception of visual function (Hubel *et al.*, 1976). Imbalance between the input from the two eyes results in reduced segregation (as after unilateral eye closure) (Hubel *et al.* 1977; Shatz and Stryker, 1978) or enhanced segregation (as after induced strabismus) (Hubel and Wiesel, 1965; Van Sluyters and Levitt, 1980). Reduced input (bilateral eye closure) or TTX administration into both eyes also results in reduced segregation (Stryker, 1981). Thus, at the cortical level, the process is much more dependent on the patterned activity of visual afferents than appears to be the case subcortically.

Two issues still remain unresolved with respect to ocular segregation. One concerns the hypopigmentation effects seen in albino animals and Siamese, and other breeds of cats. It has been suggested that hypopigmentation of the optic stalk may lead to disordering of optic axons in the nerve, misrouting in the optic chiasm, and possible aberrant patterns of segregation in the subcortical visual centers (Guillery, 1974; Silver and Sapiro, 1981; Strongin and Guillery, 1981). In embryonic Siamese kittens it appears that the uncrossed optic pathway is reduced from earliest time and has a limited distribution within the lateral geniculate nucleus (Shatz and Kliot, 1982). In albino rats, it appears that while the transient uncrossed pathway may be slightly lighter than normal, it does distribute over the same area as normal and has only slightly fewer cells contributing to it prior to the period of limitation. During that time many more cells are lost from the uncrossed pathway than is normal (Land *et al.*, 1981; Bunt *et al.*, 1983). Whether the cells lost have axons inappropriately placed in the nerve and whether the differences between rat and cat might reflect different rates of expression of the same fundamental mechanism remain unexplained.

The other major point of interest is how the patterns of segregation have a functional correlate and yet, at the subcortical level at least, are established before function is possible.

E. Parallel Centripetal Pathways

A number of classes of morphologically defined ganglion cells have been identified in the retina, and it is apparent that cells of each class respond differentially to specific visual stimuli (Rodieck and Brening, 1983). Axons of these classes show different patterns of distribution in the subcortical visual centers. In these centers, cells may be identified that shape the response properties of the retinal cell classes, suggesting that cells of a particular class in the retina make preferential or unique connections with corresponding central cells. The separation of channels appears to be maintained in the pathway from the lateral geniculate nucleus to the cortex and even in intracortical circuitry within area 17 (Lund, 1981).

Studies on histogenesis suggest that particular cell classes are generated at different times (Walsh *et al.*, 1983). Within the retina, the dendrites of ganglion cells overlap considerably when viewed in a tangential plan. However, there is no overlap between dendrites of a single cell class (Peichl and Wassle, 1981; Wassle *et al.*, 1981). This suggests that during development there may be a mutual interaction between the dendrites of cells of one class but not between those of different classes. The competitive restraint on dendritic growth is evidenced by lesion studies in which ganglion cells located over part of the retina are eliminated (Perry and Linden, 1982). Under such circumstances, there is an extension of dendrites from intact to ganglion cell-free retina.

In the lateral geniculate nucleus, while the individual cell classes may be recognized at early times, it is apparent that their maturation occurs during the period of sensitivity to visual stimulation and it can be modified by manipulation of visual input. Studies on maturation of retinogeniculate axons indicate that while the terminal field of one class (X cells) becomes restricted during this time, that of a second class (Y cells) enlarges (Sur *et al.*, 1984). There is evidence from deprivation studies to suggest that there may be a mutual interaction between axon fields of each class (Sur *et al.*, 1982). This is in marked contrast to events in the retina.

Nothing is yet known of any potential plasticity of development of these separate channels at the cortical level.

F. Interdependence between Parts of the Visual System

It is clear from the foregoing discussion that the overall organization of any part of the visual system can develop in relative isolation from other parts of the system. This is seen in the most extreme form after transplantation to ectopic locations (Lund *et al.*, 1983), but is also found after many of the manipulations described in the previous sections of this review. However, the subtleties of organization of any one part that are necessary prerequisites for normal function are clearly dependent on the integrity and activity of the other parts of the system. Such interdependence falls into two main patterns—those in which more peripheral parts of the system influence the development of more central structures and those in which parallel pathways converging on a region interact with one another. Somewhat surprisingly, examples are hard to find of centrifugal influences on development except those mediated by retrograde processes.

With regard to peripheral central interactions, the absence of an optic projection at early developmental times results in more substantial cell death in the primary optic centers such as the superior colliculus (DeLong and Sidman, 1962) and lateral geniculate nucleus (Polyak, 1957; Kalil, 1980). It also leads to a more diffusely organized geniculocortical pathway (Jeffery, 1984), to changes in dendritic spine numbers of cortical pyramidal cells in area 17 (Globus and Scheibel, 1967; Valverde, 1968), and to abnormal development of the callosal pathway between the visual areas.

Some of these effects can be replicated not only by removing the end organ but also by depriving the animals of appropriate sensory function or by silencing activity in the optic nerve. In general, functional deprivation is more effective in inducing effects in animals such as cat and monkey in which development is somewhat more protracted (see Chapter 10, this volume).

The interaction between converging systems occurs between the inputs from each eye, both with respect to the distribution of the primary afferents in the subcortical centers as well as between the terminals of axons driven by one or another eye in the cortex. Again, this has a major impact on the distribution of callosal connections. Imbalances in the interaction may be induced by unilateral eye removal and by local lesions of one eye (Lund and Lund, 1973, 1976). In addition, in animals such as cat and monkey, many of these effects can be replicated by unilateral restriction of vision and may also be modified by induced strabismus. In addition to the parallel interactions seen between like pathways, there is also interaction between convergent inputs of different origin. Thus, eye removal in young rats results in expansion of both the corticotectal pathway (Lund and Lund, 1973; Lund *et al.*, 1973) and the crossed parabigeminal pathway (Stevenson and Lund, 1982).

The interdependencies reviewed here may serve an important function in providing the basis for corrections of minor optical misalignments or the suppression of inappropriate information relayed by an eye damaged during development. They may also reflect the strategy necessary for establishing a complex organization when the genetic program may not be sufficiently precise to define the level of detail necessary for optimal functioning. In all cases the flexibility ceases once the particular maturational event that forms the basis for it has been completed.

V. Discussion

Much of the earlier work on development of the mammalian visual system focused on those events that were demonstrable in postnatal life, in particular those events modulated by visual function. With the advent of techniques suitable for examining the early development of visual pathways, it is now possible to gain a broader view of the developmental process and to examine common themes that recur at different stages. This chapter has emphasized the recent studies of early development of visual connections.

There are two features that characterize many of the stages of development of the visual system. The first is that there are periods of development during which a particular process is not absolutely determined and may be modified, but beyond which external perturbations may have little or no impact on the organization of projections. The principle was first shown with respect to the critical period for visual deprivation effects but subsequent work has shown that it is not restricted to that situation. The second feature that has attracted particular attention is the relative imprecision of early connections and the subsequent pruning mechanism that leads to the restricted adult patterns.

The transient exuberant pathways are seen at all levels, from the highly abnormal retinoretinal and occipitocorticospinal projections to the more subtle diffuse early retinotectal pathways. It is extremely difficult to estimate what proportion of the early projections are anomalous and eliminated for that reason. Approximately 40% of the ganglion cells die during a restricted period of development. However, there is no guarantee that all of these cells die because of inadequate or inappropriate axonal distribution, nor, conversely, is it necessary that elimination of an abnormal connection be always accompanied by cell death. Of the approximately 200,000 optic axons in an immature rat, only 2,000 or so appear to contribute to the anomalous uncrossed retinotectal pathway. Whether the loss of the other 70,000 cells in the early postnatal period is related to improvement of the resolution of the topographic map or is completely unrelated to the restriction of exuberant pathways remains to be seen.

One underlying mechanism that promotes cell death seems to be the lack of an appropriate target. This may be the case for the retinoretinal projection and it certainly seems to be true for the occipitocorticospinal pathway since that can be preserved by placing tectum on the spinal cord (Sharkey *et al.*, 1984). Such a concept is supported by *in vitro* study (McCaffery *et al.*, 1982) and by experiments in which the retina develops *in vivo* in a structure in which it is unable to gain access to appropriate targets (McLoon and Lund, 1984). In other examples, the pruning results from an interaction between two inputs. This can occur between visual and other sensory afferents as is the case of the early optic projection to nucleus ventralis posterior of the thalamus (Frost, 1981). It can also occur between the optic projections from each eye or between axons of ganglion cells of the same eye serving different parts of the visual map. Finally, it may occur between pathways that correlate projections, as between callosal and visual afferent projections. It would appear that in each of these examples, the effect is to limit the opportunity of the transient axon or cell to make sufficient synaptic connections to survive.

Thus, rather than there simply being a generalized sustaining factor provided by the target region, there may also be a requirement for specific synaptic connections to permit survival. Those circumstances in which there is loss of certain axonal branches with maintenance of others require a more local control than that generally attributed to growth factors. However, there are examples showing local changes in axonal growth patterns induced by local application of growth factors (Gundersen and Barrett, 1979). Whether such events can happen in a central neuropil, how specific they might be, and whether they may function to cause retraction rather than growth are important but as yet unresolved issues. Whether availability of growth factors is the only mechanism whereby axons and cells are lost is not known. Other options have been discussed and these may be important in the present setting.

The mechanisms whereby one axonal population exerts an influence on another, leading to segregation or suppression, are still far from clear although the current work in the visual system seems likely to resolve that issue before too long. In the larger literature on the effects of selective visual deprivation, it is clear that appropriate sensory function is an important determinant in balancing the relative territory occupied by interacting pathways. Many of the events described in this chapter, however, occur before significant function is possible; the eyes are yet to be exposed to visual input, and in some cases the ganglion cells of the retina are without synaptic input. Despite this, there are suggestions that function may play a role in the interactive processes leading to segregation and suppression of early exuberant pathways. Thus, in parallel studies in nonmammalian vertebrates, the use of the sodium channel blocker TTX results in somewhat less precise retinoptic maps in regenerating pathways (Meyer, 1983; Schmidt and Edwards, 1983) an in failure of segregation between axons arising from each eye (Meyer, 1982). Comparable experiments in postnatal kittens show that cortical ocular dominance patterns do not develop normally, and in young rats that the uncrossed retinotectal pathway does not restrict its terminal field as much as usual (Fawcett *et al.*, 1983). While these results appear supportive of the idea that relative activity per se plays a part in the capacity of one axon population to dominate another, there is a concern over interpretation of the results. It is apparent that TTX not only blocks conduction but, either directly or indirectly, affects axoplasmic transport (Edwards and Grafstein, 1984) and, in addition, results in anatomical changes in axon terminals in the lateral geniculate nucleus (Archer *et al.*, 1982; Kalil *et al.*, 1983). It is conceivable, therefore, that the lack of activity is not the important event but, rather, secondary factors involving metabolic and growth-related events. To resolve this issue, it will be necessary to override the effect of TTX with direct electrical stimulation.

The mechanisms suggested for activity-related elimination of synapses follow those suggested by Hebb (1949) in which synapses functioning in synchrony are enhanced. This has been further elaborated by Stent (1973) and Changeux and Danchin (1976) (see also Chapter 9, this volume). The corollary of Hebb's original proposal is particularly important here—namely, that synapses firing out of synchrony are eliminated. Furthermore, this process of elimination can occur only during a limited period of development. There are two important points, however. First, the anomalous projections must form synapses. This certainly occurs in the lateral geniculate nucleus and for a few axons in the superior colliculus. It is less likely for the callosal projection since the exuberant pathway does not appear to gain access to the supragranular layers where the heaviest density of synapses is usually found. The second point is that pathways that have yet to function must be electrically active and that adjacent cells on the retina should fire in synchrony more often than more distantly located cells. To apply similar mechanisms to the correlation of callosal projections to activity in the visual pathway, we would need to make the further condition that there was a gradient of activity across each retina, centering from the region of representation of the vertical meridian of the visual field.

There are two further and somewhat more general points that may be made from the work reviewed here. The first derives from the observations that the chick shows a transient uncrossed optic pathway which, unlike that of mammals, is completely eliminated during maturation but which can be preserved by appropriate conditions (such as eye removal) (Raffin and Reperant, 1975). It may be possible that developmentally imprecise pathways like the uncrossed optic system can serve as a reservoir for adaptive or evolutionary change in neural connections without the need for a mutation directed specifically at the pathway concerned.

The second point is that while the pruning process involved in refining connections in the subcortical visual system of all mammals studied occurs before natural visual function is possible, the relative timing of comparable processes at the cortical level varies according to the animal. In general, in animals with a short developmental period, cortical events occur at similar times to subcortical ones. When the development is protracted, as in cats or monkeys, the pruning process occurs somewhat later in development. As a result, it falls within the time period when it is modifiable by natural function. Genetic controls that influence the timing of developmental events may then be extremely important in defining the way in which inputs may modulate central organization and functioning.

Taken together, these two points suggest ways in which an imprecise developmental mechanism has become the substrate for a significant biological system and for a degree of adaptability to changes in the environment in its broadest sense in which the neurons of the visual system develop.

VI. Conclusions

This review has attempted to correlate some of the recent literature on early development of connections in the mammalian visual system and to draw from this some general principles and implications that will be important for future studies. While the final resolution of the mechanisms underlying plasticity in the developing visual system remains to be achieved, the recent studies on mammals have allowed a very different perspective from Sperry's proposal (1963) that neurons "must carry some kind of individual identification tags, presumably cytochemical in nature, by which they are distinguished one from another almost, in many regions, to the level of the single neuron; and further, that the growing fibers are extremely particular when it comes to establishing synaptic connections, each axon linking only with certain neurons to which it becomes selectively attached by specific chemical affinities."

Acknowledgments

We thank Barbara Bartoldi for typing the manuscript. Preparation of this chapter was supported in part by National Institutes of Health grants EY05308 and EY05283.

References

Archer, S. M., Dubin, M. W., and Stark, L. A. (1982). *Science* **217**, 743–745.
Berman, N. E., and Payne, B. R. (1983). *Brain Res.* **274**, 201–212.
Bishop, P. O., Kozak, W., Levick, W. R., and Vakkur, G. J. (1962). *J. Physiol. (London)* **163**, 503–539.
Bunt, S. M. (1982). *J. Comp. Neurol.* **206**, 209–226.
Bunt, S. M., and Lund, R. D. (1981). *Brain Res.* **211**, 399–404.
Bunt, S. M., Horder, T. J., and Martin, K. A. C. (1979). *In* "Developmental Neurobiology of Vision" (R. D. Freeman, ed.), pp. 331–343. Plenum Press, New York.
Bunt, S. M., Lund, R. D., and Land, P. W. (1983). *Dev. Brain Res.* **6**, 149–168.
Caviness, V. S., Jr., and Frost, D. O. (1983). *J. Comp. Neurol.* **219**, 182–202.
Caviness, V. S., Jr., and Sidman, R. L. (1973). *J. Comp. Neurol.* **148**, 141–152.
Caviness, V. S., Jr., and Yorke, C. H., Jr. (1976). *J. Comp. Neurol.* **170**, 449–460.
Changeux, J. P., and Danchin, A. (1976). *Nature (London)* **264**, 705–712.
Constantine-Paton, M., and Law, M. I. (1978). *Science* **202**, 639–641.
Constantine-Paton, M., and Reh, T. (1983). *Soc. Neurosci. Abstr.* **9**, 760.
Cragg, B. G. (1972). *Invest. Ophthalmol.* **11**, 377–384.
Cusick, C. G., and Lund, R. D. (1979). *Soc. Neurosci. Abstr.* **5**, 624.
Cusick, C. G., and Lund, R. D. (1981). *Brain Res.* **214**, 239–259.
Cusick, C. G., and Lund, R. D. (1982). *J. Comp. Neurol.* **212**, 385–398.

DeLong, G. R., and Sidman, R. L. (1962). *J. Comp. Neurol.* **118**, 205–223.
Dowling, J. E., and Boycott, B. (1966). *Proc. R. Soc. London Ser. B* **166**, 80–111.
Drager, U. C. (1981). *J. Comp. Neurol.* **201**, 555–570.
Edwards, D. L., and Grafstein, B. (1984). *Brain Res.* **299**, 190–194.
Elberger, A. J., Smith, E. L., III, and White, J. M. (1983). *Neurosci. Lett.* **35**, 19–24.
Enroth-Cugell, C., and Robson, J. G. (1966). *J. Physiol. (London)* **185**, 517–552.
Fawcett, J. W., O'Leary, D. D. M., and Cowan, W. M. (1983). *Soc. Neurosci. Abstr.* **9**, 856.
Frost, D. O. (1983). *Soc. Neurosci. Abstr.* **9**, 26.
Gaze, R. M., Keating, M. J., Szekely, G., and Beazley, L. (1970). *Proc. R. Soc. London Ser. B* **175**, 107–147.
Globus, A., and Scheibel, A. (1967). *Exp. Neurol.* **18**, 116–131.
Guillery, R. W. (1974). *Sci. Am.* **230**, 44–54.
Gundersen, R. W., and Barrett, J. N. (1979). *Science* **206**, 1079–1080.
Hebb, D. O. (1949). "The Organization of Behavior." Wiley, New York.
Hinds, J. W., and Hinds, P. L. (1974). *Dev. Biol.* **37**, 381–416.
Hsiao, K. (1984). *J. Neurosci.* **4**, 368–373.
Hsiao, K., Sacks, G. M., and Schneider, G. E. (1984). *J. Neurosci.* **4**, 359–367.
Hubel, D. H., and Wiesel, T. N. (1965). *J. Neurophysiol.* **28**, 1041–1059.
Hubel, D. H., Wiesel, T. N., and LeVay, S. (1976). *Cold Spring Harbor Symp. Quant. Biol.* **40**, 581–589.
Hubel, D. H., Wiesel, T. N., and LeVay, S. (1977). *Philos. Trans. R. Soc. London Ser. B* **278**, 377–409.
Hughes, W. F., and McLoon, S. C. (1979). *Exp. Neurol.* **66**, 587–601.
Innocenti, G. M. (1981). *Science* **212**, 824–827.
Innocenti, G. M., and Frost, D. O. (1979). *Nature (London)* **280**, 231–234.
Innocenti, G. M., and Frost, D. O. (1980). *Exp. Brain Res.* **39**, 365–375.
Innocenti, G. M., Fiore, L., and Caminiti, R. (1977). *Neurosci. Lett.* **4**, 237–242.
Ivy, G. O., and Killackey, H. P. (1981). *J. Comp. Neurol.* **195**, 367–389.
Ivy, G. O., Akers, R. M., and Killackey, H. P. (1979). *Brain Res.* **173**, 532–537.
Jeffery, G. (1984). *Dev. Brain Res.* **13**, 257–263.
Jeffery, G., and Perry, V. H. (1982). *Dev. Brain Res.* **2**, 176–180.
Jeffery, G., Cowey, A., and Kuypers, H. G. J. M. (1981). *Exp. Brain Res.* **44**, 34–40.
Jeffery, G., Arzymanow, B. J., and Lieberman, A. R. (1984). *Dev. Brain Res.* **14**, 135–138.
Jensen, K. F., and Killackey, H. P. (1984). *Proc. Natl. Acad. Sci. U.S.A.* **81**, 964–968.
Kaiserman-Abramof, I. R., Graybiel, A. M., and Nauta, W. J. H. (1975). *Soc. Neurosci. Abstr.* **1**, 102.
Kaiserman-Abramof, I. R., Graybiel, A. M., and Nauta, W. J. H. (1980). *Neurosci.* **5**, 41–52.
Kalil, R. (1980). *J. Comp. Neurol.* **189**, 483–524.
Kalil, R. E., Dubin, M. W., Scott, G. L., and Stark, L. A. (1983). *Soc. Neurosci. Abstr.* **9**, 24.
Keating, M. J. (1977). *Philos. Trans. R. Soc. London Ser. B* **278**, 277–294.
Koppel, H., and Innocenti, G. M. (1983). *Neurosci. Lett.* **41**, 33–40.
Kuwabara, T., and Weidman, T. A. (1974). *Invest. Ophthalmol.* **13**, 725–739.
Land, P. W., and Lund, R. D. (1979). *Science* **205**, 698–700.
Land, P. W., Hargrove, K., Eldridge, J., and Lund, R. D. (1981). *Soc. Neurosci. Abstr.* **7**, 141.
Law, M. I., and Constantine-Paton, M. (1981). *J. Neurosci.* **1**, 741–759.
Lazar, G. (1973). *J. Anat.* **116**, 347–355.
Lemmon, V., and Pearlman, A. L. (1981). *J. Neurosci.* **1**, 83–93.
LeVay, S., and Stryker, M. P. (1979). *In* "Aspects of Developmental Neurobiology" (J. A. Ferrendelli, ed.), pp. 83–98. Society for Neuroscience, Bethesda, Maryland.
LeVay, S., Hubel, D. H., and Wiesel, T. N. (1975). *J. Comp. Neurol.* **159**, 559–576.

LeVay, S., Stryker, M. P., and Shatz, C. J. (1978). *J. Comp. Neurol.* **179**, 223-244.

Lund, J. S. (1981). *In* "The Organization of the Cerebral Cortex" (F. O. Schmitt, F. G. Worden, G. Adelman, and S. G. Dennis, eds.), pp. 105-124. MIT Press, Cambridge, Massachusetts.

Lund, R. D. (1965). *Science* **149**, 1506-1507.

Lund, R. D. (1966). *J. Anat.* **100**, 51-62.

Lund, R. D. (1978). "Development and Plasticity of the Brain." Oxford Univ. Press, New York.

Lund, R. D., and Lund, J. S. (1971). *Science* **171**, 804-807.

Lund, R. D., and Lund, J. S. (1973). *Exp. Neurol.* **40**, 377-390.

Lund, R. D., and Lund, J. S. (1976). *J. Comp. Neurol.* **169**, 113-154.

Lund, R. D., and Mitchell, D. E. (1979a). *Brain Res.* **167**, 172-175.

Lund, R. D., and Mitchell, D. E. (1979b). *Brain Res.* **167**, 176-179.

Lund, R. D., and Mustari, M. J. (1977). *J. Comp. Neurol.* **173**, 289-306.

Lund, R. D., Cunningham, T. J., and Lund, J. S. (1973). *Brain Behav. Evol.* **8**, 51-72.

Lund, R. D., Mitchell, D. E., and Henry, G. H. (1978). *Brain Res.* **144**, 169-172.

Lund, R. D., Land, P. W., and Boles, J. (1980). *J. Comp. Neurol.* **189**, 711-720.

Lund, R. D., McLoon, L. K., McLoon, S. C., Harvey, A. R., and Jaeger, C. B. (1983). *In* "Nerve, Organ, and Tissue Regeneration: Research Perspectives" (F. J. Seil, ed.), pp. 303-323. Academic Press, New York.

Lund, R. D., Chang, F-L. F. and Land, P. W. (1984). *Dev. Brain Res.* **14**, 139-142.

McCaffery, C. A., Bennett, M. R., and Dreher, B. (1982). *Exp. Brain Res.* **48**, 377-386.

McLoon, S. C. (1982). *Science* **215**, 1418-1420.

McLoon, S. C., and Lund, R. D. (1980a). *Exp. Brain Res.* **40**, 273-282.

McLoon, S. C., and Lund, R. D. (1980b). *Brain Res.* **197**, 491-495.

McLoon, S. C., and Lund, R. D. (1982). *Exp. Brain Res.* **45**, 277-284.

McLoon, S. C., and Lund, R. D. (1984). *Dev. Brain Res.* **12**, 131-135.

Maxwell, E., and Land, P. W. (1981). *Anat. Res.* **199**, 165A.

Meyer, R. L. (1982). *Science* **218**, 589-591.

Meyer, R. L. (1983). *Dev. Brain Res.* **6**, 193-198.

Mooney, R. D., Rhoades, R. W., and Fish, S. E. (1984). *Exp. Brain Res.* **55**, 9-25.

Oberdorfer, M., Miller, N., and Silver, J. (1981). *Invest. Ophthalmol.* **20**, 174.

Olavarria, J., Malach, R., Lee, P., and Van Sluyters, R. C. (1983). *Invest. Ophthalmol. Suppl.* **24**, 9.

O'Leary, D. D. M., Stanfield, B. B., and Cowan, W. M. (1981). *Dev. Brain Res.* **1**, 607-617.

O'Leary, D. D. M., Fawcett, J. W., and Cowan, W. M. (1984). *Soc. Neurosci. Abstr.* **10**, 464.

Peichl, L., and Wassle, H. (1981). *Proc. R. Soc. London Ser. B* **212**, 139-156.

Perry, V. H., and Linden, R. (1982). *Nature (London)* **297**, 683-685.

Pettigrew, J. D., and Konishi, M. (1976). *Nature (London)* **264**, 753-754.

Polyak, S. L. (1957). "The Vertebrate Visual System." Univ. of Chicago Press, Chicago.

Potts, R. A., Dreher, B., and Bennett, M. R. (1982). *Dev. Brain Res.* **3**, 481-486.

Raffin, J. P., and Reperant, J. (1975). *Arch. Anat. Microsc. Morphol. Exp.* **64**, 93-111.

Rakic, P. (1977). *Philos. Trans. R. Soc. London Ser. B* **278**, 245-260.

Rakic, P. (1981). *Science* **214**, 928-931.

Rhoades, R. W., and Dellacroce, D. D. (1980). *Brain Res.* **202**, 189-195.

Rhoades, R. W., Mooney, R. D., and Fish, S. E. (1984). *Exp. Brain Res.* **56**, 92-105.

Rodieck, R. W., and Brening, R. K. (1983). *Brain Behav. Evol.* **23**, 121-164.

Rothblat, L. A., and Hayes, L. L. (1982). *Brain Res.* **246**, 146-149.

Rutishauser, U., (1984). *Nature (London)* **310**, 549-554.

Sachs, G. M., and Schneider, G. E. (1984). *J. Comp. Neurol.* **230**, 155-167.

Schmidt, J. T., and Edwards, D. L. (1983). *Brain Res.* **269**, 29-39.

Schneider, G. E. (1973). *Brain Behav. Evol.* **8**, 73-109.

Schneider, G. E. (1984). *Soc. Neurosci. Abstr.* **10**, 467.

Schneider, G. E., and Jhaveri, S. (1984). *Soc. Neurosci. Abstr.* **10**, 467.

Sengelaub, D. R., and Finlay, B. L. (1982). *J. Comp. Neurol.* **204**, 311–317.

Sharkey, M. A., Steedman, J. G., Lund, R. D., and Dom, R. M. (1984). *Bull. S.C. Acad. Sci.* **46**, 122.

Shatz, C. J. (1983). *J. Neurosci.* **3**, 482–499.

Shatz, C. J., and Kliot, M. (1982). *Nature (London)* **300**, 525–529.

Shatz, C. J., and Rakic, P. (1981). *J. Comp. Neurol.* **196**, 287–307.

Shatz, C. J., and Stryker, M. P. (1978). *J. Physiol. (London)* **281**, 267–283.

Sidman, R. L. (1961). *In* "The Structure of the Eye" (G. K. Smelser, ed.), pp. 487–505. Academic Press, New York.

Silver, J. (1984). *J. Comp. Neurol.* **223**, 238–251.

Silver, J., and Sapiro, J. (1981). *J. Comp. Neurol.* **202**, 521–538.

So, K.-F., Schneider, G. E., and Frost, D. O. (1978). *Brain Res.* **142**, 343–352.

Sperry, R. W. (1963). *Proc. Natl. Acad. Sci. U.S.A.* **50**, 703–710.

Sretavan, D., and Shatz, C. J. (1984). *Nature (London)* **308**, 845–848.

Stelzner, D. J., Bohn, R. C., and Strauss, J. A. (1981). *J. Comp. Neurol.* **201**, 299–317.

Stent, G. S. (1973). *Proc. Natl. Acad. Sci. U.S.A.* **70**, 997–1001.

Stevenson, J. A., and Lund, R. D. (1982). *J. Comp. Neurol.* **207**, 191–202.

Strongin, A. C., and Guillery, R. W. (1981). *J. Neurosci.* **1**, 1193–1204.

Stryker, M. (1981). *Soc. Neurosci. Abstr.* **7**, 842.

Stuermer, C. A. O. (1984). *J. Comp. Neurol.* **229**, 214–232.

Stuermer, C. A. O., and Easter, S. S., Jr. (1984). *J. Neurosci.* **4**, 1045–1051.

Sur, M., Humphrey, A. L., and Sherman, S. M. (1982). *Nature (London)* **300**, 183–185.

Sur, M., Weller, R. E., and Sherman, S. M. (1984). *Nature (London)* **310**, 246–249.

Tusa, R. J., Rosenquist, A. C., and Palmer, L. A. (1979). *J. Comp. Neurol.* **185**, 657–678.

Tusa, R. J., Palmer, L. A., and Rosenquist, A. C. (1981). *In* "Cortical Sensory Organization" (C. N. Woolsey, ed.) pp. 183–214. Humana Press, Englewood Cliffs, New Jersey.

Udin, S. B. (1983). *Nature (London)* **301**, 336–338.

Udin, S. B., and Keating, M. J. (1981). *J. Comp. Neurol.* **203**, 575–594.

Valverde, F. (1968). *Exp. Brain Res.* **5**, 274–292.

Van Sluyters, R. C., and Levitt, F. B. (1980). *J. Neurophysiol.* **43**, 686–699.

Walsh, C., Polley, E. H., and Hickey, T. L. (1983). *Nature (London)* **302**, 611–614.

Wassle, H., Peichl, L., and Boycott, B. B. (1981). *Proc. R. Soc. London Ser. B* **212**, 157–175.

Weidman, T. A., and Kuwabara, T. (1968). *Arch. Ophthalmol.* **79**, 470–484.

Wise, S. P., and Jones, E. G. (1976). *J. Comp. Neurol.* **168**, 313–344.

5

Do Neurotransmitters, Neurohumors, and Hormones Specify Critical Periods?*

JEAN M. LAUDER and HELMUT KREBS
Laboratory of Developmental Neurobiology
University of North Carolina School of Medicine
Chapel Hill, North Carolina

*Dedicated to the memory of Professor Dalbir Bindra.

Developmental NeuroPsychobiology

I. An Hierarchical Model for the Specification of Critical Periods

The term *critical period*, as used in this chapter, refers to that segment of neurogenesis when specific ontogenic processes are vulnerable to perturbation by external insults or are sensitive to internal influences required for their development. The critical period concept originates largely from behavioral experiments involving imprinting in birds and early handling in mammalian neonates (see reviews by Gottlieb, 1976; Scott *et al.*, 1974; and Thorpe, 1963) indicating that these processes can only occur during a circumscribed portion of postnatal development. Although the determinants of critical periods have been most thoroughly examined in the developing visual system with regard to the effects of monocular and binocular deprivation on the functional development of visual cortical neurons (Hubel and Wiesel, 1970; see review by Lund, 1978; Singer, this volume), there is good evidence that other aspects of nervous system ontogeny, such as the sexual differentiation of the brain (see reviews by Harlan *et al.*, 1979; Toran-Allerand, this volume) are also sensitive to external and internal influences during both the pre- and postnatal periods (Goldman, 1980; Gottlieb, 1973).

The extent of a critical period varies for each organizational system (Scott *et al.*, 1974) and depends on the presence of such fundamental neurogenic events as cell proliferation, neuronal differentiation, gliogenesis, myelinogenesis, and synaptogenesis (see Cowan, 1979). This chapter seeks to add a new dimension to the concept of critical period in which a hierarchy of internal regulatory signals, provided by neurotransmitters, neurohumors (neuronal substances such as peptides) and hormones is superimposed upon basic ontogenic events (Fig. 1). It is suggested that these internal humoral and hormonal signals can mediate the effects of external influences on developmental processes, a function that may either protect or place these events at risk depending on the nature of the influence and its mechanism of action. It is hoped that this speculative review will stimulate new thought on the subject of "critical periods" and further stress the idea that the hormonal and humoral milieu may play a key role in the coordination of temporal events during nervous system development.

II. Neurotransmitters and Neurohumors as Developmental Signals

A. Roles for Neurotransmitters and Neurohumors in Development

The hypothesis that particular neurotransmitters or neurohumors might have developmental functions prior to the onset of neurotransmission itself derives from their presence in primitive organisms and during key phases

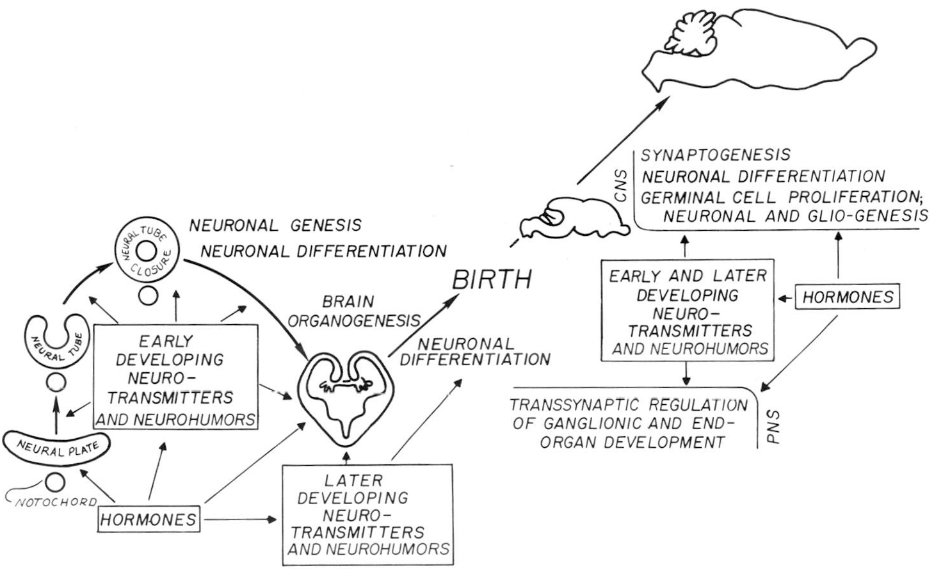

Fig. 1. Hierarchical Model for the specification of critical periods by neurotransmitters, neurohumors, and hormones. It is hypothesized that particular neurotransmitters (neurohumors) play different roles in neurogenesis depending on when and where they appear in the developing nervous system. Hormones may act either through these humoral substances or directly on underlying neurogenic processes.

of embryogenesis and neurogenesis in higher organisms (Table I). Proposed nontransmission functions for these neurohumoral agents include control of cell division and morphogenetic cell movements during early phases of embryogenesis, neural tube closure, palate formation, myoblast differentiation, regulation of cell differentiation during formation of the central nervous system (CNS), as well as involvement in such processes as metamorphosis, morphogenetic, and regenerative processes in lower animals. In the peripheral nervous system (PNS; Table II), synapse-dependent roles for neurotransmitters and neurohumors involve the influence of preganglionic cholinergic neurons on the development of postanglionic sympathetic (adrenergic) or parasympathetic (cholinergic) neurons, which in turn influence the development of their end organs. These interactions (1) involve the neurotransmitter of the preganglionic neurons, acetylcholine, and its receptors; (2) appear to require the presence of synapses, and (3) are under the regulation of descending central innervation of the preganglionic cells (by axons of unknown transmitter content). These peripheral events are referred to as anterograde transsynaptic regulation. Retrograde transsynaptic

Table I

Nontransmission Roles for Neurotransmitters and Neurohumors in Development

Phase of development	Location	Organism	Neurotransmitter or neurohumor	Proposed function	References
Cleavage, gastrulation	Fertilized egg and early zygote	Sea urchin, Fish, Amphibian, Chick, Rat, Mouse	Acetylcholine, Norepinephrine, Epinephrine, Dopamine, 5-hydroxytryptamine, (serotonin), 5-methoxytryptamine	Control of cell division and morphogenetic cell movements	Baker, 1965; Burden and Lawrence, 1973; Buznikov, 1980; Buznikov and Shmukler, 1981; Buznikov *et al.*, 1964, 1968, 1970, 1972; Deeb, 1972; Emanuelsson, 1974; Gustafson and Toneby, 1970, 1971; Pienkowski, 1977; Toneby, 1977
Metamorphosis	Larvae	Abalone	GABA	Control of settling on algal surface and metamorphosis	Morse *et al.*, 1979
Morphogenesis, growth, and regeneration	Head region, nerve and interstitial cells, cilia	Planaria, Hydra, Tetrahymena	Acetylcholine, Serotonin, Norepinephrine, Dopamine, Substance P	Promotion of morphogenetic and regenerative functions in protozoans and flatworms	Blum, 1970; Franquinet, 1979; Müller *et al.*, 1977; Taban and Cathieni, 1979; Rodríguez and Renaud, 1980
Formation of neural tube and gut, torsion and flexure of the embryo, myogenesis	Notochord, Neural plate, Neural tube, Gut, Yolk sac, Allantois	Frog, Chick, Rat, Mouse	Acetylcholine, Norepinephrine, Dopamine, Serotonin	Control of neural tube closure, cell division and cell differentiation, morphogenetic cell movements, myoblast differentiation	Allan and Newgreen, 1977; Boucek and Bourne, 1962; Burack and Badger, 1964; Caston, 1962; Cochard *et al.*, 1978; Emanuelsson and Palén, 1975; Filogamo *et al.*, 1978; Gérard *et al.*, 1978; Gershon *et al.*, 1979; Ignarro and Shideman, 1968a,b; Kirby and Gilmore, 1972; Lawrence and Burden, 1973; Palén *et al.*, 1979; Schlumpf and Lichtensteiger, 1979; Schowing *et al.*, 1977; Sims, 1977; Strudel *et al.*, 1977a,b; Teitelman *et al.*, 1978; Wallace, 1979

Palate formation	Palatal shelves	Mouse	Acetylcholine, Serotonin, GABA	Control of morphogenetic cell movements, cell migration	Kujawa and Zimmerman 1978; Clark *et al.*, 1980; Venkatasubramanian *et al.*, 1980; Wee *et al.*, 1979, 1980; Zimmerman and Wee, 1984
Brain organogenesis, retinal development	Neurons	Axolotl, Frog, Rat, Mouse, Rabbit, Monkey, Human	Acetylcholine, GABA, Norepinephrine, Dopamine, Serotonin	Control of cell division, neuronal and glial differentiation, cell migration, synaptogenesis, other "trophic functions"	Ahmad and Zamenhof, 1978; Bartels, 1971; Cadilhac and Pons, 1976; Chronwall and Wolff, 1980; Gromova *et al.*, 1983; Golden, 1972, 1973; Hanson *et al.*, 1984; Haydon *et al.*, 1984;; Lauder and Bloom, 1974; Lauder and Krebs, 1976, 1978a,b; Lauder *et al.*, 1980, 1985b; Levitt and Rakic, 1979; Lewis *et al.*, 1977a; Maeda and Dresse, 1969; Olson *et al.*, 1973; Olson and Seiger, 1972; Patel *et al.*, 1979a; Schlumpf *et al.*, 1977, 1980; Specht *et al.*, 1978a,b; Taber Pierce, 1973; Tennyson *et al.*, 1972, 1973, 1975; Vernadakis, 1973; Wolff *et al.*, 1979; Yamamoto *et al.*, 1980; Yew *et al.*, 1974

Table II

Synapse: Dependent Roles for Neurotransmitters and Neurohumors in Development

Development event	Location	Animal	Neurotransmitter or neurohumor	References[a]
Anterograde transsynaptic regulation of catecholaminergic enzyme development and end-organ innervation	Sympathetic ganglia (e.g., superior cervical ganglion)	Rat	Acetylcholine	Black, 1973; Black *et al.*, 1971, 1972; Black and Geen, 1973, 1974; Black and Mytilineou, 1976a,b; Hendry, 1975a; Otten and Thoenen, 1976
Retrograde transsynaptic regulation of preganglionic cholinergic terminal development	Sympathetic ganglia e.g., superior cervical ganglion)	Rat	NGF (norepinephrine)	Dibner *et al.*, 1977; Dibner and Black, 1976; Hendry, 1975b; Hendry and Thoenen, 1974
Central regulation of preganglionic neuron development	Sympathetic ganglia (e.g., superior cervical ganglion)	Rat	Unknown	Black *et al.*, 1976
Central regulation of end-organ innervation by postganglionic neuron	Sympathetic ganglia (e.g., superior cervical ganglion)	Rat	Unknown	Mytilineou and Black, 1978
Anterograde transsynaptic regulation of cholinergic enzyme development	Parasympathetic ganglia (e.g., ciliary ganglion	Chick	Acetylcholine	Burt and Narayanan, 1976; Chiappinelli *et al.*, 1976, 1978; Sorimachi and Kataoka, 1974

[a]For reviews, see Black, 1978; Black and Patterson, 1980; Giacobini, 1975, 1978, 1979.

regulation of preganglionic (cholinergic) development has also been documented. This requires the integrity of the end organ, nerve growth factor (NGF) produced by the end organ, and possibly norepinephrine, the transmitter of the postganglionic cell, although the last point has yet to be proven.

The key difference between these two categories of humorally mediated events is that synapses appear not to be required in the first category (Table I), whereas in the second (Table II), they form an integral part of the mechanism whereby the neurotransmitter or humoral agent exerts its influence. These two categories are therefore designated as *nontransmission* and *synapse-dependent* roles for neurotransmitters and neurohumors in development. In the first case, the critical period for these events presumably occurs prior to or during synaptogenesis, in contrast to the latter wherein synaptogenesis is a prerequisite for and apparently a determinant of the critical period. In the remainder of this section we will discuss in detail some of the nontransmission roles for these substances in CNS development, including the possible relationship to critical periods for brain organogenesis.

Logically, the ability of any factor to exert an influence on development depends on its being in the right place at the right time in the ontogenic continuum. With regard to neurotransmitters and neurohumors, an obvious first step in analyzing their ability to play particular developmental roles is to demonstrate their presence in those regions of the developing nervous system where and when key ontogenic events are taking place. Figure 1 incorporates this idea with respect to so-called early and late developing neurotransmitters and neurohumors. The basis for these temporal designations is contained in Figure 2. This latter scheme is derived from information concerning the "birthdays" of certain neuronal populations in the rat brain (time of neuronal genesis, time of origin, day of last cell division), correlated with morphochemical data regarding their neurotransmitter or neurohumoral content and the time at which this content can first be demonstrated. With this correlative scheme we can begin to infer the possible involvement of different neurotransmitters and neurohumors in various developmental processes. For example, only a few of these substances are present early enough to be concerned with the timing of "birthdays" of their target neurons in the prenatally developing rat brain: (1) the monoamines norepinephrine (NE), dopamine (DA) and serotonin (5-HT), (2) GABA and (3) the peptides somatostatin (SS), substance P (SP) and neurophysin (NP), all of which appear around E13–14 (Fig. 2). The other agents, including vasopressin (VP), vasoactine intestinal peptide (VIP), enkephalin (ENK), and lutenizing hormone-releasing hormone (LH-RH), do not appear until after the completion of neuronal genesis in most brain regions. Another interesting pattern is the appearance of early developing neurotransmitters and neurohumors shortly after the genesis of the neurons that contain them, in contrast to the relatively long lag period between the formation of other

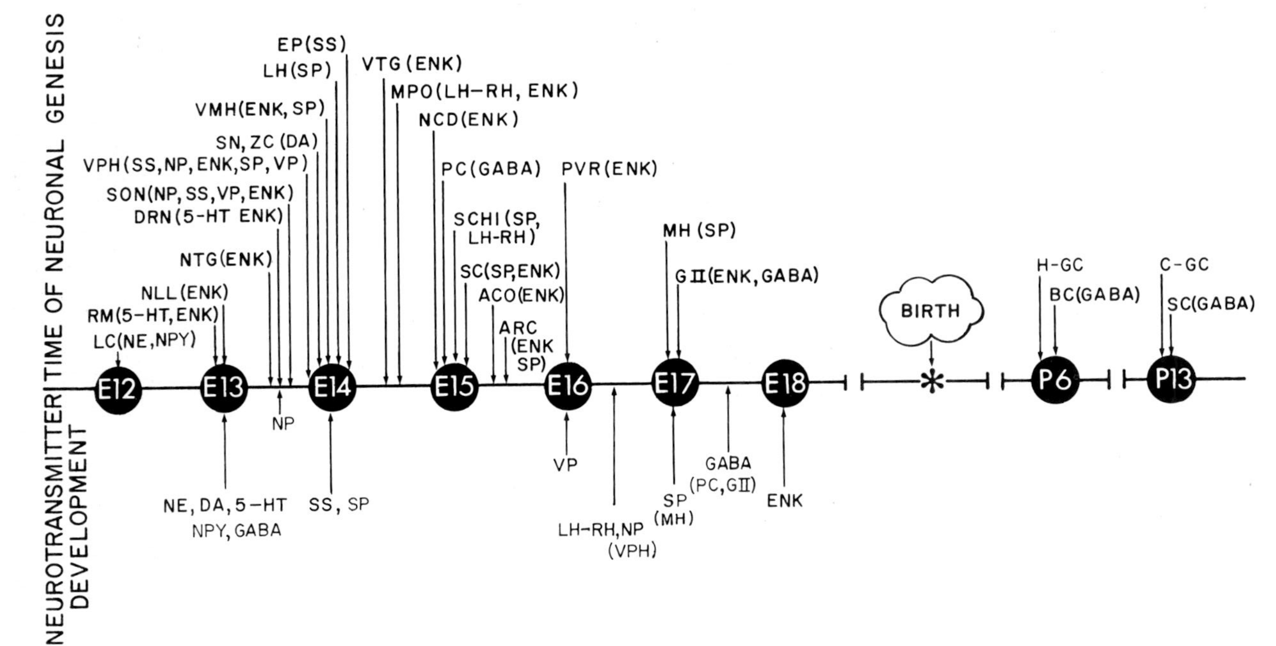
TIME OF NEURONAL GENESIS
NEUROTRANSMITTER DEVELOPMENT
E12
E13
E14
E15
E16
E17
E18
BIRTH
P6
P13
LC(NE,NPY)
RM(5-HT,ENK)
NLL(ENK)
NTG(ENK)
DRN(5-HT ENK)
SON(NP,SS,VP,ENK)
VPH(SS,NP,ENK,SP,VP)
SN,ZC (DA)
VMH(ENK,SP)
LH(SP)
EP(SS)
VTG(ENK)
MPO(LH-RH, ENK)
NCD(ENK)
PC(GABA)
SCHI(SP, LH-RH)
SC(SP,ENK)
ACO(ENK)
ARC (ENK SP)
PVR(ENK)
MH (SP)
G II(ENK, GABA)
H-GC
BC(GABA)
C-GC
SC(GABA)
NE, DA, 5-HT
NPY, GABA
NP
SS, SP
VP
LH-RH,NP (VPH)
SP (MH)
GABA (PC,G II)
ENK

neuronal populations and the late-developing neurotransmitters and neurohumors they contain. In several instances evidence for transient expression of a particular neurotransmitter characteristic has been found. This is true for certain cells of the gut that transiently express a noradrenergic phenotype (Cochard *et al.*, 1978; Teitelman *et al.*, 1978) as well as for neuroepithelial cells in the neural tube of the forebrain that appear to transiently contain tyrosine hydroxylase (Specht *et al.*, 1978;a,b). These transient neurotransmitters may play some developmental role in the cells which contain them or in other cells with which they come into contact.

B. Monoamines and the Process of Neurulation

The development of the central nervous system begins with the formation of the neural plate. Within hours of the induction of the neural ectoderm by the chorda–mesoderm, neural folds elevate at the lateral edges of the neural plate. As the neural folds approximate near the midline, they fuse forming

---

Fig. 2. Relationship of neuronal genesis (time of origin, last cell division) to first appearance of neurotransmitter (neurohumor) in the rat brain. Note that early-developing neurotransmitters and neurohumors (NE, DA, 5-HT, SS, NP) can be detected before or soon after the genesis of the neurons which will contain them, whereas those developing later (SP, ENK, LH-RH, GABA) are detectable only after a lag period of several days following the cessation of cell proliferation in the neuronal populations with which they are associated. Abbrevations: *Neuronal populations:* LC, locus coeruleus; RM, nuc. raphe magnus; NLL, nuc. lat. lemniscus; NTG, d. teg. nuc. Gudden; DRN, d. raphe nuc.; SON, supraoptic nuc.; VPH, nuc. paraventricularis hypothalamus; SN, ZC, substantia nigra, zona compacta; VMH, ventromedial nuc. hypothalamus; LH, lat. habenula; EP, entopeduncular nuc.; VTG, vent. tegmental nuc. Gudden; NCD, d. cochlear nuc.; PC, cerebellar Purkinje cells; Schi., suprachiasmatic nuc.; SC, superior colliculus; ACo, amygdala cortical nuc.; ARC, arcuate nuc.; PVR, nuc. periventricularis magnocellularis thalamus, MH, medial habenula nuc.; GII, cerebellar Golgi II neurons; BC, cerebellar basket cells, SC, cerebellar stellate cells; H-GC, hippocampal granule cells; C-GC, cerebellar granule cells. *Neurotransmitters or neurohumors:* NE, norepinephrine; DA, dopamine; 5-HT, serotonin; SS, somatostatin; NP, neurophysin; NPY, neuropeptide; SP, substance P; ENK, enkephalin; LH-RH, lutenizing hormone–releasing hormone; GABA, γ-aminobutyric acid; VP, vasopressin. *References: Time of neuronal genesis:* Altman and Bayer, 1978a, b, c; Anderson, 1978; Brückner *et al.*, 1975; Ifft, 1972; Nornes and Morita, 1979; Lauder and Bloom, 1974; Lauder and Krebs 1976, 1978. *Neurotransmitter (neurohumoral) content of neuronal populations:* (NE, DA, 5-HT): Ungerstedt, 1971; Swanson and Hartman, 1975; Steinbusch *et al.*, 1978; (SS): Finley *et al.*, 1981a; Finley and Petrusz, 1979; (NP): Sokol *et al.*, 1976; Vandesande *et al.*, 1974; Weindl *et al.*, 1978; (SP) Ljungdahl *et al.*, 1978; (ENK): Finley *et al.*, 1981b; Hökfelt *et al.*, 1977; Sar *et al.*, 1978; (LH-RH): Ibata *et al.*, 1979; Sétáló *et al.*, 1976; (GABA): McLaughlin *et al.*, 1974; Ribak *et al.*, 1978; Schon and Iversen, 1972; (VP): Sokol *et al.*, 1976; Swaab *et al.*, 1975; Weindl *et al.*, 1978. *First appearance of Neurotransmitter or neurohumoral content:* (NE, DA, 5-HT): Lauder and Bloom, 1974; Lauder *et al.*, 1980; Lidov and Molliver, 1980; Olson and Seiger, 1972; Specht *et al.*, 1978a,b; (SS): Bugnon *et al.*, 1977a,b; McDonald *et al.*, 1984; Shiosaka *et al.*, 1981; (NP): Bugnon *et al.*, 1977a,b; Silverman *et al.*, 1980; Sinding *et al.*, 1980; (VP): Boer *et al.*, 1980; Gash *et al.*, 1980; (SP, ENK): Inagaki *et al.*, 1982; Pickel *et al.*, 1979; (LH-RH): Daikoku *et al.*, 1984; Paden and Silverman, 1979; (GABA): Lauder *et al.*, 1985a; McLaughlin *et al.*, 1974; NPY: Woodhams *et al.*, 1985.

a continuous neural tube. Neurulation, the process of the formation of the neural tube, advances in both rostral and caudal directions from the initial point of contact of the neural folds and is complete with the closure of the rostral and caudal neuropores.

During this early period of neurogenesis, catecholamines and serotonin (5-HT) have been detected in chick embryos and implicated in the initial phases of nervous system development. In whole embryo extracts, the catecholamines NE and DA have been biochemically measured as early as day one of incubation (Ignarro and Shideman, 1968a,b), whereas extracts of the notochord have revealed the presence of 5-HT as well as catecholamines (Strudel *et al.*, 1977a,b). Various embryonic structures, discussed below, have been shown to accumulate (and possibly synthesize) these amines during embryonic morphogenesis (Kirby and Gilmore, 1972; Lawrence and Burden, 1973; Wallace, 1979, 1982). This early presence of monoamines may influence the development of the chick embryo neuraxis, as illustrated by the teratological effects of drugs which interfere with the metabolism of these transmitter substances (see Section V for details). For example, alterations in catecholamine metabolism result in defects in neural tube closure and failures in the development of embryonic torsion and flexure (Lawrence and Burden, 1973). Certain drugs that interfere with 5-HT synthesis, release, and receptor interactions also disturb the processes of blastoderm growth, primitive streak formation, neurulation, brain formation, and somatogenesis (Palén *et al.*, 1979; Jurand, 1980). The locations of these disturbances resulting from perturbations in monoamine metabolism may be related to the specific embryonic sites that concentrate the amines during morphogenesis.

Wallace (1979, 1982) has examined sites concentrating either 5-HT or NE in chicks during neurulation by culturing embryos with these compounds and following the development of amine accumulaton in various structures by fluorescence histochemistry or immunocytochemistry. Serotonin accumulation was found in specific portions of the developing brain within the floor plate of the mesencephalon and caudal (otic level) myelencephalon soon after the neural tube had closed in this region. Similarly, sclerotome cells of somites concentrated 5-HT soon after the somites began to exhibit morphological differentiation. In addition, caudal segments of the developing spinal cord and adjacent notochord concentrated 5-HT in the region of the neural plate actively involved in closure. As progression of closure advanced caudally, the 5-HT accumulation observed in the floor plate of the neural tube and notochord also advanced caudally, remaining in the region of the closing caudal neuropore.

Over the same period, NE concentration initially appeared throughout the developing neural plate, with regions of increased accumulation found in progressively more caudal portions of the neural tube in apparent spatial and temporal coordination with the advance of neural tube closure. As neurulation neared completion, the capacity of the neural tube to concentrate NE likewise ceased, again in a rostral to caudal sequence, with those areas initially completing closure losing the ability first. With a similar pattern, notochordal accumulation of NE was observed beneath regions of the developing brain and spread progressively into trunk portions of the chord over the period during which the neural tube in the spinal cord completed closure.

A comparison of embryonic sites concentrating either 5-HT or NE (Fig. 3) showed major differences in location during neurulation. Although an overlap in certain regions of the brain floor plate appeared temporarily, almost mutually exclusive patterns of accumulation were observed in the region of the caudal neuropore (Fig. 3A,B) as well as throughout the rostrocaudal extent of the notochord and within various components of the developing somites (Fig. 3E,F).

These findings raise the possibility that 5-HT and NE may play different roles in chick neurogenesis, possibly acting in concert to coordinate events leading to neural tube closure, brain organogenesis and somite differentiation, although such actions are speculative at present. The notochord may constitute a site of synthesis of monoamines in the early chick embryo (Kirby and Gilmore, 1972; Allan and Newgreen, 1977; Wallace, 1979, 1982), although yolk material has also been shown to contain catecholamines during early embryonic periods (Ignarro and Shideman, 1968a,b). Moreover, the malformations produced by drugs which interfere with 5-HT metabolism are associated with a delayed degradation of yolk granules which are thought to contain 5-HT and its precursor L-tryptophan (Emanuelsson, 1974). These granules, which are taken up into the neural plate and tube during neurulation (Santander and Cuadrado, 1976), could be a source of 5-HT for the developing nervous system in addition to the notochord. In the rat embryo, the yolk sac itself may be a source of monoamines for the early developing nervous system (Schlumpf and Lichtensteiger, 1979).

Recently, we have undertaken similar studies using the method of whole mammalian embryo culture (New *et al.*, 1973; Sadler and Kochhar,1976; Sadler and New,1981) to examine sites of 5-HT accumulation and synthesis in the developing mouse embryo, since we felt that this would provide a better animal model with respect to the human analogy than the chick (Lauder and Sadler, in preparation). To date, we have examined embryos in two different age groups: (1) 7–9 somites (9–9.5 days of gestation) and (2) 45–50 somites (12–12.5 days of gestation) for their ability to accumulate immunoreactive 5-HT after incubation in either 5-HT itself, 5-hydroxytryptophan (5-HTP;

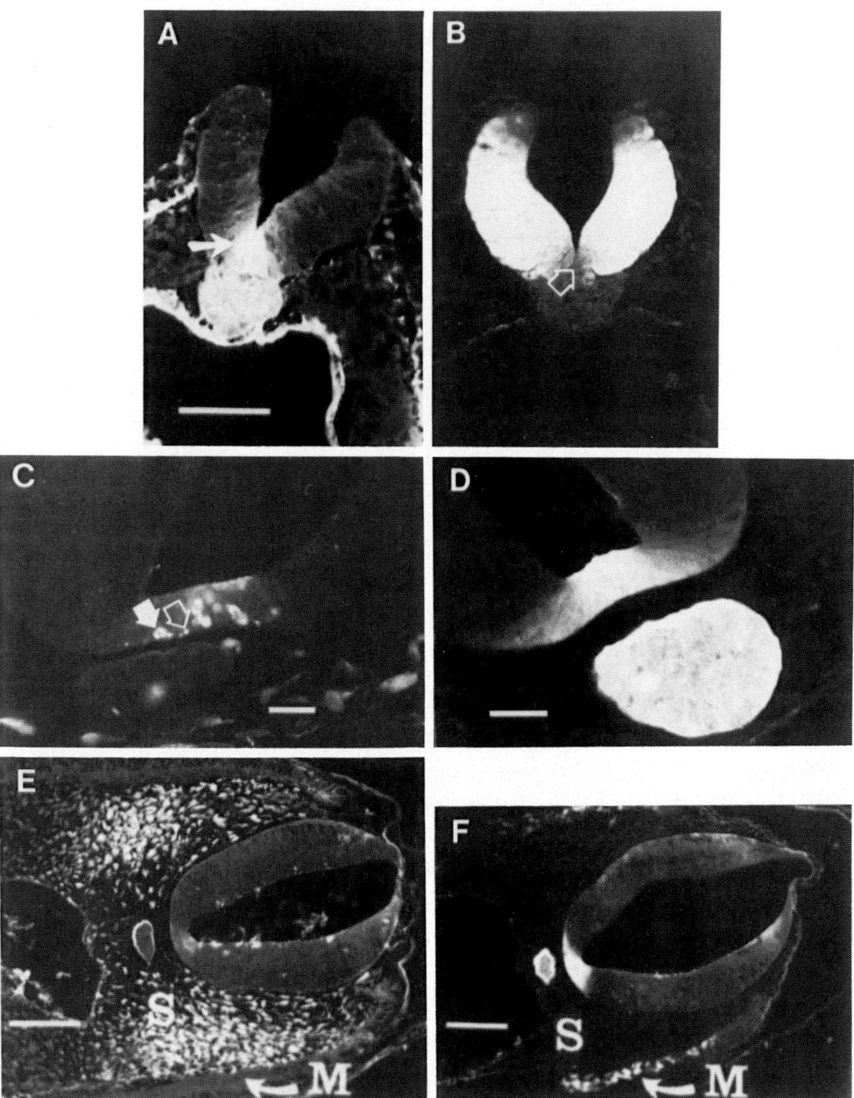

Fig. 3. Comparison between sites of serotonin (5-HT) and norepinephrine (NE) accumulation in the chick embryo following incubation *in vitro* with either of these compounds (25 μM). (A,C,E) Sites of 5-HT accumulation. (B,D,F) Sites of NE accumulation. Note the different sites of accumulation and appearance of 5-HT and NE fluorescence at the same anatomical level of the neuraxis (A,B: caudal neuropore, stage 10 embryo; C,D: cervical level, stage 15 embryo; E,F rostral somite level, stage 15 embryo). S, Sclerotome; M, myotome. From Wallace (1979), by permission. *Bars:* (A,B,E,F) 50 μm; (C,D) 20 μm.

the immediate precursor of 5-HT) or L-tryptophan (L-trp; the initial precursor of 5-HT), in the presence of nialamide (an inhibitor of monoamine oxidase) and L-cysteine (L-cys, an anti-oxidant). After incubation in medium containing these substances (at concentrations of 10^{-5}–$10^{-7}M$) for 4 hours, embryos were fixed in 4% paraformaldehyde and processed for paraffin embedding, sectioned and stained for immunocytochemistry using an antiserum to 5-HT (Wallace *et al.*, 1982).

In the youngest embryos (7–9 somites), in which the neural tube is still in the process of closure, no immunoreactivity was ever seen in the notochord or in the floor plate of the neural tube as in the chick embryo, described above. However, a prominent site of 5-HT localization after either 5-HT or 5-HTP incubation was the hindgut (Fig. 4) which is located adjacent to the caudal neural tube in a similar configuration to the notochord in the chick. Thus, considering the small size of the notochord in the mouse embryo, it is possible that the hindgut performs a function in this species which in the chick is performed by the notochord, although the nature of this function is unknown. Interestingly, another major site of 5-HT immunoreactivity was seen in the heart in these young mouse embryos. This site was also visible after incubation in L-trp, indicating that the heart could constitute a site of 5-HT synthesis from its initial precursor. However, the possibility must also be considered that 5-HT was synthesized elsewhere from L-trp and then taken up by the heart. The finding of a major site of 5-HT accumulation in the heart, nevertheless, raises the possibility that some heart malformations could have a drug-related etiology, if correlated with the use of 5-HT interactive drugs in pregnancy.

In the older embryos (45–50 somites), (Fig. 5) in which brain organogenesis is proceeding and neuronal genesis has begun, a number of brain-related as well as non-neural sites of 5-HT immunoreactivity were observed after incubation with 5-HT or 5-HTP. Only the non-neural sites were seen after L-trp, however. These non-neural sites, most of which are also present after 5-HT or 5-HTP, include the head ectoderm, the otocyst, palate, the dorsal aspect of the tongue, the roof of the mouth and lining of the mandibular and pharyngeal arches. In the brain, sites seen after 5-HT or 5-HTP (but not with L-trp) include the roof of the diencephalon and epiphysis, as well as certain neuronal populations in the brainstem and diencephalon. Those in the brainstem do not stain with an antiserum to tyrosine hydroxylase, indicating that they are not catecholaminergic neurons being loaded non-specifically with 5-HT or 5-HTP. Some of the cells in the diencephalon do appear to be catecholaminergic, although not all cells in a given group both accumulate 5-HT and stain for tyrosine hydroxylase. In addition, and perhaps rather significantly, in all animals examined, including controls not incubated

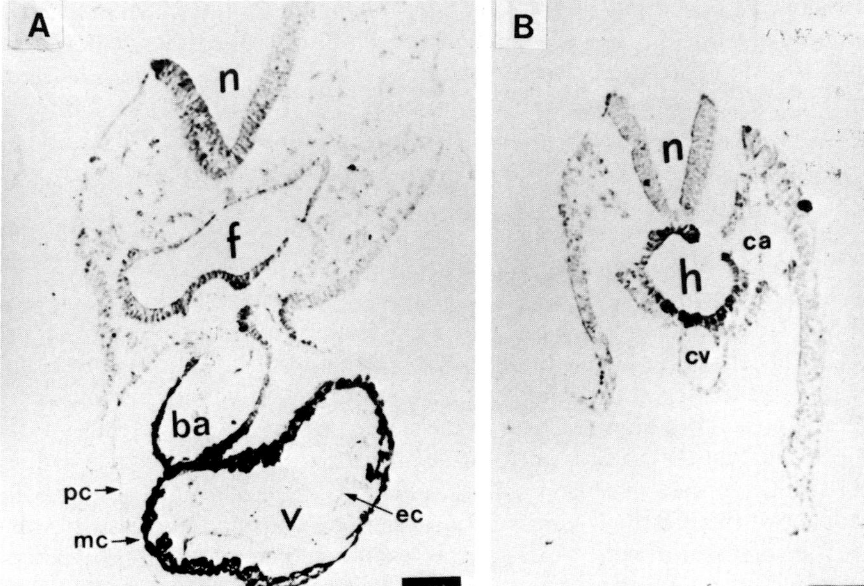

Fig. 4. Sites of accumulation of 5-HT in the 7-12 somite mouse embryo in culture. (A) Heart; (B) hindgut. Embyros were cultured according to the method of Sadler and Kochhar (1976) for 4 h in the presence of 5-HT ($10^{-5}M$), L-cys ($10^{-5}M$), and nialamide ($10^{-5}M$). These same sites were seen when embryos were incubated in 5-hydroxytryptophan (5-HTP, $10^{-5}M$), indicating that they may contain the enzyme aromatic amine decarboxylase, which converts 5-HTP to 5-HT. The heart is also seen with L-tryptophan, suggesting that it may contain the entire pathway for 5-HT synthesis, although uptake of 5-HT synthesized elsewhere is also a possibility. (A) n, Neural tube; f, foregut; ba, bulbus arteriosus; v, ventricle; pe, pericardium; mc, myocardium; ec, endocardium. (B) h, Hindgut; cv, caudal vein; ca, caudal artery. Bars = 100 μm. Anti-5-HT immunocytochemistry. From Lauder and Sadler, in preparation.

in 5-HT or its precursors, but exposed to nialamide and L-cys, numerous small, darkly staining 5-HT immunoreactive cells are seen throughout the mesenchyme and often within the brain itself. These cells, by the criteria of their size, distribution and immunoreactive properties (Combs *et al.*, 1965) have tentatively been identified as mast cells. Their presence in the head mesenchyme and brains of these embryos is interesting, since they could provide a potential source of 5-HT for uptake into the sites we have visualized by our *in vitro* incubation method.

The presence of discrete regions of the developing brain, heart, gut, ear, palate, tongue, mouth, and ectoderm with the ability to accumulate 5-HT itself or perhaps to synthesize it from its precursors raises interesting issues with regard to the possibility of teratogenic effects of psychoactive drugs

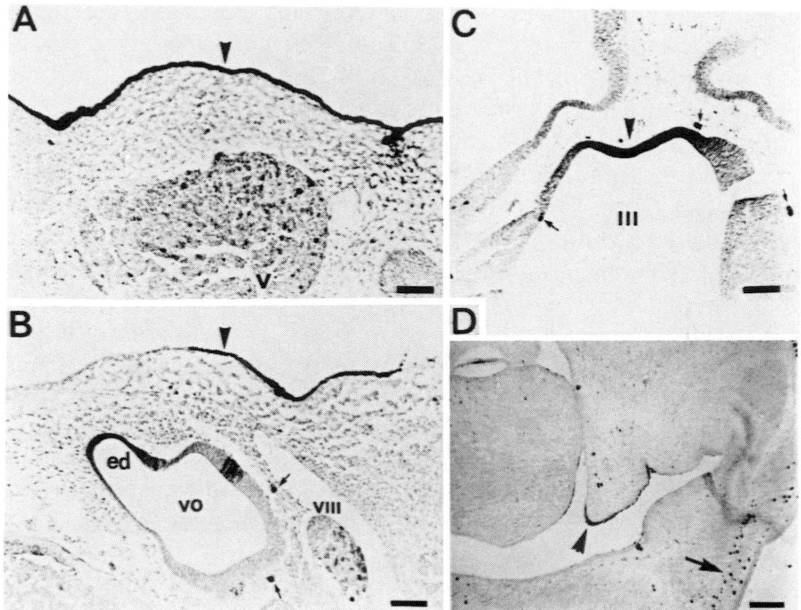

Fig. 5. Sites of accumulation of 5-HT in the 45–50 somite (A-C) and E14.5 (D) mouse embryo in culture. (A) Head ectoderm; (B) head ectoderm and otocyst (ear); (C) Roof of the diencephalon (epithalamus, epiphysis); (D) brainstem (myelencephalon). Embryos cultured as described in Fig. 1. All of these sites are also seen with 5-HTP, but L-tryptophan only produces immunoreactivity in the head ectoderm and otocyst. (A) Arrowhead, head ectoderm; V, ganglion of the Vth cranial nerve (semilunar or Gasserian ganglion). (B) Arrowhead, head ectoderm; small arrows, mast cells; VIII, ganglion of the VIIIth cranial nerve (acoustic ganglion); ed, endolymphatic duct of the otocyst; vo, vestibular part of otocyst. (C) Arrowhead, roof of the diencephalon (epithalamus, epiphysis); III, third ventricle; small arrows, mast cells. (D) Palate, E14.5. Arrowhead, site of 5-HT uptake in palatal shelf epithelium; arrow, mast cells. Bars = 100 μm. Anti-5-HT immunocytochemistry. (A-C) From Lauder and Sadler, in preparation; (D) from Lauder and Zimmerman, in preparation.

which have the ability to alter the uptake, synthesis, metabolism or receptor interaction of the neurotransmitter 5-HT.

C. Serotonin as a Differentiation Signal during Brain Organogenesis

Our studies concerning the development of monoamine neurons in the embryonic rat brain have demonstrated that these are early developing neurotransmitter systems which begin to differentiate just following neural tube closure (see Fig. 2 and Lauder and Bloom, 1974, 1975). Since most other cells are still dividing, this raises the possibility that these transmitters might influence the differentiation of other neurons. Based on these temporal correlations and the results of pharmacologic studies (described below), we have proposed that 5-HT acts as a differentiation signal for specific populations of neurons in the embryonic brain.

In pharmacologic experiments (Lauder and Krebs, 1976,1978), we have tested the hypothesis that 5-HT neurons influence certain populations of neuroepithelial cells to stop dividing and begin to differentiate as neurons. In these studies, the drug *p*-chlorophenylalanine (*p*CPA), which inhibits 5-HT synthesis (Aghajanian *et al.*, 1973; Deguchi *et al.*, 1973; Koe and Weissman, 1968), was administered to pregnant rats beginning on embryonic day 8 (E8). This maternal treatment significantly inhibits 5-HT synthesis in the embryo (Lauder *et al.*, 1981), particularly in developing 5-HT neurons (Lauder *et al.*, 1985), and produces prolonged proliferation (delayed differentiation) of specific populations of neuroepithelial cells in the embryonic brain, all of which give rise to neurons innervated by 5-HT axons in the adult (5-HT target cells). These results have recently been correlated with the spatio-temporal patterns of 5-HT axonai growth in the embryonic brain (Lauder *et al.*, 1982a; Wallace and Lauder, 1983). These results indicate that the effects of *p*CPA on embryonic neuronal differentiation are dependent on the presence of developing 5-HT axons, suggesting that these *p*CPA effects are specific to 5-HT neurons and their target cells. We also found early neurogenesis in many brain regions of injected controls compared to uninjected controls, which may indicate an effect of the chronic stress of daily injections on neurogenesis.

In the context of these *p*CPA experiments, we have examined interactions between developing 5-HT neurons and proliferating neuroepithelial cells in the neural tube (Lauder *et al.*, 1982a,b) using immunocytochemical methods with an antiserum to 5-HT-hemocyanin conjugates (Wallace *et al.*, 1982). As shown in Figure 6, such interactions exist in a variety of different configurations. For example, developing 5-HT axons pass through the marginal zone of the neural tube where they are able to contact the apical processes of dividing neuroepithelial cells (Fig. 6A,B). Likewise, varicose processes of 5-HT neurons pass directly into the neuroepithelium where they appear to contact dividing cells (Fig. 6E). Sometimes even the cell bodies of 5-HT neurons are in contact with proliferating cells in the immediately adjacent neuroepithelium (Fig. 6C,D). In addition, certain populations of 5-HT neurons appear to contact the cerebrospinal fluid directly, raising the possibility that 5-HT could also act as a diffuse humoral signal in the embryonic brain.

Recently, we have completed studies in collaboration with Dr. Patricia Whitaker-Azmitia (Whitaker-Azmitia and Lauder, 1985) which have demonstrated changes in the number of 5-HT receptors in the forebrain and

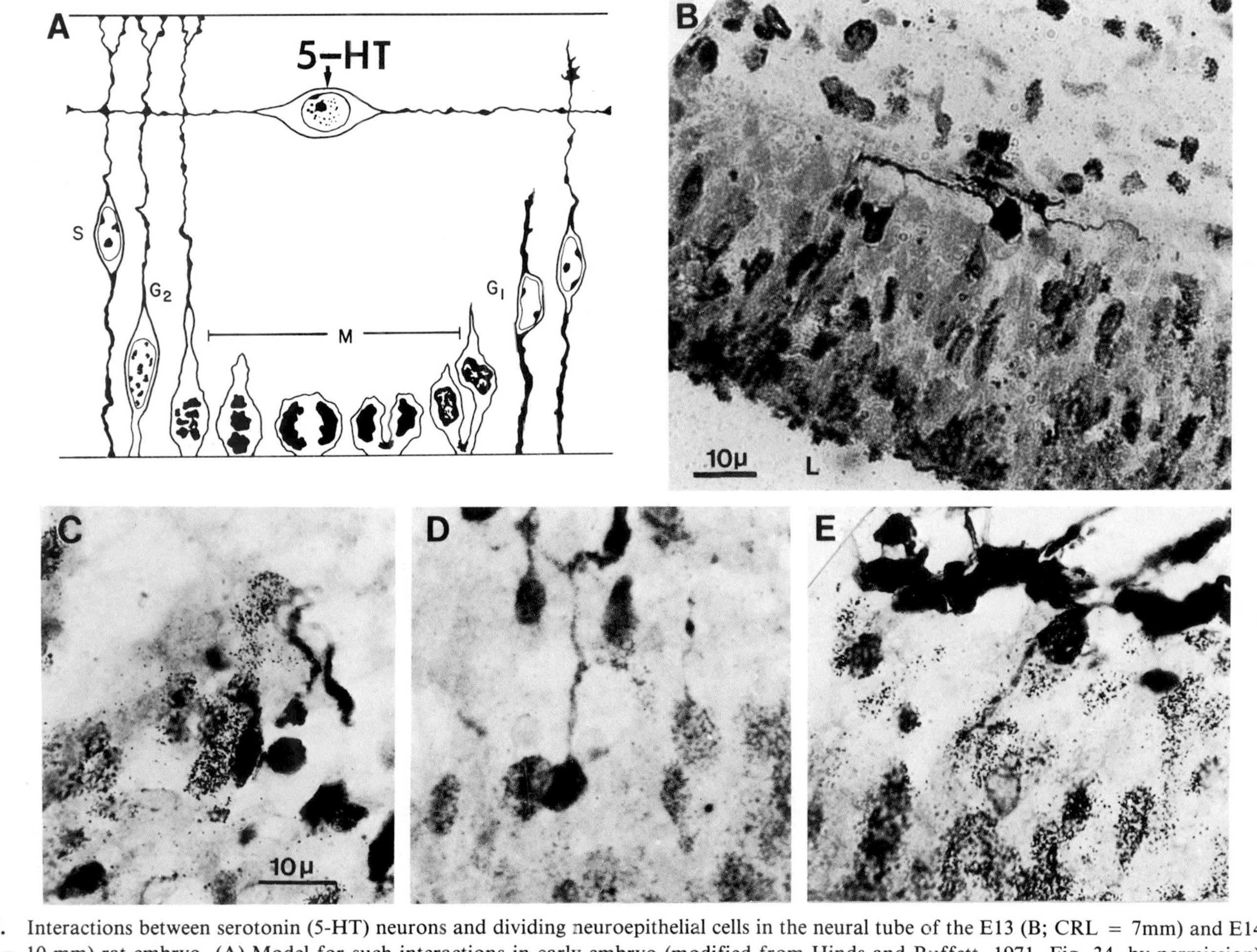

Fig. 6. Interactions between serotonin (5-HT) neurons and dividing neuroepithelial cells in the neural tube of the E13 (B; CRL = 7mm) and E14 (C-E; CRL = 10 mm) rat embryo. (A) Model for such interactions in early embryo (modified from Hinds and Ruffett, 1971, Fig. 34, by permission). Note similarity to (B). From Lauder, 1982b.

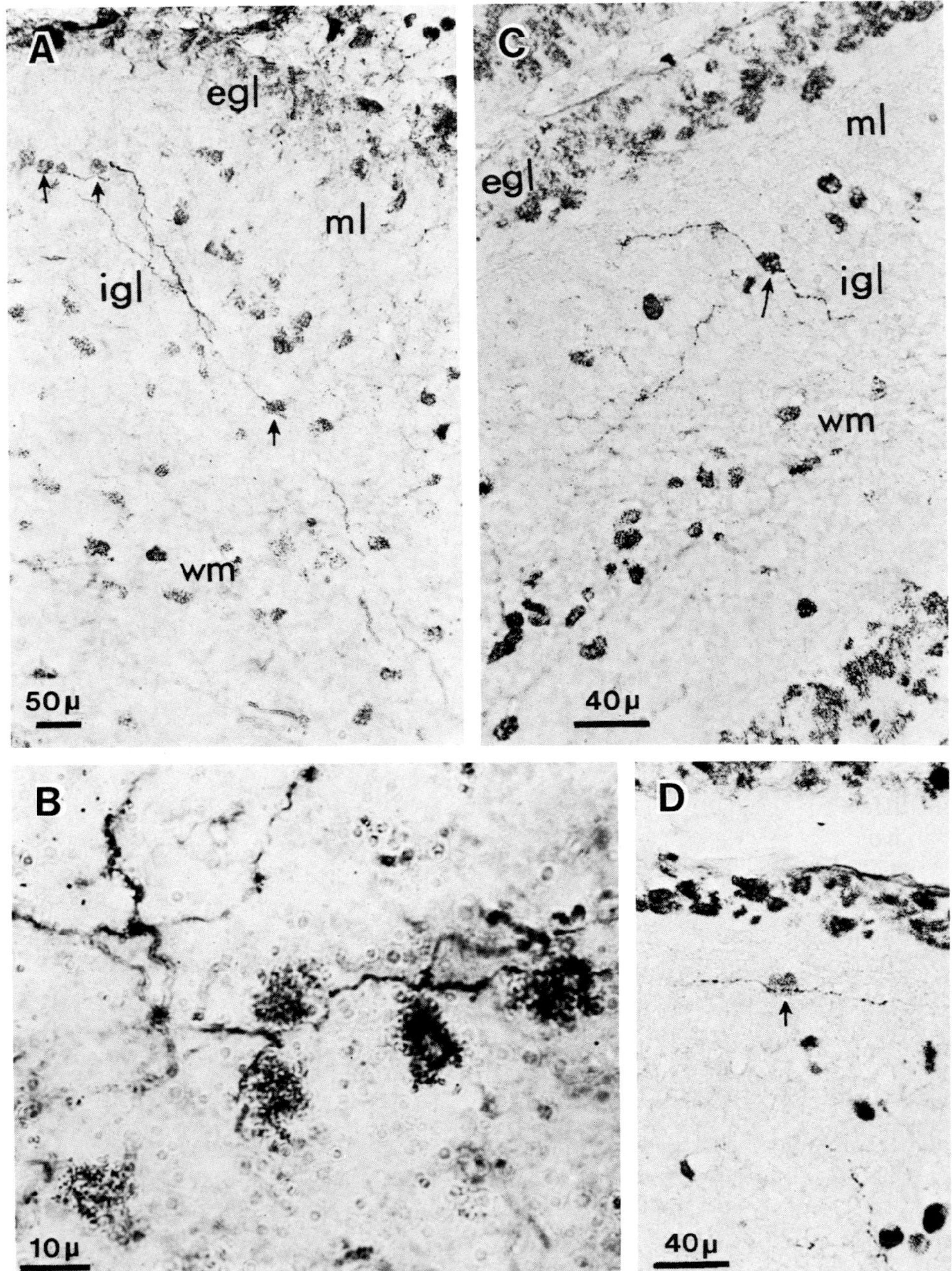
A
egl
ml
igl
wm
50μ
C
ml
egl
igl
wm
40μ
B
10μ
D
40μ

brainstem of offspring of pCPA-treated mothers, indicating that 5-HT neurons may regulate the development of their receptors and that the effects of 5-HT neurons on the differentiation of their target cells could be receptor-mediated.

After birth, axons of 5-HT neurons continue to develop and grow into most regions of the neonatal rat brain. At this time, however, most cell proliferation is related to the genesis of glia or small interneurons. In these regions (e.g., cerebellum, hippocampus), 5-HT axons are frequently seen in association with such neuronal and glial precursor cells (see Figs. 7 and 8), raising the possibility that they could exert an epigenetic influence on postnatal neuro- or gliogenesis.

Recently, we have begun to explore the possible influences of 5-HT on the *in vitro* differentiation of neurons and glia from developing rat brain using dissociated cell cultures of embryonic 5-HT neurons and their target cells. The dopaminergic (DA) neurons of the substantia nigra appear to be one of the earliest targets of 5-HT axons in the developing brain, as shown by double immunocytochemistry with antisera to 5-HT and tyrosine hydroxylase (Fig. 9A). Using dissociated cell cultures of 5-HT and DA neurons from embryonic day 14 (E14) rat brain, we have examined the effects of co-culture on the morphological and biochemical differentiation of these neurons, using immunocytochemistry (Fig. 9B,C) and uptake of [^{3}H]5-HT or [^{3}H]DA (Fig. 9D) to label 5-HT and DA neurons, respectively. As shown in Table III, co-culture has a greater effect on the morphological differentiation of DA neurons than on 5-HT neurons. This could be interpreted to mean that 5-HT neurons in these cultures influence the differentiation of DA neurons, as predicted by our hypothesis and their *in vivo* ontogenic relationship (Fig. 9A), although other possibilities must also be considered. No effect of co-culture on the biochemical differentiation of these DA neurons has yet been demonstrated, however, as measured by the amount of uptake of [^{3}H]DA/DA neuron, even though there are clearly effects on the morphological complexity of these cells. Experiments are still in progress to analyze these effects, including the use of other biochemical and molecular markers of differentiation such as tyrosine hydroxylase activity and *in situ* hybridization to localize and quantitate mRNA for this protein. We are also examining the effects of 5-HT added directly to the culture medium on the differentiation of DA neurons.

Fig. 7. Interactions between serotonergic (5-HT) axons and dividing glial precursors (glioblasts) in the presumptive IGL and in the ML of the P-2 (A,B) and P-9 (C,D) rat cerebellum. Abbreviations as in Fig. 7. (B): Presumptive IGL. Note the length of the 5-HT axon shown in A (——), which appears to interact with several glioblasts (arrows). Combined anti 5-HT immunocytochemistry, and [^{3}H]thymidine autoradiography. From Lauder *et al.*, 1982b.

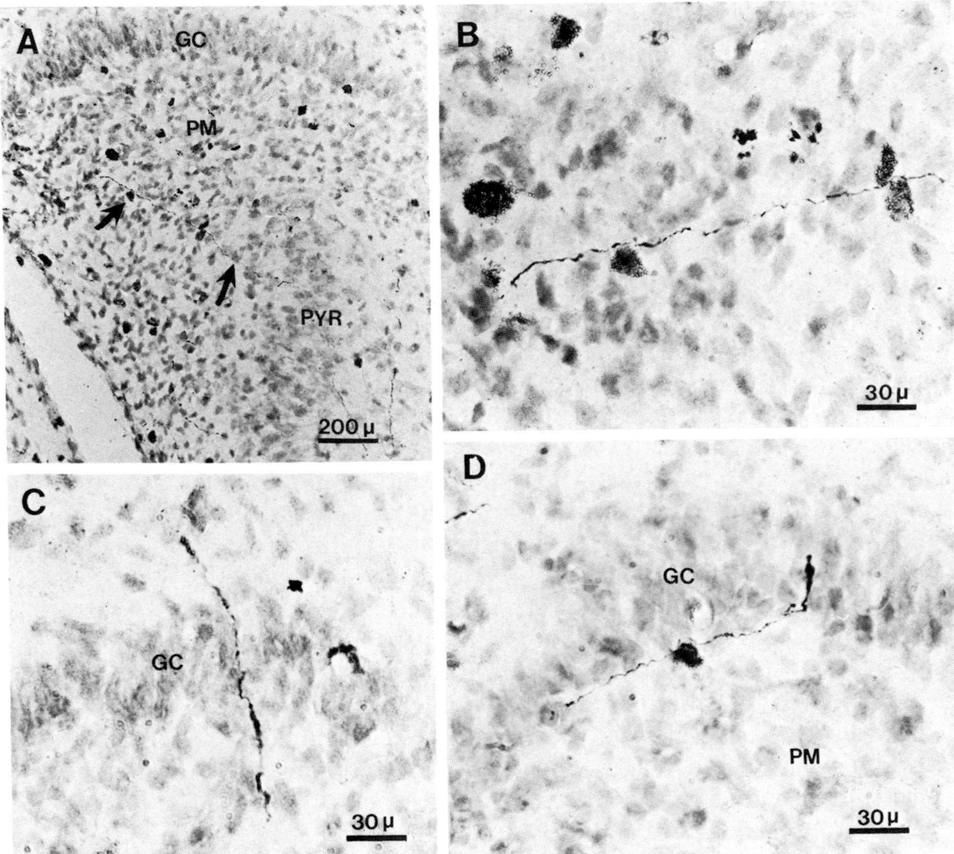

Fig. 8. Serotonergic (5-HT) axons in the ventral hippocampus of the rat at 2 days postnatal (P-2). GC, Granule cell layer; PM, polymorph cell layer; Pyr, pyramidal cell layer. (B): 5-HT axon designated by arrow in (A) that appears to be interacting with three labeled cells. Combined immunocytochemistry, [^{3}H]thymidine autoradiography. From Lauder *et al.*, 1982b.

In addition, we are exploring the possible influence of 5-HT on glial cell proliferation in monolayer cultures of astrocytes from postnatal rat brain. In studies where we have used combined [^{3}H]thymidine autoradiography (to label proliferating gial cells) and anti-5-HT + anti-GFAP (glial fibrillary acidic protein) immunocytochemistry, we have demonstrated that 5-HT neurons readily contact proliferating astrocytes in these cultures. We are now engaged in studies where 5-HT is added to the culture medium and quantitative measurements of [^{3}H]thymidine incorporation are made to assess the effects of 5-HT on glial cell proliferation. These studies are still in progress.

Thus, it appears that these *in vitro* model systems may be useful for exploring effects of 5-HT on the proliferation and differentiation of target

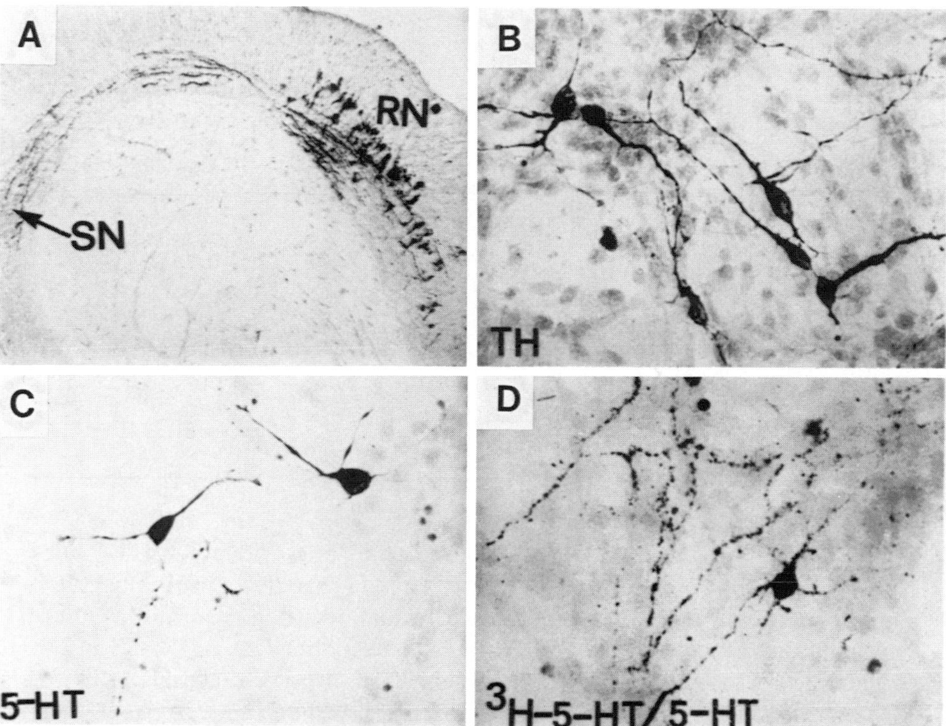

Fig. 9. Serotonergic (5-HT) and dopaminergic (DA) neurons from the raphe (RN) and substantia nigra (SN) of the E14 rat embryo (A: sagittal section) grown in dissociated cell culture for 4 days (B-D). B: SN neurons immunocytochemically stained with anti-tyrosine hydroxylase (TH); C: RN neurons stained with anti-5-HT; D: 5-HT neuron stained with anti-5-HT and labeled with ^{3}H-5-HT (combined immunocytochemistry and autoradiography). From Lauder, Konig, Han and Lieth, in preparation.

neurons and glial cells. These systems may also prove useful for examination of the direct effects of various drugs, hormones and humoral agents on the differentiation of 5-HT neurons and their interactions with other neurons or glia.

In addition to our own work, other *in vitro* studies have recently appeared which lend support to the working hypothesis that 5-HT plays roles in development prior to its use in neurotransmission. In the first of these (Gromova *et al.*, 1983), 5-HT was added to the medium of organotypic hippocampal cultures from newborn rats which were grown for 2–35 days *in vitro*. In these experiments, 5-HT appeared to stimulate neuropil development, axon myelination and synaptogenesis. In addition, more neurons in these cultures exhibited spontaneous activity and did so earlier than in untreated cultures.

The second study, carried out by Kater and coworkers (Haydon *et al.*, 1984)

Table III

Effects of Co-Culture of E14 Raphe (R) and Substantia Nigra (SN) on Morphological Differentiation of 5-HT and Dopamine (DA) neurons After 6 Days *in Vitro* (6 DIV)

Ratio of co-culture / single cult.	R+SN / R (5-HT)	R+SN / SN (DA)	Co-culture effect SN / Co-culture effect R
Absolute field surface area	1.7	3.0	1.8
Relative field surface area	1.3	2.3	1.8
Individual segment length	1.1	1.5	1.4
Total segment length	1.4	2.0	1.4
Number of segments	1.1	1.4	1.3
Distance from cell center to segment start point	1.1	1.4	1.3
Mean morphol. effects of co-culture	1.3	1.9	1.5
% increase	30%	90%	50%

on cultured snail neurons has provided compelling evidence for a role of 5-HT in the control of growth cone motility and synaptogenesis. In this study, 5-HT was applied to growth cones of individual, identified neurons in culture using a micropipette. Only certain neurons responded to 5-HT and did so by retracting their growth cones. It was also demonstrated that 5-HT inhibited the formation of electrical synapses by these cells. Importantly, it has recently been demonstrated that neurotransmitters can be released from growth cones of cultured neurons (Hume *et al.*, 1983; Kidokoro and Yeh, 1982; Young and Poo, 1983). Thus, these experiments suggest that 5-HT, if released from growth cones of 5-HT neurons, could alter the motility of growth cones of other 5-HT sensitive cells. More recently, this group has shown that the development of these 5-HT-sensitive cells is altered if 5-HT is depleted in the snail embryo during the period when 5-HT neurons are normally interacting with their target cells (Goldberger *et al.*, in preparation; reviewed by Meinertzhagen, 1985). These studies, therefore, provide direct evidence that 5-HT neurons control the differentiation of their target cells in the snail, in agreement with our more indirect evidence in the rat.

Thirdly, studies from the laboratories of Azmitia and Whitaker-Azmitia have recently shown that growth of cultured 5-HT neurons from embryonic rat brain is inhibited by either 5-HT itself, or by the 5-HT receptor agonist 5-methoxytryptamine, presumably by the stimulation of autoreceptors on these 5-HT neurons (Davila-Garcia *et al.*, 1985). These studies are in agreement with the observation of Kater and co-workers that 5-HT inhibits growth cone activity and neuritic growth in 5-HT neurons in the leech, discussed above.

Further studies are in progress by the Kater group to examine the effects of other transmitters such as GABA and norepinephrine on growth cone motility and synaptogenesis in leech neuronal cultures. Preliminary evidence (Kater, personal communication) suggests that these substances each have their own specific effects on different types of snail neurons such that some inhibit while others stimulate growth cone motility. This suggests that different neurotransmitters, released from growth cones, could produce a mutual orchestration of cell-cell interactions during neurite outgrowth and synaptogenesis, thus participating in the construction of those neuronal circuitries in which they will later function in neurotransmission.

D. GABA and Brain Development

Another neurotransmitter that has been implicated in developmental functions is γ-aminobutyric acid (GABA). In a study by McLaughlin *et al.* (1974), immunoreactivity to the GABA synthesizing enzyme, glutamate decarboxylase (GAD), was reported in the growing neurites of the developing cerebellar molecular layer where it was associated with small synaptic vesicles prior to the onset of synaptogenesis. A similar presynaptic development of GABA has been reported in the embryonic spinal cord (Reitzel *et al.*, 1979). The presence of GAD, and presumably GABA, in growth cones raises the possibility that it might also affect growth cone motility as 5-HT has been shown to do, discussed above.

Prenatal development of the GABAergic system in the rat brain has recently been studied using an antiserum to GABA-glutaraldehyde-hemocyanin conjugates, which is specific for GABAergic neurons (Lauder *et al.*, 1985). In this study, the GABA system was found to differentiate prior to most other transmitters, including the monoamines. Thus, by embryonic day 13 (E13) a well-developed fiber network exists in the brainstem, mesencephalon and diencephalon, including a large projection in the posterior commissure and adjacent areas on the surface of the mesencephalon and tectum (Fig. 10). GABAergic cell bodies begin to appear on E14 in the frontal pole of the cerebral cortex. By E16, they are visible in all regions of cortex where they are located above and below the cortical plate and in the outer part of layer I. Also contained in the outer part of layer I is a dense fiber plexus which stains intensely for GABA. These fibers may be part of the first contingent of cortical afferents to invade the telencephalic vesicle, an event which may be a stimulus for the beginning of neuronal differentiation in this region (König and Marty, 1981; Marin-Padilla, 1984).

Interestingly, the trajectories taken by growing GABAergic fibers in the brainstem, mesencephalon and diencephalon at E13 and at subsequent stages of development are coincident with regions of both monoaminergic and

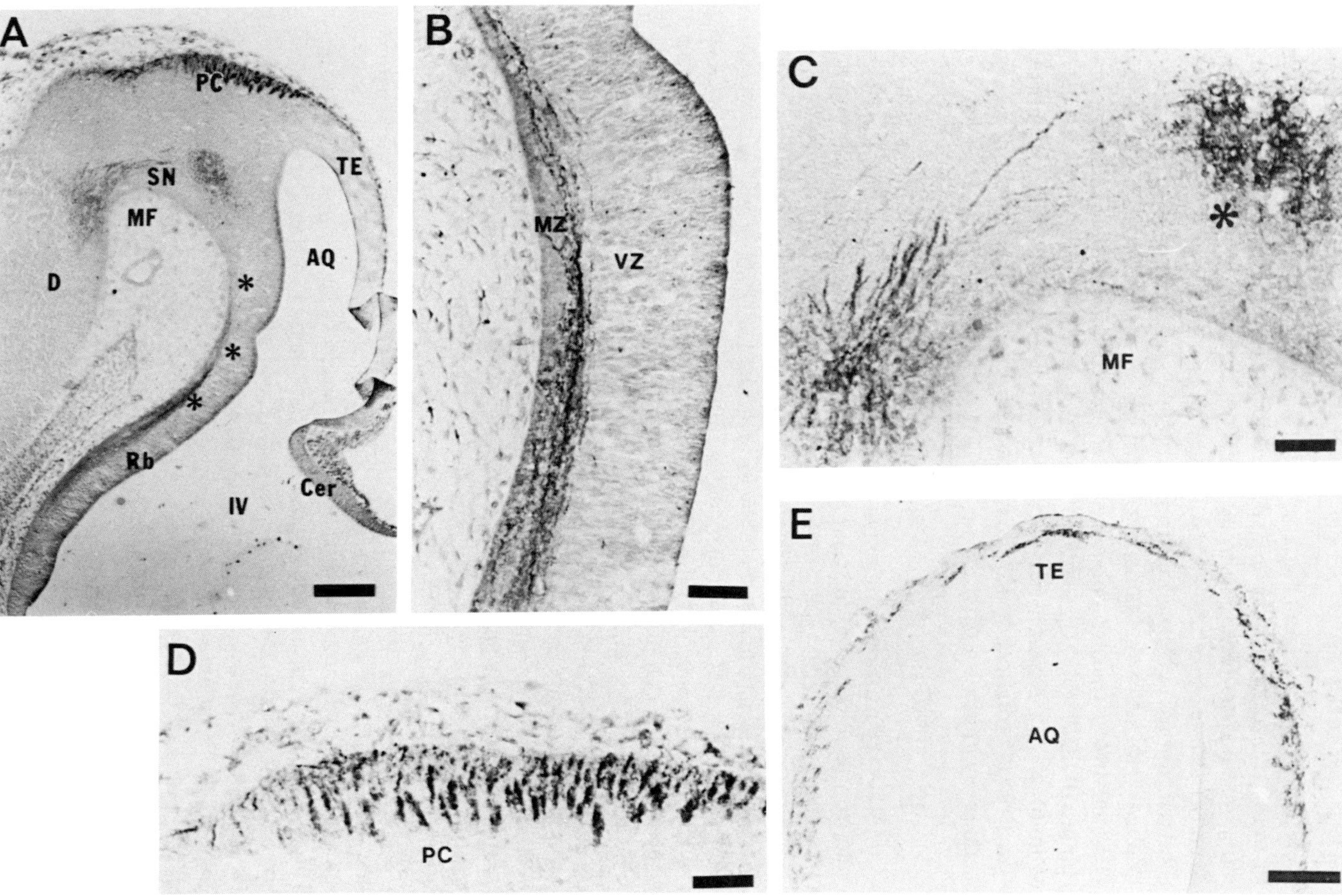
A
PC
SN
TE
MF
AQ
D
*
*
*
Rb
IV
Cer
B
MZ
VZ
C
*
MF
D
PC
E
TE
AQ

peptidergic differentiation and appear to correspond to recently reported patterns of benzodiazepine receptor development (Schlump *et al.*, 1983). These receptors, which are thought to be functionally linked to the GABA receptor (Johnston *et al.*, 1984), appear to develop slightly later than the GABAergic innervation in these same regions (Lauder *et al.*, 1985).

Since evidence from other studies suggests that GABA can stimulate neuronal differentiation and receptor development *in vitro* (Eins *et al.*, 1983; Hansen *et al.*, 1984; Meier *et al.*, 1983; Spoerri and Wolff, 1981), and may be involved in the plasticity of neuronal connections in adulthood (Houser *et al.*, 1983; Wolff *et al.*, 1978), the early differentiation of the GABAergic system may indicate a trophic role for GABA in brain development, possibly involving receptors for this neurotransmitter or related substances.

Preliminary studies in the cerebellum suggest another possible role for GABA in neurogenesis (J. M. Lauder and P. Petrusz, unpublished observations). Using an antiserum to GAD, we have studied the development of cerebellar Purkinje cells. This antibody produces a Golgi-like image with appropriate fixation and dilutions. In coronal (transverse) sections, in which Purkinje cells are viewed end-on due to their planar geometry and orientation, we noticed in 10-day-old animals that the tips of these Purkinje cell dendrites seem to penetrate into the EGL, which contains proliferating and nonproliferating germinal cells giving rise to stellate, basket, and granule cells (Fig. 11). If verified at the ultrastructural level, this could place GABA-ergic Purkinje cell dendrites in an ideal location to influence the differentiation of EGL cells. Since the stellate and basket cells are also thought to be GABA-ergic, whereas the granule cells are not, it is tempting to speculate that GABA might in some way influence cells to become one type or another. An alternative explanation, however, is that this represents nonspecific staining of Bergmann glial processes, an issue which must still be resolved.

E. Summary

In this section we have discussed a scheme whereby certain neurotransmitters and neurohumors present at appropriate times and places during pre- and postnatal neurogenesis could exert regulatory influences on an underlying matrix of fundamental ontogenic events. In this sense, these

Fig. 10. GABA immunoreactivity in the E13 rat brain. (A) sagittal section through the region of developing 5-HT neurons in the raphe nuclei [asterisks designate area shown in (B) at higher magnification]. Note fibers ascending from more caudal (spinal cord) levels. (D) Some of these fibers turn dorsally at the level of the mesencephalic flexure to join the posterior commissure and (E) cross over the surface of the tectum, while (C) others continue rostrally through the anlage of the substantia nigra and into the diencephalon (cluster in lower left section). D, Diencephalon; MF, mesencephalic flexure; SN, substantia nigra; PC, posterior commissure; TE, tectum; AQ, aqueduct of Sylvius; IV, fourth ventricle; Cer, cerebellum; Rb, rhombencephalon; MZ, marginal zone; VZ, ventricular zone. Bars: (A) 200 μm; (B-D) 50 μm; (E) 100 μm. From Lauder *et al.*, 1985a.

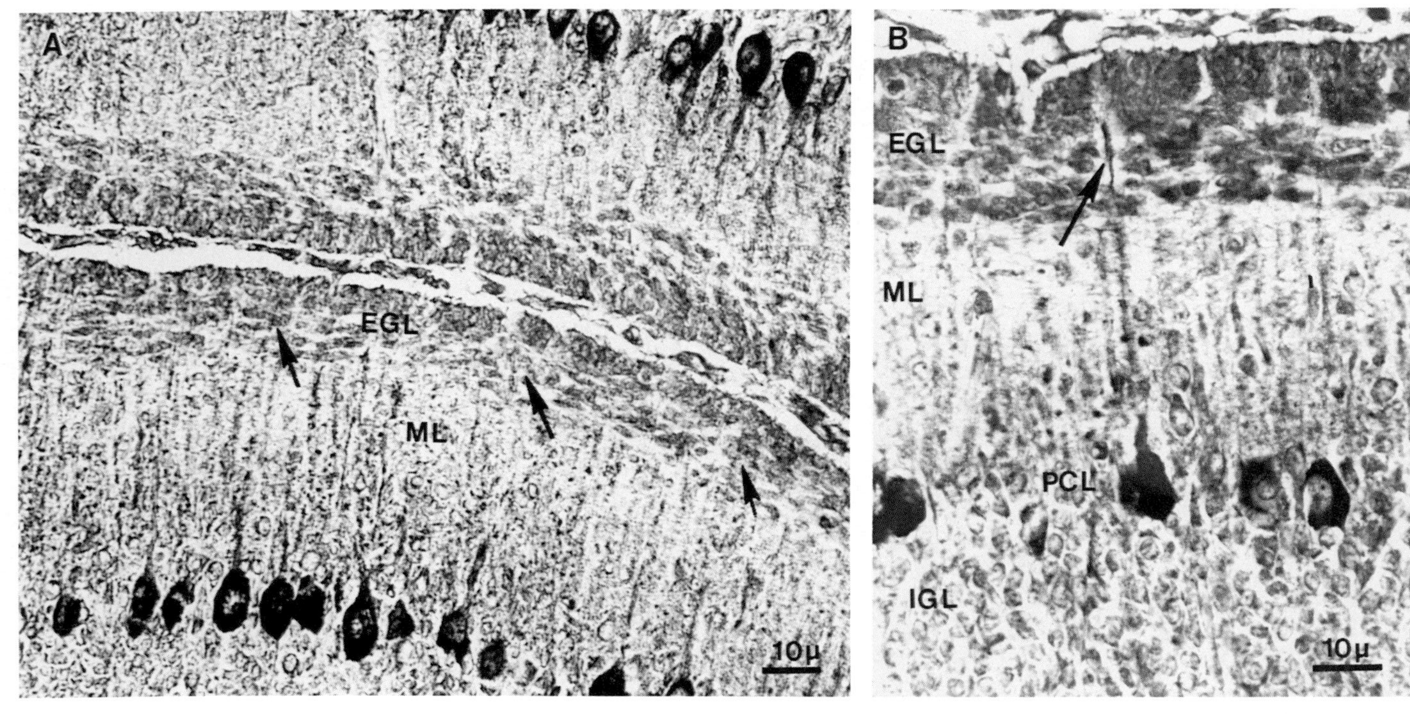

Fig. 11. Immunocytochemistry of glutamate decarboxylase (GAD) in the rat cerebellum at 10 days postnatal (P-10). Note the staining of what appear to be the distal dendrites of Purkinje cells (arrows), which penetrate the external granular layer (EGL). ML, Molecular layer, PCL; Purkinje cell layer; IGL, internal granular layer.

humoral agents are suggested to play key roles in the specification of critical periods by acting both as internal "clocks" to coordinate the timing of developmental processes and as substrates on which other influences such as hormones, drugs, stress, and environmental toxicants could act. In this scheme, neurotransmitters and neurohumors are viewed as an integral link between internal and external factors which influence critical periods for nervous system development.

III. Hormones as Temporal Regulators of Postnatal Neurogenesis

In the hierarchical framework for the humoral and hormonal specification of critical periods illustrated in Fig. 1, hormones are depicted as the most generalized influence on the underlying substrata of ontogenic events and humoral regulatory influences. In this section we discuss the experimental evidence for hormonal regulation of neurogenic events, and in the following section explain how hormones may also influence such processes indirectly through their effects on other developmental signals such as neurotransmitters and neurohumors.

The importance of the hormonal milieu for neurogenesis in the neonatal rat cerebellum and hippocampus has been the focus of a series of studies that have investigated the effects of altered thyroid states or elevated corticosteriod levels on various aspects of cell proliferation and neuronal differentiation. These studies have provided evidence that proper levels of corticosteroids and thyroid hormones are necessary for normal rates of germinal cell proliferation, cessation of cell division, formation of neurons from precursor cells, axonal and dendritic growth, neuronal migration, and formation of the correct number and types of synaptic relationships. Moreover, the different effects of these two hormones on such neurogenic processes suggest that they may act in concert to control the timing of brain development during postnatal ontogeny. The well-established and important role of sex hormones in brain differentiation is the subject of another chapter in this volume (Toran-Allerand) and is therefore not discussed here.

A. Thyroid Hormones as Differentiation Signals

Clinical studies in the 1930s made clear the necessity of thyroid hormones for normal brain growth in humans (Kerley, 1936). Shortly thereafter, more detailed studies in the rat demonstrated the profound influence of altered thyroid states on general somatic and neural development during a critical period of postnatal life consisting of the first few weeks after birth (Salmon, 1936, 1941; Scow and Simpson, 1945; Hamburg and Vicari, 1957; Eayrs, 1961).

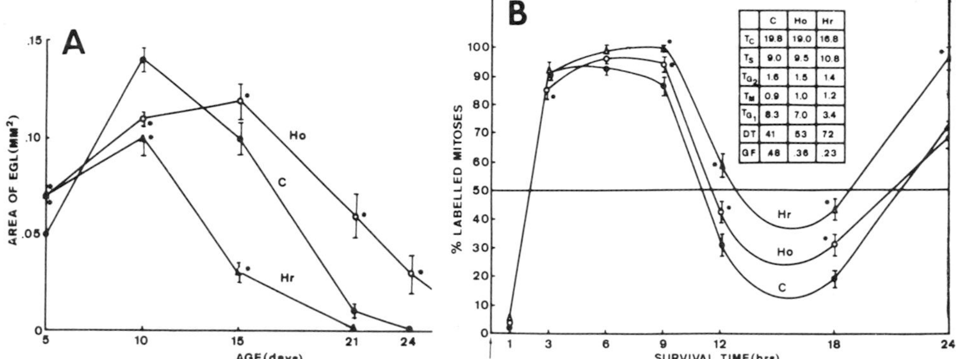

Fig. 12. Effects of hypo- (Ho) and hyperthyroidism (Hr) on cell proliferation in the rat cerebellum. (A) Area of the external granular layer (EGL; pyramis, sagittal plane). (B) Cell cycle (percent labeled mitoses; multiple survival [^{3}H]thymidine autoradiography; pyramis, coronal plane). From Lauder, 1977, by permission of Elsevier.

1. Cell Proliferation in the Cerebellum

In the newborn rat, the external granular layer (EGL), located on the surface of the developing cerebellum, consists of a proliferating population of cells that increase in number during the first postnatal week. This is followed by progressive cessation of cell proliferation and the movement of differentiating neurons from this layer to their final destinations in the molecular layer (ML; stellate and basket cells) and internal granular layer (IGL; granule cells; Altman, 1969).

In 1965, Legrand first reported the retarded growth and disappearance of the EGL in neonatal hypothyroidism. This prompted us to initiate detailed investigations into the influence of altered thyroid states on various aspects of cerebellar development (Nicholson and Altman, 1972a,b,c). In these studies, we confirmed the effects of hypothyroidism on EGL development and found that hyperthyroidism exerted opposite effects leading to accelerated growth of the EGL followed by an early decline and disappearance of this germinal layer (Fig. 12A).

Autoradiographic studies using [^{3}H]thymidine demonstrated that these effects result from premature cessation of cell proliferation in the hyperthyroid EGL and a prolonged period of proliferation in hypothyroidism. Both conditions affect the number of cells formed such that an overall cell deficit results from hyperthyroidism, whereas hypothyroidism causes differential effects on the production of particular classes of EGL-derived neurons. Similar results have been obtained in biochemical (Balázs *et al.,* 1971) and morphological studies (Clos and Legrand, 1973).

To understand these results in terms of cell proliferation and acquisition in the developing cerebellum, cell cycle analyses were conducted in neonatal hypo- and hyperthyroid rats (Lewis *et al.*, 1976; Patel *et al.*, 1979b; Lauder, 1977). Our study (Lauder, 1977), carried out at 10 days postnatal, demonstrated a significantly shorter cell cycle in the EGL of hyperthyroid animals which appeared to result from a shorter G_1 phase (Fig. 12B). (The G_1 phase is that part of the cell cycle following mitosis when neuronal precursor cells make the decision to stop dividing and differentiate into neurons or to reenter the cell cycle for another round of proliferation.) We also observed a decreased rate of cell acquisition in the EGL of these animals, presumably reflecting the premature cessation of cell proliferation.

The shorter cell cycle in our hyperthyroid rats could explain both the increased number of proliferating cells observed in hyperthyroid neonates and the progressive decrease in EGL cell production after the first postnatal week in such animals (Fig. 12A; Nicholson and Altman, 1972a; Legrand *et al.*, 1976; Patel *et al.*, 1976b). This interpretation is based on the hypothesis that EGL cells are required to complete a fixed number of cell cycles prior to ceasing cell division permanently, a point that they would reach more rapidly in such animals.

Hypothyroidism, which produces obvious retardation effects on the growth and disappearance of the EGL (Fig. 12A), paradoxically was not found to reduce the rate of cell proliferation (lengthen the cell cycle), in agreement with Lewis *et al.* (1976). However, we did observe effects on the duration of mitosis itself, a phenomenon that could easily be masked using the type of cell cycle analysis employed in both this study and our own, which is based on the mitotic index. Others have also reported an elevated mitotic index in hypothyroid animals (Rabié *et al.*, 1979). These results indicate that in the EGL of the hypothyroid neonate some cells spend an excess amount of time in mitosis. Further, other results from our study suggest that some of these cells complete extra cycles in a part of the EGL (the subproliferative zone), where most cells are normally postmitotic. These combined effects help to explain the retarded and prolonged growth of the EGL during the first three weeks after birth.

2. *Granule Cell Migration and Parallel Fiber Development in the Cerebellum*

Cerebellar granule cells, the largest contingent of cells formed from the EGL, initiate neurite outgrowth after permanently withdrawing from the cell cycle and passing to the inner part of the EGL (the subproliferative zone). Two axonal processes are first extended parallel to the pial surface, followed sometime later by the emergence of a third, vertically oriented process that descends into the ML towards the IGL. The cell body of the granule cell

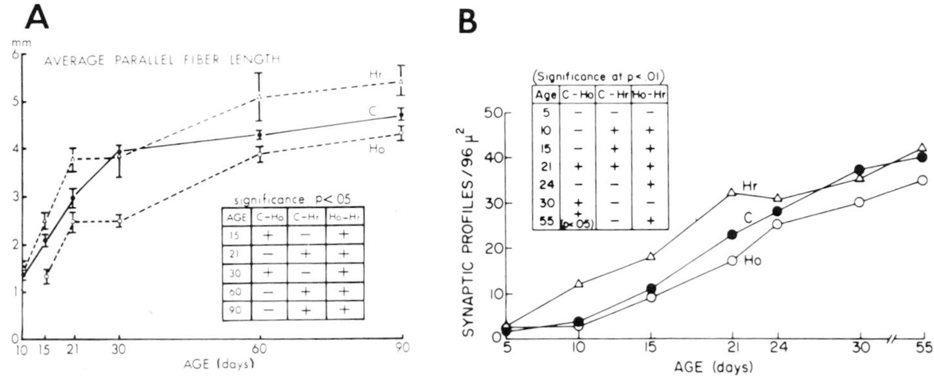

Fig. 13. Effects of hypo- (Ho) and hyperthyroidism (Hr) on axonal growth and synaptogenesis in the molecular layer of the rat cerebellum during the postnatal period. (A) Parallel fiber growth, lobules III-V, coronal plane, lesioning and degeneration method. From Lauder, 1978, by permission of Elsevier. (B) Density of synaptic profiles, pyramis, sagittal plane, ethanolic phosphotungstic acid (E-PTA) method. Revised from Nicholson and Altman, 1972c, by permission of AAAS.

migrates through the ML to the IGL, either by actively moving in an ameoboid fashion (Rakic, 1971) or by translocation of the nucleus within the descending part of the T-shaped parallel fiber growing ahead of it (Altman, 1975).

We studied the effects of altered thyroid states on granule cell migration at 10 days postnatal, including (1) the movement of cells within the EGL (from the proliferative to subproliferative zones), (2) their exit from the EGL into the ML, (3) the rate of their migration to the IGL (Lauder, 1979), and (4) the growth and development of the parallel fibers themselves (Lauder, 1978). Hypothyroidism significantly reduced the rate of granule cell migration following the exit of these cells from the EGL, but did not affect their rate of movement within the EGL or the speed of their exit into the ML. Likewise, this treatment significantly retarded the growth of parallel fibers resulting in a permanent deficit in their length (Fig. 13A).

In contrast, hyperthyroidism accelerated the exit of cells from the EGL into the molecular layer, as well as the rate of granule cell migration to the internal granular layer, but retarded movement of cells within the EGL. Parallel fiber growth was also accelerated, particularly during the time of peak granule cell migration (Fig. 13A). Moreover, the rate of acceleration of granule cell migration was proportional to the increased rate of parallel fiber growth. These observations are supported by the study of Rabié *et al.*, (1979) in which replacement therapy with excess thyroxine in hypothyroid animals also appears to enhance granule cell migration.

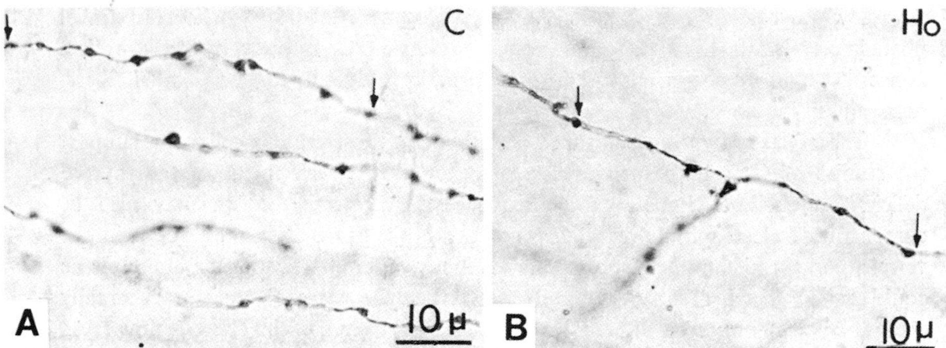

Fig. 14. Parallel fiber varicosities (sites of synapses with Purkinje cell dendrites) as visualized using the rapid Golgi method in 60-day-old control (C) and hypothyroid (Ho) animals. Arrow-to-arrow distance, 50 μm. Note the deficit in varicosities in (B). From Lauder, 1978, by permission of Elsevier.

These studies emphasize the close relationship between parallel fiber growth and granule cell migration in the developing cerebellum and suggest that changes in the rate of migration in hypo- and hyperthyroidism could be secondary to an influence on parallel fiber growth. Furthermore, they lend support to the model of granule cell migration involving translocation of the nucleus within the growing parallel fiber, where this neuritic growth provides the motive force for the migration process.

3. Synaptogenesis in the Cerebellar Cortex

Studies of synaptogenesis in the ML of the developing cerebellum (Fig. 15B) have demonstrated that hypothyroidism significantly retards the developmental increase in density of synapses after 15 days of age, a deficit that remains into adulthood (Nicholson and Altman, 1972b,c; Rebière and Dainat, 1976; Rebière and Legrand, 1970). In addition, an even greater reduction is found in total synapses, since the area of the ML is also reduced. Deficits in the amount of synaptosomal protein in the hypothyroid cerebellum can be restored by replacement therapy with thyroxine, if it is given early (Rabié and Legrand, 1973), but such replacement therapy cannot reverse the synaptic deficit if given after the first 4 weeks following birth (Rabié *et al.*, 1979), indicating a critical period for the dependence of synaptogenesis on proper levels of thyroid hormone.

The number of synaptic varicosities along individual parallel fibers in hypothyroid animals is also greatly reduced, both in terms of the number of sites per unit length as well as in the total number of sites/parallel fiber (Fig. 16; Lauder, 1978). These findings are of particular interest in light of the reports by Lewis *et al.* (1976) and Rabié *et al.*, (1979) that such animals exhibit increased cell death in the IGL, since this could be related to the reduced capacity of their granule cells to make synapses with Purkinje cells (Rebière and Legrand, 1972).

Other defects in the synaptic organization of the hypothyroid cerebellum include the persistence of basket cell axon–Purkinje cell somatic spine synapses after the time of their normal disappearance, decreased density of basket cell axon terminals, increased numbers of Purkinje cell axon collaterals and climbing fibers (Rebière and Dainat, 1976), persistence of climbing fiber–Purkinje cell somatic spine synapses (Hajós *et al.*, 1973), and retarded development of glomeruli (sites of synaptic interactions between mossy fibers, and granule and Golgi cells; Rabié *et al.,* 1977).

In hyperthyroidism, both synaptogenesis and parallel fiber growth are accelerated during the first three weeks after birth (Fig. 15; Nicholson and Altman, 1972b,c; Lauder, 1978). This results in a normal density of synapses in the molecular layer by 24 days, despite profound deficit in the number of granule cells (Nicholson and Altman, 1972a). Since there is no increase in the number of synaptic varicosities per unit length of parallel fiber in these animals, it appears that each parallel fiber succeeds in making more than the normal number of synapses by virtue of its accelerated rate of growth and increased length (up to 1.5 mm greater than controls in adults; Fig. 13A). Nevertheless, there is a drastic reduction in the total number of synapses in the molecular layer (Nicholson and Altman, 1972b,c), since the area of this lamina is significantly smaller, partly as a result of the deficit in the total number of parallel fibers.

4. *Axonal Growth in the Hippocampus*

The granule cells of the hippocampus, like those of the cerebellum, are mainly generated postnatally. For this reason, we have investigated the effects of altered thyroid states on the growth of their axons, the mossy fibers, which synapse with the dendrites of pyramidal cells (Lauder and Mugnaini, 1977, 1980).

Hyperthyroidism, initiated at birth, produces an ectopic bundle of mossy fibers beneath the pyramidal cell layer (Fig. 15B–G) that is not present in normal animals (Fig. 15A). This effect is dose dependent (Fig. 15D–G), has a critical period during the first 3 postnatal weeks, and is permanent (since the ectopic fibers persist long after the treatment has been terminated; Fig. 15C). These ectopic fibers may also be functional since the normal postsynaptic specializations (large spines or "excrescences") usually associated with mossy fiber terminals on apical pyramidal cell dendrites are found on

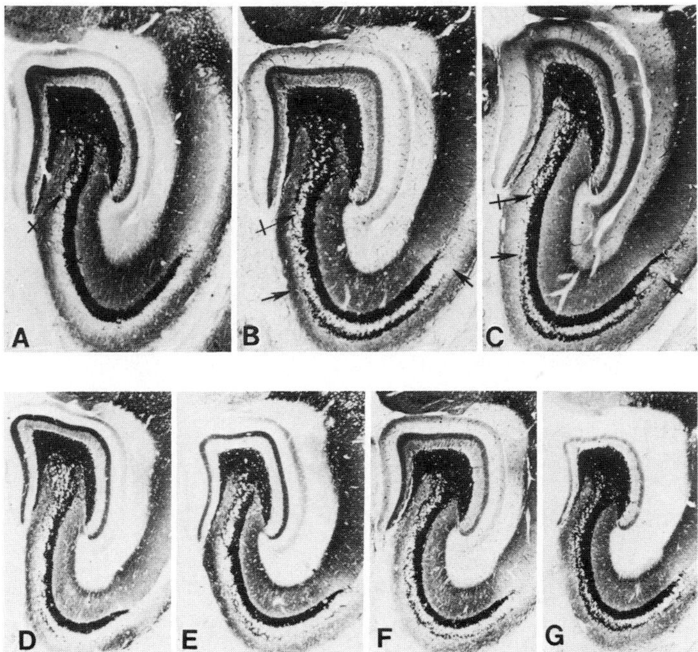

Fig. 15. (A–C) Normal (A) and ectopic hippocampal mossy fibers (uncrossed arrows; B, C) in 30-(A,B) and 90-(C) day-old animals injected with saline (A) or 10 μg 1-thyroxine from days 0 to 15 postnatal (B,C) and stained with the Timm silver sulphide method (mid-septo-temporal level of the horizontally sectioned hippocampus). (D–G) Dose-related effects of daily injections of 5 (D), 8 (E), 10 (F), or 15 (G) μg of 1-thyroxine from days 0 to 15 postnatal on the production of ectopic mossy fibers (mid-septo-temporal level of the horizontally sectioned hippocampus stained by the Timm silver sulphide method). Bars = 1mm. From Lauder and Mugnaini, 1980, by permission of S. Karger.

the basal dendrites of these cells in association with the ectopic mossy fibers and terminals (Fig. 16), which appear normal on electron microscopic examination (Fig. 17).

Hypothyroidism does not alter the distribution of hippocampal mossy fibers, but does reduce the longitudinal areal growth and volume of the developing hippocampus. This may be related to a decreased rate of granule cell acquisition (Rabié *et al.*, 1979).

Together with our parallel fiber study, these results indicate that excess thyroid hormone is capable of stimulating axonal growth in two postnatally developing brain regions, the cerebellum and hippocampus. This may indicate a general ability of thyroid hormone to promote axonal growth throughout

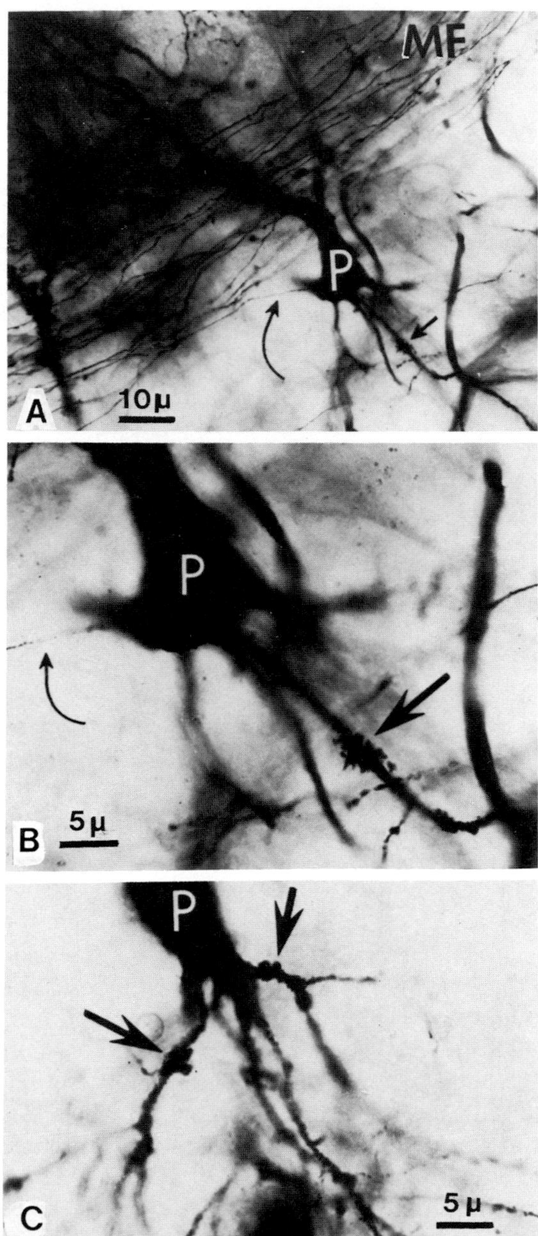

Fig. 16. Rapid Golgi impregnation of a pyramidal cell (P) from field CA3 of Ammon's horn (horizontally sectioned hippocampus at the mid-septo-temporal level) in a hyperthyroid animal (15 μg l-thyroxine, 0–15 days postnatal), with an ectopic excrescence on its basal dendrite (straight arrow). (B) The same pyramidal cell at higher magnification showing the excrescence in more

the postnatally developing brain, where such growth is most active. Alternatively, thyroid hormone may only promote the growth of certain types of axons, such as those of postnatally forming microneurons.

5. *Summary*

These studies demonstrate the necessity of proper levels of thyroid hormone during the postnatal period for normal germinal cell proliferation, formation and migration of neurons, axonal and dendritic growth, and synaptogenesis. It appears that a major function of thyroid hormone in the developing brain may be to act as a timing mechanism for coordination of events involved in the formation of correct cell numbers and the construction of appropriate neuronal circuitry.

B. Corticosteroids as Regulators of Cell Proliferation

Neonatal treatment of rats and mice with corticosteroids decreases brain weight, inhibits DNA synthesis, and reduces DNA content (cell number) in both the cerebrum and cerebellum (Balázs and Cotterrell, 1972; Cotterrell *et al.,* 1972; Howard, 1968; Clos *et al.,* 1975). Unlike treatment with thyroid hormone (Balázs *et al.,* 1971), however, these cell proliferation effects are rapid in onset and essentially reversible. Corticosteroids also may retard later aspects of neurogenesis as evidenced by deficits in development of dendritic spines in cerebral cortex and delayed ontogeny of a variety of electrophysiological and behavioral parameters (Howard, 1968; Clos *et al.,* 1974; Shapiro, 1968; Salas and Shapiro, 1970; Vernadakis and Woodbury, 1971).

These indications that corticosteroids affect cell proliferation and neuronal maturation prompted us to study the effects of postnatal hydrocortisone acetate (HCA) administration on cell proliferation and neuron formation in the cerebellar cortex and hippocampus (Bohn and Lauder, 1978, 1980; Bohn, 1980; for review see Lauder and Bohn, 1980).

1. *Cell Proliferation in the Cerebellum*

Injection of rats with 200 mg HCA on days 1–4 postnatal markedly reduced the growth of the EGL and the mitotic index within this layer during the treatment period (Fig. 18A,B). However, a rebound in cell proliferation

---

detail (straight arrow). (C) Another pyramidal cell from CA3b of the same hyperthyroid animal with more basal dendritic excrescence (arrows). These postsynaptic specializations are presumably the sites of synapses with the ectopic mossy fibers shown in Fig. 17. From Lauder and Mugnaini, 1980, by permission of S. Karger.

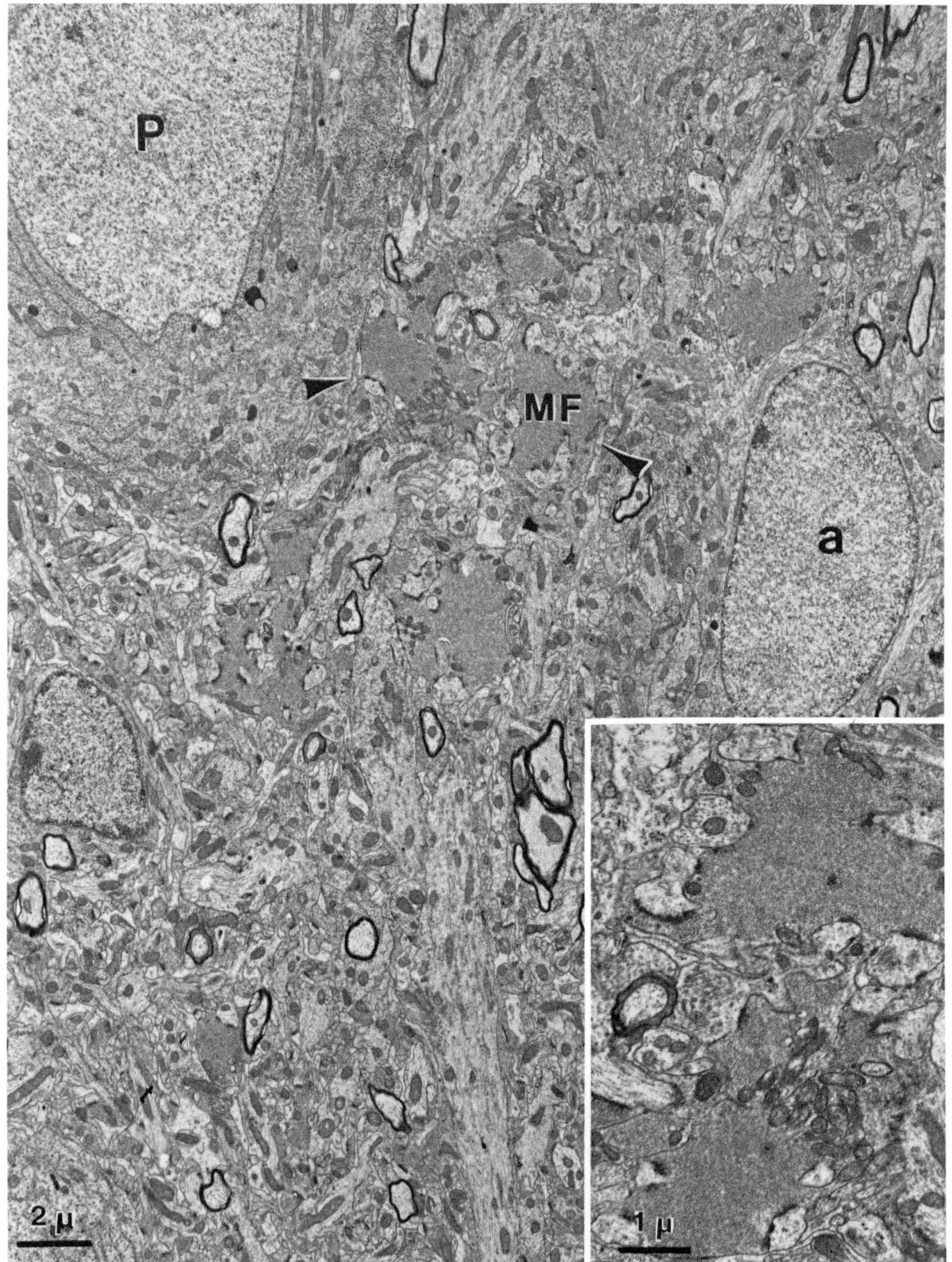

Fig. 17. Hippocampal mossy fiber terminals (MF) synapsing with basal dendritic excrescences in stratum oriens of CA3b in a 30-day-old hyperthyroid animal (10 μg 1-thyroxine, days 0-15 postnatal). P, Pyramidal cell; a, Astrocyte. Bar, 2 μm. Inset: enlargement of synaptic boutons indicated by arrows. Bar, 1μm. From Lauder and Mugnaini, 1980, by permission of S. Karger.

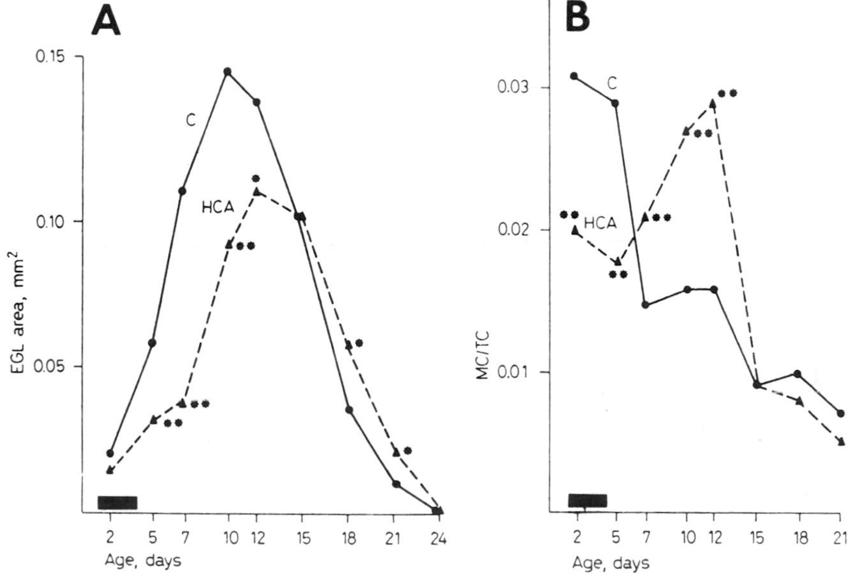

Fig. 18. Effects of hydrocortisone acetate (ACA, days 1–4; bar) on (A) areal development of the cerebellar external granular layer (EGL) and (B) the mitotic index in the EGL. MC/TC, = mitotic cells/total cells ($p < .05$); **, $p < .01$ (multiple F test). Pyramis, sagittal plane. Modified from Bohn and Lauder, 1978, by permission of S. Karger.

occurred following termination of the treatment, as shown by the elevated mitotic index (Fig. 18B). In spite of this rebound, the total numbers of cells produced by the EGL did not fully recover and the time of disappearance of the EGL was unchanged (Fig. 18A). As a consequence, total numbers of all EGL-derived cells (granule, stellate, and basket cells) were reduced to varying degrees.

During the treatment period, a certain proportion of granule cells (as well as basket and stellate cells) ceased proliferation early, whereas during the period of rebound following termination of the hormone treatment, fewer EGL cells stopped dividing.

We interpret these results to mean that HCA treatment reversibly inhibits cell proliferation in the EGL. However, for a small proportion of cells, possibly those approaching their last cell divisions, this inhibition causes them to cease proliferation prematurely. To test this, we injected HCA during a later part of the postnatal period (days 7–18) in an attempt to "catch" more granule cells at this critical point. As predicted, this later HCA treatment produced early final cell divisions for a much larger proportion of granule cells. Furthermore, the disappearance of the EGL was accelerated, presumably reflecting a more permanent inhibition of cell proliferation.

2. *Granule Cell Genesis in the Hippocampus*

In rats treated with HCA on days 1–4 postnatal, granule cell proliferation in the hilus of the dentate gyrus was suppressed during and immediately following treatment, particularly in the ventral hippocampus (Bohn, 1980). As in the cerebellum, a rebound of cell proliferation occurred following termination of the treatment, but only after a delay of 1 or 2 days, which was most pronounced in the ventral hippocampus.

This more severe and less reversible inhibition of cell proliferation in the hippocampus, as compared to the cerebellum, might be due to the higher concentrations of steroid receptors in hippocampal granule cells (McEwen *et al.*, 1975; Waremburg, 1975). In addition, these receptors may have a higher binding capacity in the ventral hippocampus at birth due to a spatiotemporal gradient in their development (Table VI).

In contrast to the effects of early HCA treatment on granule cell genesis in the cerebellum, no early last cell divisions of granule cells were observed in the hippocampus. This could be due to the fact that at the time the treatment was given, only a small number of granule cells were being generated. This would mean that only these few cells would be nearing their last cell divisions and thus be vulnerable to permanent inhibition of proliferation by HCA. Additionally, the delayed, but strong rebound in cell proliferation after termination of HCA treatment could explain the delayed cessation of cell proliferation in the majority of granule cells in these animals. It would be interesting to see if later HCA injections could "catch" a larger proportion of cells at the point where they might be induced to stop dividing prematurely, as observed in the cerebellum.

C. Thyroid Hormones and Corticosteroids as Joint Coordinators of Postnatal Neurogenesis

The studies summarized above and reviewed elsewhere (Lauder and Bohn, 1980) indicate that thyroid hormones and corticosteroids have different effects on postnatal neurogenic events such as germinal cell proliferation, neuronal genesis, and differentiation. Corticosteroid treatment produces a reversible inhibition of cell proliferation, except in cells nearing their last divisions, whereas excess thyroid hormone stimulates cell proliferation just after birth, but results in early cessation of cell division later in the post natal period. In addition, thyroid hormones seem to regulate the rate of later aspects of neuronal differentiaiton such as cell migration, axonal and dendritic growth, and synaptogenesis.

In the neonatal rat, plasma levels of thyroid hormones and corticosteroids are changing and receptors for these hormones are developing (Tables IV–VI). Moreover, ontogeny of the diurnal corticosterone rhythm is reportedly

Table IV

Brain Thyroid Hormone Receptors: Developmental Characteristics

Nuclear receptors

1. Receptors present in fetal brain (E20) with a *binding capacity* approximating that in the adult (Schwartz and Oppenheimer, 1978)
2. *Density* of cerebral T_3 receptors is high at birth, then falls to adult levels by the end of second week (Naidoo *et al.*, 1978; Valcana and Timiras, 1978)

   In whole brain (including cerebellum) peak occurs at 2 days postnatal, then declines (Schwartz and Oppenheimer, 1978)
3. T_3 *binding capacity* of cerebral cells peaks at 2 days postnatal, then falls to adult levels by 4 weeks

   In cerebellum, binding capacity does not peak until 14 days postnatal (Schwartz and Oppenheimer, 1979)
4. *Affinity constant* of cerebral T_3 binding peaks between 6 and 20 days postnatal, then falls (Dozin-van Roye and De Nayer, 1979; Valcana and Timiras, 1978)
5. Development of nuclear T_3 in whole brain occurs prior to increase in serum T_3 levels (Schwartz and Oppenheimer, 1978)

Mitochondrial receptors

Only demonstrable in brains of rats 12 days postnatal or younger (Sterling *et al.*, 1978)

Nonmitochondrial cytosol receptors

Affinity for T_3 highest at day 10 in rat cerebellum (peak of EGL growth) (Geel, 1977)

influenced by different thyroid states (Lengvári *et al.,* 1977; Meserve and Leathem, 1974; Poland *et al.,* 1979; Taylor and Lengvári, 1977). Based on the effects of thyroid hormones and glucocorticoids on postnatal neurogenesis, and the apparent relationship between development of the thyroid- and adrenal-pituitary axes, it is hypothesized that the changing ratio

Table V

Concentration of Thyroid Hormones and Their Metabolites in Brain and Plasma: Developmental Characteristics

1. *Higher levels of* $[^{131}I]T_3$ in brains of neonates than adults (Cohan *et al.*, 1967)
2. *Serum concentration* T_4:
   a. Peaks at day 2 postnatal, falls, then rises to a second peak on days 14 to 16 (Samel, 1968)
   b. Rises to a peak on day 16 (Clos *et al.*, 1974)
3. *Serum concentration* T_3 rises to adult levels by day 12 (Schwartz and Oppenheimer, 1978)
4. *Rate of metabolism* T_3 most rapid, days 11–23 postnatal, falling to low levels thereafter (Naidoo and Timiras, 1979)
5. *Formation of TRIAC* (triiodothyroacetic acid) from T_4 and deiodination of T_3 occurs mainly during early postnatal period (role for TRIAC in brain development?) (Naidoo and Timiras, 1979)

Table VI

Hippocampal Glucocorticoid Receptors: Developmental Characteristics

1. *Concentration* of *cytosol receptors* for corticosterone rises from low levels at birth to adult levels by 32 days (Olpe and McEwen, 1976)
2. *Binding capacity* of *nuclear receptors* for corticosterone is much lower during first postnatal week than in adult; hippocampus first shows selective nuclear binding at day 9 (peak of granule cell genesis; Schlessinger *et al.*, 1975) (i.e., nuclear binding exceeds whole tissue levels; Turner, 1978)
3. *Gradients* of nuclear *binding capacity* ([^{3}H]corticosterone autoradiography) match developmental gradients for pyramidal and granule *cell genesis* (Turner, 1978)

of endogenous levels of corticosteroids and thyroid hormones and the development of their respective receptors during the first few weeks after birth may constitute important coordinating influences on the rates of cell proliferation and neuronal differentiation in the cerebellum and hippocampus. In this sense, such an hormonal timing mechanism could help to regulate the duration of "critical periods" for postnatal brain development.

IV. Hormonal–Humoral Interactions

Hormonal influences on neurotransmitter systems are well documented in the adult nervous system, particularly with respect to the monoamines. Such relationships have also been found in the developing peripheral and central nervous systems. Together with indications that certain neurotransmitters or neurohumors are able to exert their developmental influences, these hormonal–humoral interactions form the basis for our model of the specification of critical periods by neurotransmitters, neurohumors and hormones (Fig. 1). In this section we discuss evidence for such hormonal–humoral interactions with respect to thyroid hormones, glucocorticoids, and the monoamines.

A. Thyroid–Monoamine Relationships

Thyroid–monoamine interactions during the postnatal period have been inferred from the effects of neonatal hypo- or hyperthyroidism on monoamine levels, the activity of their synthesizing enzymes, their metabolism, and the morphological development of the neurons that contain them, both in the periphery (Gresik, 1976; Lau and Slotkin, 1979) and the CNS. In addition, serotonin (5-HT) and norepinephrine (NE) have been reported in the thyroid gland, where they could play a functional role (Gershon and Nunez, 1976).

In the developing frog brain, proper levels of thyroid hormone are required for the formation of monoaminergic neurons in the preoptic recess organ during metamorphic climax as well as for the ontogeny of monoamine terminals in the median eminence (Kikuyama *et al.,* 1979). Moreover, this regulation seems to require an intact pituitary, since hypophysectomy of tadpoles at the tail-bud stage prevents replacement therapy with thyroxine from ameliorating the deficits caused by depriving such animals of their thyroidal primordium. Moreover, in experiments with intraocular grafts of fetal rat substantia nigra (dopamine) and raphe (5-HT) have demonstrated thyroxine is required for normal neurite outgrowth from these monoamine neurons as demonstrated using the Falch-Hillarp fluorescence histochemical method (Granholm and Seiger, 1981; Granholm *et al.,* 1984).

Although it is not known whether thyroid hormone is required for the actual formation of monoamine neurons in all vertebrates, it is clear that thyroid state can influence the metabolism of monoamines in the postnatally developing rat brain. In an extensive review, Singhal *et al.,* (1975) summarized data indicating that proper levels of thyroid hormone are required for the normal synthesis and degradation of 5-HT, and dopamine (DA) in the postnatally developing rat brain. In later work, they demonstrated dose-dependent effects of replacement therapy with triiodothyronine (T_3) on this metabolism, indicating a possible direct influence of thyroid hormones on the postnatally developing monoamine systems (Rastogi *et al.,* 1976) as well as regional differences in the effects of hypo- and hyperthyroidism (Rastogi and Singhal, 1976, 1979). Of further note is the theory of Dratman (1978) that T_3 actually participates in catecholamine metabolism in nerve terminals by acting either as a false neurotransmitter or as a precursor for catecholamine synthesis by virtue of its derivation from and close structural relationship to tyrosine.

Serotonin, on the contrary, has been reported to be increased throughout the brain of the neonate made hypothyroid with propylthiouracil, whereas neonatal hyperthyroidism is reported to have no effect on brain levels of 5-HT. The increased 5-HT levels in hypothyroidism are interpreted to derive from a decreased turnover rate as well as an enhanced reuptake into synaptic terminals (Schwark and Keesey, 1975, 1978). Relevant to these findings are the reports by Vaccari that neonatal hypothyroidism leads to decreased ^{3}H-imipramine binding (5-HT uptake sites) in cerebral cortex and striatum (Vaccari, 1985) and that such hypothyroid rats are behaviorally hyporeactive to loading with the 5-HT precusor 5-HTP (Vaccari 1982). In addition, neonatal hypothyroidism appears to lead to reduced MAO-A activity (Vaccari *et al.,* 1983).

Thus, it appears that neonatal thyroid states do affect developing monoamine systems in various ways. Moreover, the ability of hypo- and hyperthyroidism to affect these transmitters occurs during a critical period of postnatal development within the first 3 weeks after birth, since replacement therapy with T_3 does not reverse changes caused by hypothyroidism after this time (Schwark and Keesey, 1978; Rastogi and Singhal, 1978, 1979).

B. Corticosteroid–Monoamine Relationships

An intact adrenal gland is required for normal ontogeny of the monoamine systems in both the brain (see below) and periphery (Bohn *et al.,* 1981) during a critical period of postnatal development. During the first few postnatal weeks in the rat, corticosteroids in plasma and brain fall from relatively high levels at birth to low levels around days 6 to 9. This "trough period" is then followed by a rise to adult levels between days 9 and 21 (Sze *et al.,* 1976). Sze and co-workers have demonstrated that the developmental rise in the 5-HT synthesizing enzyme tryptophan hydroxylase (TPH), normally occurring during this same period (9–21 days postnatal), requires an intact adrenal gland (Sze *et al.,* 1976; Sze, 1976, 1980). Further, replacement therapy with corticosterone restores the levels of TPH to normal during this period. Finally, corticosterone given to intact neonatal rats increases TPH activity, but only during a restricted critical period (6–12 days). These results indicate that normal TPH activity in the developing rat brain is dependent on the changing postnatal patterns of plasma and brain corticosteroids which rise from low levels at the end of the first postnatal week to adult levels by the end of the third week.

Effects of corticosteroids on tyrosine hydroxylase activity during postnatal development have likewise been demonstrated, indicating that the activity of this catecholamine synthesizing enzyme in brain (Diez *et al.,* 1977) or superior cervical ganglion (Markey and Sze, 1981) can also be altered by corticosteroid or adrenalectomy injections during a critical period. Moreover, corticosteroid injections during this period increase brain catecholamine content (Costa *et al.,* 1974). Of particular interest in this regard is the finding that the cells of the locus coeruleus bind [^{3}H]corticosterone during the postnatal period and adulthood, suggesting that glucocorticoids could play a direct role in the development and functional activity of this noradrenergic cell group (Towle *et al.,* 1982).

C. Thyroid–Corticosteroid Interactive Effects on Developing Monoamine Systems

Interactive effects between thyroid hormones and corticosteroids on the postnatally developing monoamine systems are not surprising in view of the interrelationships of these two hormonal systems discussed in Section III C.

Such interactions raise the possibility that changing levels of these hormones and their receptors in the postnatally developing brain could serve to coordinate spatio-temporal relationships between different neurogenic events.

With regard to the relationship of such interactions to developing monoamine systems, Lengvári *et al.*, (1980) have demonstrated that thyroxine (T_4) given on days 8 to 10 postnatal increased NE content in the hypothalamus and mesencephalon between days 16 and 21, increased DA content in the hypothalamus, and decreased DA levels in the mesencephalon during the same period. This is in contrast to the effects of corticosterone administration from 8 to 10 days, which decreases the NE content of the hypothalamus and mesencephalon as well as hypothalamic DA, but increases DA levels in the mesencephalon. Most interestingly, thyroxine given on days 8 to 10 in conjunction with corticosterone counteracts all of the effects of the corticosterone alone.

It is perhaps noteworthy in this regard that neonatal hyperthyroidism decreases the activity of tyrosine hydroxylase and the catecholamine content in the adrenal gland, whereas hypothyroidism increases both parameters (Gripois *et al.*, 1980). As mentioned previously, the diurnal corticosterone rhythm is also changed by these altered thyroid states (Lengvári *et al.*, 1977; Meserve and Leathem, 1974; Poland *et al.*, 1979; Taylor and Lengvári, 1977). Thus, it seems likely that a balance between circulating levels of thyroid hormones and corticosteroids during the neonatal period may be central to the proposed coordinating action of these hormones on the changing spatio-temporal relationships during postnatal brain development.

V. Drugs, Stress, and Other Environmental Influences on the Humoral and Hormonal Milieu of the Developing Nervous System: Implications for Their Teratogenic Effects

A. Neural Tube Defects

In a detailed review, Jurand (1980) summarized the various malformations occurring in mouse embryos whose mothers had been treated with different psycho-active drugs, mainly the tricyclic anti-depressants, during pregnancy. While the drug doses are comparatively higher than those normally administered to the human female, the effects may be significant when viewed in the context of the sites of 5-HT accumulation we have described in the mouse embryo (see Section II B), as well as the sites of 5-HT and NE uptake previously found in the chick embryo.

For example, in the study of Jurand, desmethylimipramine, a NE uptake inhibitor, and imipramine, a 5-HT uptake inhibitor, both produced exencephaly, kinking of the spinal cord and hydromyelia. Triimipramine, chlorimipramine, amitriptyline and chlorpromazine all caused cranioraschisis and myeloschisis (failure of neural tube closure in the regions of the rostral and caudal neuropores, respectively) with amitriptyline, a 5-HT uptake blocker, being most potent in causing cranioraschisis and myeloschisis. It should also be noted that in a human study (Idänpään-Heikkilä and Saxén, 1973) a number of women who received amitriptyline or imipramine and chloropyramine together during the first or last trimester of pregnancy gave birth to children with neural tube defects or craniofacial abnormalities. Of further interest with regard to these apparent effects of amitriptyline is the report that this drug promotes the release of 5-HT from mast cells in the rat (Theoharides *et al.,* 1982), while at the same time inhibiting the release of histamine. Thus, one possible route of action of this tricyclic antidepressant on embryogenesis, at least in the mouse embryo, could be by the release of 5-HT from mast cells in the vicinity of the sites of 5-HT accumulation we have observed, while at the same time blocking its uptake into these sites.

In studies on chick embryos, Palén *et al.,* (1979) have analyzed the effects of 5-HT interactive drugs or 5-HT itself on neural tube development. They found that 5-HT antagonists, 5-HT synthesis inhibitors, or MAO-inhibitors placed in the incubation medium of embryos in culture produced morphogenetic defects of one kind or another, including distrubances in blasto-dermal expansion, primitive streak, somite and heart formation, neural tube closure, and brain formation. Interestingly, 5-HT itself was the most potent teratogen of those studied.

Although we cannot draw any clearcut conclusions from such teratogenic studies, it is worth noting that exogenously administered monoamine transmitters or drugs that alter their synthesis, action, or metabolism can produce malformations in the same regions of the neuraxis where Wallace (1979, 1983), Lawrence and Burden (1973), and Kirby and Gilmore (1972) reported the appearance of monoamine histofluorescence either endogenously or following incubation of embryos with these substances. Although far from definitive, such correlative information raises the possibility that during normal neurogenesis certain regions of the neuraxis concentrate monoamines provided by some extraneural source such as the yolk, yolk sac, mast cells, or adjacent tissues such as the notochord or gut. These events could be critical for normal neural tube development and provide a mechanism whereby monoamine-interactive drugs could interfere with this early neurogenic process.

B. Brain Organogenesis and Behavior

We have already discussed the effects of maternal administration of a 5-HT synthesis inhibitor, *p*-chlorophenylalanine (pCPA), during the second trimester of pregnancy on the time of neuronal genesis in the embryonic rat brain (Section II B). Although no similar studies yet exist using clinically relevant drugs, it is possible that such drugs could also interfere with the genesis of 5-HT target neurons, if they changed levels of 5-HT or its ability to interact with its target cells. Such studies are planned for the future and will include biochemical and immunocytochemical analyses of the effects of such treatments on 5-HT levels and the development of 5-HT neurons within the embryonic brain.

Studies employing monoamine depleting drugs such as 6-hydroxydopamine (6-HODA) and reserpine have been carried out in postnatal animals by other laboratories. Although they are not conclusive due to the possibility of direct cytotoxic effects, Patel, Lewis, and co-workers (Lewis *et al.,* 1977a; Patel *et al.,* 1977) found that administration of reserpine from birth until 11 days postnatal produced changes in [^{3}H]thymidine incorporation in forebrain and cerebellum, decreased glial cell proliferation in the subependymal layer of the lateral ventricle, and caused a decreased mitotic index in the EGL of the cerebellum. Noting degenerating cells in such regions of cell proliferation, however, these workers interpreted their results as being due largely to toxic side effects. In a more recent study (Patel *et al.,* 1979a) using lower doses of reserpine, such side effects were in part ruled out, and a dose-dependent decrease in DNA synthesis was still observed. Therefore, these studies lend support to the hypothesis that monoamines may play a role in the control of postnatal brain cell proliferation. This possibility is further enhanced by our findings of interactions between the axons of 5-HT neurons and dividing glial or neuronal precursors in the postnatally developing cerebellum and hippocampus (Figs. 7–8).

In other studies employing 6-HODA administration to neonates, defects in granule cell migration, large deficits in granule cell numbers, abnormal foliation and fissurization, and disorientation of Purkinje cell dendritic trees have been found in the developing cerebellum (Berry *et al.,* 1980; 1981; Lovell, 1981; Sievers *et al.,* 1981). Again, it is not clear to what degree the effects of this treatment are the result of cytotoxicity of the drug itself rather than resulting from the removal of noradrenergic axons. However, Lovell (1981) used much lower doses of 6-HODA than did other investigators and no evidence of cytotoxicity could be found. Although it is possible that some Purkinje cell dendritic tree anomalies could be due to the absence of a large part of the granule cell population, the effects on the granule cells themselves are interesting, especially if it can be demonstrated that NE fibers are in locations appropriate for such effects during cerebellar development. If their distribution is at all analogous to that observed in our 5-HT studies (Section II,A; Fig. 7), the possibilities for such interactions would be strengthened.

It should be noted in this regard that in their study of the NE innervation of the chicken cerebellum, Mugnaini and Dahl (1975) described a rich plexus of fibers in both the IGL and ML. If NE fibers are present in these same locations in the rat or in contact with EGL during granule cell formation and migration, the effects of 6-HODA described above, might be more easily interpretable. Further, Lidov and Molliver (1979) reported that injection of the rat fetus with 6-HODA on day 17 of gestation produced some abnormalities in cell migration in cerebral cortex, such that distinct foci of ectopic neurons were found, although no other changes could be detected and this abnormality could be the result of drug toxicity. Moreover, these 6-HODA injections were done several days after the arrival of NE fibers in the cortex, which may be too late to produce clearcut effects on earlier cortical neurogenic events (Schlumpf *et al.,* 1980).

There is also evidence for a trophic influence of NE fibers on postnatal dendritic growth. Maeda *et al.,* (1974) reported that neonatal ablation of the locus coeruleus resulted in an immature dendritic pattern for cerebral cortical pyramidal cells of layer VI in animals sacrificed as young adults. In another study, Amaral *et al.,* (1975) found that the dendrites of modified pyramidal cells in field CA4 of the hippocampus were abnormally long in adult rats treated with 6-HODA as neonates. However, these results were not confirmed by Wendlandt *et al.,* (1977). Based on these and other data, Berry *et al.,* (1980) speculated that monoamine fibers may promote dendritic outgrowth prenatally. Finally, Pettigrew and associates (Katsamatsu and Pettigrew, 1979; Katsamatsu *et al.,* 1979) have reported that plasticity of the developing cat visual cortex seems to be influenced by NE, since 6-HODA treatment of newborn kittens prevents the shift in ocular dominance normally seen after neonatal monocular deprivation (see chapter by Singer, this volume). Microperfusion with NE restores this plasticity in the immediate vicinity of the perfusion. Microperfusion of NE also restored this plasticity in monocularly deprived animals who presumably were past the critical period for ocular dominance shifts. See recent review by Berger and Verney (1984).

Vorhees *et al.,* (1979) analyzed the potential of a number of clinically relevant drugs to act as behavioral teratogens when given to rats during the second and third trimesters of pregnancy. Of these drugs, fenfluramine, an anorectic, caused a variety of behavioral anomalies including hyperactivity and delayed development of swimming, as well as a small decrease in brain weight. Since administration of this drug decreases 5-HT levels in the adult brain (Duhault and Boulanger, 1977), it might have a similar effect in the developing brain. If so, this treatment could be analogous to our study indicating maternal pCPA administration affected development of 5-HT target cell populations (Section II,B). It is conceivable that the behavioral consequences of pCPA and fenfluramine might both be related to alterations in the developing 5-HT system. Coyle *et al.,* (1976) stressed the point that psychoactive drugs seem to be the most potent of the behavioral teratogens. This may be due to their interference with the development of the same neurotransmitter system they interact with in the adult, the monoamines, especially if these transmitters act as developmental signals.

C. Effects of Drugs on Developing Monoamine Systems

When given to the developing rodent, certain *5-HT agonists* or *antagonists* produce changes in brain levels of 5-HT in these animals as neonates or adults. For example, regional changes in 5-HT concentration and depletability by pCPA have been noted in adult offspring of rats treated with imipramine during gestation and the suckling period. A similar effect was also found after identical treatment with methylamphetamine, chlorpromazine, or phencyclidine (Tonge, 1974). Likewise, increases in whole brain 5-HT levels occur in 15-day-old rats treated with cyproheptadine or methysergide during the second postnatal week (Hole, 1972a) as well as in newborn offspring of rats treated with imipramine on days 18 to 21 of gestation (Nair, 1974). Moreover, chlorpromazine administered on days 1 to 6 postnatal also elevates brain 5-HT levels at 20 or 60 days (Taub and Peters, 1978). Permanent decreases in brain 5-HT have been found in adult offspring of mothers treated with LSD on days 1–2, 9–12, 13–15, or 18–21 of gestation, with the most pronounced effects occuring in offspring from mothers treated during the earliest periods. These effects on 5-HT levels could reflect changes in the metabolic machinery for 5-HT synthesis or degradation as well as changes in the number of 5-HT terminals. Such alterations could have consequences for brain development, if 5-HT acts as an ontogenic signal for neuro- or gliogenesis.

Effects of maternal treatment with these drugs on the brain weights of offspring have been reported (Hole, 1972a,b; Hoff, 1976). Especially noteworthy in the present context is the finding that LSD treatment of pregnant mice on day 12 or 18 of gestation produces increased and decreased weight of the mesencephalon and pons–medulla, respectively. This could be significant in light of our findings regarding the effects of maternal pCPA treatment on cell proliferation in the embryonic brain (Section II,B).

The problem of effects of *addictive drugs* such as the opiates on prenatal brain development has been documented in animals (Zagon and McLaughlin, 1977, 1983). Often, addicted mothers are maintained on methadone during pregnancy in an attempt to avoid opiate use. Studies (McGinty and Ford, 1980; Slotkin *et al.,* 1979) have indicated that treatment of the pregnant rat with methadone during the last trimester can retard brain growth, decrease synaptosomal uptake of monoamines, alter catecholamine synthesis, and reduce levels of catecholamines in forebrain regions (with no effects on brain stem levels). These results indicate decreased numbers of monoamine (particularly catecholaminergic) nerve terminals in the brains of offspring of methadone-treated mothers, accompanied by an overall retardation in brain development.

The effects of catecholamine-depleting drugs and antagonists administered to the pregnant rat on receptor development in the offspring have also been explored (Rosengarten and Friedhoff, 1979). In this study, α-methyl-*p*-tyrosine (α-MT) or haloperidol given to the mother from days 8 to 20 of gestation decreased the specific binding of [^{3}H]spiroperidol in the caudate nucleus of offspring and decreased sensitivity to apomorphine in the production of stereotyped behavior. Assuming that α-MT blocks dopamine synthesis and haloperidol acts as a dopamine antagonist in the embryonic caudate, these results suggest that dopamine may be an important factor in the embryonic development of receptors. A similar situation might exist with respect to 5-HT and its receptors.

For an excellent summary of interactions of drugs with developing neurotransmitter systems, see Yanai (1984).

D. Stress and Developing Monoamine Systems

Moyer *et al.,* (1978) reported that stressing the pregnant rat produces changes in catecholamine levels in discrete brain regions of the offspring that are different for males and females and appear to correlate with changes in sexual behavior in the two sexes, especially the feminizing effect of stress on male offspring.

Giulian *et al.,* (1974) found that a single disruptive experience (cold stress) on day 1 after birth changes the activity of rats in an open field and decreases their reaction to handling stress. These behavioral changes are associated with elevated 5-HT levels in the forebrain and midbrain during the third postnatal week. Interestingly, these alterations seem to occur during a critical period, since treatment of animals on day 10 does not produce these effects. Adrenal involvement has been ruled out by adrenalectomy or corticosterone administration, neither of which seem to influence the effects of this stressful experience on 5-HT levels.

This latter finding is of particular interest, since Griffin *et al.,* (1978) have shown that neonatally stressed rats exhibit a decreased capacity for DNA synthesis in the cerebellum for a short period following the stressful experience. Since corticosteroid involvement appears to have been ruled out in the above study, it is tempting to speculate that such an effect could be related to increased levels of 5-HT in the cerebellum which might inhibit cell proliferation. The effects of "stress" on development of 5-HT target cells in the embryonic brain is discussed in Section II,B.

VI. Summary and Conclusions

In this chapter we have discussed the concept of *critical period* in the context of an hierarchical scheme of neural development in which neurotransmitters, neurohumors, and hormones orchestrate the scenario of fundamental pre- and postnatal neurogenic events. The literature reviewed provides evidence for the involvement of neurotransmitters in neural tube formation and brain organogenesis, for the hormonal regulation of cell proliferation and differentiation during postnatal neurogenesis and for key hormonal–humoral interactions in the developing nervous system. The possibility that drugs and stress act as behavioral teratogens primarily through their effects on this hormonal–humoral milieu is also discussed with respect to the known critical periods for such behavioral alterations.

It appears likely, therefore, that neurotransmitters, neurohumors, and hormones may play integral roles in the regulation of those fundamental ontogenic events that provide the basis for critical periods in nervous system development. In this sense it seems reasonable to propose that this hormonal–humoral milieu helps to "specify critical periods" in terms of both their temporal framework and their sensitivity to external stimuli or environmental insults.

Acknowledgments

Supported by NIH grants NS-13481, NS-15706, and NS-09904. We would like to thank Pamela Henderson and Anthony DiNome for technical assistance and Kathrinn Plemmons for typing the manuscript. We are also grateful to Dr. James Wallace for helpful discussions and written contributions to portions of this manuscript. Jean Lauder is the recipient of NIH Research Career Development Award, NS-00507.

References

Aghajanian, G. K., Kuhar, M. J., and Roth, R. H. (1973). *Brain Res.* **54,** 85–101.
Ahmad, G., and Zamenhof, S. (1978). *Life Sci.* **22,** 963–970.
Allan, I. J., and Newgreen, D. F. (1977). *Am. J. Anat.* **149,** 413–421.
Altman, J. (1969). *J. Comp. Neurol.* **136,** 269–294.
Altman, J. (1975). *J. Comp. Neurol.* **163,** 427–448.
Altman, J., and Bayer, S. A. (1978a). *J. Comp. Neurol.* **182,** 945–972.
Altman, J., and Bayer, S. A. (1978b). *J. Comp. Neurol.* **182,** 973–994.
Altman, J., and Bayer, S. A. (1978c). *J. Comp. Neurol.* **182,** 995–1016.
Amaral, D. G., Foss, J. A., Kellogg, C., and Woodward, D. J. (1975). *Soc. Neurosci. Abstr.* **1,** 789.
Anderson, C. H. (1978). *Brain Res.* **154,** 119–122.

Baker, P. C. (1965). *Acta Embryol. Morphol. Exp.* **8**, 197-204.
Balázs, R., and Cotterrell, M. (1972). *Nature (London)* **236**, 348-350.
Balázs, R., Kovács, S., Cocks, W. A., Johnson, A. L., and Eayrs, J. T. (1971). *Brain Res.* **25**, 555-570.
Bartels, W. (1971). *Z. Zellforsch. Mikrosk. Anat.* **116**, 94-118.
Berger, B. and Verney, C. (1984). *In* "Monoamine Innervation of Cerebral Cortex", pp.95-125. Alan R. Liss, N.Y.
Berry, M., McConnell, P., and Sievers, J. (1980). *Curr. Top. Dev. Biol.* **15**, 67-101.
Berry, M., Sievers, J., and Baumgarten, H. G. (1981). *Prog. Brain Res.* **53**, 65-92.
Black, I. B. (1973). *J. Neurochem.* **20**, 1265-1267.
Black, I. B. (1978). *Annu. Rev. Neurosci.* **1**, 183-214.
Black, I. B., and Geen, S. C. (1973). *Brain Res.* **63**, 291-302.
Black, I. B., and Geen, S. C. (1974). *J. Neurochem.* **22**, 301-306.
Black, I. B., and Mytilineou, C. (1976a). *Brain Res.* **101**, 503-521.
Black, I. B., and Mytilineou, C. (1976b). *Brain Res.* **108**, 199-204.
Black, I. B., and Patterson, P. H. (1980). *Curr. Top. Dev. Biol.* **15**, 27-40.
Black, I. B., Hendry, I. A., and Iversen, L. L. (1971). *Brain Res.* **34**, 229-240.
Black, I. B., Hendry, I. A., and Iversen, L. L. (1972). *J. Neurochem.* **19**, 1367-1377.
Black, I. B., Bloom, E. M., and Hamill, R. W. (1976). *Proc. Natl. Acad. Sci. U.S.A.* **73**, 3575-3578.
Blum, J. J. (1970). *In* "Biogenic Amines as Physiologic Regulators" (J. J. Blum, ed.), pp. 95-118. Prentice-Hall, Englewood Cliffs, New Jersey.
Boer, G. J., Buijs, R. M., Swaab, D. F. and DeVries G. J. (1980). *Peptides* **1**, 203-209.
Bohn, M. C. (1980). *Neuroscience* **5**, 2003-2012.
Bohn, M. C., and Lauder, J. M. (1978). *Dev. Neurosci. (Basel)* **1**, 250-266.
Bohn, M. C., and Lauder, J. M. (1980). *Dev. Neurosci. (Basel)* **3**, 81-89.
Bohn, M. C., Goldstein, M., and Black, I. B. (1981). *Dev. Biol.* **82**, 1-10.
Boucek, R. J., and Bourne, B. B. (1962). *Nature (London)* **193**, (4821), 1181-1182.
Brückner, G., Mares, V., and Biesold, D. (1975). *J. Comp. Neurol.* **166**, 245-256.
Bugnon, C., Fellmann, D., and Bloch, B. (1977a). *Cell Tissue Res.* **183**, 319-328.
Bugnon, C., Bloch, B., Lenys, D., and Fellmann, D. (1977b). *Brain Res.* **137**, 175-180.
Burack, W. R., and Badger, A. (1964). *Fed. Proc. Fed. Am. Soc. Exp. Biol.* **23**, 561.
Burden, H. W., and Lawrence, I. E. (1973). *Am. J. Anat.* **136**, 251-257.
Burt, A. M., and Narayanan, C. H. (1976). *Exp. Neurol.* **53**, 703-713.
Buznikov, G. A. (1980). *In* "Neurotransmitters, Comparative Aspects" (J. Salánki and T. M. Turpaev, eds.), pp. 7-29. Akadémiai Kiadó, Budapest.
Buznikov, G. A., and Shmukler, Yu.B. (1981). *Neurochem. Res.* **6**, 55-68.
Buznikov, G. A., Chudakova, I. V., and Znezdina, N. D. (1964). *J. Embryol. Exp. Morphol.* **12**, 563-573.
Buznikov, G. A., Chudakova, I. V., Berdysheva, L. V., and Vyazmina, N. M. (1968). *J. Embryol. Exp. Morphol.* **20**, 119-128.
Buznikov, G. A., Kost, A. N., Kucherova, N. F., Mndzhoyan, A. L., Suvorov, N. N., and Berdysheva, L. V. (1970). *J. Embryol. Exp. Morphol.* **23**, 549-569.
Buznikov, G. A., Sakharova, A. V., Manukhin, B. N., and Markova, L. V. (1972). *J. Embryol. Exp. Morphol.* **27**, 339-351.
Cadilhac, J., and Pons, F. (1976). *C. R. Seances Soc. Biol. Ses Fil.* **170**, 25-31.
Caston, J. D. (1962). *Dev. Biol.* **5**, 468-482.
Chiappinelli, V., Giacobini, E., Pilar, G., and Uchimura, H. (1976). *J. Physiol.* **257**, 749-766.
Chiappinelli, V., Fairman, K., and Giacobini, E. (1978). *Dev. Neurosci (Basel)* **1**, 191-202.
Clark, R. L., Venkatasubramanian, K., Wolf, J. M., and Zimmerman, E. F. (1980). *J. Cell Biol.* **87**, 114a.
Clos, J., and Legrand, J. (1973). *Brain Res.* **63**, 450-455.
Clos, J., Crepel, F., Legrand, C., Legrand, J., Rabié, A., and Vigouroux, E. (1974). *Gen. Comp.*

Endocrinol. **23,** 178–192.
Clos, J., Selme-Matrat, M., Rabié, A., and Legrand, J. (1975). *J. Physiol. (Paris)* **70,** 207–218.
Cochard, P., Goldstein, M., and Black, I. (1978). *Proc. Natl. Acad. Sci. U.S.A.* **75**(6), 2986–2990.
Cohan, S., Ford, D., and Rhines, R. K. (1967). *Acta Neurol. Scand.* **43,** 11–32.
Combs, J. W., Lagunoff, D., and Benditt, E. P. (1965). *J. Cell Biol.* **25,** 577–592.
Costa, M., Eränkö, O., and Eränkö, L. (1974). *Brain Res.* **67,** 457–466.
Cotterrell, M., Balázs, R., and Johnson, A. L. (1972). *J. Neurochem.* **19,** 2151–2167.
Cowan, W. M. (1979). *In* "The Neurosciences Fourth Study Program" (F. O. Schmitt and F. G. Worden, eds.), pp. 59–79. MIT Press, Cambridge, Massachusetts.
Coyle, J., Wayner, M. J., and Singer, G. (1976). *Pharmacol. Biochem. Behav.* **4,** 191–200.
Daikoku, S., Maki, Y., Olamura, Y., Tsuruo, Y., and Hisano, S. (1984). *Int. J. Dev. Neurosci.* **2,** 113–120.
Davila-Garcia, M. I., Alvarez, P., Whitaker-Azmitia, P., Towle, A. C., and Azmitia, E. C. (1985). *Soc. Neurosci. Abst.* **11,** 601.
Deeb, S. S. (1972). *J. Exp. Zool.* **181,** 79–86.
Deguchi, T., Sinha, A. K., and Barchas, J. D. (1973). *J. Neurochem.* **20,** 1329–1336.
Dibner, M. D., and Black, I. B. (1976). *Brain Res.* **103,** 93–102.
Dibner, M. D., Mytilineou, C., and Black, I. B. (1977). *Brain Res.* **123,** 301–310.
Diez, J. A., Sze, P. Y., and Ginsburg, B. E. (1977). *Neurochem. Res.* **2,** 161–170.
Dozin-van Roye, B., and De Nayer, Ph. (1979). *Brain Res.* **177,** 551–554.
Dratman, M. B. (1978). *In* "Hormonal Proteins and Peptides," Vol. 6, pp. 205–271. Academic Press, New York.
Duhault, J., and Boulanger, M. (1977). *Eur. J. Pharmacol.* **43,** 203–205.
Eayrs, J. (1961). *Growth* **25,** 175–189.
Eins, S., Spoerri, P. E., and Heyder, E. (1983). *Cell Tiss. Res.* **229,** 457–460.
Emanuelson, H. (1974). *Wilhelm Roux's Arch. Dev. Biol.* **175,** 253–271.
Emanuelson, H., and Palén, K. (1975). *Wilhelm Roux's Arch. Dev. Biol.* **177,** 1–17.
Filogamo, G., Peirone, S., and Sisto Daneo, L. (1978). *In* "Maturation of Neurotransmission" (A. Vernadakis, E. Giacobini, and G. Filogamo, eds.), pp. 1–9. Karger, Basel.
Finley, J. C. W., and Petrusz, P. (1979). *Anat. Rec.* **193,** 538.
Finley, J. C. W., Maderdrut, J. L., Roger, L. J., and Petrusz, P. (1981a). *Neuroscience* **6,** 2173–2192.
Finley, J. C. W., Maderdrut, J. L., and Petrusz, P. (1981b). *J. Comp. Neurol.* **198,** 541–565.
Franquinet, P. R. (1979). *J. Embryol. Exp. Morphol.* **51,** 85–95.
Gash, D., Sladek, C., and Scott, D. (1980). *Brain Res.* **181,** 345–355.
Geel, S. E. (1977). *Nature (London)* **269,** 428–430.
Gérard, A., Gérard, H., and Dollander, A. (1978). *Compte Rendus Acad. Sci. (Paris) Ser. D,* **286,** 891–894.
Gershon, M. D., and Nunez, E. A. (1976). *Mol. Cell Endocrinol.* **5,** 169–180.
Gershon, M. D., Teitelman, G., Rothman, T. P., Joh, T. H., and Reis, D. J. (1979). *Soc. Neurosci. Abstr.* **5,** 334.
Giacobini, E. (1975). *J. Neurosci. Res.* **1,** 315–331.
Giacobini, E. (1978). *In* "Maturation of Neurotransmission" (A. Vernadakis, E. Giacobini, and G. Filogamo, eds.), pp. 41–64. Karger, Basel.
Giacobini, E. (1979). *In* "Neural Growth and Differentiation" (E. Meisami and M. A. B. Brazier, eds.), pp. 153–167. Raven, New York.
Giulian, D., McEwen, B. S., and Pohorecky, L. A. (1974). *Proc. Natl. Acad. Sci. U.S.A.* **71,** 4106–4110.
Golden, G. S. (1972). *Brain Res.* **44,** 278–282.
Golden, G. S. (1973). *Dev. Biol.* **33,** 300–311.
Goldman, A. S. (1980). *In* "Drug and Chemical Risks to the Fetus and Newborn," (R. H. Schwartz and S. J. Yaffe, eds.), pp. 9–31. Alan R. Liss, New York.

Gottlieb, G. (1973). *In* "Studies on the Development of Behavior and Nervous System: Behavioral Embryology" (G. Gottlieb, ed.), Vol. 1, pp. 3–45. Academic Press, New York.
Gottlieb, G. (1976). *In* "Neural and Behavioral Specificity" (G. Gottlieb, ed.), pp. 24–54. Academic Press, New York.
Granholm, A.-C., and Seiger, Å. (1981). *Neuroscience Letters* **22,** 279–284.
Granholm, A.-C., and Siegel, R. A., and Seiger, Å. (1984). *Int. J. Dev. Neurosci.* **2,** 337–345.
Griffin, W. S. T., Woodward, D. J., and Chanda, R. (1978). *Brain Res. Bull.* **3,** 365–367.
Gripois, D., Klein, C., and Valeus, M. (1980). *Biol. Neonate* **37,** 165–171.
Gromova, H. A., Chubakova, A. R., Chumasov, E. I., and Konovalov, H. V. (1983). *Int. J. Dev. Neurosci.* **1,** 339–349.
Gustafson, T., and Toneby, M. (1970). *Exp. Cell Res.* **62,** 102–117.
Gustafson, T., and Toneby, M. (1971). *Am. Sci.* **59,** 452–462.
Hajós, F., Patel, A. J., and Balázs, R. (1973). *Brain Res.* **50,** 389–401.
Hamburg, M., and Vicari, E. (1957). *Anat. Rec.* **127,** 302.
Hanson, G. H., Neier, E., and Schousboe, A. (1984). *Int. J. Dev. Neurosci.* **2,** 247–257.
Harlan, R. E., Gordon, J. H., and Gorski, R. A. (1979). *Rev. Neurosci.* **4,** 31–71.
Haydon, P. G., McCobb, D. P., and Kater, S. B. (1984). *Science* **226,** 561–564.
Hendry, I. A. (1975a). *Brain Res.* **90,** 235–244.
Hendry, I. A. (1975b). *Brain Res.* **94,** 87–97.
Hendry, I. A., and Thoenen, H. (1974). *J. Neurochem.* **22,** 999–1004.
Hoff, K. M. (1976). *Gen. Pharmacol.* **7,** 395–398.
Hökfelt, T., Elde, R., Johansson, O., Terenius, L., and Stein, L. (1977). *Neurosci. Lett.* **5,** 25–31.
Hole, K. (1972a). *Dev. Psychobiol.* **5,** 157–173.
Hole, K. (1972b). *Eur. J. Pharmacol.* **18,** 361–366.
Houser, C. R., Lee, M., and Vaughn, J. E. (1983). *J. Neurosci.* **3,** 2030–2042.
Howard, E. (1968). *Exp. Neurol.* **22,** 191–208.
Hubel, D. H., and Wiesel, T. N. (1970). *J. Physiol. (London)* **206,** 419–436.
Hume, R. I., Role, L. W., and Fischbach, G. D. (1983). *Nature* **305,** 632–635.
Ibata, Y., Watanabe, K., Kinoshita, H., Kubo, S., Sano, Y., Sin, S., Hashimura, E., and Imagawa, K. (1979). *Cell Tissue Res.* **198,** 381–395.
Idänpään-Heikkilä, J., and Saxén, L. (1973). *Lancet* **ii,** 282–284.
Ifft, J. D. (1972). *J. Comp. Neurol.* **144,** 193–204.
Ignarro, L. J., and Shideman, F. E. (1968a). *J. Pharmacol. Exp. Ther.* **159,** 38–48.
Ignarro, L. J., and Shideman, F. E. (1968b). *J. Pharmacol. Exp. Ther.* **159,** 49–58.
Inagaki, S., Sakanaka, M., Shiosaka, S., Senba, E., Takatsuke, K., Takagi, H., Kawai, Y., Minagawa, H., and Tahyama, M. (1982). *Neuroscience* **7,** 251–277.
Johnston, G. A. R., Allan, R. D., and Skerritt, J. H. (1984). *In:* Handbook of Neurochemistry Vol. 6, 2nd edition. (A. Lajtha, ed.), pp. 213–237, Plenum, New York.
Jurand, A. (1980). *Dev. Growth Differ.* **22,** 61–70.
Katsamatsu, T., and Pettigrew, J. D. (1979). *J. Comp. Neurol.* **185,** 139–162.
Katsamatsu, T., Pettigrew, J. D., and Ary, M. (1979). *J. Comp. Neurol.* **185,** 163–182.
Kerley, C. G. (1936). *Endocrinology (Baltimore)* **20,** 611.
Kidokoro, Y., and Yeh, E. (1982). *Proc. Natl. Acad. Sci.* **79,** 6727–6731.
Kikuyama, S., Miyakawa, M., and Arai, Y. (1979). *Cell Tissue Res.* **198,** 27–33.
Kirby, M. L., and Gilmore, S. A. (1972). *Anat. Rec.* **173,** 469–478.
Koe, B. K., and Weissman, A. (1968). *Adv. Pharmacol.* **6B,** 29–47.
König, N., and Marty, R. (1981). *Biblithaca Acta* **19,** 152–160.

Kujawa, M. J., and Zimmerman, E. F. (1978). *Teratology* **17,** 29A.

Lau, C., and Slotkin, T. A. (1979). *J. Pharmacol. Exp. Ther.* **208,** 485–490.

Lauder, J. M. (1977). *Brain Res.* **126,** 31–51.

Lauder, J. M. (1978). *Brain Res.* **142,** 25–39.

Lauder, J. M. (1979). *Dev. Biol.* **70,** 105–115.

Lauder, J. M., and Bloom, F. E. (1974). *J. Comp. Neurol.* **155,** 469–481.

Lauder, J. M., and Bloom, F. E. (1975). *J. Comp. Neurol.* **163,** 251–264.

Lauder, J. M., and Bohn, M. C. (1980). *In* "Progress in Psychoneuroendocrinology" (F. Brambilla, G. Racagni, and D. deWied, eds.), pp. 603–620. Elsevier, Amsterdam.

Lauder, J. M., Han, V. K. M., Henderson, P., Verdorn, T., and Towle, A. C. (1985a). *Neuroscience* (in press).

Lauder, J. M., and Krebs, H. (1976). *Brain Res.* **107,** 638–644.

Lauder, J. M., and Krebs, H. (1978a). *Dev. Neurosci. (Basel)* **1,** 15–30.

Lauder, J. M., and Krebs, H. (1978b). *In* "Maturation of Neurotransmission" (A. Vernadakis, E. Giacobini, and G. Filogomo, eds.), pp. 171–180. Karger, Basel.

Lauder, J. M., and Mugnaini, E. (1977). *Nature (London)* **268,** 335–337.

Lauder, J. M., and Mugnaini, E. (1980). *Dev. Neurosci. (Basel)* **3,** 248–265.

Lauder, J. M., Wallace, J. A., and Krebs, H. (1980). *In* "Progress in Psychoneuroendocrinology" (F. Brambilla, G. Racagni, and D. deWied, eds.), pp. 539–556. Elsevier, Amsterdam.

Lauder, J. M., Sze, P. Y., and Krebs, H. (1981). *Dev. Neurosci. (Basel)* **4,** 291–295.

Lauder, J. M., Petrusz, P., Wallace, J. A., DiNome, A., Wilkie, M. B., and McCarthy, K. (1982a). *J. Histochem. Cytochem.* **30,** 788–793.

Lauder, J. M., Towle, A. C., Patrick, K., Henderson, P., and Krebs, H. (1985b). *Dev. Brain Res.* **20,** 107–114.

Lauder, J. M., Wallace, J. A., Krebs, H., Petrusz, P., and McCarthy, K. (1982b). *Brain Res. Bull.* **9,** 605–625.

Lawrence, I. E., Jr., and Burden, H. W. (1973). *Am. J. Anat.* **136,** 199–208.

Legrand, J. (1965). *C.R. Hebd. Seances Acad. Sci.* **261,** 544–547.

Legrand, J., Selme-Matrat, M., Rabié, A., Clos, J., and Legrand, C. (1976). *Biol. Neonate* **29,** 368–380.

Lengvári, I., Branch, B. J., and Taylor, A. N. (1977). *Neuroendocrinology* **24,** 65–73.

Lengvári, I., Branch, B. J., and Taylor, A. N. (1980). *Dev. Neurosci. (Basel)* **3,** 59–65.

Levitt, P., and Rakic, P. (1979). *Soc. Neurosci. Abstr.* **5,** 341.

Lewis, P. D., Patel, A. J., Johnson, A. L., and Balázs, R. (1976). *Brain Res.* **104,** 49–62.

Lewis, P. D., Patel, A. J., Béndek, G., and Balázs, R. (1977). *Brain Res.* **129,** 299–308.

Lidov, H. G. W., and Molliver, M. E. (1979). *Soc. Neurosci. Abstr.* **5,** 341.

Lidov, H. G. W., and Molliver, M. E. (1980). *Soc. Neurosci. Abstr.* **6,** 351.

Ljungdahl, A., Hökfelt, T., Nilsson, G., and Goldstein, M. (1978). *Neuroscience* **3,** 945–976.

Lovell, K. L. (1981). *Soc. Neurosci. Abstr.* **7,** 555.

Lund, R. D., (1978). "Development and Plasticity of the Brain," pp. 253–284. Oxford Univ. Press, London and New York.

McEwen, B. S., Gerlach, J. L., and Micco, P. J. (1975). *In* "The Hippocampus" (R. L. Isaacson and K. C. Pribram, eds.), Vol. 1, pp. 285–322. Plenum, New York.

McGinty, J. F., and Ford, D. H. (1980). *Dev. Neurosci. (Basel)* **3,** 224–234.

McLaughlin, B. J., Wood, J. G., Saito, K., Barber, R., Vaughn, J. E., Roberts, E., and Wu, J.-Y. (1974). *Brain Res.* **76,** 377–391.

Maeda, T., and Dresse, A. (1969). *Acta Neurol. Belg.* **69,** 5–10.

Maeda, T., Tohyama, M., and Shimizu, N. (1974). *Brain Res.* **70,** 515–520.

Markey, K. A., and Sze, P. Y. (1981). *Dev. Neurosci.* **4,** 267–272.
Marin-Padilla, M. (1984). *In* "Cerebral Cortex," (A. Peters and E. G. Jones, eds.), pp. 447–478. Plenum, New York.
Meier, E., Drejer, J., and Schousboe, A. (1983). *In* "CNS Receptors from Molecular Pharmacology to Behavior. (P. Mandel and F. DeFeudis, eds.), pp. 47–58. Raven Press, New York.
Meinertzhagen, I. A. (1985). *Nature* **313,** 348–349.
Meserve, L. A., and Leathem, J. H. (1974). *Proc. Soc. Exp. Biol. Med.* **147,** 510–512.
Morse, D. E., Hooker, N., Duncan, H., and Jensen, L. (1979). *Science* **204,** 407–410.
Mugnaini, E., and Dahl, A.-L. (1975). *J. Comp. Neurol.* **162,** 417–432.
Müller, W. A., Mitze, A., Wickhorst, J.-P., and Meier-Menge, H. M. (1977). *Wilhelm Roux's Arch. Dev. Biol.* **182,** 311–328.
Mytilineou, C., and Black, I. B. (1978). *Brain Res.* **158,** 259–268.
Naidoo, S., and Timiras, P. S. (1979). *Dev. Neurosci. (Basel)* **2,** 213–224.
Naidoo, S., Valcana, T., and Timiras, P. S. (1978). *Am. Zool.* **18,** 545–552.
Nair, V. (1974). *In* "Drugs and the Developing Brain" (A. Vernadakis and N. Weiner, eds.), pp. 171–197. Plenum, New York.
New, D. A. T., Coppola, P. T., and Terry, S. (1983). *J. Reprod. Fert.* **35,** 135–138.
Nicholson, J. L., and Altman, J. (1972a). *Brain Res.* **44,** 13–23.
Nicholson, J. L., and Altman, J. (1972b). *Brain Res.* **44,** 25–36.
Nicholson, J. L., and Altman, J. (1972c). *Science* **176,** 530–532.
Nornes, H. O., and Morita, M. (1979). *Dev. Neurosci. (Basel)* **2,** 101–114.
Olpe, H.-R., and McEwen, B. S. (1976). *Brain Res.* **105,** 121–128.
Olson, L., and Seiger, Å. (1972). *Z. Anat. Entwicklungsgesch.* **137,** 301–316.
Olson, L., Boreus, L. O., and Seiger, Å. (1973). *Z. Anat. Entwicklungsgesch.* **139,** 259–282.
Otten, U., and Thoenen, H. (1976). *Neurosci. Lett.* **2,** 93–96.
Paden, C. M., and Silverman, A. J. (1979). *Soc. Neurosci. Abstr.* **5,** 454.
Palén, K., Thorneby, L., and Emanuelson, H. (1979). *Wilhelm Roux's Arch. Dev. Biol.* **187,** 89–103.
Patel, A. J., Béndek, B., Balázs, R., and Lewis, P. D. (1977). *Brain Res.* **129,** 283–297.
Patel, A. J., Bailey, P., and Balázs, R. (1979a). *Neuroscience* **4,** 139–143.
Patel, A. J., Lewis, P. D., Balázs, R., Bailey, P., and Lai, M. (1979b). *Brain Res.* **172,** 57–72.
Pickel, V. M., Sumal, K. K., Miller, R. J., and Reis, D. J. (1979). *Soc. Neurosci. Abstr.* **5,** 174.
Pienkowski, M. M. (1977). *Anat. Rec.* **189**(3), 550.
Poland, R. E., Weichsel, M. E., Jr., and Rubin, R. T. (1979). *Horm. Metab. Res.* **11,** 222–227.
Rabié, A., and Legrand, J. (1973). *Brain Res.* **61,** 267–278.
Rabié, A., Favre, C., Clavel, M. C., and Legrand, J. (1977). *Brain Res.* **120,** 521–531.
Rabié, A., Favre, C., Clavel, M. C., and Legrand, J. (1979). *Brain Res.* **161,** 469–479.
Rakic, P. (1971). *J. Comp. Neurol.* **141,** 283–312.
Rastogi, R. B., and Singhal, R. L. (1976). *Life Sci.* **18,** 851–868.
Rastogi, R. B., and Singhal, R. L. (1978). *Naunyn-Schmiedeberg's Arch. Pharmacol.* **304,** 9–13.
Rastogi, R. B., and Singhal, R. L. (1979). *Psychopharmacology (Berlin)* **62,** 287–293.
Rastogi, R. B., Lapierre, Y., and Singhal, R. L. (1976). *J. Neurochem.* **26,** 443–449.
Rebière, A., and Dainat, J. (1976). *Acta Neuropathol.* **35,** 117–129.
Rebière, A., and Legrand, J. (1970). *C. R. Acad. Sci. Ser. D* **274,** 3581–3584.
Rebière, A., and Legrand, J. (1972). *Arch. Anat. Microsc. Morphol. Exp.* **61,** 105–126.
Reitzel, J. L., Maderdrut, J. L., and Oppenheim, R. W. (1979). *Brain Res.* **172,** 487–504.
Ribak, C. E., Vaughn, J. E., and Saito, K. (1978). *Brain Res.* **140,** 315–332.
Rodriguez, N., and Renaud, F. L. (1980). *J. Cell Biol.* **85,** 242–247.
Rosengarten, H., and Friedhoff, A. J. (1979). *Science* **203,** 1133–1135.
Sadler, T. W., and Kochar, D. M. (1976). *J. Embryol. Exp. Morphol.* **36,** 273–281.

Sadler, T. W., and New, D. A. T. (1981). *J. Embryol. Exp. Morphol.* **66**, 109–116.
Salas, M., and Shapiro, S. (1970). *Physiol. Behav.* **5**, 7–12.
Salmon, T. N. (1936). *Proc. Soc. Exp. Biol. Med.* **35**, 489–491.
Salmon, T. N. (1941). *Endocrinology (Baltimore)* **29**, 291–296.
Samel, M. (1968). *Gen. Comp. Endocrinol.* **10**, 229–234.
Santander, R. G., and Cuadrado, G. M. (1976). *Acta Anat.* **95**, 368–383.
Sar, M., Stumpf, W. E., Miller, R. J., Chang, K.-J., and Cuatrecasas, P. (1978). *J. Comp. Neurol.* **182**, 17–38.
Schlumpf, M., and Lichtensteiger, W. (1979). *Anat. Embryol.* **156**, 177–187.
Schlumpf, M., Richards, J. G., Lichtensteiger, W., and Mohler, H. (1983). *J. Neurosci.* **3**, 1478–1487.
Schlumpf, M., Shoemaker, W. J., and Bloom, F. E. (1977). *Soc. Neurosci. Abstr.* **3**, 361.
Schlumpf, M., Shoemaker, W. J., and Bloom, F. E. (1980). *J. Comp. Neurol.* **192**, 361–376.
Schon, F., and Iversen, L. L. (1972). *Brain Res.* **42**, 503–507.
Schowing, J.,Sprumont, P., and Van Toledo, B. (1977). *C. R. Comp. Rendus Acad. Sci. (Paris)* **171**(6), 1163–1166.
Schwark, W. S., and Keesey, R. R. (1975). *Res. Commun. Chem. Pathol. Pharmacol.* **10**, 37–50.
Schwark, W. S., and Keesey, R. R. (1978). *J. Neurochem.* **30**, 1583–1586.
Schwartz, H. L., and Oppenheimer, J. H. (1978). *Endocrinology (Baltimore)* **103**, 943–948.
Scott, J. P., Stewart, J. M., and DeGhett, V. J. (1974). *Dev. Psychobiol.* **7**, 489–513.
Scow, R. O., and Simpson, M. E. (1945). *Anat. Rec.* **91**, 209–226.
Sétáló, G., Vigh, S., Schally, A. V., Arimura, A., and Flerkó, B. (1976). *Brain Res.* **103**, 597–602.
Shapiro, S. (1968). *Gen. Comp. Endocrinol.* **10**, 214–228.
Sievers, J., Berry, M., and Baumgarten, H. (1981). *Brain Res.* **207**, 200–208.
Shiosaka, S., Takatsuke, K., Sakanaka, M., Inagaki, S., Takagi, H.,Senba, E., Kawai, Y., and Tahyama, M. (1981). *J. Comp. Neurol.* **202**, 115–124.
Silverman, A.-J., Goldstein, R., and Gadde, C. A. (1980). *Peptides* **1**, 27–44.
Sims, J. J. (1977). *J. Comp. Neurol.* **173**, 319–336.
Sinding, C., Robinson, A. G., Seif, S. M., and Schmid, P. G. (1980). *Brain Res.* **195**, 177–186.
Singhal, R. L., Rastogi, R. B., and Hrdina, P. D. (1975). *Life Sci.* **17**, 1617–1626.
Slotkin, T. A., Whitmore, W. L., Salvaggio, M., and Seidler, F. J. (1979). *Life Sci.* **24**, 1223–1230.
Sokol, H. W., Zimmerman, E. A., Sawyer, W. H., and Robinson, A. G. (1976). *Endocrinology (Baltimore)* **98**, 1176–1188.
Sorimachi, M., and Kataoka, K. (1974). *Brain Res.* **70**, 123–130.
Specht, L. A., Pickel, V. M., Joh, T. H., and Reis, D. J. (1978a). *Soc. Neurosci. Abstr.* **4**, 386.
Specht, L. A., Pickel, V. M., Joh, T. H., and Reis, D. J. (1978b). *Brain Res.* **156**, 315–321.
Spoerri, P. E., and Wolff, J. R. (1981). *Cell Tiss. Res.* **218**, 567–579.
Sterling, K., Lazarus, J. H., Milch, P. O., Sakurada, T., and Brenner, M. A. (1978). *Science* **201**, 1126–1129.
Strudel, G., Meiniel, R., and Gateau, G. (1977a). *Compte Reudus Acad. Sci. Ser. D* **284**, 1097–1100.
Strudel, G., Recasens, M., and Mandel, P. (1977b). *Compte Reudus Acad. Sci. Ser. D* **284**, 967–969.
Swaab, D. F., Nijveldt, F., and Pool, C. W. (1975). *J. Endocrinol.* **67**, 461–462.
Swanson, L. W., and Hartman, B. K. (1975). *J. Comp. Neurol.* **163**, 467–506.
Sze, P. Y. (1976). *Adv. Biochem. Psychopharmacol.* **15**, 251–265.
Sze, P. Y. (1980). *Dev. Neurosci. (Basel)* **3**, 217–223.
Sze, P. Y., Neckers, L., and Towle, A. C. (1976). *J. Neurochem.* **26**, 169–173.
Taban, C. H., and Cathieni, M. (1979). *Experientia* **35**, 811–812.
Taber Pierce, E. (1973). *Prog. Brain Res.* **40**, 53–65.
Taub, H., and Peters, D. A. V. (1978). *Gen. Pharmacol.* **9**, 97–100.

Taylor, A. N., and Lengvári, I. (1977). *Neuroendocrinology* **24,** 74–79.
Teitelman, G., Joh, T. H., and Reis, D. J. (1978). *Brain Res.* **158,** 229–234.
Tennyson, V. M., Barrett, R. E., Cohen, G., Cote, L., Heikkila, R., and Mytilineou, C. (1972). *Brain Res.* **46,** 251–285.
Tennyson, V. M., Mytilineou, C., and Barrett, R. E. (1973). *J. Comp. Neurol.* **149,** 233–258.
Tennyson, V. M., Mytilineou, C., Heikkila, R., Barrett, R. E., Cole, L., and Cohen, G. (1975). *In* "The Golgi Centennial Symposium" (M. Santini, ed.), pp. 449–464. Raven, New York.
Theoharides, T. C., Bondy, P. K., Tsakalos, N. D., and Akenase, P. W. (1982). *Nature* **5863,** 229–231.
Thorpe, W. H. (1963). "Learning and Instinct in Animals," 558 pp. Harvard Univ. Press, Cambridge, Massachusetts.
Toneby, M. (1977). *In* "Functional Aspects of 5-Hydroxytryptamine and Dopamine in Early Embryogenesis of Echinoidea and Asteroidea." University of Stockholm, Stockholm.
Tonge, S. R. (1974). *Life Sci.* **15,** 245–249.
Towle, A. C., Sze, P. Y., and Lauder, J. M. (1982). *Dev. Neurosci.* **5,** 458–464.
Turner, B. B. (1978). *Am. Zool.* **18,** 461–475.
Ungerstedt, U. (1971). *Acta Physiol. Scand. Suppl.* **367,** 1–48.
Vaccari, A. (1982). *Eur. J. Pharmacol.* **82,** 93–95.
Vaccari, A. (1985). *Br. J. Pharmacol.* **84,** 773–778.
Vaccari, A., Biassoni, R., and Timiras, P. S. (1983). *J. of Neurochem.* **40,** 1019–1025.
Valcana, T., and Timiras, P. S. (1978). *Mol. Cell Endocrinol.* **11,** 31–41.
Vandesande, F., DeMey, J., and Dierickx, K. (1974). *Cell Tissue Res.* **151,** 187–200.
Venkatasubramanian, K., Clark, R. L., Wolff, J. M., and Zimmerman, E. F. (1980). *J. Cell Biol.* **87,** 55a.
Vernadakis, A. (1973). *Prog. Brain Res.* **40,** 231–243.
Vernadakis, A., and Woodbury, D. M. (1971). *Influence Horm. Nerv. Syst., Proc. Int. Soc. Psychoneuroendocrinol. 1970, 1971,* pp. 85–97.
Vorhees, C. V., Brunner, R. L., and Butcher, R. E. (1979). *Science* **205,** 1220–1224.
Wallace, J. A. (1979). Ph.D. Thesis, University of California, Davis.
Wallace, J. A. (1982). *Am. J. Anat.* **165,** 261–276.
Wallace, J. A., and Lauder, J. M. (1983). *Brain Res. Bull.* **10,** 459–479.
Wallace, J. A., Petrusz, P., and Lauder, J. M. (1982). *Brain Res. Bull.* **9,** 117–129.
Warembourg, M. (1975). *Brain Res.* **89,** 61–70.
Wee, E. L., Babiarz, B. S., Zimmerman, S., and Zimmerman, E. F. (1979). *J. Embryol. Exp. Morphol.* **53,** 75–90.
Wee, E. L., Phillips, N. J., Babiarz, B. S., and Zimmerman, E. F. (1980). *J. Embryol. Exp. Morphol.* **58,** 177–193.
Weindl, A., Sofroniew, M. V., and Schinko, I. (1978). *In* "Neurosecretion and Neuroendocrine Activity" (W. Bargmann, A. Oksche, A. Polenov, and B. Scharrer, eds.), pp. 312–319. Springer Publ., New York.
Wendlandt, T. J., Crow, T. J., and Stirling, R. V. (1977). *Brain Res.* **125,** 1–9.
Whitaker-Azmitia, P. and Lauder, J. M. (1985). *Soc. Neurosci. Abstr.* **11,** 601.
Woodhams, P. L., Allen, Y. S., McGovern, J., Allen, J. M., Bloom, S. R., Balázs, R., and Polak, J. M. (1985). *Neuroscience* **15,** 173–202.
Wolff, J. R., Joo, F., and Dames, W. (1978). *Nature* **274,** 72–74.
Yanai, J., ed. (1984). "Neurobehavioral Teratology," Elsevier, New York.
Yew, D. T., Ho, A. K. S., and Meyer, D. B. (1974). *Experientia* **30,** 1320–1322.
Young, S. H., and Poo, M.-M. (1983). *Nature* **305,** 634–637.
Zagon, I. S., and McLaughlin, P. J. (1977). *Pharmacology* **15,** 276–282.
Zagon, I. S., and McLaughlin, P. J. (1983). *Science* **221,** 227–229.
Zimmerman, E. F., and Wee, E. L. (1984). In: *Current Topics in Developmental Biology/ Palate Development* (E. F. Zimmerman, ed.), pp. 37–63, Academic Press, New York.

6

Sexual Differentiation of the Brain

C. DOMINIQUE TORAN-ALLERAND
Center for Reproductive Sciences
and Department of Neurology, and Department of Anatomy and Cell Biology
Columbia University College of Physicians and Surgeons
New York, New York

I. Introduction

There are clear sex differences in the neural control of endocrine function and in a wide variety of reproductive and nonreproductive behavioral patterns of most vertebrate species. It is generally believed that these differences in brain function are the result of the gonadal hormonal environment during a specific period of central nervous system development. The gonadal hormones exert what is referred to as an inductive or organizational influence on the developing brain that results in the differentiation of a broad spectrum of neural functions congruent with the genetic sex. This phenomenon is termed sexual differentiation of the brain.

Sexual differentiation is thought to result from exposure of an undifferentiated and bipotential brain to testicular androgens during a restricted "critical" time period of neural development when the tissue is sufficiently plastic to respond permanently and irreversibly to the hormones and after which it is refractory or responds in a reversible manner (for reviews, see MacLusky and Naftolin, 1981; McEwen, 1983). Although I focus principally on the rodent, the most extensively studied animal model of sexual differentiation, the basic phenomena have been shown to exist in vertebrates in general. It should always be kept in mind, however, that, although the underlying principles of hormonal action are probably valid across species, the neural functions that may be dimorphic, the brain regions involved, the timing of the hormone-sensitive period, and even the very hormone or hormones responsible may vary widely.

The most extensively studied and most clear-cut sex-specific differences in brain function are those concerned with reproductive physiology and with sexual or mating behavior. The female rodent exhibits an endogenous rhythm of sexual behavior and of pituitary and gonadal hormone release that is absent in the adult male. Evidence indicating that the brain rather than peripheral endocrine organs is altered in rodents includes the fact that a male pituitary transplanted under a female brain responds cyclically, while the converse does not occur (Harris and Jacobson, 1952). The phenomenon of sexual differentiation in the rodent actually consists of two processes, both of which are dependent on androgen, but with differential hormonal sensitivities and different critical periods. One, termed *masculinization*, refers to the development and potentiation of male characteristics. The other, termed *defeminization*, relates to suppression of the female characteristics. The defeminization component of this process does not occur in higher mammals for unknown reasons (McEwen, 1983). This difference serves as a good example of the great importance of considering the variability existing across species.

During the critical period, which in the rodent may extend from the late fetal through the first few postnatal days, the expression of the intrinsic rhythmicity of the brain is permanently and irreversibly suppressed by the normal (in the case of the male) or experimental (in the case of the female) exposure to androgens that both masculinize and defeminize postpubertal sexual behavior (enhancement of mounting behavior; reduction in the capacity to exhibit lordosis) and the noncyclic or tonic pattern of gonadotropin and gonadal hormone release. While masculinization and defeminization of brain function in both rodent sexes follow exposure to androgen, *feminization* (gonadotropin and gonadal hormone surges and lordosis behavior) is generally considered to emerge relatively passively in the absence of such hormonal induction and to represent the expression of the brain's presumed intrinsic or unmodified pattern of neural organization. The sensitive period for masculinization appears to precede that for defeminization, at least in the rat (Ward, 1969).

II. Possible Mechanisms of Steroid Action

The permanent and irreversible nature of the steroidal effects has strongly suggested that androgens may influence the organization of the central nervous system (CNS) through alterations in neuronal genomic expression. How androgens elicit these effects is not fully known. Considerable evidence has accumulated, however, from a variety of experiments in the rodent that local intraneuronal conversion of aromatizable androgens to the active steroid, estradiol-17β, and the subsequent binding of estradiol to putative intracellular (intraneuronal) receptors in such target regions as the hypothalamus, preoptic area (POA), and the amygdala may be requisite events for the initiation of masculinization of the brain. These regions of the adult brain have all been implicated in the neural control of reproductive function and sexual behavior and contain in the fetal and neonatal rodent of both sexes (as in the adult) high affinity, distinct androgen- and estrogen-binding macromolecules ("receptors"), as shown by [^{3}H]steroid autoradiography and by receptor assay as well as high levels of aromatizing enzymes (for reviews see MacLusky and Naftolin, 1981; McEwen, 1983).

In non-CNS tissues of adult vertebrates estrogen has been thought to act through binding to specific intracellular receptor sites, which subsequently interact with the cell nucleus (Gorski and Gannon, 1976) and initiate changes in nucleic acid and protein synthesis (Hudson *et al.*, 1970; Salaman, 1974). Studies by Welshons *et al.* (1984), however, suggest that the estrogen receptor is entirely intranuclear in intact cells and that the generally accepted model of intracytoplasmic binding followed by nuclear translocation of the receptor–steroid complex is erroneous, an artifact of extraction.

The importance of estradiol has been shown in a number of ways. Intrahypothalamic implants of testosterone or estradiol have been found to be equally effective in eliciting masculinization (Christensen and Gorski, 1978). The nonaromatizable androgens such as 5α-dihydrotestosterone (DHT), on the other hand, appear to be ineffective (Luttge and Whalen, 1970; McDonald and Doughty, 1974) or only partially effective (Gerall *et al.*, 1975). Testosterone-induced (McDonald and Doughty, 1973; Doughty and McDonald, 1974; McEwen *et al.*, 1977; Vreeburg *et al.*, 1977) and estrogen-induced (Doughty *et al.*, 1975) masculinization can be blocked by antiestrogens that compete with estradiol for the receptor sites. The masculinizing effects of both endogenous and exogenous testosterone, furthermore, have been attenuated by aromatizing enzyme inhibitors (Booth, 1977; McEwen *et al.*, 1977; Vreeburg *et al.*, 1977). And finally, in androgen-insensitive male rats and mice with the genetic defect Tfm (testicular feminizing mutation), in which androgen receptors are deficient in numbers (10–15% of normal) (Naess *et al.*, 1976), the brain is nonetheless normally masculinized and defeminized with respect to those estrogen-dependent neural functions, since aromatase activity and the estrogen receptors are quite normal (Wieland and Fox, 1981).

III. Sexual Dimorphism of Brain Structure

Brain lesions studies have indicated that destruction of the medial preoptic area and anterior hypothalamus (AH) results, in most cases, in complete failure to mount, thrust, or intromit in males (Heimer and Larsson, 1966, 1967; Singer, 1968). The most important afferents to this area appear to be those from the medial forebrain bundle (MFB) (Scouten *et al.*, 1980). Implantation of testosterone in this area activates sexual behavior in castrated male rats (Davidson, 1966). In contrast, the POA–AH is not essential for lordosis in females; lesions thereof either facilitate (Powers and Valenstein, 1972) or have no effect (Singer, 1968) on lordosis, as is also the case with MFB lesions (Hitt *et al.*, 1970). Rather, lesions in the ventromedial hypothalamus (VMH) (Law and Meagher, 1958) and cuts through lateral afferent connections to this region (Malsbury *et al.*, 1979) suppress lordosis. Implantation of estrogen in the VMH is more effective than in other hypothalamic areas in inducing lordosis in ovariectomized female rats (Rubin and Barfield, 1980). The POA is, however, crucial to cyclic endocrine patterns in female rodents; lesions of this area or knife cuts separating it from the medial basal hypothalamus–pituitary region prevent gonadotropin surges and subsequent ovulation (Halasz and Gorski, 1967). Thus studies of neural substrates of sex dimorphism have concentrated largely on these regions.

The neural substrates for the steroidal effects and the morphogenetic basis for sexual differentiation are less well understood. A variety of studies suggest that the critical period for steroid sensitivity may be related to the maturational state of neurons. Neonatal hyper- (Phelps and Sawyer, 1976) and hypo-thyroidism (Kikuyama, 1969) shortens and prolongs respectively the susceptibility of neonatal female rats to androgenization (see chapter by Lauder, this volume). Reier *et al.* (1977) and Lawrence and Raisman (1980) have shown in the rat POA a close correlation between ultrastructural maturation of its neurons and neuropil and the period of androgen sensitivity;

the major phase of differentiation appears to occur after the critical period, with synaptogenesis extending for at least a week beyond the maximum estimate of the critical period for behavioral and physiological organization by gonadal hormones. Similarly, the major phase of POA dendritic proliferation follows the period of sensitivity to testosterone in the hamster (Hsu *et al.*, 1978). A crucial issue has been whether or not the gonadal steroids can directly influence the neural substrate by inducing structural changes through cellular interactions.

Obviously, some of the effects of testosterone or estradiol need not be mediated by gross anatomical changes. Since steroidal effects are believed to be mediated through specific receptors, their primary action may rather be at the molecular level as a result of genomic (transcriptional) or posttranscriptional events. Hormones may alter cellular responsiveness, perhaps through modulations of enzymes controlling synaptic transmission (Ani *et al.*, 1980; Arimatsu *et al.*, 1981; Luine *et al.*, 1975) or by alterations in membrane properties. It has also been suggested that androgenization may result from interference with maturational or metabolic aspects of the receptor system (Vertés and King, 1971).

It is reasonable to consider, however, that sex differences in neurally controlled functions might be expressed also in the structural dimorphism of the relevant brain regions. That the gonadal hormones may both influence and alter the direction of neuronal differentiation has been supported by increasingly numerous examples of steroid- or sex-dependent structural dimorphism in such physiologically significant, steroid-receptor-containing regions of the adult CNS as the hypothalmus, the POA, the amygdala, the hippocampus, the septum, habenula, and the spinal cord of mammals, the vocal control nuclei and their motor effector units of song birds and frogs, and the autonomic ganglia of the cat, mouse and rat. Some morphological sex differences and consequences of exposure to steroid are summarized in Table I (see also Toran-Allerand, 1984; Arnold and Gorski, 1984; and Chapter 14, this volume). They may be broadly characterized as differences in the neuronal cell body and its organelles, including neuronal numbers and size (Gorski *et al.*, 1978; Breedlove and Arnold, 1980; Gurney and Konishi, 1980; Gurney, 1981; Hannigan and Kelley, 1981; Jordan *et al.*, 1982; Ifft, 1964; Pfaff, 1966; Dörner and Staudt, 1968, 1969a,b; Bubenik and Brown, 1973; Gregory, 1975; Hellman *et al.*, 1976; Staudt and Dörner, 1976; Güldner, 1976), in the length of dendrites (Hammer and Jacobson, 1984), in the numbers of dendritic spines (Meyer *et al.*, 1978; Ayoub *et al.*, 1982) in dendritic distribution (Greeenough *et al.*, 1977; Ayoub *et al.*, 1982) in the morphology (King, 1972; Ratner and Adamo, 1971; Güldner, 1982), and

Table I

Steroid-Dependent Structural Dimorphism in the Vertebrate Central Nervous System[a]

Cytological difference	Region	Animal
Neuronal numbers	POA,[b] amygdala, spinal cord, vocal centers, neuromuscular effector units	Rat, mouse, songbird, frog
Neuronal size	POA, ventromedial hypothalamus, amygdala, habenula, hippocampus, cerebral cortex, vocal centers, neuromuscular effector units	Rat, mouse, monkey, songbird, frog
Dendritic length/branching	POA, suprachiasmatic nucleus, vocal centers	Rat, hamster, monkey, frog, songbird
Dendritic spines	POA, hippocampus, vocal centers	Rat, monkey, songbird
Numbers of synapses	Arcuate, suprachiasmatic nuclei	Rat
Type of synapses	Suprachiasmatic nuclei	Rat
Synaptic organization	POA, arcuate nuclei, amygdala	Rat
Synaptic organelles	Arcuate, suprachiasmatic nuclei, amygdala	Rat
Axonal density	Hippocampus (sympathetic); septum and habenula (vasopressingergic)	Rat
Regional nuclear volume	POA, spinal cord, amygdala, vocal centers	Rat, mouse, frog, songbird
Volume of neural structures	Cerebral cortex	Rat
	?Corpus callosum	Human

[a]For references, see text. Modified from Toran-Allerand (1984).
[b]POA, Preoptic area.

topographic distribution (organization) (Matsumoto and Arai, 1981a; Raisman and Field, 1973) of synapses, in the regional density of the vasopressinergic (DeVries *et al.*, 1981, 1983) and sympathetic innervation (Loy and Milner, 1980) of the septum and hippocampus respectively, and in the volume of certain neural structures such as the human corpus callosum (Delacoste-Utamsing and Holloway, 1982). It should also be noted that some steroid effects on brain morphology have been reported to occur after the traditionally regarded perinatal critical period has ended. Dramatic increases in synapse density in the arcuate nucleus and amygdala, but not POA, have been reported after estrogenic treatment that may be quite brief in adolescent rats (Clough and Rodriguez-Sierra, 1983). This emphasizes the point made earlier that there is no single critical period but probably one for each neural function, whose timing may range from the perinatal period to puberty, as seen here. In addition, effects of adult hormone manipulation on morphological measures have been reported in gerbils and canaries (Commins and Yahr, 1984; DeVoogd and Nottebohm, 1981), and in the ovariectomized, deafferented female rat (Matsumoto and Arai, 1981b). In the case of the adult, whether in hormone-sensitive regions that normally exhibit structural plasticity in adulthood such as the telencephalic song nuclei of the male canary (DeVoogd and Nottebohm, 1981), or in mammalian brain regions that have been deafferented, estrogen-responsive neurons may exhibit responses that are normally expressed during development, a type of recapitulation of the ontogenetic situation.

The precise sexually dimorphic functions subserved by these morphological differences are largely unknown. Their cytological features (soma, axons, dendrites, synapses) as well as the electrophysiological demonstration of dimorphism in the synaptic connectivity between the amygdala and POA–AH regions (Dyer *et al.*, 1976) have led to the hypothesis that sex-specific differences in neural connectivity or circuitry may form the substrate for sexual differentiation. Such morphological observations in the adult, however, merely represent the final results of the steroidal effects. They tell little about the underlying morphogenetic or cellular mechanisms that produced them.

IV. The Possible Role of α-Fetoprotein

Although it is generally accepted that androgen is necessary for differentiation of the male rodent brain, little is known about the steroid requirements, if any, for the normal development of the female brain, since androgen alone is generally viewed as the determining factor. It has been largely assumed that the perinatal female brain is protected from excessive estrogenization (i.e., masculinization) by the high level of perinatal estrogens through the extracellular sequestration by binding to α-fetoprotein (AFP) (McEwen *et al.*, 1975). AFP is a specific α_1 globulin and the major plasma protein of embryonal and fetal sera and the amniotic fluid of numerous vertebrate species, including man (Ruoslahti and Engvall, 1978). Synthesized in the yolk sac and liver, it is estrogen- (but not androgen-) binding in the rodent (Raynaud *et al.*, 1971).

The current view of AFP's biological role with respect to the developing rodent brain is that this protein retards by mass action the interaction of estradiol with its intracellular receptors (McEwen *et al.*, 1975), estradiol's lower affinity for AFP ($K_D = 10^{-8}\ M$) than for its receptor ($K_D = 10^{-11}\ M$) (Linkie and LaBarbera, 1979) being offset by AFP's extremely high concentration in fetal and newborn plasma (1–6 mg/ml) (Ruoslahti and Engvall, 1978). Support for this hypothesis has been suggested by MacLusky *et al.* (1979a,b) who found that while the estrogen receptors of the fetal and neonatal male rodent brain are always occupied by endogenous estradiol of presumed aromatized androgen origin, such nuclear binding was not detectable in the female brain. However, the possible limits in sensitivity of the assay used should also been taken into consideration.

The existence of high levels of estrophilic AFP, equally present in the fetal and neonatal rodent of both sexes, has raised questions regarding its possible involvement in the differentiation and growth of estrogen target organs. The work of Raynaud (1971) suggests that AFP controls uterine uptake of estradiol, since injection of a synthetic estrogen with little affinity for AFP exerted a more pronounced uterotrophic effect than estradiol. Other studies (Alvarez and Ramirez, 1970; Nunez *et al.*, 1979) however, have shown that injected estradiol is readily concentrated in the uterus of immature rats. Soto and Sonnenschein (1980), moreover, have suggested that AFP per se may be a specific inhibitor of cell multiplication of estrogen-sensitive tumor cells. It is not clear whether estrogen enters the cell alone or bound to AFP.

It has been proposed that at low steroid concentrations cellular entry of estrogen into the immature rat uterus may be carrier-mediated in contrast to the diffusion seen with high levels (Milgrom *et al.*, 1973). Several studies have shown an intracellular pool of AFP of unknown functional significance in soluble extracts of fetal and neonatal mouse and rat brain (Attardi and Ruoslahti, 1976; Palpinger and McEwen, 1973). Its intraneuronal localization has been confirmed by immunocytochemistry within the CNS of a wide variety of developing avian and mammalian species, including primates (see Toran-Allerand, 1984, for review). The discrete intracytoplasmic localization within the neurons raises questions regarding its possible active role in neuronal differentiation and development.

V. The Possible Importance of Estrogen Per Se

Speculation that estrogen may have a more fundamental and general role in the development of the CNS of both sexes is increasing. In both the neonatal male and female rodent, estrogen-concentrating neurons (Sheridan, 1979; Gerlach *et al.*, 1983) and high levels of estrogen receptors (Barley *et al.*, 1974; MacLusky *et al.*, 1979a,b; Friedman *et al.*, 1983) have been demonstrated in the cerebral cortex *only* during the first two weeks of postnatal life. Estrogen receptors have also been demonstrated in the cerebellar cortex of the immature (22–27 days old) mouse of both sexes (Fox, 1977). In the adult female rodent, the greatest concentration of estradiol receptors is localized in the hypothalamus–POA and the amygdala and such receptors are virtually absent from the cerebral cortex (Eisenfeld, 1970; Pfaff and Keiner, 1973; Zigmond and McEwen, 1970). In neonatal rats and mice of both sexes, however, equal amounts of high-affinity estradiol receptors are found in *both* the hypothalamus and cerebral cortex at a level similar to that of the adult female hypothalamus (MacLusky *et al.*, 1976; Friedman *et al.*, 1983). The significance of the cerebral and cerebellar receptors is unknown. Barley *et al.* (1974) have questioned whether estradiol itself might not be involved in an unspecified "organizational" effect on the brain. Fox (1975), furthermore, has proposed that relative concentrations or ratios between both androgens and estrogens may be important in both sexes as a mechanism for the hormonal effects. It is not known whether or not the estrogen receptors of the cerebral and cerebellar cortex are at all functional, since their occupancy by endogenous estrogen has not yet been shown, perhaps due to AFP's preventing its entry into the cell (MacLusky *et al.*, 1979b).

Tangential evidence, however, supports the hypothesis that the estradiol receptors in the neonatal brain may mediate developmental effects of endogenous estrogen as well as those of testosterone. The ontogenetic patterns of various morphological, biochemical and physiological aspects of the neonatal period, viewed in relation to one another, would argue for an active role for estrogen in CNS development. MacLusky *et al.* (1976, 1979b) and Friedman *et al.* (1983) have shown that neonatal estradiol binding by the cerebral cortex is maximal around day 10 after birth, the cortical binding sites declining precipitously thereafter between days 10 and 15 to virtually undetectable levels. Thus the cortical receptors differ from the hypothalamic estradiol receptors, which are maintained throughout adulthood. Of great significance in this regard is that the cerebral and cerebellar cortex of the rodent, like the hypothalamus–POA are developing rapidly during the postnatal period. For example, dendritic extension in the visual cortex is maximal between approximately days 3 and 7 (Juraska and Fifkova, 1979a) and the rate of synaptogenesis rises beginning at about day 7 (Juraska and Fifkova, 1979b). Mylelination and other critical aspects of morphological, biochemical and functional differentiation develop during the second and third postnatal weeks (Bass *et al.*, 1969a,b), a period that coincides with the maximal presence of the estradiol receptors. Of particular note is that, unlike the hypothalamus–POA in which steroid sensitivity is characteristic of the adult as well, the cortical estradiol receptors are present *only* during neural differentiation. In the postnatal cerebellum, moreover, estradiol receptors appear to be localized primarily to the granule cells (Fox, 1977) during the third week, a major period of neuronal differentiation, during which granule cell genesis ceases and synapses form at high rates during and after their migration from external to internal granular layer (Altman, 1967).

In the rodent, during this same postnatal period, AFP decreases linearly from its very high concentration at birth (half-life of 24 h) to trace levels (0.01% of fetal levels) at weaning (Vannier and Raynaud, 1975), where it persists unless its synthesis is restimulated by such conditions as liver malignancy (hepatoma) in the adult (Ruoslahti and Engvall, 1978). In both neonatal male and female rats serum estrogen levels, on the other hand, after being elevated for the first 2 postnatal days (maternal origin) and then rapidly falling, increase abruptly between days 9 and 19 (with a peak around day 10) to levels that are never observed subsequently during life (Döhler and Wüttke, 1975). Declining levels of the estrogen-binding AFP thus occur in the presence of very high serum estrogen levels. This has suggested that the increasing levels of free estradiol might exert an organizational effect in estrogen receptor-containing regions of the developing CNS. These speculations find support in the observations of exogenous estrogen-induced enhancement of cortical maturation (Heim and Timiras, 1963; Heim, 1966), myelinogenesis (Curry and Heim, 1966), and cerebral amino acid concentrations (Hudson *et al.*, 1970), and of the permanent inhibitory action of estrogen on the incorporation of [^{3}H]lysine into the cerebral cortex and cerebellum (Litteria, 1977; Litteria and Thorner, 1974, 1976). (In addition, estrogen may be involved in sex differences in the response of cortical and hippocampal dendritic branching to rearing in complex versus isolated environments; see chapter by Juraska, this volume.)

Finally, the possible fundamental importance of estradiol to CNS differentiation is underlined by the absence of known genetic defects of its receptor, suggesting that such defects may not be compatible with embryonic or fetal survival. By contrast, defects of the androgen receptor are well described, such as the Tfm mutation of rodents (Ohno *et al.*, 1974; Shapiro *et al.*, 1976). This is not meant to imply, however, that androgens acting alone or in conjunction with estradiol may not also exert pre- and/or postnatal influences of their own, independent of aromatization. Putative androgen nuclear receptors have been detected in both sexes by assay in the hypothalamus, POA, and cerebellum of neonatal mice (Attardi and Ohno,

1976; Fox, 1975, 1977), and by autoradiography in the neonatal rat (Sheridan *et al.*, 1974; Breedlove and Arnold, 1980). Their specific absence has been described in the androgen-insensitive, androgen-receptor–deficient, Tfm/y mutant mouse. Breedlove and Arnold (1980) and Breedlove *et al.* (1982) have shown also that the spinal cord of the rat contains a sexually dimorphic nucleus that contains androgen receptors and not estrogen receptors and whose presence is androgen-dependent exclusively.

VI. An Experimental Approach to the Problem

Questions raised by these observations are difficult to study in the intact animal. One way of investigating the effects of the gonadal hormones on the differentiation and development of the newborn rodent brain has been to use tissue culture. Tissue culture methods offer a unique and more dynamic approach to many problems of CNS development by providing a situation that avoids many *in situ* limitations and *in vivo* constraints. Despite its isolation *in vitro*, cultured CNS tissue is only relatively free from the complex influences of its cellular and humoral environments. It may, for example, receive hormonal and other contributions from the serum and embryo extract components of the nutrient medium as well as exhibit structural and functional alterations attributable to its deafferented or nonafferented state. The existing culture systems, organotypic or explant culture, cell dissociates, and clonal cell lines differ primarily in the degree to which they vary from the *in situ* condition; organotypic culture, the system to be described, retains the closest similarity to normal CNS tissue.

These *in vitro* studies offer a novel approach for supplementing and extending *in vivo* studies in order to elucidate some cellular aspects of sexual differentiation. The organotypic or explant culture method encourages the cytologic and histologic differentiation and development of fragments of undifferentiated or poorly differentiated fetal or neonatal nervous tissue over periods of weeks and months in ontogenetic patterns and cytoarchitectonic relationships remarkably similar to those regions *in situ*. This method is based on the premise that interactions between all the cells of a region and their three-dimensional architectonic relationships are fundamental to developmental processes and to normal neural function.

Implicit in this approach is the attempt also to preserve as many cells of as heterogeneous a population as possible so as to avoid introducing the inevitable bias of selection inherent in dissociated cell culture methods. This is particularly important for the *in vitro* study of the hypothalamus, since it is a region of great developmental, anatomical, and functional heterogeneity in which subpopulations of neurons such as the steroid-receptor–bearing ones may be randomly distributed or regionally localized and hence very susceptible to loss during dissociation. Despite the limitations imposed by the method, the accessibility of hypothalamic neurons to direct experimental manipulation and direct observation under the microscope in a culture system, exhibiting remarkable structural and functional integrity, enables correlative cytologic, physiological, and biochemical studies otherwise not possible.

Tissue culture studies (Toran-Allerand, 1976, 1980a) have shown that addition of estradiol-17β or testosterone to the nutrient medium of explants of the hypothalamus–POA of the newborn mouse of either genetic sex induces an accelerated and progressively intense growth of neuronal processes (neurites) that is restricted to specific regions of specific hypothalamic explant levels. At least some of the enhanced growth reflects increased dendritic arborization within the explant, based upon quantitative analyses of Golgi-stained cultures (Toran-Allerand *et al.*, 1983). This steroid response is most marked in the POA and infundibular–premamillary regions and is characterized in its extreme by the formation of extensive neuritic arborizations or plexuses of very fine fibers, as shown in Fig. 1A and B. There is a close correlation between the pattern and regional localization of this response and the presence and topographic distribution of nuclear estrogen receptors as shown by [^{3}H]estradiol autoradiography in Fig. 2 (Toran-Allerand *et al.*, 1980). Localized areas of steroid responsive neuritic arborizations appear to emanate only from regions containing labeled cells. Nonresponsive regions contain few or no labeled cells. Relatively few fibers appear to contribute to the plexuses. This pattern of the response and its regional localization have suggested that steroids may induce neuritic branching perhaps only in neurons containing the steroid receptor.

The developmental importance and specificity of estradiol is emphasized by a suggestive dose-dependence of the response and by exposure of the cultures to nutrient medium whose serum component had been pretreated by physical, pharmacological or immunochemical means in order to reduce the availability of the steroid to the culture (Toran-Allerand, 1976, 1980a). The development of neurites in cultures so treated is retarded and only in regions previously shown to be steroid-responsive as indicated in Fig. 3. The importance of estrogen derived from aromatization as a component of the testosterone effect is suggested by the apparent failure of the nonaromatizable androgen DHT (Toran-Allerand, 1980a) to induce a response and by the characteristic neuritic growth elicited by exogenous testosterone in cultures derived from the androgen-insensitive (receptor-deficient) Tfm/y mouse

(C. D. Toran-Allerand, unpublished results). That local aromatization alone, however, may not be sufficient for the action of testosterone and that there may exist an additional requirement for estradiol other than that derived from aromatization is further suggested by the observation that in the absence of estrogen, testosterone, added at levels previously shown to be stimulatory, is without any visible inductive effect (Fig. 4) (Toran-Allerand, 1980a).

Moreover, while the functions of the estrogen receptors of the cerebral cortex has been seriously questioned, since it has not been possible to demonstrate occupancy of these receptors by endogenous estrogen (MacLusky *et al.*, 1979a), tissue culture studies suggest the contrary. Organotypic cultures of the perinatal mouse anterior cingulate/frontal cortex have been shown to contain estrogen receptors by means of [^{3}H]estradiol autoradiography (Fig. 5) and to respond to estradiol by a marked enhancement of radial neuritic growth, which, unlike the hypothalamus–POA, shows little tendency to arborize (Fig. 6). Estradiol has also been shown to stimulate the incorporation of [^{3}H]fucose and/or [^{3}H]leucine into the glycoproteins and proteins of cultures of the hypothalamus–POA (N. J. MacLusky, C. D. Toran-Allerand, and B. S. McEwen, unpublished results) as well as those of the cerebral cortex (N. J. MacLusky and C. D. Toran-Allerand, unpublished results).

VII. How Might Hormones Change Growth Patterns and Neuronal Circuits?

These tissue culture observations may be viewed in terms of some general principles of CNS differentiation in order to provide clues toward an understanding of the possible morphogenetic basis for sexual differentiation. Differentiation and development of the CNS are based on well-ordered sequences of interlocking phenomena whose timing is critical, as other chapters have noted. Temporal alterations may have profound physiological consequences that may be expressed morphologically in such subtle ways as by alterations in the patterns of axonal growth, dendritic differentiation, and synaptic organization. Many studies on the ontogeny of neural circuits throughout the CNS have shown the morphogenetic and temporal importance of the afferent fiber input to the dendritic differentiation and synaptic organization of target neurons (Ramon y Cajal, 1910–1911; Morest, 1969; Kornguth and Scott, 1972). Dendritic development often appears to be induced by and partially dependent on its afferent axonal input (e.g., see Chapter by Clopton, this volume). Since the amount of postsynaptic space

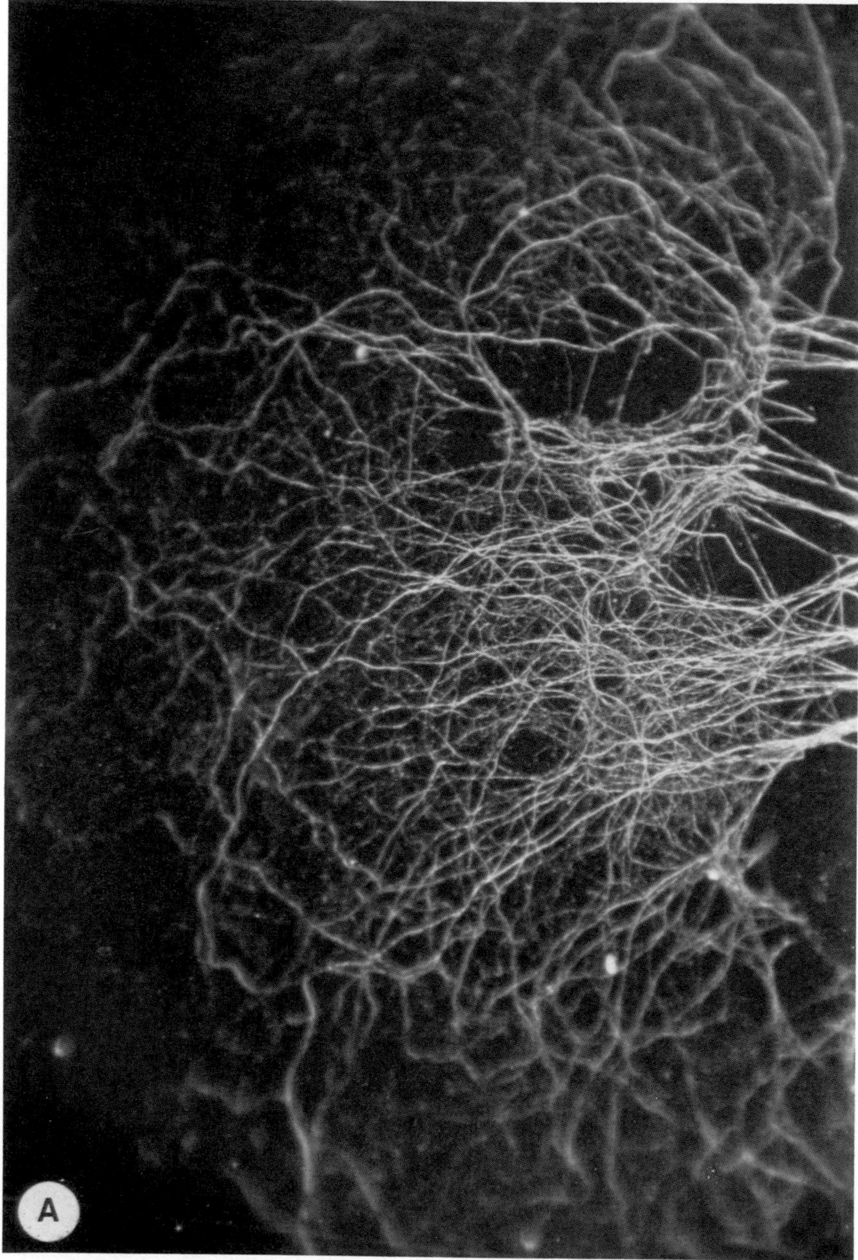

Fig. 1. Steroid-responsive neuritic outgrowth. Homologous pair of explants from the preoptic area, 13 days *in vitro*. (A) Control (horse serum, endogenous estradiol 200 pg/ml). (B) 50 ng/ml in horse serum. Morphological concomitant of a dose–response effect. The surface areas of

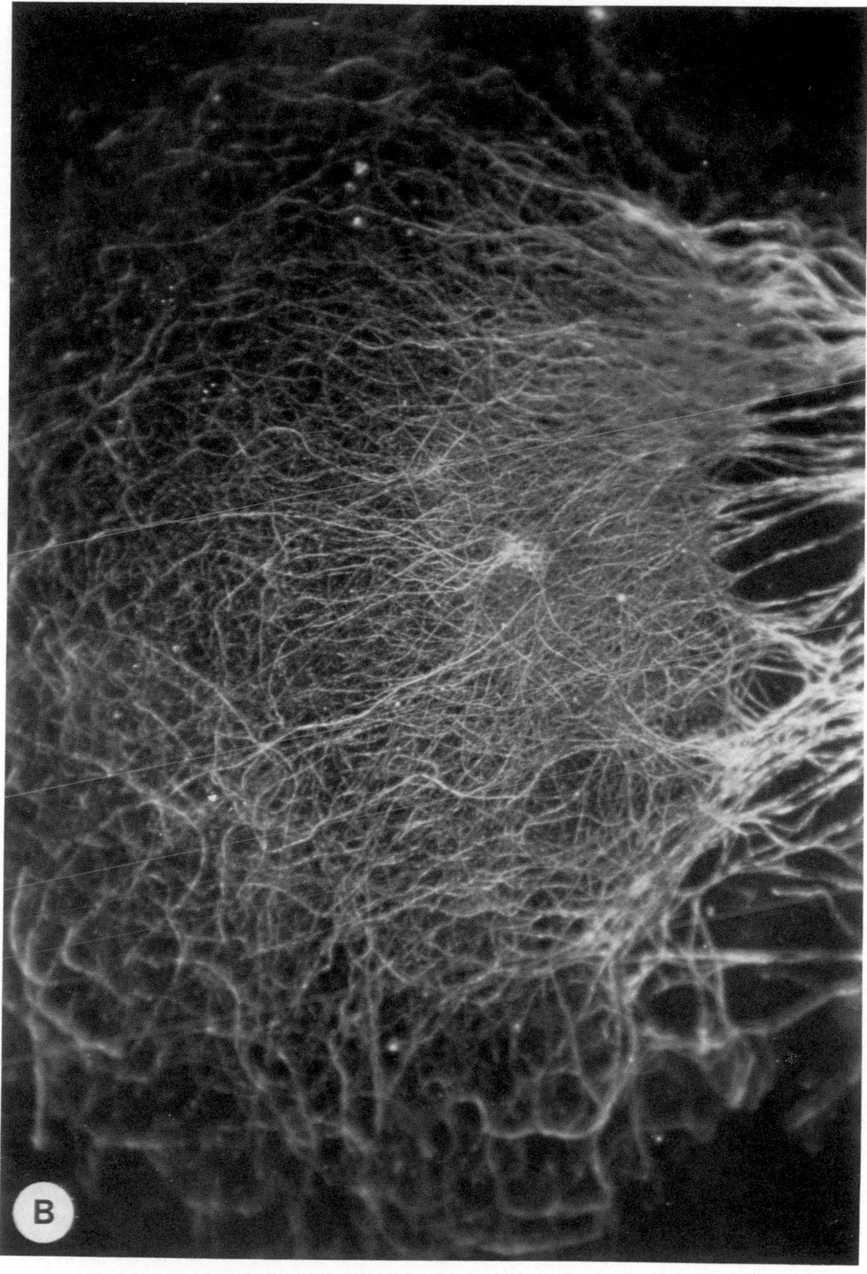

the neuritic arborizations do not differ significantly but the differences in unit density are striking, suggesting steroidal induction of neuritic branching. Holmes' silver impregnation, dark field, ×25. From Toran-Allerand (1980a).

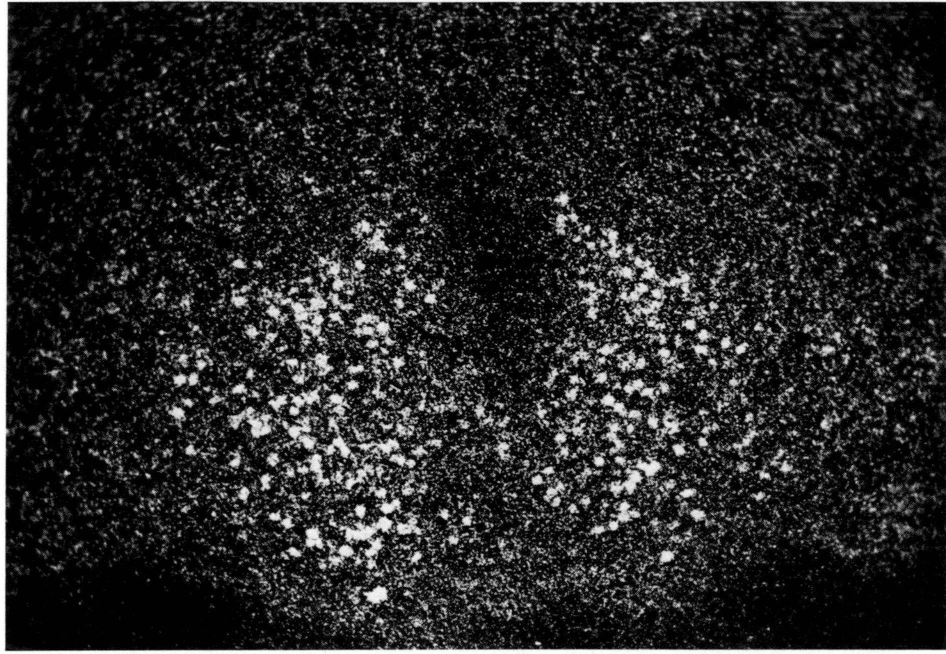

Fig. 2. Autoradiogram of cells of the newborn mouse preoptic area exposed to [^{3}H]estradiol on the seventh day *in vitro*. Dark field, ×40. From Toran-Allerand *et al.* (1980).

available appears to be limited and relatively constant for each neuronal type, the distribution of synaptic sites among different groups of converging axons appears to be determined competitively on a temporally restricted basis (Gottlieb and Cowan, 1972). Even dendrites appear to compete for their afferents (Perry and Linden, 1982). Thus primacy of dendritic maturation, axonal arrival, and synapse formation may heavily influence the pattern of connections.

The *in vitro* observations show that explant regions containing estrogen receptors appear to respond to varying levels of estrogen by variations in both the rate and extent of the growth of their neurites. This suggests that steroidal trophic influences on neuritic growth patterns and/or on neuronal survival may play a role in the morphogenesis of sexual differentiation by so influencing afferent axonal growth, terminal arborization, synapse formation, and dendritic differentiation as to result in fundamentally different, sex-specific patterns of neural circuitry. If, as has been shown in various CNS regions, the topographic distribution of synapses reflects the differential growth rates of specified axons during development, then one might postulate that differences in neuronal responsiveness to steroids might well result in differences in the growth of axons and in the mode and stability of their termination. Thus, if one were to assume that enhanced growth of neurites and/or augmented neuronal survival might represent the primary or initial responses of steroid-sensitive neurons to estrogen and/or androgen (depending upon the animal species), then the morphogenetic consequences would necessarily exert pervasive effects on all subsequent aspects of target neural differentiation. Such a cascading sequence of events summarized in Fig. 7, could well result in both sexually differentiated circuits and in the observed structural dimorphism of the adult brain referred to earlier (Toran-Allerand, 1984). For example, the increased numbers of neurites, whether resulting from trophic influences on neuritic growth (neurite-promoting) or on enhanced neuronal survival (neuronotrophic) (Toran-Allerand, 1984), would result theoretically in both more axons and more dendrites. As a consequence of the former, there would be an increase in terminal arborizations and more potential synaptic contacts. These increased synaptic contacts might serve to preserve (stabilize) more or different synapses (i.e., increase the number and type of surviving synapses and influence their topographic distribution and hence the pattern of organization). In addition, these contacts might result in enhanced target neuron survival, expressed as a gross or selective increase in cell numbers. Thus the pattern of synaptic organization, and possibly even the numerical distribution of cell types, may be modulated by steroid acceleration of neuritic outgrowth. Sexual dimorphism of brain structure may perhaps be viewed, therefore, as the ultimate morphogenetic consequence of steroid-induced differences in the rate and extent of neuronal development and of neuritic interactions in select, steroid-sensitive neuronal populations.

The tissue culture studies also suggest that the absence of androgen may not be sufficient for the emergence of the female pattern of neural organization. Masculine and feminine patterns may *both* require active induction by estrogen. The possible source of such estrogen in the rodent is intriguing. As mentioned above, the developing rodent brain, particularly that of the female, is generally assumed to be completely protected from potential masculinization by the high perinatal levels of estrogens through their extracellular sequestration by binding to AFP.

In the rodent AFP is present throughout neural development at concentrations that are sufficiently high to presumably bind all available circulating estrogens. The intraneuronal localization of the estrophilic AFP (Toran-Allerand, 1980b) however, as shown in Fig. 8, may have profound implications for the process of sexual differentiation of the rodent brain. Its intracellular presence forces one to reconsider the extent to which the developing rodent brain is actually protected from exposure to estrogens by the extracellular AFP. If one considers that the intraneuronal AFP is probably derived from endocytosis and internalization of the extracellular plasma protein since neurons lack the mRNA for AFP (Schacter and Toran-Allerand, 1982), then neuronal uptake of the estrogen-binding proteins must bring estradiol *into* the cell. Since there is a difference of several orders of magnitude

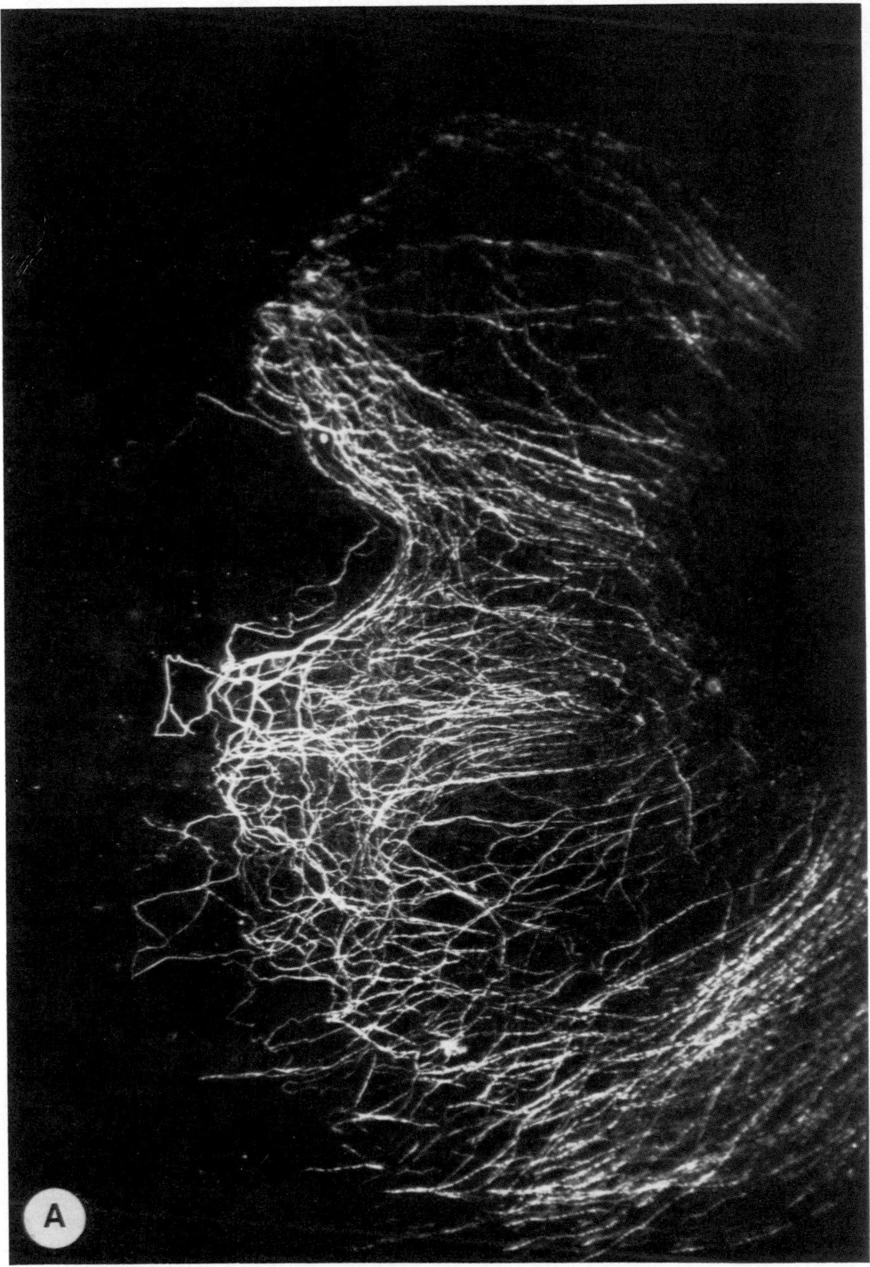

Fig. 3. Homologous pair of explants from the preoptic area, 22 days *in vitro*, exposed to normal serum containing either (A) antibodies to bovine serum albumin (BSA) or (B) antibodies to estradiol-17β/BSA. Note the striking reduction in neuritic outgrowth following physiological

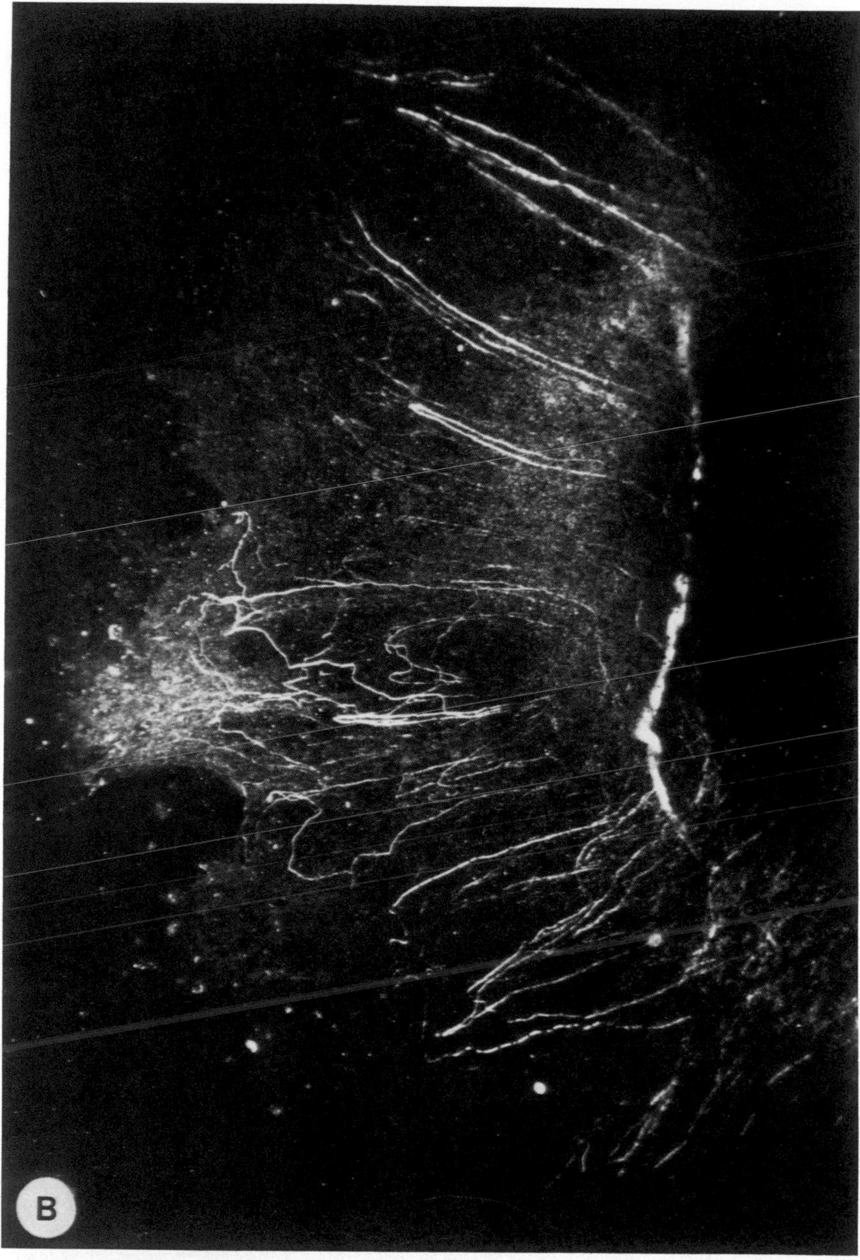

inactivation of the estradiol. Note also the typical looping back of the fibers toward the explant, the characteristic pattern seen in many regions of the CNS *in vitro*. Holmes', dark field, ×15.8. From Toran-Allerand (1980a).

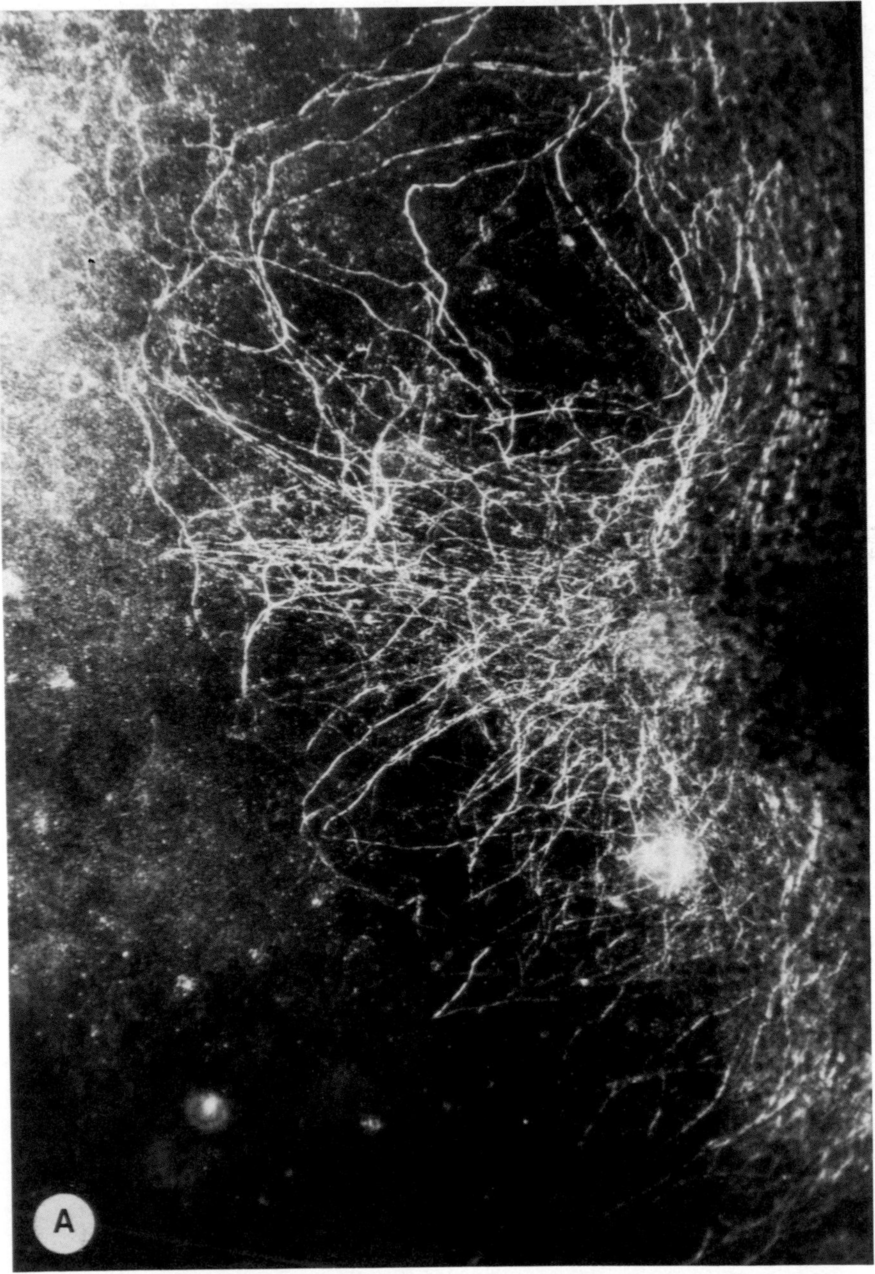

Fig. 4. The response to testosterone in the absence of estradiol. (A) Control [estradiol-deficient serum (Sephadex LH-20)]; (B) testosterone (5 μg/ml in Sephadex LH-20 treated serum).

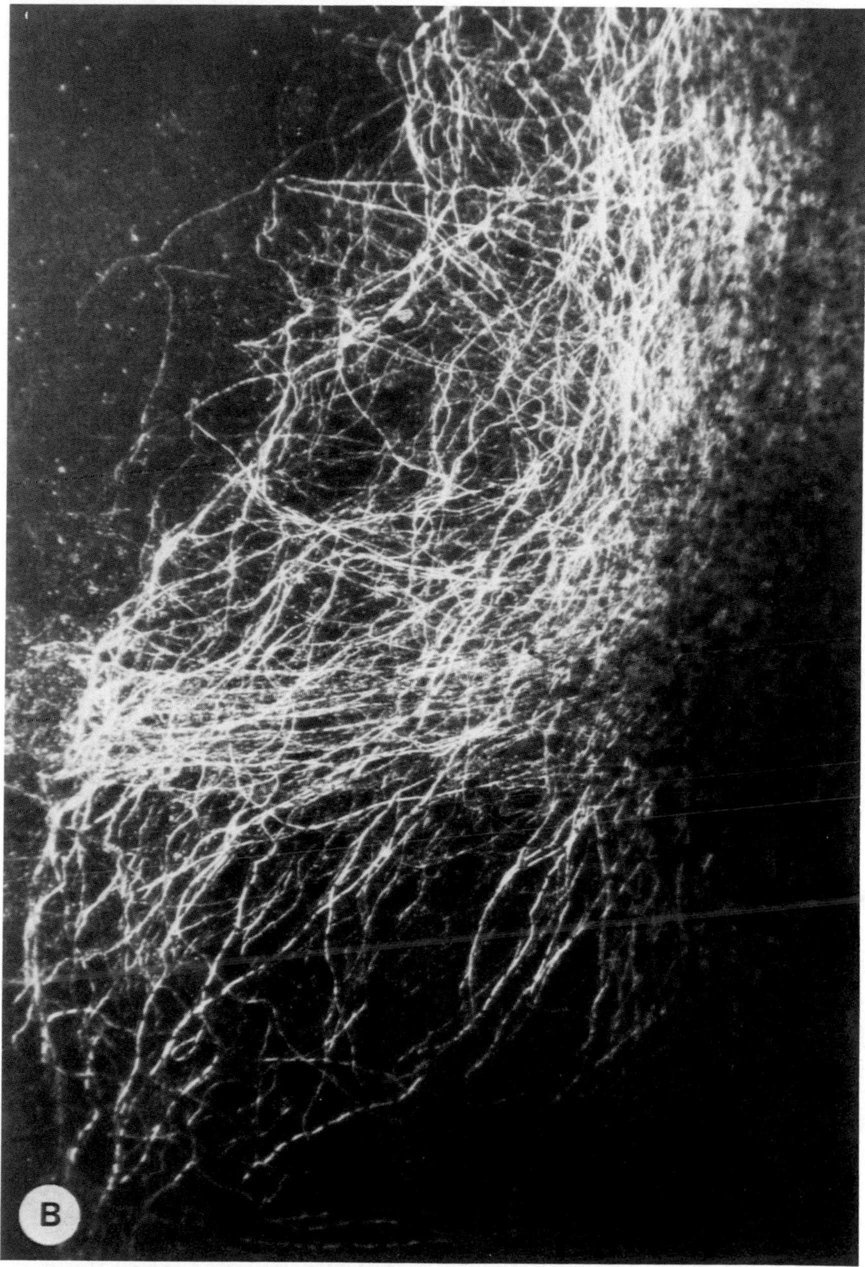

Homologous explant pairs from the preoptic area, 20 days *in vitro*. Note the absence of a significant neuritic response in the homolog exposed to testosterone alone. Holmes', dark field, ×125. From Toran-Allerand (1980a).

between the affinity constants for the binding of estradiol by AFP and by the estrogen receptor (Savu *et al.*, 1981), intracellular dissociation of the plasma protein–estradiol complexes, resulting from this greater affinity of estradiol for its receptor, could liberate the steroid for subsequent receptor binding. Such a mechanism might thus provide target neurons of both sexes with the intracellular source of estradiol of nonandrogen origin that was postulated earlier. This also suggests that in the rodent, at least, AFP may have an important role as mediator or modulator of the intraneuronal transport of estrogen and the other biologically active substances that bind to them. For example, rat, mouse, and human AFP both bind endogenous, nonestrogen lipid ligands that behave as true competitors for both the AFP-bound estrogens (Nunez *et al.*, 1979) and the estrogens bound to a uterine cytosol. The majority of these lipids have been identified as polyunsaturated, long-chain free fatty acids of which arachidonic, docohexaenoic, and docosatetraenoic acids appear to predominate (Pineiro *et al.*, 1979). These have been shown (Crawford and Sinclair, 1972; Sinclair and Crawford, 1972) to have critical importance to brain development in mammals, including man. Maximal incorporation of these fatty acids into the brain of the postnatal rat occurs around days 10 to 12 (Pineiro *et al.*, 1979), a period of rapid cerebral cortical development involving neuronal growth and differentiation and the onset of myelinogenesis (Bass *et al.*, 1969a,b; Jacobson, 1963), which coincides with the presence of the estrogen receptor.

Although the female neural phenotype is assumed to result normally from a female genotype *and* the absence of excessive androgen exposure, the tissue culture findings suggest a mechanism by which estradiol per se could exert a positive developmental influence on the CNS. As shown in Fig. 9 for example, in the genetic female or neonatally castrated male, exposure only to low trophic (submasculinizing) levels of estradiol, which originate perhaps from the intracellular dissociation of the AFP/estradiol complexes, might result in a specific pattern of neural organization. The superimposition of intraneuronal aromatization of testosterone to estradiol, on the other hand, could produce a more concentrated estrogenic effect, and this resultant additive stimulus to neuritic development or change in the balance of ratio of testosterone too estradiol (Fox, 1975) might induce a different or what would be termed a male pattern of neural differentiation.

The possible importance of ovarian estrogen in the ontogeny of female neural patterns is further supported by various studies showing the importance of the postnatal and prepubertal ovary for feminization of both reproductive (Gerall *et al.*, 1973; Hendricks and Duffy, 1974; Hendricks and Weltin, 1976) and nonreproductive (Stewart and Cygan, 1980) behaviors. Furthermore, in neonatally gonadectomized males or females, low, submasculinzing levels of estradiol have been shown to lead to the development of the high levels

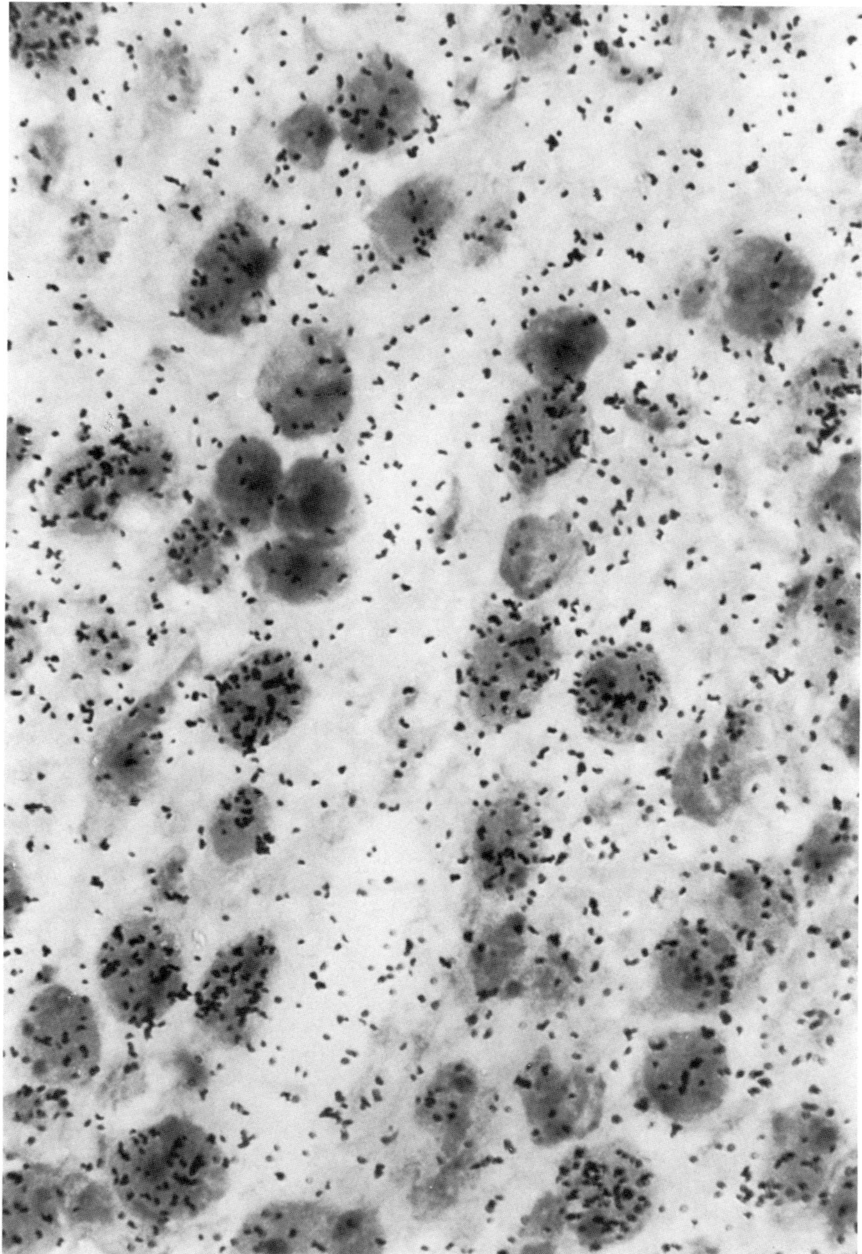

Fig. 5. Autoradiogram of cells in a culture of the newborn mouse anterior cingulate cortex exposed to [^{3}H]estradiol on the eighth day *in vitro*. Radioactivity is seen as silver grains concentrated over the nuclei of large cells, which presumably are neurons. ×252.

Fig. 6. Estrogen-induced enhancement of neuritic growth in living homologous cultures of the newborn mouse anterior cingulate cortex, 9 days *in vitro*. (A) Control (horse serum, estradiol,

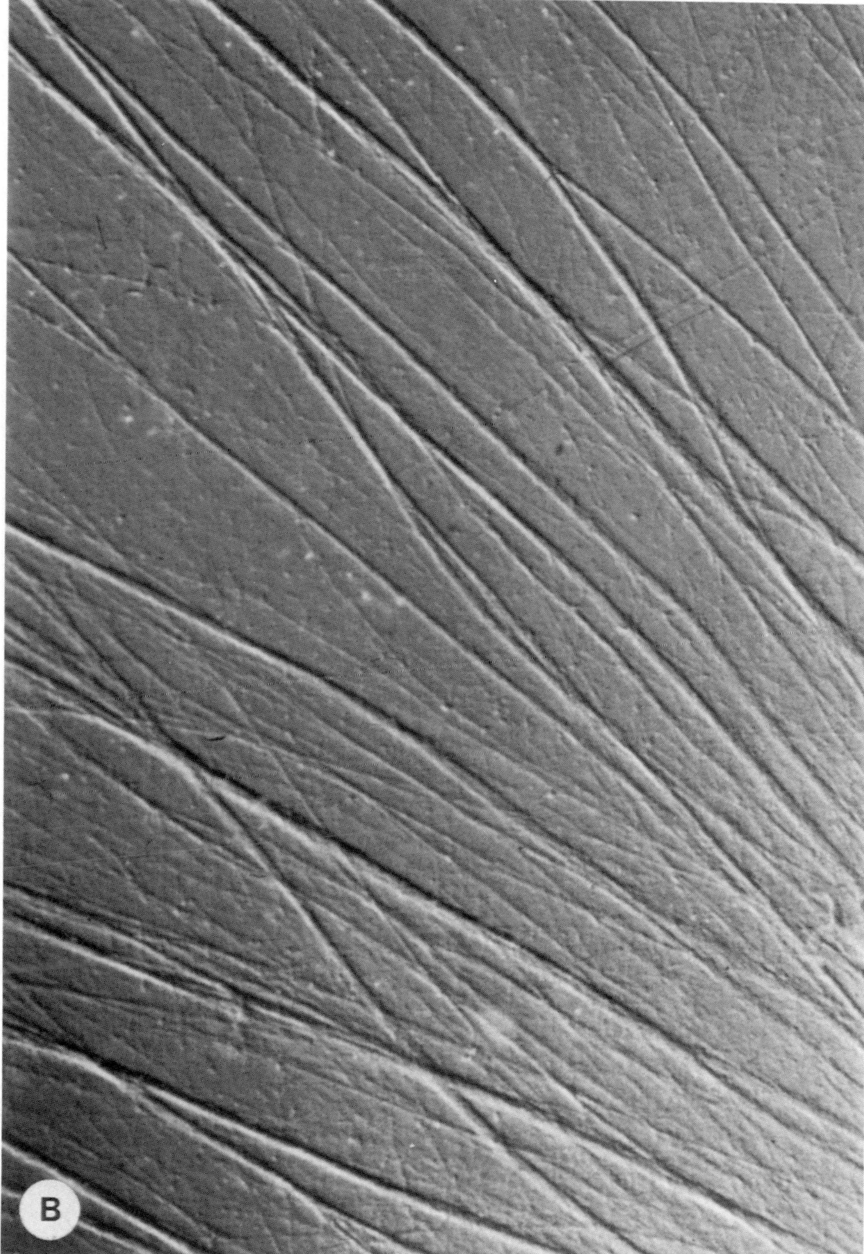

200 pg/ml). (B) Estradiol 50 ng/ml in horse serum. Note that unlike the response in the POA (Fig. 1B), the responsive cortical neurites are radial and very long, with little or no tendency to arborize extensively. Nomarski, ×25.2.

of open field activity that characterize the adult female. The development of sex differences in cerebral cortical thickness of the rat is also influenced by ovariectomy (Diamond *et al.*, 1979), and postnatal but not adult ovariectomy has been shown to influence (masculinize) the sex difference in cerebral asymmetry (Diamond *et al.*, 1981). In addition, postnatal exposure to the estrogen antagonist Tamoxifen resulted both in the inhibition of the feminine form of the sexually dimorphic nucleus and defeminization without masculinization of both gonadotropin regulation and female sexual behavior (Hancke and Döhler, 1980), effects that were blocked by concomitant of "low" doses of estradiol. An alternative explanation for the absence of masculinization in these experiments takes into account the known temporal differences in the "critical" periods for many aspects of masculinization (largely prenatal) and of defeminization, a largely postnatal phenomenon (see McEwen, 1983, for discussion).

VIII. Hormonal Effects: Direct or Indirect?

What is not yet clear but obviously fundamental to an understanding of steroidal effects on the developing CNS is whether the response of neurites to estradiol is mediated by the hormone directly or indirectly, via the intermediary of secreted growth or trophic factors of neuronal and/or neuroglial origin, as diagrammed in Fig. 10 (Toran-Allerand, 1984). Since it is still not possible to relate estrogen-responsive neurites to their cell bodies of origin, they may be considered as originating from either estrogen-receptor–containing neurons *per se* or from other (nonreceptor) neurons, responding to estrogen-stimulated growth factors. Experiments in collaboration with Karl Pfenninger and Leland Ellis (Toran-Allerand *et al.*, 1984, and in preparation) suggest the latter possibility and provide support for the hypothesis that estradiol's effect on the developing CNS may be in part indirect and involve interactions of the steroid with an endogenous growth (neurite-promoting) factor.

Concurrent exposure of cultures of the hypothalamus, POA, and cingulate cortex of the E-17 mouse and of the cingulate cortex and olfactory bulb of the E-18 rat to both estradiol and high (10–50 μg/ml) concentrations of insulin results in a massive outgrowth of neurites restricted to regions containing estrogen receptors. The response far exceeds that seen with either hormone alone, and on a polylysine substratum consists of a dense halo of fine radial neuritic outgrowth resembling that elicited by nerve growth factor in cultures of peripheral ganglia.

The superphysiological concentrations of insulin required suggest that the factor(s) involved may not be insulin but a closely related molecule(s) such

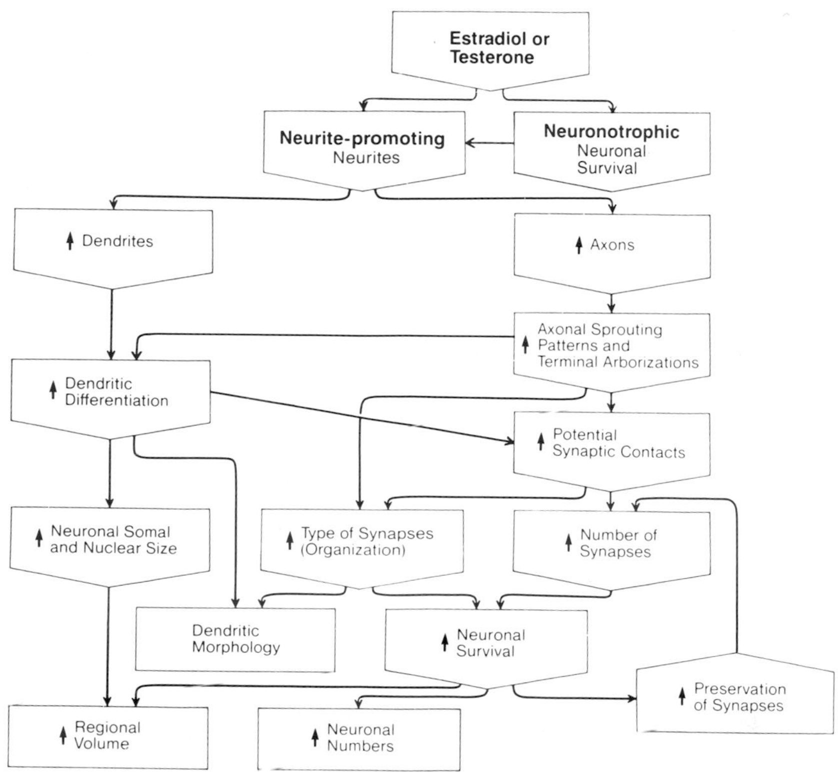

Fig. 7. Diagrammatic representation of the "cascade hypothesis." The morphogenetic consequences of early stimulation of neuritic growth by steroid could elicit a cascade of cellular events affecting subsequent aspects of target neural differentiation to result in the sexual dimorphism of brain structures. Modified from Toran-Allerand (1984).

as one or more of the insulinlike growth factors (IGF), a member of the family of somatomedins (Daughday *et al.*, 1972; Zapf *et al.*, 1981; Lenoir and Honegger, 1983). Insulin and/or IGF have been shown in cultures of the CNS and peripheral nervous system to stimulate DNA synthesis (Lenoir and Honegger, 1983; Raizada *et al.*, 1980) and to promote neuronal survival (Snyder and Kim, 1980) and neurite extension (Bhat, 1983). Moreover, both insulin and IGF binding sites, are widely distributed throughout the CNS. Insulin receptors, at least, are reported to be particularly concentrated in the hypothalamus, POA, cortex, and olfactory bulbs (Baskin *et al.*, 1983; Underhill *et al.*, 1982), all regions also rich in estrogen receptors. Both insulin (Rosenzweig *et al.*, 1980) and IGF (Binoux *et al.*, 1981; D'Ercole *et al.*, 1980) have been shown to be synthesized locally by the fetal brain.

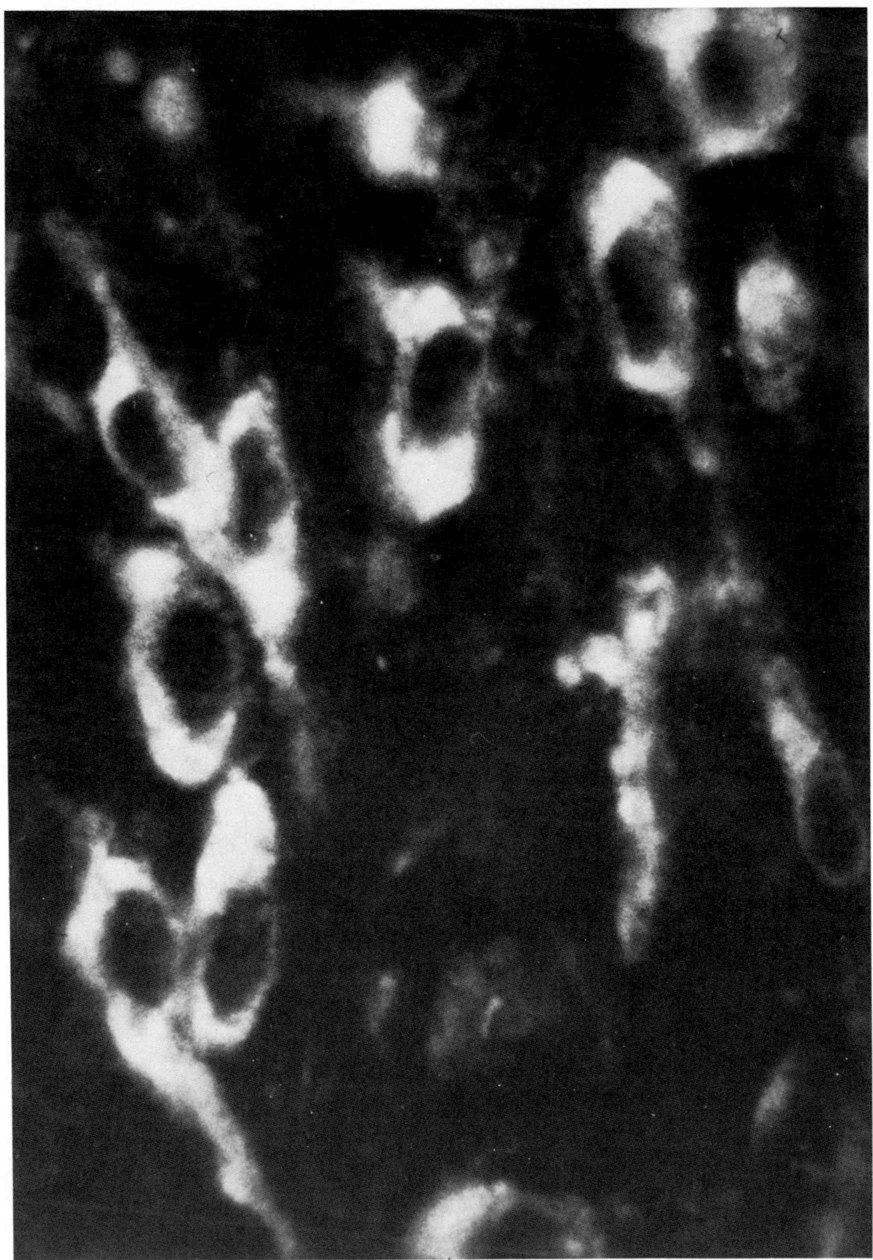

Fig. 8. Localization of immunoreactive mouse AFP in neurons of the newborn male mouse hypothalamus. Rhodamine immunofluorescence, ×200. From Toran-Allerand (1984).

The apparent synergism between estradiol and insulin with respect to neuritic growth suggests that estradiol's effect may be indirect and involve modulation of the synthesis and/or release of growth factors such as IGF, the regulation (up or down) of their receptors, or that it may even influence other aspects such as target cell binding affinities and degradation rates of IGF.

IX. *In Vitro Veritas*?

It is often difficult to relate *in vitro* findings to the *in vivo* state, since many variables that do not normally exist *in situ* may be introduced by the culture situation. There is considerable evidence from a wide range of tissue culture studies, however, that the responses of CNS tissue *in vitro* are not anomalies of culture per se but represent, rather, the intrinsic potential of nervous tissue elicited or modified by the very abnormalities of the environment. Thus while the cellular nature of the responses of the culture to steroids (i.e., the focal enhancement of neuritic outgrowth) represents quite likely a response intrinsic to CNS tissue, the specific pattern of the response, (i.e., the extensive neuritic arborizations themselves) may well result from both de- or nonafferentation, as well as the absence of suitable targets.

Ramon y Cajal (1910–1911) has shown, for example, that during the early stages of dendritic differentiation of the Purkinje cell, multiple undifferentiated primary dendritic processes are formed from various parts of the cell body. He suggested that these were formed at points of least resistance, attributing them to induction by the cell body proper (?genomic). Ramon y Cajal further proposed that the definitive pattern of dendritic arborization depended on the imminent or actual contact of the afferent axon, followed by resorption of the extraneous branches and increasing complexity of the remaining ones. Thus the early neuritic response to steroids *in vitro* may well reflect this early response of undifferentiated neurons to the steroid hormones with none of the subsequent modifying influences of afferentation. Furthermore, Reier *et al.* (1977) have shown in the POA of the rat of both sexes an abrupt maturation of neurons of the fifth postnatal day, maximal maturation by the tenth day, and few subsequent qualitative changes. The major period of synaptogenesis, in addition, does not start until *after* the tenth day, suggesting that in this region the onset of both structural and functional maturation coincides with the end phase of the critical period. The 31% estrogen-induced increase in the number of complete first-order dendritic branches emanating from the cell body proper (Toran-Allerand *et al.*, 1983), which has not been observed *in vivo* (Hammer and Jacobson, 1984), may perhaps reflect the unopposed action of estrogen on early dendritic

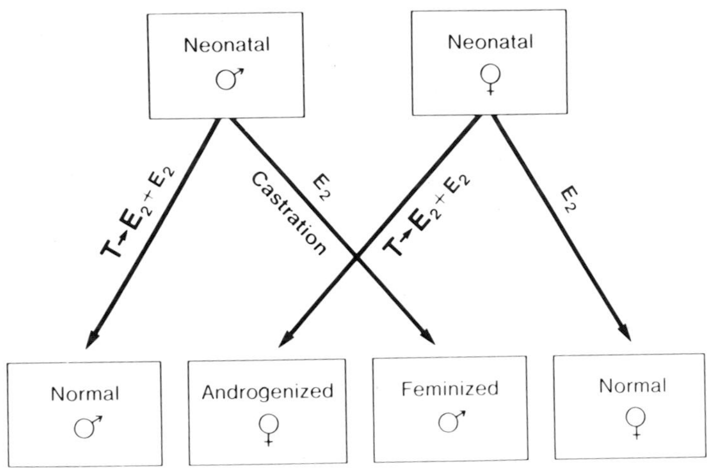

Fig. 9. Proposed model for sexual differentiation of the brain and the requirement for both estradiol and testosterone. Modified from Toran-Allerand (1981).

differentiation arrested at that stage of development by the absence of afferentation. Such a view is supported by observations *in vivo* in which, in contrast, Hammer and Jacobson (1984) found that only dendritic length was greater in males than in females (a parameter similarly responsive to steroids *in vitro*), while the number of primary and terminal branches were similar in both sexes.

The preferential involvement of first-order branching as a consequence of exposure to estrogen is extremely interesting. Both primary dendritic differentiation (exemplified by first order branching), which contributes to the basic dendritic phenotype, and the early undifferentiated protoplasmic somatic dendritic processes (referred to by Ramon y Cajal) are probably "intrinsically" (genomically) determined. Higher order branching, in contrast, appears to be more dependent on neuronal interactions via afferent axonal inputs that define the complexity of the ultimate dendritic arborizations. If hormonal regulation of gene expression is an important factor in the process of differentiation of other target tissues of estrogen (O'Malley and Birnbaumer, 1978 for review), then this selective involvement of first order branching in the estrogen treated cultures would provide evidence supporting such a postulated genomic effect for estrogen in the CNS as well.

Moreover, in neurons whose cell surface specificity (Gottlieb and Glaser, 1980) may have been modified by steroid hormone exposure (one possible consequence of steroidal effects on gene expression), the absence of both

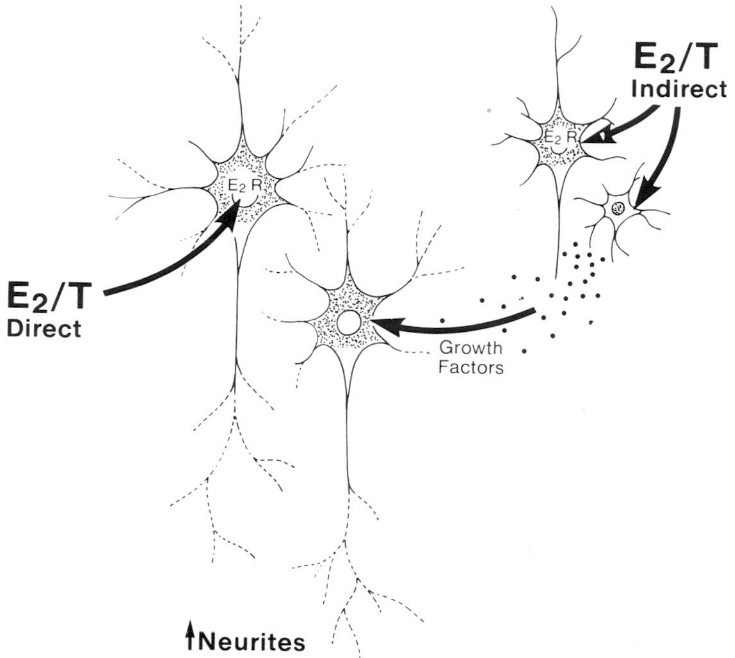

Fig. 10. Diagrammatic representation of the hypothesized neurite-promoting effect of estradiol (E_2) and testosterone (T). The gonadal steroids may exert their effects directly on the genome via the intranuclear estrogen receptor (E_2R) or indirectly via the intermediary of other hormone-sensitive neurons and/or neuroglia that might themselves secrete neurite-promoting growth factors whose targets need not necessarily be estrogen receptor-containing. From Toran-Allerand (1984).

afferentiation and targets of similar specificity may elicit (collateral) neuritic sprouting of the type seen classically following denervation (Lynch and Cotman, 1975). That this excessive neuritic growth may represent an unfulfilled extended search for similarly specified targets is suggested by their ultimate disintegration after 10 to 14 days *in vitro* (C. D. Toran-Allerand, unpublished observation), in much the same fashion that neurites from retinal explants degenerate with time unless cocultured with target tectal tissue (Smalheiser *et al.*, 1981).

It has been suggested that the *in vitro* observations may well reflect cellular features of the critical period *in vivo* and that a direct effect of steroids on neuritic growth *does* underlie the *in vivo* hormonal response. Stanley and Fink (1984) have shown that in the hypothalamus–POA of the rat during the critical period for sexual differentiation, the brains of males and androgenized females have an increased capacity by mRNA translation to synthesize a protein with the molecular and electrophoretic characteristics of tubulin, as compared to normal females, and that the content of a protein of similar weight and characteristics is 30% higher in males than in females at birth. Since tubulin is a major neuronal cytoskeletal constituent whose polymerization is involved in neuritic growth and differentiation, these findings in conjunction with the subsequent demonstration of a striking sex difference in the hybridization of tubulin cDNA to total RNA (H. F. Stanley and J. L. Roberts, in preparation) are strongly indicative of a steroid-dependent enhancement of neuritic growth and of a sex difference in neuritic differentiation *in vivo*.

That the morphological response of the cultures to estrogen are comparable to those *in vivo* is also supported by observations of the significant enhancement of synapse formation on dendritic shafts within anterior chamber transplants of the medial amygdala exposed to masculinizing levels of estrogen *in oculo* (Nichizuka and Arai, 1982). The response of this ectopic amygdaloid tissue in what may be considered another type of culture situation is comparable to the specific and permanent increase in shaft synapses reported in the medial amygdaloid nuclei of neonatal male rats exposed to gonadal steroids *in vivo* (Nichizuka and Arai, 1981), suggesting that these were the responses of neurons intrinsic to the amygdala and not merely an anomaly of deafferentation.

Furthermore, that the nature of the morphological response of the cultures to estrogen may be a general characteristic of undifferentiated, steroid-receptor–containing nerve cells exposed to steroids is suggested by the studies of Cherbas *et al.* (1982) at Harvard on a hormone-dependent cell line of probable neural origin, derived from embryos of the fruit fly *Drosophila melanogaster*. Exposure of cells in culture from clones of undifferentiated imaginal disc cells to the insect steroid molting hormone, 20-hydroxyecdysone, results in cessation of cell division, followed by subsequent cytologic transformation. This cellular differentiation is characterized morphologically by the extension of long and elaborate cell processes that arborize extensively. This neuritic effect is specific to ecdysteroid hormones and mediated by ecdysteroid receptors, which are similar to vertebrate steroid receptors.

The effects of the gonadal steroids on undifferentiated nervous tissue represent only one of the very many extrinsic influences on axonal and dendritic growth and differentiation during development. The cytological characteristics of the responses to the gonadal steroids are *not* solely restricted or peculiar to the gonadal steroids or even to sexual differentiation but are consistent with the types of cellular responses associated with neural plasticity. Axonal growth, dendritic differentiation, dendritic spine density, and synaptogenesis, for example, are cytological features that have been shown not only to be dependent on gonadal hormones and sexually dimorphic, but which also exhibit considerable pre- and postnatal plasticity. Such disparate influences as vision (Ruiz-Marcos and Valverde, 1969; Lund and Lund, 1972; Parnavelas *et al.*, 1973; see chapter by Boothe, this volume), environmental enrichment (Holloway, 1966; Globus *et al.*, 1973; Greenough *et al.*, 1973; see chapters by Greenough and Juraska, this volume) pre- and postnatal injury (Goldman *et al.*, 1974; Goldman-Rakic, 1981), as well as other hormones such as thyroxine (Rebière and Legrand, 1972; see chapter by Lauder, this volume), all acting during restricted, "critical" developmental periods, can also modify these cytological features permanently. The neuritic responses to some of these very influences, such as brain injury, moreover, may also be modified by sex differences and gonadal hormone exposure as well (e.g., Goldman *et al.*, 1974; Loy and Milner, 1980). Sexual differentiation may thus be viewed as representing yet another facet of the much broader question of the factors and cellular mechanisms contributing to neural development and neural plasticity.

References

Altman, J. (1967). *In* "The Neurosciences: A Study Program" (G. C. Quarton, T. Meinechuk, and F. O. Schmitt, eds.), pp. 723–743. Rockefeller Univ. Press, New York.

Alvarez, E. A., and Ramirez, V. D. (1970). *Neuroendocrinology* **6**, 349–360.

Ani, M., Butterworth, P. J., and Thomas, P. J. (1980). *Brain Res.* **183**, 341–353.

Arimatsu, Y., Seto, A., and Amano, T. (1981). *Brain Res.* **213**, 432–437.

Arnold, A. P., and Gorski, R. A. (1984). *Annu. Rev. Neurosci.* **7**, 413–442.

Attardi, B., and Ohno, S. (1976). *Endocrinology (Baltimore)* **99**, 1279–1290.

Attardi, B., and Ruoslahti, E. (1976). *Nature (London)* **263**, 685–687.

Ayoub, D. M., Greenough, W. T., and Juraska, J. M. (1982). *Science* **219**, 197–198.

Barley, J., Ginsburg, M., Greenstein, B. D., MacLusky, N. J., and Thomas, P. J. (1974). *Nature (London)* **252**, 259–260.

Baskin, D. G., Porte, D., Jr., Guest, K., and Dorsa, D. M. (1983). *Endocrinology (Baltimore)* **112**, 898–903.

Bass, H., Netsky, M. G., and Young, E. (1969a). *Neurology* **19**, 258–268.

Bass, H., Netsky, M. G., and Young, E. (1969b). *Neurology* **19**, 405–414.

Bhat, N. R. (1983). *Dev. Brain Res.* **11**, 315–318.

Binoux, M., Hossenlopp, P., Lassarre, C., and Hardouin, N. (1981). *FEBS Lett.* **124**, 178.

Booth, J. E. (1977). *J. Endocrinol.* **72**, 53–54.

Breedlove, S. M., and Arnold, A. P. (1980). *Science* **210**, 564–566.

Breedlove, G. A., Jordan, C. L., and Arnold, A. P. (1982). *Brain Res.* **237**, 173–181.

Bubenik, G. A., and Brown, G. M. (1973). *Experientia* **26**, 619–621.

Cherbas, P., Savakis, C., Cherbas, L., and Koehler, M. M. D. (1982). *In* "Molecular Genetic Neuroscience" (F. O. Schmitt, S. J. Bird, and F. E. Bloom, eds.), pp. 277–288, Raven, New York.

Christensen, L. W., and Gorski, R. A. (1978). *Brain Res.* **146**, 325–340.

Clough, R. W., and Rodriguez-Sierra, J. F. (1983). *Am. J. Anat.* **167**, 205–214.

Commins, D., and Yahr, P. (1984). *J. Comp. Neurol.* **224**, 132–140.

Crawford, M. A., and Sinclair, A. J. (1972). *Lipids Malnutr. Dev. Brain, Ciba Found. Symp., 1971, 1972*, pp. 267–287.

Curry, J. J., and Heim, L. (1966). *Nature (London)* **209**, 915–916.

Daughaday, W. H., Hall, K., Raben, M. S., Salmon, W. D., Jr., Van den Brande, J. L., and Van Wyk, J. (1972). *Nature (London)* **235**, 107.

Davidson, J. M. (1966). *Endocrinology (Baltimore)* **79**, 783–794.

DeLacoste-Utamsing, C., and Holloway, R. L. (1982). *Science* **216**, 1431–1432.

D'Ercole, A. J., Appelwhite, G. T., and Underwood, L. E. (1980). *Dev. Biol.* **75**, 315–328.

DeVoogd, T. J., and Nottebohm, F. (1981). *Science* **214**, 202–204.

DeVries, G. J., Buijs, R. M., and Swaab, D. F. (1981). *Brain Res.* **218**, 67–78.

DeVries, G. J., Best, W., and Sluiter, A. A. (1983). *Dev. Brain Res.* **8**, 377–380.

Diamond, M. C., Johnson, R. E., and Ehlert, J. (1979). *Behav. Neural. Biol.* **26**, 485–491.

Diamond, M. C., Young, D., Sukwinder, S., and Johnson, R. E. (1981). *Soc. Neurosci. Abstr.* **7**, 286.

Döhler, K. D., and Wüttke, W. (1975). *Endocrinology (Baltimore)* **97**, 898–907.

Dörner, G., and Staudt, J. (1968). *Neuroendocrinology* **3**, 136–140.

Dörner, G., and Staudt, J. (1969a). *Neuroendocrinology* **5**, 103–106.

Dörner, G., and Staudt, J. (1969b). *Neuroendocrinology* **4**, 278–281.

Doughty, C., and McDonald, P. G. (1974). *Differentiation* **2**, 275–285.

Doughty, C., Booth, J. E., McDonald, P. G., and Parrott, R. F. (1975). *J. Endocrinol.* **67**, 459–460.

Dyer, R. G., MacLeod, N. K., and Ellendorf, F. (1976). *Proc. R. Soc. London Ser. B.* **193**, 421–440.

Eisenfeld, A. J. (1970). *Endocrinology (Baltimore)* **86**, 1313–1318.

Fox, T. O. (1975). *Proc. Natl. Acad. Sci. U.S.A.* **72**, 4303–4307.

Fox, T. O. (1977). *Brain Res.* **128**, 263–273.

Friedman, W. J., McEwen, B. S., Toran-Allerand, C. D., and Gerlach, J. L. (1983). *Dev. Brain Res.* **11**, 19–27.

Gerall, A. A., Dunlap, J. L., and Hendricks, S. E. (1973). *J. Comp. Physiol. Psychol* **82**, 449–465.

Gerall, A. A., McMurray, M. M., and Farrell, A. (1975). *J. Endocrinol.* **67**, 439–445.

Gerlach, J. L., McEwen, B. S., Toran-Allerand, C. D., and Friedman, W. J. (1983). *Dev. Brain Res.* **11**, 7–18.

Globus, A., Rosenzweig, M. R., Bennett, E. L., and Diamond, M. C. (1973). *J. Comp. Physiol. Psychol.* **82**, 175–181.

Goldman, P. S., Crawford, H. T., Stokes, L. P., Galkin, T. W., and Rosvold, H. E. (1974). *Science* **186**, 540–542.

Goldman-Rakic, P. S. (1981). *Prog. Brain Res.* **53**, 3–19.

Gorski, J., and Gannon, F. (1976). *Annu. Rev. Physiol.* **38**, 425–450.

Gorski, R. A., Gordon, J., Shryne, J. E., and Southam, A. (1978). *Brain Res.* **148**, 333–346.

Gottlieb, D. I., and Cowan, W. M. (1972). *Brain Res.* **41**, 452–456.

Gottlieb, D. I., and Glaser, L. (1980). *Annu. Rev. Neurosci.* **3**, 303–318.

Greenough, W. T., Volkmar, F. R., and Juraska, J. M. (1973). *Exp. Neurol.* **41**, 371–378.

Greenough, W. T., Carter, C. S., Steerman, C., and DeVoogd, T. J. (1977). *Brain Res.* **126**, 63–72.

Gregory, E. (1975). *Brain Res.* **99**, 152–156.
Güldner, F. H. (1976). *Cell Tissue Res.* **165**, 509–544.
Güldner, F. H. (1982). *Neurosci. Lett.* **28**, 145–150.
Gurney, M. E. (1981). *J. Neurosci.* **1**, 658–673.
Gurney, M. E., and Konishi, M. (1980). *Science* **208**, 1380–1383.
Halasz, B., and Gorski, R. A. (1967). *Endocrinology (Baltimore)* **80**, 608–622.
Hammer, R. P., Jr., and Jacobson, C. D. (1984). *Int. J. Dev. Neurosci.* **2**, 77–85.
Hancke, J. C., and Döhler, D.-D. (1980). *Acta Endocrinol. (Copenhagen)* **94**, 102.
Hannigan, P. C., and Kelley, D. B. (1981). *Soc. Neurosci. Abstr.* **7**, 269.
Harris, G. W., and Jacobson, D. (1952). *Proc. R. Soc. London, Ser. B.*, **139**, 263–276.
Heim, L. M. (1966). *Endocrinology (Baltimore)* **78**, 1130–1134.
Heim, L. M., and Timiras, P. S. (1963). *Endocrinology (Baltimore)* **72**, 598–606.
Heimer, L., and Larsson, K. (1966). *Brain Res.* **3**, 248–263.
Heimer, L., and Larsson, K. (1967). *Physiol. Behav.* **2**, 207–209.
Hellman, R. E., Ford, D. H., and Rhines, R. K. (1976). *Psychoneuroendocrinology* **1**, 389–397.
Hendricks, S. E., and Duffy, J. A. (1974). *Dev. Psychobiol.* **7**, 297–303.
Hendricks, S. E., and Weltin, M. (1976). *Physiol. Psychol.* **4**, 105–110.
Hitt, J. C., Hendricks, S. E., Ginsberg, S. I., and Lewis, J. H. (1970). *J. Comp. Physiol. Psychol.* **173**, 377–384.
Holloway, R., Jr. (1966). *Brain Res.* **2**, 393–396.
Hsu, C. H., Carter, C. S., and Greenough, W. T. (1978). *Soc. Neurosci. Abstr.* **4**, 115.
Hudson, D. B. Vernadakis, A., and Timiras, P. S. (1970). *Brain Res.* **23**, 213–222.
Ifft, J. D. (1964). *Anat. Rec.* **148**, 599–604.
Jacobson, J. (1963). *J. Comp. Neurol.* **121**, 5–29.
Jordan, C. L., Breedlove, S. M., and Arnold, A. P. (1982). *Brain Res.* **249**, 309–314.
Juraska, J. M., and Fifkova, E. (1979a). *J. Comp. Neurol.* **183**, 247–256.
Juraska, J. M., and Fifkova, E. (1979b). *J. Comp. Neurol.* **183**, 257–268.
Kikuyama, S. (1969). *Endocrinol. Jpn.* **16**, 269–273.
King, J. C. (1972). Ph.D. Thesis, Tulane University, New Orleans, Louisiana.
Kornguth, S. E., and Scott, G. (1972). *J. Comp. Neurol.* **146**, 61–82.
Law, T., and Meagher, M. (1958). *Science* **128**, 1626–1627.
Lawrence, J. M., and Raisman, G. (1980). *Brain Res.* **183**, 466–471.
Lenoir, D., and Honegger, P. (1983). *Dev. Brain Res.* **7**, 205–213.
Linkie, D. M., and LaBarbera, A. R. (1979). *Life Sci.* **25**, 1665–1674.
Litteria, M. (1977). *Brain Res.* **127**, 164–167.
Litteria, M., and Thorner, M. W. (1974). *Brain Res.* **80**, 152–154.
Litteria, M., and Thorner, M. W. (1976). *Brain Res.* **103**, 584–587.
Loy, R., and Milner, T. A. (1980). *Science* **208**, 1282–1284.
Luine, V. N., Khylchevskaya, R. I., and McEwen, B. S. (1975). *Brain Res.* **86**, 293–306.
Lund, J. S., and Lund, R. (1972). *Brain Res.* **42**, 21–32.
Luttge, W. G., and Whalen, R. E. (1970). *Horm. Behav.* **1**, 265–281.
Lynch, G., and Cotman, C. W. (1975). *In* "The Hippocampus" (R. L. Isaacson and K. H. Pribram, eds.), Vol. 1, pp. 123–154. Plenum, New York.
McDonald, P. G., and Doughty, C. (1973). *Neuroendocrinology* **13**, 182–188.
McDonald, P. G., and Doughty, C. (1974). *J. Endocrinol.* **61**, 95–103.
McEwen, B. S. (1983). *Int. Rev. Physiol.* **27**, 99–145.
McEwen, B. S., Plapinger, L., Chaptal, C., Gerlach, J., and Wallach, G. (1975). *Brain Res.* **96**, 400–407.

McEwen, B. S., Lieberburg, I., and Krey, L. C. (1977). *Horm. Behav.* **9**, 249-263.
MacLusky, N. J., and Naftolin, F. (1981). *Science* **211**, 1294-1303.
MacLusky, N. J., Chaptal, C., Lieberburg, I., and McEwen, B. S. (1976). *Brain Res.* **114**, 158-165.
MacLusky, N. J., Lieberburg, I., and McEwen, B. S. (1979a). *Brain Res.* **178**, 129-142.
MacLusky, N. J., Chaptal, C., and McEwen, B. S. (1979b). *Brain Res.* **178**, 143-160.
Malsbury, C. W., Marques, D. M., and Daood, J. T. (1979). *Brain Res. Bull.* **4**, 833-842.
Matsumoto, A., and Arai, Y. (1981a). *Neuroendocrinology* **33**, 166-169.
Matsumoto, A., and Arai, Y. (1981b). *J. Comp. Neurol.* **197**, 197-206.
Meyer, G., Ferres-Torres, R., and Mas, M. (1978). *Brain Res.* **155**, 108-112.
Milgrom, E., Atger, M., and Baulieu, E. E. (1973). *Biochem. Biophys. Acta* **320**, 267-283.
Morest, D. K. (1969). *Z. Anat. Entwicklungsgesch.* **128**, 290-317.
Naess, O., Haug, E., Attramadal, A., Aakvaag, A., Hansson, V., and French, F. (1976). *Endocrinology (Baltimore)* **99**, 1295-1303.
Nichizuka, M., and Arai, Y. (1981). *Brain Res.* **212**, 31-38.
Nichizuka, M., and Arai, Y. (1982). *Proc. Natl. Acad. Sci. U.S.A.* **79**, 7024-7026.
Nunez, E. A., Benassayag, C., Savu, L., Vallette, G., and Delorme, J. (1979). *In* "Carcino-Embryonic Proteins. Chemistry, Biology, Clinical application" (F. G. Lehmann, ed.), Vol. 1, pp. 171-180. Elsevier, Amsterdam.
Ohno, S., Geller, L. N., and Young, Lai, E. V. (1974). *Cell* **3**, 235-242.
O'Malley, B. W., and Birnbaumer, L. (eds.) (1978). *Recept. Hormone Action* 2.
Parnavelas, J. G., and Globus, A. and Kaups, P. (1973). *Nature (London) New Biol.* **245**, 287-288.
Perry, V. H., and Linden, R. (1982). *Nature (London)* **297**, 683-685.
Pfaff, D. W. (1966). *J. Endocrinology* **36**, 415-416.
Pfaff, D., and Keiner, M. (1973). *J. Comp. Neurol.* **115**, 121-158.
Phelps, C. P., and Sawyer, C. H. (1976). *Horm. Behav.* **7**, 331-340.
Pineiro, A., Olivito, A.-M. and Uriel, J. (1979). *C. R. Hebd Seances Acad. Sci. Ser. D* **289**, 1053-1056.
Plapinger, L., and McEwen, B. S. (1973). *Endocrinology (Baltimore)* **93**, 1119-1128.
Powers, J. B., and Valenstein, E. S. (1972). *Science* **175**, 1003-1005.
Raisman, G., and Field, P. M. (1973). *Brain Res.* **54**, 1-29.
Raizada, M. S., Yang, J. W., and Fellows, R. E. (1980). *Brain Res.* **200**, 389-400.
Ramon y Cajal, S. (1909-1911). "Histologie du système Nerveux de l'Homme et des Vertébrés" (L. Azoulay, transl.), Vol. 2. CSIC, Madrid (1952).
Ratner, A., and Adamo, N. J. (1971). *Neuroendocrinology* **8**, 26-35.
Raynaud, P. (1971). *Steroids* **21**, 249-258.
Raynaud, J. P., Mercier-Bodard, C., and Baulieu, E. E. (1971). *Steroids* **18**, 767-788.
Rebière, A., and Legrand, J. (1972). *C. R. Hebd. Seances Acad. Sci., Ser. D* **274**, 3581-3584.
Reier, P. J., Cullen, M. J. Froelich, J. S., and Rothchild, I. (1977). *Brain Res.* **122**, 415-436.
Rosenzweig, J. L., Havrankova, J., Lesniak, M. A., Brownstein, M. J., and Roth, J. (1980). *Proc. Natl. Acad. Sci. U.S.A.* **77**, 572-576.
Rubin, B. S., and Barfield, R. J. (1980). *Endocrinology (Baltimore)* **106**, 504-509.
Ruiz-Marcos, A., and Valverde, F. (1969). *Exp. Brain Res.* **8**, 284-294.
Ruoslahti, E., and Engvall, E. (1978). *Scand. J. Immunol.* 7 *Suppl.* **6**, 1-17.
Salaman, D. F. (1974). *Prog. Brain Res.* **41**, 349-362.
Savu, L., Benassayag, C., Vallette, G., Christeff, N., and Nunez, E. A. (1981). *J. Biol. Chem.* **256**, 9414-9418.
Schachter, B. S., and Toran-Allerand, C. D. (1982). *Dev. Brain Res.* **5**, 93-98.

Scouten, C. W., Burrell, L., Palmer, T., and Cegavske, C. F. (1980). *Physiol. Behav.* **25**, 237–243.
Shapiro, B. H., Goldman, A. S., Steinbeck, H. F., and Neumann, F. (1976). *Experientia* **32**, 650–651.
Sheridan, P. J. (1979). *Brain Res.* **178**, 201–206.
Sheridan, P. J., Sar, M., and Stumpf, W. F. (1974). *Am. J. Anat.* **140**, 589–593.
Sinclair, A. J., and Crawford, M. A. (1972). *J. Neurochem.* **19**, 1753–1758.
Singer, J. J. (1968). *J. Comp. Physiol. Psychol.* **66**, 738–742.
Smalheiser, N. R., Crain, S. M., and Bornstein, M. B. (1981). *Brain Res.* **204**, 159–178.
Snyder, E. Y., and Kim, S. U. (1980). *Brain Res.* **196**, 565–571.
Soto, A.M., and Sonnenschein, C. (1980). *Proc. Natl. Acad. Sci. U.S.A.* **77**, 2084–2087.
Stanley, H. F., and Fink, G. (1984). *Int. Congr. Endocrinol. 7th Abstr.* July 1–7, 1984, p. 70.
Staudt, J., and Dörner, G. (1976). *Endokrinologie* **67**, 296–300.
Stewart, J., and Cygan, D. (1980). *Horm. Behav.* **14**, 20–32.
Toran-Allerand, C. D. (1976). *Brain Res.* **106**, 407–412.
Toran-Allerand, C. D. (1980a). *Brain Res.* **189**, 413–427.
Toran-Allerand, C. D. (1980b). *Nature (London),* **286**, 733–735.
Toran-Allerand, C. D. (1981). *In* "Biogregulators of Reproduction" (H. Vogel and G. M. Jagiello, eds.), pp. 43–57. Academic Press, New York.
Toran-Allerand, C. D. (1984). *Prog. Brain Res.* **61**, 63–96.
Toran-Allerand, C. D., Gerlach, J. L., and McEwen, B. S. (1980). *Brain Res.* **184**, 517–522.
Toran-Allerand, C. D., Hashimoto, K., Greenough, W. T., and Saltarelli, M. (1983). *Dev. Brain Res.* **7**, 97–101.
Toran-Allerand, C. D., Pfenninger, K. H., and Ellis, L. (1984). *Soc. Neurosci. Abstr.* **10**, 455.
Underhill, L. H., Rosenzweig, J. L., Roth, J., Brownstein, M. J., Young, W. H. III, and Havrankova, J. (1982). *Front. Horm. Res.* **10**, 96–110.
Vannier, B., and Raynaud, J. P. (1975). *Mol. Cell. Endocrinol.* **3**, 323–337.
Vertès, M., and King, R. J. B. (1971). *J. Endocrinol.* **51**, 271–282.
Vreeburg, J. T. M., van der Vaart, P. D. M., and van der Schoot, P. (1977). *J. Endocrinol.* **74**, 375–382.
Ward, I. (1969). *Horm. Behav.* **1**, 25–36.
Welshons, W. V., Lieberman, M. A., and Gorski, J. (1984). *Nature (London)* **307**, 747–749.
Wieland, S. J., and Fox, T. O. (1981). *J. Steroid Biochem.* **14**, 409–414.
Zapf, J., Froesch, E. R., and Humbel, R. E. (1981). *Curr. Top. Cell. Regul.* **19**, 257–309.
Zigmond, R. E., and McEwen, B. S. (1970). *J. Neurochem.* **17**, 889–899.

7

Behavioral Neuroembryology: Motor Perspectives

ROBERT R. PROVINE
Department of Psychology
University of Maryland Baltimore County
Catonsville, Maryland

> I should like to work like the archeologist who pieces together the fragments of a lovely thing which are alone left to him. As he proceeds, fragment by fragment, he is guided by the conviction that these fragments are part of a whole which, however, he does not yet know. He must be enough of an artist to recreate, as it were, the work of the master, but he dare not build according to his own ideas. Above all, he must keep holy the broken edges of the fragments; in that way only may he hope to fit new fragments into their proper place and thus ultimately achieve a true restoration of the master's creation. [Spemann, 1938]

I. Introduction

The growing popularity of embryological research reflects an increased interest in developmental issues and a commitment to the developmental approach as an important, if not necessary, tactic for the analysis of

Developmental NeuroPsychobiology

neurobehavioral mechanisms. Unfortunately, our understanding of basic principles of neurobehavioral development lags far behind the rapidly growing pool of often unorganized facts. Examples abound. Efforts are scattered because of a lack of consensus about basic issues. Tantalizing observations about embryonic behavior lie buried in reports dealing with other phenomena. Anatomists seeking structural criteria for functional synapses may be unaware that the embryo was performing synaptically mediated behavior before they were able to detect synaptic structure. A report on one species does not cite related work on others. Artificial age boundaries exist; embryological studies usually end at birth when neuropsychological and developmental psychological studies begin. There is also a prenatal–postnatal dichotomy in research on human development that is reflected in, and is, in part, the product of the administrative boundaries between the departments of obstetrics and pediatrics at many medical schools. Some of these problems are being solved as the developmental research literature is being organized and reviewed in terms of major neurobehavioral themes (Tobach *et al.*, 1971; Gottlieb, 1973, 1974; Burghardt and Bekoff, 1978; Bekoff, 1981; Immelmann *et al.*, 1981; Oppenheim, 1982). However, a major advance will probably only come about when students of development adopt a naturalistic, multidisciplinary approach that brings them into more intimate contact with the embryo or developing organism. As Viktor Hamburger, a pioneer neuroembryologist and student of Spemann (author of the introductory quotation), has often stressed, the embryo is always the best teacher. We should heed its lessons.

In this spirit, the present chapter describes the development of behavior and its neural correlates. Research on the chick embryo, a representative and widely studied vertebrate embryo, receives the most attention. These data provide a basis for comparison with research findings from other vertebrate and invertebrate embryos. Most of the reviewed research concerns the prenatal period although it is understood that the major formative period for many neurobehavioral processes extends well beyond birth. An emphasis is also placed on spinal cord motor processes because of their unique contribution to embryonic behavior. Important research on the ontogeny of sensory systems, supraspinal influences and the role of environmental factors are considered in other chapters of the present volume and elsewhere (Gottlieb, 1971, 1973, 1974, 1976, 1978; Oppenheim, 1982).

II. History

The founding of the scientific study of embryonic behavior is attributed to the 19th century physiologist-child psychologist, Preyer (Hamburger, 1963; Gottlieb, 1973; Oppenheim, 1982). The birthdate of the discipline is 1885, the publication date for Preyer's monumental book, *Specielle Physiologie des*

Embryo (Preyer, 1885). One of the many important contributions of the book is the report that the chick embryo moved several days before the first reflexes could be evoked. This developmental "experiment of nature" suggested that embryonic motility is the product of spontaneous activity of the central nervous system; it occurred in the absence of sensory stimulation. The result lay dormant until pursued by Hamburger and co-workers in the 1960s (Hamburger, 1963).

The next landmark event in the neuroembryological study of behavior was the publication of George Ellett Coghill's *Anatomy and the Problem of Behavior* (1929) that summarized decades of his research on the behavioral development of the salamander *Ambystoma*. Coghill sought to describe behavioral events during embryonic development and correlate them with the appearance of neuroanatomical structure (Hamburger, 1963; Oppenheim, 1982). Coghill was one of the first to attempt to relate events in the development of behavior to the emergence of neural connections. On the basis of his research with *Ambystoma*, he proposed a general theory of behavior development that held that behavior was totally integrated at all ages and that independent movements emerged from a total pattern by a process of individuation (whole → parts). This proposition suffered a fate common to many ambitious general accounts of behavior; it extended an explanation which was appropriate for one species (*Ambystoma*) to other inappropriate ones. Later comparative analyses provided exceptions to the individuation model (Hamburger, 1963; Provine, 1980).

The best known champion of an alternative general view of behavioral development was William Windle (1940, 1944), who based his account primarily upon observations of mammalian fetuses. He suggested that complex coordinated movements developed by the integration of local reflex circuits (parts → whole). The focus on the reflex is in part due to the discovery of a close synchronization of onset of the first movements and the onset of reflexes in mammalian fetuses. The emphasis upon the reflex-integration approach ultimately contributed to a "reflexology" that was compatible with the radical environmental *zeitgeist* of the 1930s and 1940s. Because ongoing behavior was considered to be evoked, little attention was given to spontaneous movements. Indeed, the report of a spontaneous process was considered by some to be either an admission of ignorance of the causative agent or the acceptance of biological mysticism. The reflex-integration approach ultimately produced extensive catalogs of reflex development. The approach was useful for the description of aspects of normal and abnormal development but it contributed little experimentally or conceptually to the discovery of the neural mechanisms underlying ongoing embryonic movement. Later analyses ultimately demonstrated that the reflex, the key element of Windle's system, was not the basic unit of behavior in either the adult (Grillner, 1975) or the embryo (Hamburger, 1963).

Weiss (1941) made several conceptually advanced but often overlooked contributions to our understanding of how the CNS patterns motor behavior (Gallistell, 1980; Bekoff, 1981). He was also one of the first to apply the procedures of experimental embryology to the analysis of neurobehavioral problems (also see Detwiler, 1936). In a classic study, Weiss (1941) reversed the anterior–posterior axes of the buds of the forelimbs of embryonic salamanders. After development, individual limbs of the salamanders produced coordinated movements but the fore- and hind limbs worked against each other (the front legs walked "backward") and prevented normal locomotion. Weiss concluded that sensory input, that was abnormal in the operated salamanders, was not necessary for the development or performance of within- or between-limb coordination. Deafferentation studies also supported this conclusion. By exclusion, Weiss looked at the CNS for the origin of the patterned motor outflow.

Other well-known studies from the early period of behavioral embryology reported that amphibians raised in a solution of paralyzing chloretone anesthetic performed age-typical swimming movements after the solution was removed. Because swimming developed in immobilized and apparently neuroelectrically quiescent embryos raised from eggs in the chloretone solution, it did not require exercise or practice for its genesis. This line of research was initiated with frogs by the master embryologist Ross Harrison (1904) and was later replicated by Matthews and Detwiler (1926) and Carmichael (1926, 1927) with salamanders and by Fromme (1941) with frogs. (A more detailed description is provided below.) The widely known replication by Carmichael (1926) is one of the few neuroembryological studies cited in developmental psychology texts. It is ironic that Carmichael initially considered his results to support a conditioned reflex view of behavior development despite overwhelming contrary evidence. Because of rampant reflexology and environmentalism, a maturational account of behavior development was theoretically unpalatable in the 1920s and 1930s.

Another major advance in understanding behavior development was made by Hamburger and co-workers in the 1960s (Cowan, 1981). Their approach combined behavioral observation of the developing chick embryo with elegant microsurgical procedures that were virtually unknown in behavioral research. This and other behavioral and electrophysiological work considered in the next section indicated a unique *active* role for the embryonic spinal cord in the production of movement. Clearly, the embryo is not a reflex machine.

III. Embryology of Behavior and Its Neural Correlates

The first step in understanding embryonic behavior development is to describe ongoing movements. Spontaneous behavior is a conservative expression of the state of neuromuscular maturation of the embryo because some properties of the developing neuromuscular system may not be reflected in behavior because of inhibition, lack of activation or masking (Hamburger, 1973; Pollack and Crain, 1972). Ongoing, spontaneous motility is also the baseline behavior that must be considered in experimental studies of stimulation or learning. Comparative analyses of behavioral descriptions from a variety of organisms provide insights into phenomena common to developing neuromuscular systems. Unfortunately, behavioral descriptions are available for only a few commonly used vertebrate and invertebrate embryos, and comparative developmental analyses are extremely rare. Moreover, behavioral workers tend to ignore embryos and spontaneous behavior, and neurophysiologists usually examine the development of specific sensory systems or motor patterns and are not interested in general embryological problems.

The domestic chicken provides one of the most studied embryos. The eggs are easy to obtain and the embryos are large and hardy enough to permit dramatic experimental manipulation. Most researchers are in general agreement about the course of behavior development of the chick embryo from the onset of movement at 3 to 4 days of incubation until hatching at 21 days (Preyer, 1885; Visintini and Levi-Montalcini, 1939; Hamburger, 1963; Hamburger and Oppenheim, 1967; Corner and Bot, 1967; Oppenheim, 1974; Provine, 1980). The first movements are lateral flexions of the head. Other regions of the trunk become active in cephalocaudal order. Slow S-shaped flexions of the entire body appear as more body regions become active. These flexions may be initiated in any region of the axial musculature and pass either rostrally or caudally or they can be initiated in midtrunk regions and propogate in both directions simultaneously. At 6 to 7 days, the limbs begin to move, the trunk movements become more complex and the early S movements are gradually replaced by jerky generalized motility that seems to involve the movement of all body parts in irregular sequences. Despite the random appearance of the motility, intrajoint and intralimb coordination exists from early stages (Bekoff, 1976, and below). The jerky movements predominate until several days before hatching when they alternate with smooth, coordinated prehatching and hatching movements (Hamburger and Oppenheim, 1967; Provine, 1980).

The frequency and temporal distribution of motility changes during development. The frequency of embryonic movement peaks around 13 days after which it declines. Up to 13 days, movement occurs at regular 30 to 60-sec intervals. The cyclic character of movements is lost as the frequency of movement and the duration of activity periods increase until ~13 days when the embryo appears to be almost continually active.

An important observation concerning the mechanisms that underlie embryonic motility was the cited finding by Preyer (1885) that embryos moved days before a response could be evoked by exteroceptive sensory stimulation. On the basis of this observation, Preyer suggested that embryonic motility originated in endogenous processes. Additional indirect support for endogenous processes are the observations that motility is performed under relatively stable environmental conditions, that it is often cyclic, and that efforts to increase or decrease the amount of stimulation of the embryo have little effect on the amount of movement (Oppenheim, 1972).

The spontaneous, nonreflexogenic nature of chick embryonic motility was clearly demonstrated by Hamburger *et al.* (1966) in a crucial experiment. Their results are theoretically important because they deal with the historically significant issue of the impact of the environment on behavior development and their procedures were technically innovative. Hamburger *et al.* (1966) extirpated several segments of the thoracic neural tube (immature spinal cord) of 2-day embryos. This isolated the lumbosacral cord from input from the brain and rostral cord regions. Simultaneously, a second operation removed the dorsal half of the neural tube caudal to the thoracic spinal gap and the associated neural crest areas (the precursors of the dorsal root sensory ganglia), eliminating sensory input to the residual, postgap cord. The ventral root efferents remained intact. Thus, leg movements of the operated embryos were the product of neuronal discharges within the surgically isolated lumbosacral spinal cord segment that innervated them. The legs of the operated embryos were highly motile during most of the embryonic period, which indicated that the embryonic motility was both spontaneous, as suggested by the nonreflexogenic period noted by Preyer (1885), and patterned by neuronal circuitry within the residual spinal cord.

The next phase in the search for the neural mechanisms of motility involved the electrophysiological analysis of the embryonic spinal cord. These experiments were necessary to discover the bioelectric correlates of embryonic motility and clarify the correlation between neural and behavioral events.

Electrical recordings from the spinal cords of intact, curarized (Provine *et al.*, 1970; Provine, 1972a) and freely moving embryos (Ripley and Provine, 1972; Provine, 1973) confirmed the predictions from the behavioral findings and for the first time directly established the *neurogenic* basis of embryonic motility in the chick. Massive polyneuronal burst discharges were identified within the ventral cord region (Provine *et al.*, 1970) shown by Hamburger *et al.* (1966) to produce motility in deafferented embryos. The burst discharges were synchronized with embryonic movements in motile embryos as young as 4.5 days; the bursts were synchronized with motor nerve discharges in older curarized preparations (Ripley and Provine, 1972). When a burst occurred, the embryo moved; embryonic movements never occurred in the absence of a simultaneous burst. Simultaneous multiple electrode recordings indicated that a given burst occurred almost simultaneously along the rostrocaudal axis of the ventral two-thirds of the cord (Provine, 1971). Therefore, many motor and interneurons participate in the discharges. The variable intervals (± 100 msec or more) between bursts occurring at different ventral cord loci suggested that the bursts were initiated at rostral, caudal or midcord loci and propagate rapidly throughout the remainder of the cord (Provine, 1971). The finding of a variable locus of initiation is consistent with behavioral and electrophysiological evidence that typical motility or its electrical correlate continued to occur following cervical or midthoracic spinal transections (Provine and Rogers, 1977) and that the early S movements may be initiated in any body region. The discovery that motility is neurogenic permits inferences concerning neural development on the basis of behavior, and vice-versa. The movements are clearly not the product of *myogenic* activity produced by the spontaneous contraction of the immature muscles.

The details of how the spinal cord burst discharges are related to the patterning of motor outflow to the muscles must await simultaneous electrical recordings from the spinal cord and specific, identified muscles or nerves. Only limited evidence is currently available. Electromyographic (EMG) records from specific muscles of the chick hind limb indicate that patterned outflow is present from early stages (Bekoff, 1976, and below). Because there seem to be more region-specific differences in cord motor outflow than in cord bursts, the cord bursts may be driving local pattern generators or may be the summed activity of many different pattern generating circuits that are simultaneously active along the cord's rostrocaudal extent. Work is needed to determine which of these is the correct alternative. Unfortunately, the investigators examining spinal cord and EMG activity have pursued different questions. Provine (1973; Provine and Rogers, 1977) observed cord burst phenomena to establish the neurogenic nature of motility and the locus of burst (movement) initiation, and to describe the development of intracord communication. In contrast, Bekoff (1976) took advantage of the high resolution of the EMG technique to describe the development of patterned motor outflow (motor coordination) and, by inference, the spinal cord circuitry that produced it. Knowledge that a cord burst is synchronized with the twitch of an unspecified muscle is clearly inadequate for this task. Conversely, the EMG data by themselves do not predict the dynamics or significance of the ubiquitous cord burst discharges.

Landmesser and O'Donovan (1984) provide data that help to reconcile the above findings of transregionally coherent cord bursts (Provine) with that of muscle-specific motor outflow (Bekoff). However, further discussion of this matter requires a brief consideration of the structure of spinal cord bursts discharges. As noted by Provine (i.e., 1972, 1973; Provine and Rogers, 1977), cord bursts typically begin with a high amplitude "initiating" discharge that was usually, but not always, followed by a longer, lower amplitude and more variable "afterdischarge." Simultaneous recordings from two or more spinal cord sites indicated that the initiating burst component occurred almost simultaneously along the cords rostro-caudal extent; regional differences in burst activity were confined to the afterdischarge that varied in amplitude, duration and shape of the discharge envelope. An almost identical pattern of electromyographic (EMG) discharge was reported in the hindlimb muscles of the chick embryo by Landmesser and O'Donovan (1984). They observed that EMG bursts "were often initiated by a short duration, high amplitude discharge that occurred synchronously in all muscles studied" (p. 189). "Following the occurrence of a synchronous discharge...muscles were immediately silenced for a variable period of time after which a more prolonged burst occurred" (p. 194). They observed further that most differences in the activation sequence of muscles occurred during the afterdischarge, not the ubiquitous, high amplitude, initiating burst. Alternating activity of antagonistic muscles followed a synchronous, initiating burst. When considered together, the work by Provine on spinal cord and Landmesser, O'Donovan, and Bekoff on muscle activation sequences suggests that the cord initiating discharge is transmitted to all limb muscles; the more variable cord afterdischarge is correlated with the motor outflow to specific limb muscles. Perhaps the afterdischarge is the product of spinal motor pattern generating circuits. This interpretation is consistent with the development of the afterdischarge between 6 and 7 days of embryonic age (Provine, 1972, 1973) when the limb rapidly develops and begins to move (Provine, 1980). Before 7 days, cord bursts consisted exclusively of the short "initiating" type.

Landmesser and O'Donovan (1984) did not consider the relationship between their EMG findings and the remarkably similar results on spinal cord burst discharges that were published over a decade earlier (i.e., Provine, 1972, 1973; Stokes and Bignall, 1974). For example, Landmesser and O'Donovan (1984) announced the discovery of the "initiating-afterdischarge" burst configuration that "has not been described previously in mature animals" (p. 189), or "at this stage of development *in ovo*" (p. 193). While their claimed priority may hold within the narrow province of EMG activity, their neglect of the related cord burst activity that is a demonstrated motility correlate leads to an unfortunate discontinuity within an already small research literature.

Data concerning neurobehavioral development of the kind discussed above has been useful to anatomists and others who wish to perform "developmental dissections." For example, by examining progressively younger embryos, it is possible to establish the minimal structural criteria for a functional synapse (Foelix and Oppenheim, 1973). Some investigators have chosen to ignore potentially useful behavioral data. Several anatomists describing the development of spinal and myoneural synapses have noted the appearance of synapses at stages many days after the embryo has been moving and showing evidence of rostrocaudal integration.

Descriptions of embryonic behavior development in a variety of species provide a basis for evaluating the generality of the chick data and suggest phenomena common to the developing neuromuscular system.

Most of the embryos that have been observed move when left undisturbed. The ongoing motility of the snapping turtle is much like that of the chick. Early S waves break up into total body movements during which the activity of various body parts seem uncoordinated (Decker, 1967). The behavior of both the turtle (Decker, 1967) and lizard embryo (Hughes *et al.*, 1967) is cyclic. This and the identification of a prereflexogenic period in both the turtle and lizard embryos suggest that ongoing motility remains spontaneous even after the onset of reflexes.

The teleosts and urodels complicate comparative analyses because, as noted in the discussion of Coghill, their behavior remains smooth and integrated throughout development. Ongoing motility has been described in both the salamander (Coghill, 1929) and the toadfish (Tracy, 1926), and a definite prereflexogenic period has been demonstrated in the latter.

Electrophysiological and behavioral studies by Stehouwer and Farel (1980) demonstrate that endogenous spinal cord pattern generating circuits are responsible for bullfrog tadpole swimming behavior. Relevant work on teleosts has been contributed by Eaton *et al.* (1977) and Pollack and Crain (1972). The swimming of these aquatic embryos resembles the early S movements of the chick embryo, suggesting the involvement of similar mechanisms.

Evidence that frog swimming, and by inference the motor circuitry that produces it, does not require exercise or practice for its genesis was provided by Harrison (1904). Harrison reared frog eggs in a solution of the paralyzing anesthetic chloretone. The drug immobilized the developing embryos and may have prevented, altered, or reduced CNS neuroelectric activity. (The latter possible drug effects have never been evaluated in this preparation.) After removal of the paralyzing chloretone, the tadpoles swam as well as untreated controls of the same age. A similar result was obtained by Carmichael (1926, 1927) in a replication with salamanders. A subsequent, more detailed analysis by Fromme (1941) confirmed that the basic swimming movements matured in chronically immobilized frog embryos but found that they swam more slowly than controls. An analysis by Haverkamp (cited in Oppenheim, 1982) seems to reconcile these somewhat different findings. He detected a short-term, transient decrement in the swimming speed of chloretone reared frog and salamander embryos that was probably due to the presence of residual chloretone. In addition, salamanders but not frogs exhibited morphological abnormalities that decreased swimming performance. The latter effect probably accounts for the long-term swimming decrement in the chloretone reared salamanders that were described by Matthews and Detwiler (1926). The swimming decrements that Fromme reported in frog embryos immediately after the return of movement were probably due to the lingering effect of chloretone. The basic conclusion that the neuromuscular machinery of amphibian swimming develops autonomously still stands.

Most work with the mammalian fetus has focused on reflexogenesis (Hamburger, 1963; Oppenheim, 1982; Bekoff, 1981). Narayanan *et al.* (1971) provided one of the few analyses of ongoing mammalian fetal motility. The rat fetus performs intermittent movements which are jerkier and less well coordinated than those of the adult. The nonreflexogenic character of these movements has not been experimentally verified but they occur independently of at least one possible source of stimulation, uterine contractions. Two leading investigators of human fetal development, Hooker (1952) and Humphrey (1969), largely ignored ongoing motility while attending to reflex development. Edwards and Edwards (1970) contributed an analysis of spontaneous motility of the human fetus that was based upon maternal reports. The presence of cyclic motility with a period of about a minute was detected in late human fetuses by Robertson *et al.* (1982) who used motion detectors placed on the mother's abdomen. Thelen (1979) provided valuable descriptions of ongoing, stereotypic movements in human infants. Her observations are a starting point for the developmental analysis of motor coordination in humans.

Although the motility of the mammalian fetus has not been studied in detail, preliminary evidence suggests that its behavior differs from other vertebrate forms in at least two respects: mammals do not have a prereflexogenic period, and their behavior has a relatively late onset in regard to body development and neuronal maturation. Only a few neurophysiological investigations of direct relevance to mammalian motility development have been reported (Naka, 1964a,b; Saito, 1979; Scheibel and Scheibel, 1970, 1971). The technical problems of fetal research are formidable. Perhaps new ultrasonic imaging procedures (Birnholz, 1981) will provide a noninvasive means of observation and advances in *in utero* microsurgery

(Taub, 1976), and electrophysiology will finally permit the experimental analyses of the neuronal mechanisms of movement production in mammals.

The sparse literature on the behavior development of invertebrate embryos is reviewed by Provine (1977, 1981c), Berrill (1973) (also see the invertebrate chapters in Burghardt and Bekoff, 1978). Only the cockroach *Periplaneta americana* has been studied in a manner comparable to the chick (Provine, 1976a, 1977, 1981c). The frequency of cockroach embryonic movement increases to a midincubation peak and then declines shortly before hatching as with the chick and other vertebrates. The proportion of multisegmental relative to unisegmental movements increases during development, especially during the last few days of incubation when multisegmental movements become the most common kind. The character of the movements of the cockroach embryo are difficult to interpret. Because of the softness of the cuticle (exoskeleton), only minute muscle twitches and contractions are visible before hatching. The cockroach embryo is a soft, yolk-filled bag, and its limp, flexible legs lie folded beneath the body. This provides a natural movement deprivation condition. Electromyographic studies of specific muscles would be necessary to determine whether the apparently random twitches of the embryo are an early expression of posthatching behavior such as "walking." However, unmistakable examples of intersegmentally coordinated ventilation (dorsal–ventral flattening of the abdomen) and occasional, isolated eclosion/hatching movements are performed during the last third of incubation. The cockroach performs fewer spontaneous movements than vertebrate embryos. It is not clear whether the lower motility levels are due to a low level of spontaneous neuronal firing (Clark and Cohen, 1979), to low intersegmental additivity of excitation of the segmented ganglionic chain (the vertebrate spinal cord is a continuous column), to neuromuscular immaturity, or to some yet unknown process. Cursory observations of spontaneous motility of the grasshopper (Bentley *et al.*, 1979) suggests that behavior similar to that of the cockroach may be present in at least one other orthopteran embryo.

IV. Development of Specific Motor Patterns

Relatively more contemporary research has been done on the development of specific movements such as walking, flying, hatching, and eclosion than on the generalized embryonic movements described in the previous section. An advantage of this approach is that one can focus on specific, often cyclic and therefore easily quantified movements. Much research on the development of specific motor patterns is strongly influenced by neurophysiological investigations of the neural basis of normal movement in behaving adult vertebrates and invertebrates (Grillner, 1975; Herman *et al.*, 1976; Wilson, 1961). Selected movements of the chick, mammals, and invertebrates are considered below.

Electromyographic (EMG) studies of motile chick embryos by Bekoff (1976; Bekoff *et al.*, 1975) describe motor development in terms of activation sequences of specific muscles. The quantitative EMG procedure offers much greater temporal and spatial resolution of the patterning of motor outflow than was provided by the earlier behavioral observations. For example, despite earlier failures to detect coordinated behavior, Bekoff (1976) observed *intrajoint* coordination in the alternating EMG activity of ankle antagonists by 7 days of incubation. This is soon after the onset of leg motility and before the onset of reflex responses by the leg. A gradual increase in the precision of temporal patterning of motor outflow was observed during incubation, especially between 17 days and hatching (21 days). The early, prereflexive onset of reciprocal activation indicates that the muscle contraction sequence is produced by a centrally generated motor program that does not require sensory input for either its occurrence or maturation.

Bekoff (1976) also examined the development of *interjoint* coordination as reflected in the coactivation of knee and ankle extensors. Interjoint coordination was detected by 9 days, the earliest age observed. As with intrajoint activation, the precision of interjoint coordination gradually increased until hatching.

Few studies consider the development of *interlimb* (between-limb) coordination. Provine (1980) described the development of visually observed movement synchronization of bilateral (wing–wing and leg–leg) and homolateral limb pairs (wing–leg) in the chick embryo using a statistical measure of coincidence. At early stages, movements of bilateral limb pairs were no more synchronized than those of homolateral limb pairs; a wing was no more likely to move synchronously with the contralateral wing than with the homolateral leg. During incubation, the synchronization between bilateral limb pairs increased while that between homolateral limb pairs decreased, especially after 15 to 17 days (hatching at 21 days). Thus, an increase in within-girdle limb synchronization is correlated with a decrease in between-girdle limb synchronization. By the time of hatching, bilateral leg movements that resemble walking and wing movements that resemble flapping are present.

These results suggest that interlimb coordination develops gradually. They may also reconcile the finding of early intrajoint and interjoint coordination by Bekoff (1976) with the visual impression of jerkiness and lack of coordination of embryonic movement reported by behavioral investigators

(Hamburger and Oppenheim, 1967). Perhaps the uncoordinated appearance of most embryonic movement is due to the absence or low degree of interlimb coordination. The considerable coincident movement of homolateral (between-girdle) limb pairs during all but the final days of incubation is consistent with this conclusion (Provine, 1980). [Behavior that requires bilateral coordination may even appear relatively late in human development (Provine and Westerman, 1979).]

Elegant and technically difficult transplantation studies by Straznicky (1963) and Narayanan and Hamburger (1971) suggest that the circuitry coordinating the legs in the walking pattern or the wings in the flapping pattern is located in the adjacent spinal cord regions. This was shown by substituting brachial for lumbosacral and lumbosacral for brachial segments of the spinal cord of $2\frac{1}{2}$day chick embryos. At embryonic stages, wings and legs, both of which were innervated by brachial spinal segments, moved together more frequently than unoperated controls (Narayanan and Hamburger, 1971). The posthatching behavior of both groups of chicks was striking (Straznicky, 1963; Narayanan and Hamburger, 1971). Chicks with double brachial cords moved the wings and legs together as in normal wing-flapping. Chicks whose wings were innervated by transplanted lumbosacral cord moved in an alternating sequence as in normal walking. These results indicate both that limb-specific movement is produced by circuits within the adjacent spinal cord segments and that the circuits cannot be modulated by the appendages that they innervate. Therefore, the neural mechanisms that produce these movements appear to be irreversibly determined at the time of the operation shortly after the closure of the neural tube.

The study of wing-flapping in birds offers a rare opportunity to examine the developmental continuity of a bilaterally coordinated behavior in a vertebrate. Wing-flapping is also convenient to study because it can be elicited in the absence of flight. For example, wing-flapping may be evoked in birds which cannot fly because of immaturity or surgical modification. Walking is more difficult to study because the experimental animal may be unable to support itself against gravity and walk without artificial supports and a treadmill. [For mammals, swimming offers similar advantages (Fentress, 1972; Bekoff and Trainer, 1979).] Birds with mutations of various neural (Nicolai, 1976), muscular (Provine, 1983), wing (Sawyer, 1982) and feather-related (Provine, 1981b) components of the flight motor system are also available for genetic study and the analysis of pathogenesis.

As noted earlier, in the chick, spontaneous wing motility appears toward the end of the first week of embryonic development and the bilateral synchronization of wing movements increased during incubation, especially during the week before hatching. Low-amplitude, bilaterally synchronous "flutters" shortly before hatching (Provine, 1980) and drop-evoked wing-flapping within hours after hatching (Provine, 1981a) indicate that the circuitry that bilaterally coordinates wing-flapping was established during the prenatal period. However, the wing motor system continues to develop postnatally (Provine, 1981a). The amplitude and rate of wing-flapping increase through 2 to 3 weeks after hatching. The increased frequency of stereotypic limb movements during development has been reported in both vertebrates (Bekoff and Trainer, 1979) and invertebrates (Altman, 1975; Kutsch, 1971). Flight, which becomes possible after the development of wing feathers, has its onset between 7 and 9 days in the chick.

The maturation of the wing motor system after hatching does not depend on wing-flapping practice or functional significance because normal adult rates of wing-flapping are achieved in chicks whose wings were immobilized from the day of hatching until immediately before testing at 13 days. In addition, normal rates of wing-flapping or its muscular correlate were present in chicks that experienced bilateral wing amputation on the day of hatching (Provine, 1979) and in mutant, featherless chicks that were flightless because they lacked feathers (Provine, 1981b). These results are consistent with the finding that normal patterns of "grooming" develop in mice whose forelimbs were amputated shortly after birth (Fentress, 1973).

The above developmental studies indicate that the avian flight motor system remained intact during relatively short periods of movement restriction or disuse during the lifetime of the individual (i.e., Provine, 1981a). The neural mechanism for flight also survived the multigenerational artificial selection for high egg or meat yields during the domestication of chickens because white leghorn chickens (domestic egg layers) and Cornish × rock chickens (domestic meat producers) flap at rates similar to the red jungle fowl, their presumed ancestor (Provine *et al.*, 1984). Although domestication had little affect on the neural mechanism of wing-flapping, it did have a dramatic effect on flight performance. For example, the vigorous wing-flapping of the large, heavy Cornish × rock meat chicken was futile because the bird had an unfavorable ratio of wing area to body weight. Comparative analyses of birds that have been flightless for millenia suggest that the motor circuitry for wing-flapping is amenable to change only under appropriate conditions and over sufficiently long intervals of phylogenetic history. For example, emus and penguins, both of which are flightless birds that are presumed to have evolved from flighted forms, have different wing motor behavior than flying birds (Provine, 1984). The penguin and emu both lack the drop-evoked wing-flapping response, and the emu performs no spontaneous, bilaterally synchronized wing-flapping. The penguin retained wing-flapping, which it exercises during submarine "flight." Taken collectively, the results from emus and penguins suggest that a behavior that has no adaptive significance may be lost.

Although central motor programs probably orchestrate the basic patterning of motor outflow to the muscles, sensory factors must play a role in fine tuning wing movement to the ever changing patterns of body attitude, wing position, wind velocity, and wind direction that occur during flight. A role for sensory input in the modulation of motor output to the wings of pigeons is suggested by Trendelenberg's (1906) finding that a bilateral but not a unilateral brachial rhizotomy eliminated wing-flapping and flight (also Narayanan and Malloy, 1974). A role for feather-related sensory input is suggested by Gewecke and Woike (1978), who reported that receptors associated with the breast feathers are involved in the initiation, maintenance, and modulation of wing-flapping in tethered siskins flying in a wind tunnel. However, drop-evoked wing-flapping in featherless chickens (Provine, 1981b) argues against the necessity of a flap-releasing or-evoking role for breast feathers in drop-evoked chicken flight. The featherless chickens also performed drop-evoked wing-flapping when deprived of visual stimulation by masking the eyes or by dropping in a darkened room. Thus, the wing-flapping was probably initiated by drop-activated vestibular processes.

The observation of spontaneous, bilaterally synchronous "flutters" in embryos at late incubation stages and drop-evoked wing-flapping in chick hatchlings suggests that a bilateral coordinating mechanism develops prenatally. One experimental approach to the study of bilateral coordination uses the technique of *induced bilateral asymmetry*, in which the onset of bilateral coordination is defined as the age when the amputation, immobilization, or weighting of one wing first influences the drop-evoked flapping of the contralateral wing (Provine, 1982). If the wings continue to beat synchronously under such conditions of increased or decreased loading, a mechanism that utilizes movement-produced feedback to coordinate bilaterally symmetrical wing movements *actively* is indicated. Bilaterally symmetrical wing-flapping at a slower than normal rate was performed by chicks with unilaterally weighted or immobilized wings. Normal rates of bilaterally symmetrical flapping were performed by the unilateral amputees. These results were obtained in 3- to 5-day-old chicks, the earliest ages examined. (Flight develops at 7 to 9 days.) Therefore, a mechanism that acts across the body midline to synchronize wing flapping develops preflight.

Hatching is a coordinated, stereotyped motor act performed by all birds (Oppenheim, 1973). Chick hatching is accomplished by back thrusts of the head and beak that chip the shell. The back thrusts are performed simultaneously with alternating leg movements that propel the chick in a counterclockwise direction as viewed from the blunt end of the egg. When the shell has been cracked around approximately two-thirds of its circumference, the shell cap breaks off and the chick emerges (Hamburger and Oppenheim, 1967; Provine, 1972a). Reinforcing the shell increases the duration of the "read out" of the stereotyped rotatory movement sequence while weakening the shell shortens it (Provine, 1972a). Therefore, a shell-related feedback mechanism is involved that maximizes the probability that hatching movements will be performed long enough to permit escape from the shell and yet terminates the movements when hatching is completed. (A similar feedback mechanism is involved in cockroach hatching and eclosion behavior discussed below.) Straightening of the neck that occurs when the chick is released from the confines of the shell may "turn off" the hatching behavior sequence (Corner *et al.*, 1973; Bekoff and Kauer, 1980). Bending the neck of a hatchling back into the prehatch posture produces a resumption of the hatching movements.

Brown (1915) and Windle and Griffin (1931) report alternating interlimb coordination patterns that resemble postnatal walking in kitten fetuses. Bekoff and Lau (1980) used videotape analyses to identify interlimb coordination in rat fetuses. The latter investigators found coordination between bilateral (homologous) but not homolateral limb pairs. This suggests that the neural circuits that coordinate the movement of bilateral limb pairs develop before those of homolateral limb pairs. Coordinating mechanisms for both bilateral and contralateral limb pairs are present in the rat by the day after birth as indicated by the movement of the four limbs during swimming (Bekoff and Trainer, 1979).

Taub and associates have used deafferentation procedures to study limb motor development in the monkey (Taub, 1976). When bilateral forelimb deafferentation was performed on the day of birth by bilateral dorsal rhizotomy, infant monkeys were able to walk, reach for, and clasp objects. Similar motor capacity was demonstrated after birth by monkeys that received bilateral forelimb deafferentation two-thirds of the way through gestation (Taub, 1976). Thus, neither spinal reflexes nor local somatosensory feedback is necessary for the development of several types of forelimb movements by the monkey. However, the presence of obvious deficits in the fine control and timing of movements by deafferented monkeys (they were clumsy) suggests that exteroceptive and proprioceptive cues play an important role in the development performance of motor skill.

Although little is known about the general behavioral embryology of invertebrates, the developmental analysis of specific movements is becoming common. Bentley and Hoy (1970) demonstrated typical flight and stridulation (song) patterns in the muscles of nymphal crickets several instars (molts) before the full development of the wings necessary for the successful performance of these behaviors. Therefore, the neuronal circuits responsible for the production of the motor patterns are developed without benefit of exercise or practice. Cockroach walking (Provine, 1977) and wing-flapping in butterflies (Petersen *et al.*, 1957), moths (Kammer and Kinnamon, 1979), and locusts (Kutsch, 1971) are also successfully performed at the first opportunity without the benefit of practice.

Davis and Davis (1973) demonstrated an independence of central and peripheral factors during development in the lobster. Swimmerets were extirpated prior to differentiation in newly hatched larvae. Normal patterns of motor outflow remained in motoneurons that would ordinarily innervate the swimmerets. Therefore, the ontogeny of motor output is not dependent on feedback normally received from peripheral muscles or sense organs.

Hatching and eclosion (shedding of the cuticle) are commonly studied invertebrate behaviors (Wigglesworth, 1972; Truman, 1971; Provine, 1976a, 1977, 1981c; Hughes, 1980). Hatching and eclosion of the cockroach occur simultaneously and involve identical cyclic, caudal-to-rostral peristaltic wave movements; the embryo sheds its cuticle as it surges out of the oötheca, the communal egg case (Provine, 1976a, 1977). All 12–16 cockroach embryos occupying a given oötheca eclose and hatch simultaneously. A communal effort is required to pry open the "spring-loaded" mouth of the oötheca. The initiation and performance of eclosion–hatching movements may require hormonal priming (or triggering) of an endogenous central motor program as in the silk moth (Truman and Sokolove, 1972). However, the synchronization of effort of the many embryos in an oötheca that is necessary for successful hatching requires mutual tactile triggering (Provine, 1976a, 1977). Cockroach hatching–eclosion is recapitulated by individual nymphs on the occasion of each molt until adulthood.

Once cockroach hatching–eclosion behavior is initiated, the caudal-to-rostral waves are maintained on both sides of various midabdominal ligations. The waves on each side of the ligation are asynchronous. Thus, any abdominal ganglion is capable of initiating a wave in its own body segment and triggering activity in the next most rostral abdominal ganglion. The ligation experiment also indicates that the caudalmost abdominal ganglion initiates the eclosion–hatching sequence of the chain. The duration of the "read out" of the eclosion–hatching motor program is determined in part by a sensory feedback mechanism involving the cuticle. Once eclosion–hatching is initiated in an embryo, the length of time that the program runs can be increased by gluing the cuticle to the body. Conversely, eclosion–hatching is immediately terminated if the cuticle is quickly removed by the experimenter (Provine, 1976a). The adaptation to environmental cues insures that that eclosion-hatching behavior sequence will accomplish its task and then be terminated.

V. Tissue Culture Approaches to Neurobehavioral Problems

A key issue in behavioral embryology as well as in developmental biology concerns the evaluation of the relative contributions of intrinsic and extrinsic influences on development. The technique of tissue culture has emerged as one of the most important means to attack such problems. Tissue culture is the ultimate "simple system." Cultures are prepared by explanting small pieces of tissue or single cells into a nutrient medium and incubating them. Large pieces of tissue usually do not survive because of problems with diffusion of materials into and out of the tissue. The tissue culture technique provides the investigator with control of the tissues environment without the worry of the blood–brain barrier that exists *in vivo*. With this environmental control comes the task of providing suitable physiological conditions for a tissue. If satisfactory conditions are not found, the results are equivocal. A negative result may indicate either the absence of a critical developmental influence, a potentially useful finding, or a problem with the tissue culture technique. Tissue culture procedures are most powerful when *in vitro* preparations model developmental processes *in vivo*. Only then does a true control group exist against which the effects of the experimental manipulations *in vitro* can be compared. Procedures and advances in the tissue culture of the vertebrate nervous system have been reviewed by Crain (1974), Nelson (1975), and Letourneau (Chapter 2, this volume). Invertebrate work is considered by Provine (1981c) and Provine *et al.* (1976).

Studies have effectively demonstrated that neurites grow out and synapses form in the "deprived" environment of the culture dish even when bioelectrical activity is pharmacologicallly blocked (Crain, 1974) (but see Bergey *et al.*, 1981). Such studies provide our strongest evidence that circuit formation proceeds in nonspiking neural tissue. In this sense, they are an important extension of the scientific debate concerning the role of extrinsic factors in neurobehavioral development that was first pursued experimentally by Harrison (1904) and others with anesthesized amphibian embryos. A shortcoming of most relevant tissue culture experiments is that the specificity of connections that are formed *in vitro* is often unknown because the inputs and outputs of various parts of the CNS (i.e., dorsal and ventral roots) or clearly defined neural populations are usually not present. Bioelectrical discharges are maintained and increase in complexity in cultures. In some cases, bioelectrical activity in cultures mimics activity *in vivo*. For example, burst discharges occur in cultures of the embryo spinal cord that resemble the bursts that are correlated with spontaneous motility in the chick embryo spinal cord *in vivo* (Provine and Rogers, 1977). Unfortunately, such comparisons are tentative because the *in vitro* and *in vivo* studies are done by different researchers in different labs using different techniques to investigate different problems. This lack of cooperation is especially problematic for the *in vitro* workers who need data from *in vivo* systems to evaluate the adequacy of their "model" systems. The justifiable excitement over the potential of various models has contributed to the neglect of the phenomena to be modeled. It is ironic that in some labs the tissue culture "tool" has become the scientific problem.

Fewer technical problems are encountered with isolated (perfused) preparations in which pieces of tissue are maintained for short periods *in vitro*, often for only the duration of a recording session. Isolated preparations dispose of animal movement, respiration, hemorrhaging, and other maintenance problems that may make some problems difficult, if not impossible, to pursue *in vivo*. Investigators have taken advantage of the unique properties of isolated cord (Saito, 1979; Landmesser and O'Donovan, 1984; Velumian, 1981) and cord–muscle preparations (Landmesser, 1978) to pursue problems of behavioral relevance. These extremely useful *in vitro* preparations are not typically considered to be "tissue cultures" because of the short-term nature of the experiments and the absence of long-term maintenance or development of the explanted tissue. In contrast with long-term cultures, short-term preparations often have anatomical features (i.e., dorsal or ventral roots) that permit electrical recordings to be made from identified neuronal populations. Such preparations have been most effective in establishing the presence of specific neuronal connections or circuits. In the future, they may effectively model more complex, spontaneous processes in the embryonic nervous system.

VI. Conclusions and Future Priorities

A. Functions of Embryonic Motility

General conclusions about the contribution of embryonic motility to development and adaptation are difficult to derive because of species differences and the many possible effects that early movement may have during the lifespan of the organism (Oppenheim, 1981). The discussion that follows does not review this complex topic but summarizes representative ideas and findings about the adaptive significance of prenatal motility. Prenatal respiratory and hatching movements have obvious survival value and will not be considered. Other references to specific examples of the adaptive significance of embryonic behavior are scattered throughout this chapter.

The most conservative possibility is that embryonic motility is only an epiphenomenon, a passive by-product of neuromuscular development that plays no role in ontogenesis. However, to prove the epiphenomenon hypothesis is to prove the null hypothesis; it cannot be done. What can be done is to evaluate the magnitude of the effect of motility on some dimensions of behavior over some portion of the lifespan.

A major theme running through this chapter is that many basic locomotor systems mature in the absence of sensory input, practice and other sources of environmental "instruction." This contrasts with the frequent finding of plasticity or environmentally induced "enrichment" or "deprivation" effects in sensory systems. The many cases of "motor precocity" also suggest that motor systems develop without the benefit of their inputs. There are also numerous physiological (Bruce and Tatton, 1980; Stein *et al.*, 1980) and anatomical (Foelix and Oppenheim, 1973) precedents for the early development of efferent relative to afferent processes. This has been referred to as a "retrograde" developmental sequence (Oppenheim, 1982).

On a different level of analysis, embryonic behavior plays an important role in the development of the chick skeletomuscular system. The immobilization of the embryo for 2 days with curare produces muscle and joint abnormalities (Drachman and Sokoloff, 1966; Oppenheim *et al.*, 1978). Apparently, the constant movement of a limb during growth sculpts joints, and pathology appears in the absence of this process. The influence of immobilization on muscle is less clear and may involve the prevention of exercise or the deprivation of a trophic substance that does not move from nerve to muscle across the blocked neuromuscular synapse.

These results are compatible with the view that chick embryonic behavior is highly autonomous. Perhaps autonomous (spontaneous) motility is an adaptation to insure that the necessary amount of movement is performed to permit normal joint and muscle development. It would be maladaptive for so critical a process to fall under the control of sporadic environmental stimuli.

One of the most novel functions of embryonic motility and/or associated neuromuscular activity is the regulation of naturally occurring motoneuron cell death. The normal loss of a significant number of neurons during vertebrate ontogeny is a general and well-established phenomenon. One of the most studied cases of neuronal death involves spinal cord motoneurons of the chicken (reviewed in Oppenheim and Chu-Wang, 1983; Lamb, 1985). Depending on the spinal level, between 30 to 60% of the limb-innervating lateral motoneurons degenerate and disappear between days 5 and 10 of embryonic life. (Hatching occurs at 21 days.) An increasing amount of evidence implicates neuromuscular interactions in the regulation of motoneuron numbers: motor innervation of the limbs occurs before the onset of cell death; a reduction in size of motoneuron "targets" (i.e., the limb musculature) by embryonic microsurgery results in increased cell death; and target enlargement by grafting supernumerary limbs reduces cell death. Neuromuscular transmission is apparently a component of this regulative interaction because chronic paralysis produced by blocking neuromuscular activity with curare or related drugs during the cell death period almost entirely prevents the normal loss of spinal motoneurons. Direct electrical stimulation of the limb muscles during the cell death period accelerates the motoneuron death rate.

Corner (1977) suggested that embryonic motility is a prenatal manifestation of the neural state of rapid-eye-movement (REM) sleep. At prenatal stages, REM sleep may be expressed as overt motility because of the absence of motor inhibition that is present after birth. One source of evidence for Corner's proposition is the finding of an increasingly large proportion of REM sleep in progressively younger animals (Roffwarg *et al.*, 1966). Roffwarg *et al.* (1966) used similar sleep data to suggest a possible role for the REM sleep process in neural development.

Bekoff *et al.* (1980) proposed that there is a functional continuity between prenatal motility and postnatal play. Although prenatal motility and play involve different movements and underlying mechanisms, both may contribute to neural and muscular development and motor training. Spontaneous motility may insure that sufficient motility occurs in the impoverished sensory environment of most embryos. They suggest further that the amount of motility performed by an embryo is an index of the adaptive significance of movement for normal development of that species. Parallel arguments are made for the function of play at postnatal stages.

At postnatal stages, the organism shapes and is shaped by its sensory environment. Held and Hein (1963) provided valuable insights into how movement-produced feedback is important in fine-tuning (calibrating) movement to the environment, a process that is known as sensory–motor integration. They provided an environment in which two groups of kittens shared similar visual experiences, but only one group had access to the visual changes produced by its own movements. The group receiving visual feedback from its own movements, the "active experience" group, gradually developed good paw–eye coordination. In contrast, kittens receiving no movement-produced visual feedback, the "passive experience group," was retarded in this capacity. Thus, movement-produced, active, visual feedback is important for the development of skilled movements.

Piaget realized that movement is important in determining the environment and experience of a child and he constantly stressed the interdependence of motor and cognitive development. This interplay was central to his view of the child as experimenter, not passive metaphysician. The pursuit of these interesting issues takes us beyond the boundaries of behavioral neuroembryology. Once the basic machinery of movement is in place, the study of behavior passes on to the psychologist or ethologist who studies the organism's adaptation to the demands of postnatal life.

B. Priorities for Future Research

1. Comparative analyses of embryonic behavior in a wide variety of vertebrates and invertebrates would help define common features of developing neuromuscular systems. It is important that such studies focus on properties of movement quality, frequency, and periodicity that can be rigorously quantified and examined across species. Quantitative videotape analyses would help to avoid much of the confusion that characterized comparative efforts during the early era of behavioral embryology. The resulting behavioral descriptions would be valuable in their own right and would help to focus subsequent high resolution but narrow and labor intensive investigations of the physiological and anatomical correlates of behavioral mechanisms. Few truly comparative analyses of embryonic behavior development have ever been conducted.

2. Electromyographic analyses of activation sequences of selected muscles establish the sequence of development of intrajoint, interjoint, and interlimb coordination. Such studies would help to operationalize what is meant by "motor development" and provide a starting point for comparative analyses of embryonic motility. Electromyography also provides a useful common ground between behavioral and neurophysiological studies. Efforts at integration are important. Both behavioral and physiological investigators often forget that behavior is physiology. No one discipline has access to a higher truth.

3. A descriptive survey of patterns of ongoing bioelectrical activity from different brain regions of both behaving and curarized embryos would be a useful beginning in understanding what the embryonic brain does and how

its actions contribute to embryonic motility. The survey would help determine (a) the time of onset of electrical activity in various brain regions, (b) the character of this early activity (active–inactive, tonic–phasic, etc.), (c) the coupling between brain regions (multiple electrodes must be used), and (d) the correlation between the activity of various brain regions and movement (i.e., Are discharges correlated with the initiation, cessation, or presence of movement?). 2-Deoxy-*d*-glucose procedures may also be useful in determining highly active regions of the developing brain. Such a fishing expedition in the embryonic brain may sound like a strange undertaking to many neurophysiologists who pursue the details of already established mechanisms. However, we are almost completely ignorant of immature brain function and should therefore cast a large net in our first efforts. We should especially be on guard for unique features of embryonic brain activity (Oppenheim, 1981). The developing brain is not simply an imperfect replica or simpler version of the adult brain.

4. It is especially important that we understand when and how sensory input modulates movement during development. Most of the research presented in this chapter stressed the "precocial onset" and autonomous nature of embryonic motility. However, it is obvious that movement must eventually be adjusted to environmental demands to become adaptive. The details of this process are largely unknown. Tentative data are available concerning the timing and relative sequence of onset of sensory function in a small group of vertebrates (Gottlieb, 1971). These data are most often based on reflexes or learning-based paradigms and consider movements only as a convenient indicator of sensory function. Studies are needed of the effect of various types of stimulation on the production and control of specific movements. Videotape, electromyographic, and other neurophysiological measures will be required to detect the possibly subtle differences between the patterning of evoked and spontaneous motility.

5. The time has come finally to answer the question about the role of ongoing motility and CNS bioelectrical activity in the maturation of specific neural circuits (Changeaux and Danchin, 1976; Lamb, 1985) and put to rest over a half century of provocative half-answers and half-replications (Carmichael, 1927; Fromme, 1941). The techniques for a definitive study are at hand. Frog eggs can be reared in a paralyzing anesthetic solution such as chlorotone. In contrast to the "classical" studies, the spinal cord and brain of the experimental embryos will need to be surveyed electrophysiologically to be sure all unit electrical activity is suppressed by the anesthetic solution. After removal from the immobilizing anesthetic, the state of specific motor circuits can be evaluated with electromyographic recordings from known agonists and antagonists. Short-term blockage of bioelectrical activity starting

before and terminated after the appearance of key behaviors would reduce possible artifacts resulting from non-specific drug action. The replication of this study using the chick would be desirable to show whether the amphibian data hold for a "higher" vertebrate. The chick study would be difficult because, so far, pharmacological agents that block CNS bioelectrical activity also interfere with cardiac activity and normal development. However, a partial replication that examines the ontogeny of specific motor circuits in immobilized, curarized embryos with normal CNS bioelectrical activity would be informative.

6. In conclusion, I return to the admonition of Viktor Hamburger that "the embryo is the best teacher." We should be attentive to its lessons and resist the temptation to simply replicate classical neuroscience using the embryo. In our pursuit of facts, we must not overlook subtle but vital clues to yet undiscovered processes governing neurogenesis.

Acknowledgments

The author appreciates the helpful comments of Drs. A. C. Catania and R. W. Oppenheim on an early draft of the manuscript. Preparation of the manuscript was supported by grant HD 11973 from the National Institute of Child Health and Human Development and grant MH 36474 from the National Institutes of Mental Health.

References

Altman, J. S. (1975). *J. Comp. Physiol.* **97**, 127–142.

Bekoff, A. (1976). *Brain Res.* **106**, 271–291.

Bekoff, A. (1981). *In* "Studies in Developmental Neurobiology: Essays in Honor of Viktor Hamburger" (W. M. Cowan, ed.), Oxford Univ. Press, London and New York.

Bekoff, A., and Kauer, J. (1980). *Soc. Neurosci. Abstr.* **6**, 75.

Bekoff, A., and Lau, B. (1980). *J. Exp. Zool.* **214**, 173–175.

Bekoff, A., and Trainer, W. (1979). *J. Exp. Biol.* **83**, 1–11.

Bekoff, A., Stein, P., and Hamburger, V. (1975). *Proc. Natl. Acad. Sci., U.S.A.* **72**, 1245–1248.

Bekoff, M., Byers, J. A., and Bekoff, A. (1980). *Dev. Psychobiol.* **13**, 225–228.

Bentley, D. R., and Hoy, R. R. (1970). *Science* **170**, 1409–1411.

Bentley, D., Keshishian, H., Shankland, M., and Toroian-Raymond, A. (1979). *J. Embryol. Exp. Morphol.* **54**, 47–74.

Bergey, G. K., Fitzgerald, S. C., Schrier, B., and Nelson, P. G. (1981). *Brain Res.* **207**, 49–58.

Berill, M. (1973). *In* "Behavioral Embryology" (G. Gottlieb, ed.), pp. 141–158. Academic Press, New York.

Birnholz, J. C. (1981). *Science* **213**, 679–681.

Brown, G. T. (1915). *J. Physiol. (London)* **49**, 208–215.

Bruce, I. C., and Tatton, W. G. (1980). *Exp. Brain Res.* **39**, 411–419.

Burghardt, G. M., and Bekoff, M., eds. (1978). "The Development of Behavior: Comparative and Evolutionary Aspects." Garland, New York.

Carmichael, L. (1926). *Psychol. Rev.* **33**, 51–68.

Carmichael, L. (1927). *Psychol. Rev.* **34**, 34–47.

Changeaux, J. P., and Danchin, A. (1976). *Nature (London)* **264**, 705–712.

Clark, R. D., and Cohen, M. J. (1979). *J. Insect Physiol.* **25**, 725–731.

Coghill, G. E. (1929). "Anatomy and the Problem of Behavior." Hafner, New York.

Corner, M. A. (1977). *Prog. Neurobiol. (Oxford)* **8**, 279–295.

Corner, M. A., and Bot, A. P. C. (1967). *Prog. Brain Res.* **26**, 21–236.

Corner, M. A., Bakhuis, W. L., and van Wingerden, C. (1973). *In* "Behavioral Embryology" (G. Gottlieb, ed.), pp. 245–279. Academic Press, New York.

Cowan, W. M., ed. (1981). "Studies in Developmental Neurobiology: Essays in Honor of Viktor Hamburger." Oxford Univ. Press, London and New York.

Crain, S. M. (1974). *In* "Aspects of Neurogenesis" (G. Gottlieb, ed.), pp. 69–114. Academic Press, New York.

Davis, W. J., and Davis, K. B. (1973). *Am. Zool.* **13**, 409–425.

Decker, J. D. (1967). *Science* **157**, 952–954.

Detwiler, S. R. (1936). "Neuroembryology." Macmillan, New York. (Reprinted by Hafner, New York, 1964).

Drachman, D. B., and Sokoloff, L. (1966). *Dev. Biol.* **14**, 401–420.

Eaton, R. C., Farley, R. D., Kimmel, C. B., and Schabtach, E. (1977). *J. Neurobiol.* **8**, 151–172.

Edwards, D. D., and Edwards, J. S. (1970). *Science* **169**, 95–97.

Edwards, D. L., and Grafstein, B. (1983). *Brain Res.* **269**, 1–14.

Fentress, J. C. (1972). *In* "The Biology of Behavior" (J. A. Kiger, ed.), pp. 83–132. Oregon State Univ. Press, Corvallis.

Fentress, J. C. (1973). *Science* **179**, 704–705.

Foelix, R. F., and Oppenheim, R. W. (1973). *In* "Behavioral Embryology" (G. Gottlieb, ed.), pp. 104–139. Academic Press, New York.

Fromme, A. (1941). *Genet. Psychol. Monogr.* **24**, 219–256.

Gallistel, C. R. (1980). "The Organization of Action: A New Synthesis." Erlbaum, Hillsdale, New Jersey.

Gewecke, M., and Woike, M. (1978). *Z. Tierpsychol.* **47**, 293–298.

Gottlieb, G. (1971). *In* "The Biopsychology of Development" (E. Tobach, L. R. Aronson, and E. Shaw, eds.), pp. 67–128. Academic Press, New York.

Gottlieb, G., ed. (1973). "Behavioral Embryology." Academic Press, New York.

Gottlieb, G., ed. (1974). "Aspects of Neurogenesis." Academic Press, New York.

Gottlieb, G., ed. (1976). "Development of Neural and Behavioral Specificity." Academic Press, New York.

Gottlieb, G. (1978). "Early Influences." Academic Press, New York.

Grillner, S. (1975). *Physiol. Rev.* **55**, 247–306.

Hamburger, V. (1963). *Q. Rev. Biol.* **38**, 342–365.

Hamburger, V. (1973). *In* "Behavioral Embryology" (G. Gottlieb, ed.), pp. 52–76. Academic Press, New York.

Hamburger, V., and Oppenheim, R. W. (1967). *J. Exp. Zool.* **166**, 177–204.

Hamburger, V., Wenger, E., and Oppenheim, R. (1966). *J. Exp. Zool.* **162**, 133–160.

Haverkamp, L. Cited in Oppenheim, R. W. (1982). *Curr. Top. Dev. Biol.* **17**, 274–275.

Harris, W. A. (1981). *Ann. Rev. Physiol.* **43**, 689–710.

Harrison, R. G. (1904). *Am. J. Anat.* **3**, 197–220.

Held, R., and Hein, A. (1963). *J. Comp. Physiol. Psychol.* **56**, 872–876.
Herman, R. M., Grillner, S., Stein, P. S. G., and Stuart, D. G., eds. (1976). "Neural Control of Locomotion." Plenum Press, London and New York.
Hooker, D. (1952). "The Prenatal Origin of Behavior." Univ. of Kansas Press, Lawrence.
Hughes, A., Bryant, S. V., and Bellairs, A. d'A. (1967). *J. Zool.* **153**, 139–152.
Hughes, T. D. (1980). *Physiol. Entomol.* **5**, 47–54.
Humphrey, T. (1969). *In* "Brain and Early Behavior Development in the Fetus and Infant" (R. J. Robinson, ed.). Academic Press, New York.
Immelmann, K., Barlow, G. W., Petrinovich, L., and Main, M. B., eds. (1981). "Behavioral Development: The Bielefeld Interdisciplinary Project." Cambridge Univ. Press, London and New York.
Kammer, A. E., and Kinnamon, S. C. (1979). *J. Comp. Physiol.* **130**, 29–37.
Kutsch, W. (1971). *Z. Vgl. Physiol.* **74**, 156–168.
Lamb, A. H. (1985). *CRC Critical Rev. Clin. Neurobiol.* **1(2)**, 141–179.
Landmesser, L. T. (1978). *J. Physiol. (London)* **284**, 391–414.
Landmesser, L. T., and O'Donovan, M. J. (1984) *J. Physiol.* **347**, 189–204.
Matthews, S. A., and Detwiler, S. R. (1926). *J. Exp. Zool.* **45**, 279–292.
Naka, K. I. (1964a). *J. Gen. Physiol.* **47**, 1003–1022.
Naka, K. I. (1964b). *J. Gen. Physiol.* **47**, 1023–1038.
Narayanan, C. H., and Hamburger, V. (1971). *J. Exp. Zool.* **178**, 415–432.
Narayanan, C. H., and Malloy, R. B. (1974). *J. Exp. Zool.* **189**, 177–188.
Narayanan, C. H., Fox, M. W., and Hamburger, V. (1971). *Behaviour* **40**, 100–134.
Nelson, P. G. (1975). *Physiol. Rev.* **55**, 1–61.
Nicolai, J. (1976). *Z. Tierpsychol.* **40**, 225–243.
Oppenheim, R. W. (1972). *Proc. Int. Congr. Ornithol. 15th, 1970* pp. 283–302.
Oppenheim, R. W. (1973). In "Studies on The Development of Behavior and the Nervous System" (G. Gottlieb, ed.), Vol. 1, pp. 163–244. Academic Press, New York.
Oppenheim, R. W. (1974). *Adv. Study Behav.* 133–172.
Oppenheim, R. W. (1981). *In* "Maturation and Development: Biological and Psychological Perspectives" (K. J. Connolly and H. F. R. Prechtl, eds.), pp. 73–109. Lippincott, Philadelphia.
Oppenheim, R. W. (1982). *Curr. Top. Dev. Biol.* **17**, 257–309.
Oppenheim, R. W., and Chu-Wang, I.-W. (1983). *In* "Somatic and Autonomic Nerve-Muscle Interactions" (G. Burnstock, R. O'Brien, and G. Vrboza, eds.), pp. 57–107. Elsevier, Amsterdam.
Oppenheim, R. W., Pittman, R., Gray, M., and Maderdrut, J. (1978). *J. Comp. Neurol.* **174**, 619–640.
Petersen, B., Lundgren, L., and Wilson, L. (1957). *Behaviour* **10**, 324–339.
Pollack, E. D., and Crain, S. M. (1972). *J. Neurobiol.* **3**, 381–385.
Preyer, W. (1885). "Specielle Physiologie des Embryo." Grieben, Leipzig.
Provine, R. R. (1971). *Brain Res.* **29**, 155–158.
Provine, R. R. (1972a). *Psychonomic Sci.* **29**, 27–28.
Provine, R. R. (1972b). *Brain Res.* **41**, 365–378.
Provine, R. R. (1973). *In* "Behavioral Embryology" (G. Gottlieb, ed.), pp. 77–102. Academic Press, New York.
Provine, R. R. (1976a). *J. Insect. Physiol.* **22**, 127–131.
Provine, R. R. (1976b). *In* "Simpler Networks and Behavior" (J. Fentress, ed.), pp. 203–220. Sinauer, Sunderland, Massachusetts.
Provine, R. R. (1977). *J. Insect Physiol.* **23**, 213–220.

Provine, R. R. (1979). *Behav. Neural Biol.* **27**, 233–237.
Provine, R. R. (1980). *Dev. Psychobiol.* **13**, 151–163.
Provine, R. R. (1981a). *Dev. Psychobiol.* **14**, 279–291.
Provine, R. R. (1981b). *Dev. Psychobiol.* **14**, 481–486.
Provine, R. R. (1981c). *In* "The American Cockroach" (W. J. Bell and K. G. Adiyodi, eds.), pp. 399–423. Chapman & Hall, London.
Provine, R. R. (1982). *Dev. Psychobiol.* **15**, 245–255.
Provine, R. R. (1983). *Dev. Psychobiol.* **16**, 23–27.
Provine, R. R. (1984). *Am. Sci.* **72**, 448–455
Provine, R. R., and Rogers, L. (1977). *J. Neurobiol.* **8**, 217–228.
Provine, R. R., and Westerman, J. A. (1979). *Child Dev.* **50**, 437–441.
Sci. U.S.A. **65**, 508–515.
Provine, R. R., Seshan, K. R., and Aloe, L. (1976). *J. Comp. Neurol.* **165**, 17–30.
Provine, R. R., Strawbridge, C. L., and Harrison, B. J. (1984). *Dev. Psychobiol.* **17**, 1–10.
Ripley, K. L., and Provine R. R. (1972). *Brain Res.* **45**, 127–134.
Robertson, S. S., Dierker, L. J., Sorokin, Y., and Rosen, M. G. (1982). *Science* **218**, 1327–1330.
Roffwarg, H. P., Muzio, J. N., and Dement, W. C. (1966). *Science* **152**, 604–619.
Saito, K. (1979). *J. Physiol. (London)* **294**, 591–594.
Sawyer, L. M. (1982). *J. Embryol. Exp. Morphol.* **68**, 69–86.
Scheibel, M. E., and Scheibel, A. B. (1970). *Exp. Neurol.* **29**, 328–335.
Scheibel, M. E., and Scheibel, A. B. (1971). *Exp. Neurol.* **30**, 367–373.
Spemann, H. (1938). "Embryonic Development and Induction." Yale Univ. Press, New Haven, Connecticut.
Stehouwer, D. J., and Farel, P. B. (1980). *Brain Res.* **195**, 323–335.
Stein, B. E., Clamann, H. P., and Goldberg, S. J. (1980). *Science* **210**, 78–80.
Stokes, B. T., and Bignall, K. E. (1974). *Brain Res.* **77**, 231–242.
Straznicky, K. (1963). *Acta Biol. Acad. Sci. Hung.* **14**, 143–153.
Taub, E. (1976). *In* "Neural Control of Locomotion" (R. M. Herman, S. Grillner, P. S. G. Stein, and D. G. Stuart, eds.), pp. 675–705. Plenum, New York.
Thelen, E. (1979). *Anim. Behav.* **27**, 699–715.
Tobach, E., Aronson, L. R., and Shaw, E., eds. (1971). "The Biopsychology of Development." Academic Press, New York.
Tracy, H. C. (1926). *J. Comp. Neurol.* **40**, 253–369.
Trendelenberg, W. (1906). *Arch. Anat. Physiol, Physiol. Abt.* **30**, 1–126.
Truman, J. W. (1971). *J. Exp. Biol.* **54**, 805–814.
Truman, J. W., and Sokolove, P. G. (1972). *Science* **175**, 1491–1493.
Velumian, A. A. (1981). *Brain Res.* **229**, 502–506.
Visintini, F., and Levi-Montalcini, R. (1939). *Schweiz. Arch. Neurol. Psychiatr.* **43**, 1–45.
Weiss, P. (1941). *Comp. Psychol. Monogr.* **17**, 1–96.
Wigglesworth, V. B. (1972). "The Principles of Insect Physiology," 7th Ed. Chapman & Hall, London.
Wilson, D. M. (1961). *J. Exp. Biol.* **38**, 471–490.
Windle, W. F. (1940). "Physiology of the Fetus." Saunders, Philadelphia.
Windle, W. F. (1944). *Physiol. Zool.* **17**, 247–260.
Windle, W. F., and Griffin, A. M. (1931). *J. Comp. Neurol.* **523**, 149–188.

8

Ontogeny of the Encephalization Process

DENNIS J. STELZNER
Department of Anatomy
State University of New York Upstate Medical Center
Syracuse, New York

This chapter discusses the ontogeny of the brain's control over the spinal motor system. This increasing developmental influence of the brain is called encephalization. I concentrate mainly on the development of the integrated goal-directed responses found after birth. However, it is first necessary to give a general overview of behavioral development. The brief review below is incomplete but, I feel, represents the field at this point. Other views and more complete accounts can be found elsewhere (Carmichael, 1933; Hamburger, 1973; Oppenheim, 1981a, 1982).

Developmental
NeuroPsychobiology

I. Behavioral Development

A. Spontaneous Motility

As Provine has pointed out (Chapter 7, this volume), the spinal cord organizes the behavior of the embryo for much of its development without significant influence from descending sources, and at least in chick embryo, without sensory input. An intrinsic programmed development of the spinal motor system appears to control the early "spontaneous" motility of the embryo, both *in ovo* (Hamburger, 1963; Hamburger *et al.*, 1966) and *in utero* (Narayanan *et al.*, 1971). This behavior is not in any clear way related to the integrative adaptive behaviors found during and after hatching in the chick (Oppenheim, 1973) or after parturition in the mammal. This seeming disparity between spontaneous embryonic motility and integrated postnatal behaviors may reflect the environmental conditions during these different developmental periods. The environment *in ovo* and *in utero* is highly protected, and the behavioral role of the embryo appears largely to be passive. The behavioral development of the salamander (Coghill, 1929) and fish (Tracy, 1926) is more continuous and the behaviors generated throughout development play an active role in the survival of these species. In avian and mammalian species, although different types of sensory stimuli may cause behavioral modifications at later periods (Gottlieb, 1976; Vince, 1973), most types of sensory stimuli (particularly somesthetic) probably do not reach the CNS or do so in a very different manner than after birth. More importantly, most responses of the developing animal to such stimuli are not necessary for its immediate survival. Early spontaneous motility appears related more to the normal formation and maintenance of joints (Drachman and Sokoloff, 1966; Oppenheim *et al.*, 1978), to the maintenance of the musculature (for review, see Jacobson, 1978), or to development of motoneurons (Oppenheim *et al.*, 1978; Oppenheim, 1981b) than to the development of postnatal and posthatching behaviors. Corner (1977) has noted that the type of autogenous behavior seen *in ovo* and *in utero* may continue postnatally in the state of sleep, rather than being related to the beginnings of goal-directed responses.

B. Stimulated Embryonic Movements

The extensive data concerning embryonic development of responses to sensory stimuli have been more comprehensively reviewed elsewhere (Carmichael, 1933; Windle, 1940; Narayanan *et al.*, 1971). In most vertebrate embryos, the trigeminal area first becomes sensitive to somesthetic stimuli. The earliest response to stimulation involves neck flexion and later includes trunk and limb movements (Humphrey, 1964). Thus, local perioral stimulation later evokes a generalized movement. This early general movement has been termed a total pattern response (Coghill, 1929). Later, more discrete responses to stimulation of this area are much more variable. In many species, at about the same point in development in which perioral stimulation gives a generalized reaction, a local segmental response follows stimulation of the forelimb (Bodian, 1966; Bodian *et al.*, 1968; Windle *et al.*, 1933). Thus local reflexes may occur simultaneously with total pattern movements, and neither appears to be primary. The ease of eliciting these reflexes and the parts of the body involved in response to stimulation also appear to parallel the amount and type of spontaneous motility so that both probably use much of the same neural substrate (Narayanan *et al.* 1971). At least in the chick (Singer *et al.*, 1978; Nornes *et al.*, 1980) much of the early motility appears more related to intersegmental connections (propriospinal) than to descending connections from the brain. Humphrey (1964), in her detailed analysis of the development of responses to cutaneous stimulation in the human fetus, found that responses of the neck are seen first, followed by the trunk, etc., generally in a cephalocaudal direction. She found that proximal movements are generally seen before distal ones, the direction of response ontogeny following developmental gradients of the neuromuscular system (Pierce, 1973; Rodier, 1980). However, the areas which elicit these responses do not follow the same cephalocaudal progression. The areas first eliciting responses are those with the highest density of sensory receptors: first the perioral region, then the palms of the hands, the anal and genital areas, and the soles of the feet. Most other areas elicit responses later in development. As noted above, since these types of stimuli are not natural at this stage of development, the responses seen to them may have no adaptive significance. They do, however, give an indication both of response development and of its underlying neural substrate.

C. Goal-Directed Responses in Precocial and Altricial Species

Experimenters have used two approaches to study the ontogeny of goal-directed responses. One has been to study precocial species that are relatively mature at birth such as the sheep (Barcroft and Barron, 1939) or guinea pig (Carmichael, 1934). In these species many *in utero* responses appear to be similar to the goal-directed responses seen after birth. Study of behavioral development in precocial species has the advantage that most behavioral development occurs prenatally under constant environmental conditions. The other approach has been to study response development in altricial species born immaturely at birth such as the opossum (Langworthy, 1925; Martin *et al.*, 1978), rat (Altman and Sudarshan, 1975; Stelzner, 1971; Tilney, 1933), or cat (Langworthy, 1929), where most goal-directed behaviors are generated during the postnatal period. Anokhin (1964) has pointed out the importance of the synchrony of neural and behavioral development in these altricial species so that the responses of the animal continue to meet new demands of the postnatal environment. He theorized a program of response development for which the underlying neural substrate must mature selectively. Components of each system mature at different rates at various levels of the neuroaxis and systems extending through the neuraxis also mature at different rates. For instance, the olfactory and thermal senses are well developed at birth in most altricial mammals. The olfactory sense is important for suckling at birth (see chapter by Blass, this volume) and appears to inhibit locomotion at this time. Later it is used in returning to the nest. The newborn uses its thermal sense to guide approach to the mother or siblings and in huddling (see chapters 15 and 16, this volume). This sense becomes less important as the pup begins to regulate its body temperature and as olfaction begins to predominate in approach locomotion behaviors. Mature locomotor responses begin to appear at the time of eye opening in the rat. Another example of the developmental sequence proposed by Anokhin is the development of the scratch response in the rat (Stelzner, 1971).

1. Characteristics of the Scratch Response

In the adult rat this is an integrated response involving a hind-limb scratching very precisely and at a high rate a point on the irritated skin. It is a difficult response for an experimenter to elicit. The other hind limb supports the hindquarters during scratching, and muscles of the back and forequarters are involved. After scratching, the rat often licks the digits of the scratching limb, a component of grooming. When the spinal cord is transected, the scratch response is severely depressed initially but then recovers (Sugar and Gerard, 1940). This response depression immediately after spinal transection is called spinal shock and, as discussed below, is due to loss of supraspinal control. Scratch is one of the last responses to recover from spinal shock in the adult rat, suggesting that it is under relatively heavy descending control. Sherrington (1947) has shown that at least in carnivores, the rhythm of scratch and opposite hind limb support as well as the involvement of the muscles of the back and forelimbs during scratching are organized at the spinal level. These responses recover after spinal transection. The only components of scratching that are lost permanently are the ability to scratch the exact locus on the skin being stimulated (local sign) and the integration of the scratch response in grooming.

2. *Ontogeny of the Scratch Response*

The scratch response is primitive in the rat at birth (Stelzner, 1971). During the first 3 postnatal days there is no opposite hind-limb support, the local sign is poor, the frequency is slow (one per second), and punctate stimuli easily and repetitively elicit scratch even to inappropriate areas such as the base of the back. In essence, it is a local and primitive avoidance response to many different types of stimuli. Spinal transection has no effect during this period suggesting minimal supraspinal control. By 6 to 8 days of age the local sign improves, scratching is no longer seen to inappropriate stimuli, the response becomes much more difficult to elicit, contralateral hind-limb support begins, and the frequency increases to two or three per second. Although spinal transection does not abolish scratching at this stage, scratch does become more rudimentary and resembles the neonatal response for the first few postoperative days (Weber and Stelzner, 1977). By 15 days of age the scratch response appears mature except that it is not used in grooming until a few days later. Spinal transection at this period causes the same type of spinal shock seen in the adult. Scratching only recovers at the end of the first postoperative week and is never as easily elicitable as in the mature neonatally transected animal. In the neonatally transected rat the frequency of scratching increases with postoperative time at approximately the same pace as in unoperated littermates. All of these data are consistent with Anokhin's idea that spinal components of scratching and the descending control for scratching mature together during the ontogeny of this response.

II. Encephalization of the Spinal Motor System

Encephalization is a comparative as well as a developmental term. In higher vertebrates there is great expansion in the size of the brain and its descending control of spinal cord circuitry, culminating in the immense descending influence found in humans. It has been thought that as encephalization evolved, functions residing in lower centers (spinal cord, medulla, midbrain) were increasingly influenced by higher centers until, in certain cases, these functions were permanently "lost" from the lower centers (Fulton, 1949; Hughlings-Jackson, 1886). For instance, after decerebration, certain functions do not recover in humans (Fulton, 1949) that do recover in the rat (Woods, 1964). Similarly, the reflex ability of the isolated spinal cord is greater in the frog (Fukson *et al.*, 1980), than cat (Chambers *et al.*, 1966), than monkey (McCouch *et al.*, 1966).

Sherrington (1947) stressed the parallel between the evolution of the cerebral hemispheres and the severe depression of spinal reflexes (spinal shock) following spinal cord transection. The amount of spinal shock increases as relative forebrain size increases, lasting only momentarily in the frog (Sherrington, 1947), for a few hours in the cat (Chambers *et al.*, 1966), and for up to 1 month in the monkey (McCouch *et al.*, 1966). As noted in our analysis of the ontogeny of the scratch response, the amount of spinal shock also increases during ontogeny. As in phylogeny, the amount of spinal shock parallels the maturation of descending influences (see Section IV). In the next sections we discuss the normal ontogeny of the brain's control over the spinal cord and how removing descending influences at different developmental periods alters the encephalization process.

III. The Encephalization Process

As described earlier, the spinal motor system general develops in a rostrocaudal sequence, the forelimbs functioning earlier, and being, in most mammalian altricial species, the primary means of locomotion during the neonatal period. The brain's control over the spinal motor system develops in a reverse sequence, caudorostrally. The rhombencephalon affects the responses of the animal earlier than the midbrain, and telencephalic control develops last. This maturation of the brain and spinal cord generally follows the sequences found both in phylogeny and in the initial morphogenesis of the nervous system.

A number of different descriptive and experimental investigations have documented the encephalization process.

A. Anatomy

Neurofibrillar techniques, myelin methods, and more recent degeneration and autoradiographic methods have all shown that the development of descending motor control of the brain progresses from the medulla anteriorly. The first descending nerve tracts are from the brain stem, the medial longitudinal fasciculus, and reticulospinal, vestibulospinal, and tectospinal tracts (Langworthy, 1932; Rhines and Windle, 1941; Scheibel and Scheibel, 1964; Windle, 1970). Myelinated axons are first found in the medulla and cervical spinal cord, then in the midbrain, and finally in the forebrain (Langworthy, 1932; Windle *et al.*, 1934). Many of these pathways can be identified using neurofibrillar, degeneration, autoradiographic, or retrograde transport methods before they have any obvious behavioral effect (Donatelle, 1977; Gilbert and Stelzner, 1979; Martin *et al.*, 1978, Martin *et al.*, 1982;

Okado and Oppenheim, 1985; Van Mier and ten DonKelaar, 1985). On the other hand, nerve tracts responsible for a particular behavior may myelinate long after they have begun to function (Langworthy, 1928; Windle *et al.*, 1934).

B. Neurochemistry

Neurochemical and pharmacological experiments also show a caudorostral sequence of brain maturation. These studies have demonstrated that brainstem noradrenergic systems are present at birth in the rat and distribute in the forebrain postnatally (Coyle and Axelrod, 1972a,b; Lamprecht and Coyle, 1972; Loizow and Salt, 1970). These noradrenergic systems function at birth (Campbell *et al.*, 1969). Cholinergic systems from the forebrain (Fibiger *et al.*, 1970; Moorcroft, 1971) and serotenergic systems (Mabry and Campbell, 1974) do not affect this brain stem system until the second postnatal week. This lack of descending inhibition from the rat forebrain at birth can also be shown by ablating frontal neocortex or the hippocampus (Moorcroft, 1971; see below).

C. Lesion and Stimulation Studies

The major approach to studying the effect of encephalization on behavioral development has been to study effects of lesions at various levels of the nervous system upon postoperative behaviors. Spinal shock (response depression) or a disappearance in a certain behavior that was apparent preoperatively indicates that the pathways damaged had contributed to that behavior. When a postoperative behavioral effect is similar to that found after adult injury, the functioning of the damaged system is considered to be mature. This method of analysis does not tell whether a damaged pathway was present earlier but was nonfunctional and, as we see in Section IV, the disappearance of a particular behavior does not, necessarily, mean that the damaged pathway was exclusively responsible for it. However, it remains the method that best relates neural and behavioral development.

Development of descending pathways has also been assessed through electrical stimulation-evoked movement. This method has a number of technical difficulties. For instance, current spread is difficult to control, and it is nearly impossible to determine specific pathways responsible for the movement. In addition, a movement evoked in this way is entirely artificial; the circuitry necessary for that area to normally function in the movement may not be available until much later. However, it does indicate when a substrate for movement first becomes evident from the stimulated area.

Experiments using the lesion method or electrically evoked movements again show that the encephalization of behavioral development follows the caudorostral pattern.

1. Prenatal Studies

a. Sheep. Studies in the sheep fetus by Barcroft and Barron (1937, 1939, 1942) have shown movements evoked by electrical stimulation of the medulla on the thirty-eighth day of gestation (38DG), of the midbrain on 43DG, and of the cerebral cortex by 100DG. Barcroft and Barron also described several stages of behavioral development, studying evoked responses to snout stimulation. The fetus was removed from the amnionic sac prior to stimulation but still had an intact placental circulation. On 40DG, snout stimulation evoked jerky or twitchlike movements that involved the limbs, neck, and tail. On 50DG, the responses, instead, were sustained; a prolonged flexion of the hind limbs, extension and abduction of the forelimbs, and raising the head. By 60DG, this response was inhibited and there was little elicited or spontaneous behavior until just prior to parturition. Responses of the diaphragm and intercostal muscles similar to respiratory movements were first twitchlike (38DG) and occurred with other muscles after snout stimulation. Later (40–45DG) the rate of contraction of the diaphragm increased and the response lasted for a longer period, corresponding to the activity of other elicited movements. In the next stage, respiratory contractions continued after general muscular activity had ended (45DG). Spontaneous respiratory movements were seen from 55 to 60 DG. These movements were then suppressed along with other activity by 65DG.

Barcroft and Barron transected the brain at various levels between 40 and 70DG and replaced the fetus in the uterus. They studied these animals at 54 to 132DG, most between 1 and 13 days postoperatively. A problem with these data and other studies described below is that the response ontogeny of these chronic operates may have been changed by the lesion (see Section IV). For instance, Barcroft and Barron (1939) reported that the scratch response appeared earlier in a spinally transected fetus (70DG) than in unoperated animals (90DG), With this reservation, their results show a fascinating sequence in the ontogeny of respiration and other responses during encephalization.

No respiratory movements were seen after spinal transection (C4–C5). On the other hand, the spinally transected fetuses were very active at 70DG, an age when unoperated fetuses were unresponsive. This suggests descending inhibition of spinal circuitry at this age. A transection caudal to the pons at 70DG (10 days postoperatively) also resulted in a very active fetus, hypersensitive to most stimuli. This puts the site of descending inhibition anterior to the medulla. The development of respiration in the operated fetuses was particularly instructive. A transection at a midmedullary level resulted in a fetus of 71DG having two or three respiratory movements in response to snout stimulation, similar to an intact 38DG fetus. With a lesion anterior to the medulla at 70DG, snout stimulation evoked an increased number of respiratory contractions, comparable to a 40 to 42DG fetus. If the lesion was anterior to the pons, at 68DG or 132DG (10 days postoperatively in each case), continuous respiratory movements were seen, as in a 55 to 60DG fetus. After a lesion between the superior and inferior colliculi, a 66 to 68DG animal remained as inactive as an unoperated animal and it was difficult to elicit respiratory movements. At this age, transection at a more anterior level had no effect. After 70 to 80DG, removal of the basal ganglia and cerebral cortex caused more frequent gasping movements, suggesting that these areas had some role in inhibition of respiration by this age. These data indicate increasing encephalization of behavioral responses during ontogeny.

b. Chicks. Acute spinal cord transection at 4 or 5 or 8 or 9 days *in ovo* had no effect on spontaneous motility of the chick embryo (Oppenheim, 1975). There was no modification in the temporal pattern and rhythm of motility until 10 days *in ovo* if the cervical spinal cord was transected early in development. Only at 16 or 17 days *in ovo* was a qualitative change in the behavior of these chronic operates seen; the rotatory movements that usually began during this period were absent (Oppenheim, 1975). Sedlacek's results (1978) put the beginning of the brain's control over spontaneous motility even later. Prior to 15 days *in ovo* decapitation or spinal transection had little affect on spontaneous motility. At 15 days, there was an acute decrease in the frequency of spontaneous movements after spinal transection that returned 24 h later. By 17 to 19 days *in ovo* spinal shock was greater and there was only partial recovery 24 h later. After this period, motility after spinal transection was lower and remained at a pre-15-day level. Sedlacek also noted a progression in the brain's control of motility. Removal of the entire brain reduced the amount of motility at 14 days *in ovo*; removal of the fore and midbrain reduced motility starting at 15 days, and removal of the telencephalon had no effect until 16 days *in ovo*. Decker and Hamburger (1967) found no effect of a lesion anterior to the medulla on the duration of spontaneous activity and inactivity cycles until 17 days *in ovo*.

Oppenheim (1972) studied effects of brain ablation on prehatching and hatching behaviors. After midbrain removal at 38 to 45 h *in ovo* the normal rotatory prehatching movements at 16 or 17 days *in ovo* did not appear and hatching did not occur. With forebrain removal normal prehatching movements were found. However, the prehatching movements continued after 20 days in these operates, unlike normal animals, and hatching did not occur. Again, these data show a clear encephalization of behavioral control.

c. Rabbit. Volokhov (1961) found a similar encephalization of behavioral control in decapitated or decorticated rabbit fetuses. There was no behavioral effect of decapitation until 25DG, at which time decapitation rendered reactions similar to those normally seen before 23DG. There was no effect of decortication at 25DG. At 25 or 26DG tonic generalized movements are normally inhibited in intact fetuses and specialized motor reactions such as washing and licking appear. Decortication at 27 to 29 DG resulted in a return to the responses seen before 25 or 26DG.

2. Postnatal Studies

a. Opposum. Langworthy (1925, 1926, 1927a) decerebrated the altricial opposum at different postnatal periods. Up to 50 days of age there was no behavioral effect even immediately after decerebration. It is of note given the precautions discussed earlier, that, even though decerebration had no detectable behavioral effect, there was a forelimb response to cerebral cortex electrical stimulation as early as day 40. By 56 days of age, decerebration increased postoperative activity. At day 68 decerebration began to produce extensor rigidity and an adult level of extensor rigidity was seen after decerebration at day 76. At about this period spinal shock of adult severity was seen in the hind limbs of the opposum after midthoracic spinal transection (Martin *et al.*, 1978). Thus, encephalization and spinal shock, due to the removal of descending connections, corresponded.

b. Kittens. Langworthy (1926, 1927b) and Weed (1917) found that decerebration had little effect at birth in the kitten even though electrical stimulation of the cerebral cortex caused contralateral forelimb movement. At 9 days of age there was a transient increase in forelimb extensor tone immediately after decerebration. Decerebrate rigidity was first seen in the forelimbs at day 14 and became fully mature in both the fore and hind limbs in the ninth postnatal week.

More recently, Skoglund (1960, 1966) studied the effects of decerebration in the kitten on the development of alpha and fusimotor rigidity found in the adult cat. Damage to the anterior lobe of the cerebellum in the adult releases anterior cerebellar inhibition of the vestibular nuclei. This increases activity of descending vestibular connections onto alpha motoneurons (alpha rigidity). At birth alpha rigidity was found after anterior lobe damage. However it was not fully mature and was especially weak in the hind limbs. In the adult cat fusimotor rigidity involves release of reticulospinal facilitation of extensor fusimotor neurons caused by midcollicular decerebration. It is the type of decerebrate rigidity described by others that was discussed above. After midcollicular decerebration at birth, fusimotor rigidity was not apparent, in fact there was only a slight increase in flexor tone and an increased responsiveness of both the fore and hind limbs. By 12 to 14 days, midcollicular decerebration increased forelimb extensor tone, and at the end of the third week, there was also increased hind-limb extensor tone after this lesion. Skoglund pointed out that, as the descending reticulospinal fusimotor control developed, the IA loop of the stretch reflex within the spinal cord, which was affected by these descending connections, was also maturing. This shows that spinal circuitry was maturing at the same time as its descending control.

3. Overview

A point made earlier concerning postnatal behavioral development is worth emphasizing. Even though encephalization progresses in a caudocranial sequence, the ontogenesis of supraspinal control is not a simple one where the medulla matures first, then the midbrain, etc. Systems within the brain stem and spinal cord mature differentially and the regions within the rhombencephalon which are essential to the survival of the neonate are mature at birth (Fox, 1966; Anokhin, 1964). In the rhombencephalon, areas of the reticular formation necessary for respiration, cardiovascular control, deglutition, and vocalization are relatively mature at birth (Scheibel and Scheibel, 1964). Yet, postural reactions and the cerebellar circuitry controlling posture are still very immature. Only the rudimentary forelimb locomotor responses necessary to stay near the mother and to locate the teat are found at birth. The locomotor reactions are stimulated by contact with the face, probably through a trigeminal circuit within the brain stem (Humphrey, 1964). The vestibular portion of the VIII nerve matures relatively early as do the vestibular connections and vestibulospinal tract. The righting response and vestibular reactions are found in most altricial species during the neonatal period (Stelzner, 1971). Yet the cochlear portion of the VIII nerve and the cochlear nuclei mature late in many species (Rubel, 1978; Windle, 1970). A startle reaction to hand clap is also not apparent at birth in many altricial species (Fox, 1966; Stelzner, 1971). The dorsal column nuclei and medial lemniscus mature relatively late, along with the development of descending axons from the cerebral cortex (Langworthy, 1932; Windle, 1970). There is obviously a systems as well as a levels maturation to the encephalization process that has proven more resistant to analysis using the type of experiments described above. This type of developmental process becomes clearer when encephalization is not allowed to occur.

IV. Interference with Encephalization

As described above, the immediate effect of decerebration or spinal transection is minimal at birth in most altricial species. This lack of descending influence at birth provides the opportunity to study the development of the spinal motor system without the influence of descending control.

A. The Effect of Spinal Transection during Postnatal Development

Shurrager and Dykman (1951) were the first to examine the chronic effects of midthoracic transection in kittens and puppies. Prior to 4 weeks of age, the severe depression (spinal shock) seen immediately after this injury in adults was not found. Within hours rhythmical stepping movements developed in infant operates and extensor spasms, which characterized the chronic spinally transected adult, never developed. Shurrager and Dykman did extensive conditioning and postoperative handling of these animals and felt that this facilitated later walking in these chronic neonatal operates. A study testing this point found that exercise had no affect on the locomotor reactions that developed in spinalized kittens although some recovery of locomotion did occur in spinalized adults (12 weeks of age) when they were exercised (Smith *et al.*, 1982).

I was interested in the effect of spinal transection in neonatal operates which received no conditioning or special handling. I also wanted to know when behaviors that are not found in chronic adult operates appeared during development in neonatal operates, compared to normal response ontogeny. I chose the rat because it is behaviorally immature at birth, and development of responses in the hindquarters is very rapid, maturing by the end of the third postnatal week (Altman and Sudarsham, 1975; Donatelle, 1977; Stelzner, 1971; Tilney, 1933).

In an initial experiment (Stelzner *et al.*, 1975), the midthoracic spinal cord of neonatal (0–4 days of age) and weanling rats (21–26 days of age) was transected. The animals were given a neurological examination before surgery and at short intervals after surgery for the first 3 postoperative weeks. They were then retested at varying periods until they were sacrificed, up to a year later. Unoperated littermate controls were also tested.

1. *Effect of Spinal Transection in the Neonatal Rat*

At birth, few responses of the hindquarters were influenced by cephalic stimulation except for primitive vestibular and locomotor reactions that disappeared permanently after surgery in the neonatal operates. The only other hind-limb response affected by the surgery was the flexed posture of the proximal musculature which was not seen during the first postoperative day. Wriggling, which is a generalized long-lasting response to pinch, flexion reflexes, urogenital responses to stroking the genitalia, including postural reactions as well as urination and defecation, and the scratch response were evident immediately after recovery from anesthesia. Thus, spinal shock immediately after surgery was minimal.

During the next 3 postoperative weeks additional responses appeared in the hind limbs of neonatal operates. There were a number of similarities with normal response ontogeny, and in general, the actual time of appearance of many responses was similar to the time of appearance of these same responses in unoperated littermates. However, certain responses did not disappear as in normal response ontogeny; certain responses appeared several days earlier than normal; and in other cases there was a delay of 4 or 5 days in the appearance of responses.

In neonatal operates, bilateral hind-limb support was first noted 4 or 5 days postoperatively, whether the lesion was made at birth or at 4 days of age. Locomotor responses began in unoperated animals by 6 to 8 days of age, several days before stepping was seen in neonatal operates. On the other hand, wriggling continued throughout life in neonatal operates whereas it disappeared in normal littermates by 6 to 8 days of age. The postural reactions to stimulating the genitalia disappeared in both neonatal operates and normal littermates between 10 and 12 days of age. Also, the posture of the hind limbs of both intact littermates and neonatal operates resembled the adult posture by the end of the second postnatal week. By 15 to 21 days of age, there was a rapid development of hind-limb coordination in neonatal operates, resembling that seen in unoperated littermates 4 or 5 days earlier. During this period a waddling gait of the hind limbs, which was not coordinated with the forelimbs, occurred during forward progression. We also noted rapid replacement of a hind limb if it fell through a grid opening, hopping responses, an extensor thrust reaction if the ventral pads landed on a grid bar, and palpitations of the hind-feet on the surface of a screen. Most remarkable was a placing reaction of the hindfoot either to the bending of hairs or to slight pressure on the dorsum of the foot ("tactile" or contact placing). Except for an overall decrease in responsiveness seen with increasing survival time in neonatal operates, there were few further behavioral changes for as long as a year after surgery.

2. Effect of Spinal Transection in the Weanling Rat

Midthoracic spinal transection in weanling rats produced effects similar to those in adults (Sugar and Gerard, 1940). Few responses were seen during the first 2 or 3 postoperative days, indicating spinal shock in the weanling operate comparable to that seen in adult animals. It was necessary to evacuate the urinary bladder of these animals manually for the first 7 to 10 postoperative days. Over the last half of the first postoperative week, scratching, flexion reflexes, and brief wriggling appeared. With increasing time, choreoathetoid-like movements and flexor and extensor spasms appeared—behaviors seldom observed in neonatal operates. Other responses that are found in neonatal operates, such as hind-limb support and tactile placing, were never observed in weanling operates even with postoperative recovery periods of up to 1 year. Figure 1 shows that the posture of a chronic neonatal operate (a) and a chronic weanling operate (b), and Table I summarizes the behavioral differences in the hind limbs of adult animals spinally transected at the neonatal and weanling stages. It can be seen in Table I that a large number of responses survive or are spared by neonatal spinal transection which are lost after transection in a weanling or adult animal. I refer to these behavioral differences between infant-operates and adult-operates as *sparing of function*.

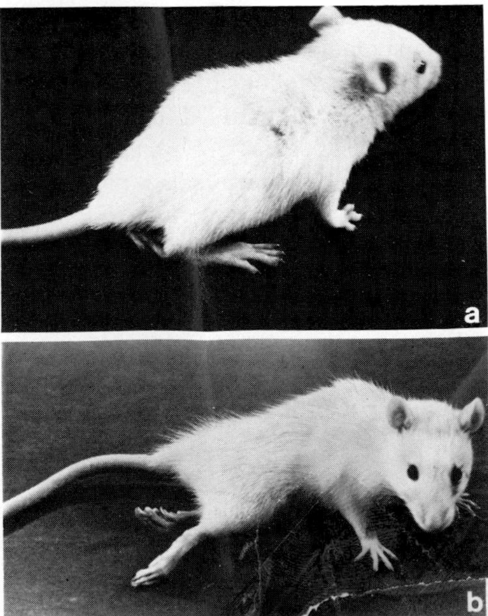

Fig. 1. A neonatal operate at the weanling stage of development (a) and a chronic weanling operate (b). From Stelzner (1982a), with permission.

Table I

Differences in Chronic Hind-limb Behaviors of Spinally Transected Neonatal and Weanling Operates

Neonatal operates	Weanling operates
1. Spinal shock minimal	1. Spinal shock pronounced
2. Wriggling, long lasting	2. Wriggling, brief
3. Spontaneous responses common: scratch, flexion reflexes but spasms infrequent	3. Spasms and choreoathetoid movements common: scratch, flexion reflexes present
4. Hind-limb support	4. —
5. Locomotor responses: stepping, hopping	5. —
6. Hind-limb replacement	6. —
7. Hind-limb palpitations	7. —
8. Extensor thrust responses	8. —
9. "Tactile placing"	9. —

3. *Stage of Maturation when Responses Are No Longer Spared after Spinal Transection*

To determine when sparing of function no longer occurs, the spinal cord of rats at several different ages (newborn, 9 days, 12 days, 15 days, 18 days, 21 days, and adult) was midthoracically transected (Weber and Stelzner, 1977). The immediate effects of surgery, the postoperative recovery and development of responses, and the chronic behaviors of the hind limbs were evaluated using a rating system from 1 (maximal response depression) to 4 (responses of a mature neonatal operate) developed from the results of our first study. Seven different classes of behavior (posture, locomotion, placing, etc.) were evaluated by two individuals who rated the behavior independently and without knowledge of the animal's age at the time of surgery or postoperative recovery time. A high correlation (r = .88) was found between the scores of the two raters. To analyze the data we also divided the postoperative recovery period into 4 stages:

Stage 1 (period of spinal shock):	The first 2 postoperative days, when responses are most depressed
Stage 2 (recovery period):	The remainder of the first postoperative week, when response recovery in the weanling transected animals is complete
Stage 3 (developmental period):	From 8 to 28 days postoperatively, when response ontogeny in the normal and neonatally transected animals is still taking place
Stage 4 (chronic period):	The period after 28 postoperative days, when responses are stable for all groups of spinally transected rats

Figure 2 presents the summed median scores of all categories of responses to give an estimate of the total amount of behavioral recovery. A fully recovered rat would have a score of 56. There are 7 different behavioral categories: a value of 4 indicates full recovery and the scores of two raters were summed. Notice the clean break in the median scores between the newborn, 9-day, and 12-day operates and those groups transected at older ages. Highly significant differences were found between animals operated at 12 days or less and animals operated at 15 days or older. During stage 1 the responses of the hind limbs of newborn, 9-day, and 12-day operates were equivalent, and there was less spinal shock than in the animals operated at 15 days or later. Recovery (stage 2) appeared greater for the 9-day and 12-day groups, since many of the behaviors which recovered in these two groups by 7 postoperative days had not yet developed in neonatal operates or their littermate controls by 7 days of age. These responses had been lost in the 9- and 12-day groups during stage 1. Thus, there was greater sparing of function and reduced spinal shock in animals spinally transected before 15 days of age.

4. *Anatomical Controls to Ensure a Complete Transection*

Previous studies of adults have shown that a small number of axons spared by a spinal transection can result in a disproportionately large amount of behavioral recovery caudal to the lesion site (Eidelberg *et al.*, 1977; Windle *et al.*, 1958). There is also the possibility in the young nervous system that axons could regenerate or grow through the transection site (Bernstein and Stelzner, 1983; Chambers, 1955; Kalil and Reh, 1979; So *et al.*, 1981). We studied the lesion site for evidence of axonal growth using several different anatomical methods. In our initial studies (Stelzner *et al.*, 1975; Weber and Stelzner, 1977), all that could be found using Nissl and myelin methods was a dural sac, up to several millimeters long (Fig. 3a–c). However, there was still the possibility that a small number of unmyelinated axons may have been unstained or were broken during tissue preparation. We then studied the lesion site in chronic neonatal operates using electron microscopy (Bernstein *et al.*, 1981). Again, although evidence of axonal growth was found on both sides of the lesion site, no axons were detected in the zone of lesion, itself. A final anatomical study used anterograde transport of [3H]proline to label corticospinal tract axons or axons of ascending nerve tracts in chronic neonatal operates (Cummings *et al.*, 1981). Again, there was no evidence that these axons crossed through or around the spinal transection. In addition, others have shown that there is no ingrowth of sympathetic peripheral axons

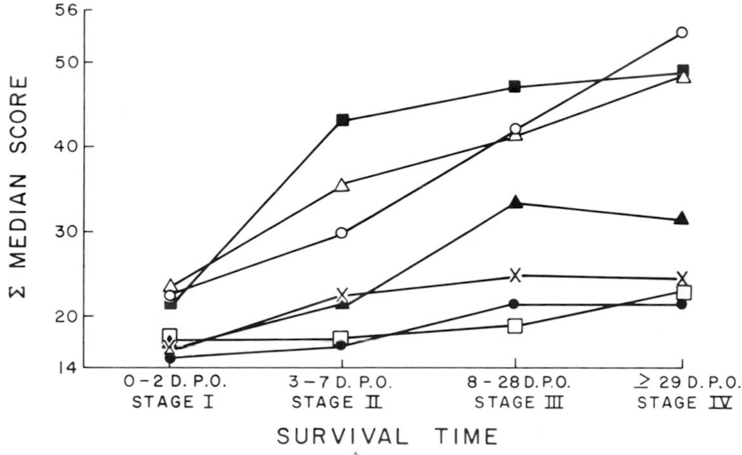

Fig. 2. Behavioral recovery at different ages following spinal transection. See text for details. (○) NB; (△) 9D; (■) 12D; (x) 15D; (▲) 18D; (●) 21D; (□) AD.

into the isolated spinal cord of neonatal operates (A. Björklund, personal communication).

During the course of these studies, we found connective tissue, and glial scarring at the lesion site in older animals (Bernstein *et al.*, 1981; Bernstein and Stelzner, 1983; Stelzner *et al.*, 1975; Weber and Stelzner, 1977). To test the possibility that these tissue reactions were related to the behavioral differences, we retransected the spinal cord in six of the neonatal operates at the weanling stage, one or two segments caudal to the original lesion. The immediate behavioral effects of this surgery were less severe than for the original lesion and the recovery greater than that found in weanling or adult operates (Stelzner *et al.*, 1975). Since there was no difference between neonatal and weanling operates in the amount of connective tissue or glial scarring at the second lesion site, we concluded that tissue reactions were not responsible for the spared behaviors. This experiment also ruled out the possibility that the age-related behavioral differences were caused by changes in connections made near the lesion site by cut ascending axons.

Together these results suggest that the mechanism responsible for sparing of function in pre-15-day operates must be occurring in the lumbosacral spinal cord, itself.

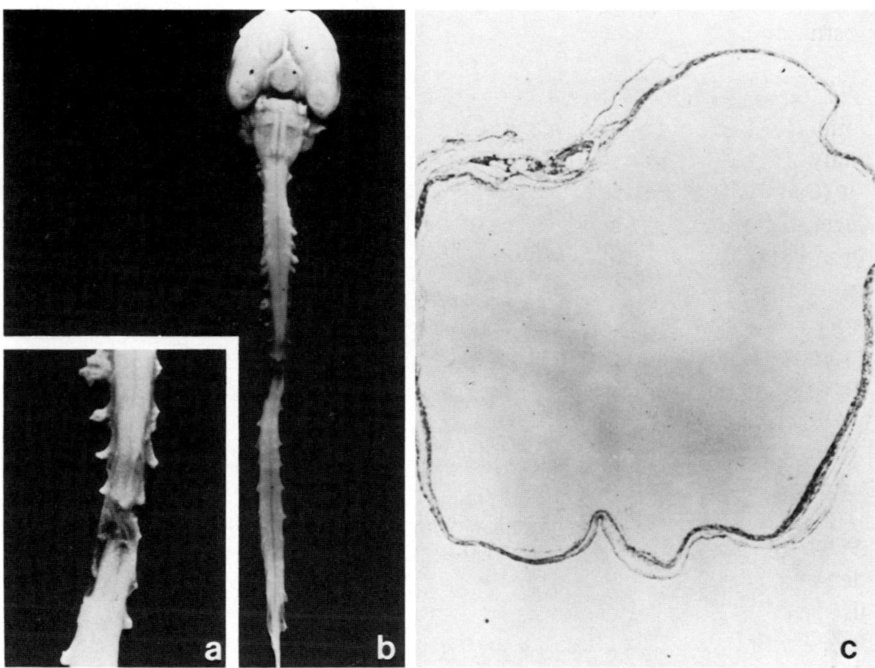

Fig. 3. (a) The site of spinal transection made in a neonatal rat that was sacrificed 30 days postoperatively. This midthoracic area is an enlargement of the area of lesion shown in (b). (c) Photomicrograph of a cresyl violet–stained transverse section of the lesion site of another chronic neonatal operate. From Stelzner (1982b), with permission.

5. Development of Dorsal Root, Descending, and Intrinsic Connections in the Lumbosacral Spinal Cord

The results described above showed an inverse relationship between the amount of spinal shock and the final development and recovery of responses in the isolated spinal cord and hindquarters. There appears to have been a permanent change in the stability of the spinal motor system at the age that spinal shock is mature. Most studies of the adult CNS have shown that the amount of spinal shock is related to the amount of influence the cut axons had on spinal circuitry preoperatively (Chambers *et al.*, 1973; Sherrington, 1947). Because spinal shock increases with age, we wanted to determine if it was related to the maturing of axonal connections at the time of injury. We asked whether there was a temporal relationship between the maturation of dorsal root, descending, and intrinsic connections and the increased spinal shock and decreased behavioral recovery which begins near the fifteenth postnatal day.

a. Dorsal Root Connections. We analyzed the development of dorsal root connections in the lumbosacral spinal cord using a lesion–degeneration analysis after intradural section of dorsal roots in the neonatal and weanling rat (Gilbert and Stelzner, 1979). As others have found, the rate of axonal degeneration in the neonate was much more rapid than in the adult. Otherwise, dorsal root connections were fully mature at birth as was previously shown by Ramon y Cajal (1911) using the Golgi technique and, more recently by Saito (1979) in his elegant electrophysiological work on the prenatal rat. In our experiment, the distribution of degeneration, including electron-dense degenerating synaptic endings, was similar in the neonatal rat to that found in the weanling or adult.

b. Descending Connections. We also analyzed the development of descending connections in the lumbosacral spinal cord (Gilbert and Stelzner, 1979). After midthoracic spinal hemisection the rate of degeneration was identical to that found after dorsal root section. However, unlike the mature distribution of dorsal root degeneration found at birth, the distribution of degeneration of descending axons was incomplete in the neonatal animal. No degeneration was found at the base of the dorsal funiculus, which is where the corticospinal tract descends in the rat, nor in the dorsal horn or intermediate nucleus of Cajal, which are projection zones of the corticospinal tract. Other investigators (Donatelle, 1977; Hicks and D'Amato, 1975; Schreyer and Jones, 1982; Wise *et al.*, 1979) have also reported that the corticospinal tract has not reached the lumbosacral spinal cord at birth in the rat. This is another indication that the encephalization process in the rat continues into the postnatal period.

Most brain stem pathways have arrived in the lumbosacral spinal cord during the neonatal period. As in other altricial species, the reticular formation is relatively mature at birth in the rat and certain descending reticulospinal axons are present (Das and Hine, 1972). Central monoamine neurons form in the brain stem before birth and most descending monoamine pathways have also been traced into the spinal cord prior to birth (Commissiong, 1983; Seiger and Olson, 1973). Rubrospinal tract axons have also descended the entire length of the spinal cord at birth (Prendergast and Stelzner, 1976; Shieh *et al.*, 1983). The vestibular reactions seen in the hind limbs while the rat pup is righting or being tilted suggest that the vestibulospinal tract is also present and functioning in the lumbosacral spinal cord at birth. Experiments using retrograde transport of horseradish peroxidase further document most brain stem nuclei project to lumbosacral spinal cord at birth (Leong *et al.*, 1984).

Our light microscopic results and the data above suggest that most descending nerve tracts are in position to form synaptic connections in the lumbosacral spinal cord of the newborn rat. However, we were surprised that no sign of electron-dense degenerating synaptic endings could be found in either the ventromedial gray (lamina VIII) or lateral gray (laminae VI–VII) after neonatal hemisection. This was true even though we used a large number of different postoperative survival periods (8h–3 days), periods when degeneration in these areas was identified using the light microscope. There are two reasons for this finding that seem likely. The appearance of synapses in a target area is often delayed for a variable period once axons have arrived near it (Commissiong, 1983; Rakic, 1977; Wise and Jones, 1976; Wise *et al.*, 1977). Thus, few synapses from descending axons may have formed at birth. Schoenfeld *et al.* (1979) have also shown that immature synaptic endings may undergo a different type of degenerative reaction so that they are destroyed before becoming electron dense. In either case, our results indicate that few mature connections from descending nerve tracts are present at birth in the lumbosacral spinal cord.

A mature distribution of degenerating axons was not found in the gray matter after thoracic hemisection until 15 days of age. This is also the period when Donatelle (1977) found that the distribution of corticospinal tract axons was mature in the lumbosacral spinal cord of the rat. Degenerating synaptic endings could be found in the ventromedial gray matter by 12 days of age but not in the lateral gray until 18 days of age. Thus, even though descending axons are found in the lumbosacral spinal cord at birth, a mature distribution is not present until 15 days of age, and mature connections appear between 12 and 18 days of age. This coincides well with our behavioral findings; this is the same period when spinal shock increases and recovery of function decreases after midthoracic spinal transection. In addition, this is the period when there is a mature reaction in the hind limbs to brain electroconvulsive shock (Vernadakis and Woodbury, 1969). Thus, descending connections appear to become functionally mature at ~15 days of age. Martin *et al.* (1978) also found an increase in spinal shock after spinal cord transection in the developing opossum at the time degenerating synaptic endings first appeared in the lumbosacral gray matter after this lesion.

c. Intrinsic Connections. Since the majority of synaptic endings in the lumbosacral spinal cord are from intrinsic sources (Gelfan, 1963), study of total synaptogenesis gives an estimate of their maturation. We used the Rasmussen technique, which is thought to preferentially stain mitochondria in the synaptic knob, to get an idea of overall synaptic development (Stelzner, 1971). This light microscopic technique shows dorsal root zones contain the greatest density of Rasmussen endfeet, suggesting that these zones are most mature at birth. This correlates well with the maturing of dorsal root connections at birth described in Section IV,A,5,a. At 14 days of age, using the density of Rasmussen endfeet as a criterion, the ventromedial and lateral portions of the intermediate gray are still immature. This also correlates well with our analysis of the maturing of descending connections discussed in Section IV,A,5,b. To get a more quantitative analysis of synaptic development, we studied the lateral gray (laminae VI–VII) of the L6 segment of the lumbosacral spinal cord using the electron microscope (Weber and Stelzner, 1980). Using a morphometric analysis, we found a large increase in the area of neuropil in this zone as well as the area of the total gray matter until 15 days of age, after which little further change was found. There was also a 50% increase in the density of axon terminals between 3 and 12 days of age and a large increase in their size, approaching the mature state, between 12 and 15 days of age. Thus, at 15 days of age, when spinal shock increases and sparing of function decreases after spinal transection, the neuropil stops growing and synaptogenesis ceases in this area of the lumbosacral gray matter. Again, the evidence indicates that the intrinsic circuitry of the lumbosacral spinal cord is developing *at the same time* descending axons are developing to influence this circuitry during the encephalization process. Moreover, the rate of synaptogenesis in the lumbosacral spinal cord is unaffected by midthoracic spinal transection at birth suggesting that, although descending connections may modify local circuitry, they are not necessary for its formation (E. D. Weber and D. J. Stelzner, unpublished observations).

6. *Summary and Conclusions*

Our work shows that the maturation of descending axons in the lumbosacral spinal cord of the rat significantly affects the response outflow of the isolated spinal cord. The increased number of responses generated by the isolated spinal cord if midthoracic transection is made prior to 15 days of age is related to (a) less spinal shock in rats spinally transected before 15 days of age, (b) the completion of synaptogenesis in the lumbosacral spinal cord by 15 days of age, (c) an immature distribution of descending axons prior to 15 days of age, (d) the maturation of synapses from descending nerve tracts between 12 and 18 days of age, and (e) the maturation of most hindlimb responses by 16 to 18 days of age. Our evidence indicates that these age differences are not related to the generation, regeneration, or sparing of descending axons passing through the transection site or to local tissue reactions at the lesion site.

B. The Effect of Decerebration

Decerebration is another example in which aborting the encephalization process results in the recovery or development of behaviors that are not seen after this same lesion in the adult. After decerebration in the kitten, behavioral responses emerged in much the same order and at approximately the same rate as in normal littermates (Bignall and Schramm, 1974). Thermoregulation, including piloerection and shivering, auditory orientation and startle reflexes, placing responses (including "tactile placing"), defensive reactions, and reflexive feeding behaviors all appeared near the appropriate age. Most of these behaviors did not recover following decerebration in the adult cat. These data, along with our own results on neonatal spinal transection, suggest that behaviors in infant operates may be intact in the adult but are permanently depressed by the decerebration.

Study of the effects of decerebration on the righting reflex supports this idea (Bignall, 1974). In the rat, righting matures rapidly between 3 and 6 days of age. At 4 days of age decerebration at the level of the rostral medulla (postmesencephalic lesion) did not impair righting immediately after surgery. If these animals were tested chronically, the success of righting was only 75–85% of the level seen in unoperated littermates by 7 days of age, and the righting response never developed to the degree seen in unoperated littermates or in animals decerebrated at a premesencephalic level. However, if a postmesencephalic lesion was made at 7 days of age, righting was completely and permanently abolished, as in the adult. Thus, the circuitry responsible for the immature righting response in postmesencephalic neonatally decerebrated rats was depressed by this lesion after 7 days of age.

C. The Effect of Cerebral Lesions

There is also evidence that the behavioral effects of ablations of the sensorimotor cortex in rats may be related to response depression. Brooks and Peck (1940) made unilateral sensorimotor cortex ablations in newborn rats and found that tactile placing never appeared in the contralateral limbs. This response normally is mature in the forelimbs at the end of the second postnatal week and in the hind limbs at the beginning of the third postnatal week (Brooks and Peck, 1940; Donatelle, 1977; Hicks and D'Amato, 1975; Stelzner, 1971; Tang, 1935). In addition, no differences in placing could be discerned when the two sides of the body were compared after the remaining sensorimotor cortex was ablated at 2 to 6 months of age. This appears to be a case where an infant lesion had the same effect as in the adult. However, Hicks and D'Amato (1975) found that tactile placing developed and was retained if the sensorimotor cortex was ablated bilaterally at birth. They also found that immature placing reactions seen at earlier ages were lost in the limb contralateral to a unilateral sensorimotor cortex ablation at about the time tactile placing is fully mature in a normal rat (17 days of age). If the remaining cortex was damaged after this time, placing responses were abolished bilaterally just as was found in the Brooks and Peck experiment described above. Together, these results indicate that maturation of the control of placing reactions by sensorimotor cortex interferes with the retention of these responses when the cortex is ablated once this control has been established. These results also suggest that this control develops bilaterally, at least after unilateral damage in the neonate. Presumably, in the experiment of Hicks and D'Amato (1975), the maturation of connections from the sensorimotor cortex interfered with the immature placing responses of the ipsilateral side of the body while facilitating tactile placing contralaterally.

A similar result was also found after sensorimotor cortex ablation in the kitten (Amassian and Ross, 1978a,b). Normally, a contact placing response is found in the forelimbs of the kitten by the end of the first postnatal week. This response improves but is not mature until the seventh postnatal week. If the sensorimotor cortex is ablated at birth, this lesion has no immediate effect on contact placing and it develops normally. If this region is removed during the following 3 weeks, contact placing is lost immediately after surgery but recovers. However, if this region is removed between 5 and 9 weeks of age, there is an increasingly severe deficit in contact placing that does not recover. Cooling the sensorimotor cortex or its projections is without effect at birth but causes a reversible deficit in contact placing during the second and third week. Cooling or destroying the thalamic projections to the sensorimotor cortex does not cause a loss in contact placing until much later in development (6–10 weeks of age), at about the time contact placing is mature. Thalamic projections to the cortex and the descent of the corticospinal tract, itself, occur much earlier in development (Wise *et al.*, 1977), showing that it is the maturation of descending connections rather than their appearance which is responsible for the suppression of this response. In this context, the ability of intracortical microstimulation to cause muscular contraction in the kitten does not mature until relatively late in development (8–9 postnatal week, Bruce and Tatton, 1980a,b), at about the time contact placing is mature and ablation of sensorimotor cortex results in its irreversible loss.

Kennard (1936, 1938, 1940, 1942) was the first to show that ablating a cortical area during development resulted in less severe behavioral deficits than if the lesion was made in the adult. She made unilateral and bilateral lesions of the precentral gyrus (motor cortex, area 4 of Brodman) and premotor cortex (area 6) in infant monkeys. These lesions resulted in few detectable motor deficits immediately after surgery and relatively minor motor disturbances when these animals matured. An adult monkey with the same lesion had profound motor paralysis immediately after surgery; without special training few voluntary movements recovered.

More recent work on adult monkeys shows that, except for a lack of response depression after surgery in the infant, few qualitative differences exist between infant operates and fully recovered adults with the same injury. First of all, the sparing of function found in infant operates was not complete. The forced grasp response that was present at the time of neonatal surgery remained until adulthood in neonatal operates, and the slow uncoordinated movements characteristic of the normal infant were still present in the mature neonatal operates (Kennard, 1936). Neither the fine motor control nor the ability to perform purposeful prehensile movements with the distal portions of the extremities ever developed in infant operates. When adult monkeys with motor cortex or pyramidal tract lesions were given extensive postoperative therapy or special training, or when the lesions were made in a number of stages (Beck and Chambers, 1970; Goldberger, 1972, 1974; Lawrence and Kuypers, 1968; Travis and Woolsey, 1956), these animals were left with deficits remarkably similar to those described by Kennard in her neonatal operates. This suggests that the differences between neonatal and adult operates in Kennard's work were a result of responses remaining depressed in the adult operate and not depressed in the infant operate. If the lesions are made in multiple stages in adult monkeys (Travis and Woolsey, 1956), these responses are never depressed and remain intact as in the infant operates. Special training can also restore these behaviors in the adult showing that the circuitry to perform them is still available.

D. Overview of Experiments Interfering with the Encephalization Process

The various data discussed in this section suggest that the circuits responsible for behaviors spared after neonatal surgery are intact but are permanently depressed after the same injury is made in the adult. Thus, as descending influences mature during the process of encephalization, they significantly affect the spinal motor system, presumably altering its functioning or circuitry. Once these descending pathways are mature, the spinal motor system is permanently modified so that certain circuits are unable to operate without these descending connections. If encephalization is not allowed to occur, both spinal reflexes and the intrinsic functioning of the spinal motor system are able to operate independently, without descending connections, This hypothesis dictates that responses such as tactile placing, or at least certain of its components, are actually present in the isolated spinal cord of the adult animal. This would explain why the tactile placing response lost after a motor cortex lesion sometimes can be reinstated by a second lesion (Bogen and Campbell, 1962). It would also follow that tactile placing and locomotor reactions found in the mature spinally transected neonate should closely resemble the normal state. Forssberg, Grillner, and colleagues (Forssberg, 1979; Forssberg *et al.*, 1974, 1980a,b) have studied locomotor responses and hopping and placing in the hind limbs of chronic neonatally transected kittens using electromyography and cinematography. These investigators concluded that the pattern of movement in chronic operates is strikingly similar to normal. The spinal mechanisms could even reproduce subtle details of walking movements found in the intact cat. They also found that the spinal placing and hopping reactions, including "tactile placing," met many of the criteria for these responses in the intact animal (but see Bregman and Goldberger, 1983a,b,c, and Smith *et al.*, 1982 for differing results).

Other evidence supports the hypothesis that sparing of function during development is related to absence of the response depression which increases with encephalization. A number of experiments have shown that behaviors thought to be lost after adult injury could be restored using different types of excitatory pharmacological agents (Amassian and Ross, 1978a; Beck and Chambers, 1970; Forssberg and Grillner, 1973; Hart, 1971; Maling and Acheson, 1946; Meyer *et al.*, 1963; Sechzer *et al.*, 1973). These data show that the circuits responsible for the behaviors which were tested such as spinal reflexes, stepping, tactile placing, visual placing, and righting still exist after lesions in the adult but are subthreshold or permanently depressed.

The question remains why initial response depression should cause certain responses to be permanently depressed in animals operated as adults that are not depressed in infant operates. Two possibilities are equally supported by the data at this time and both may be partially correct. I have discussed these possibilities at length elsewhere (Stelzner, 1982b), so I mention them only briefly here.

One possibility is that the greater response depression found in the mature animal is related to permanent changes in the physiology or morphology of connections remaining subsequent to neural injury. These changes found after injury in the adult CNS either may not occur or may occur to a lesser degree if the lesion is made before encephalization is complete. This hypothesis says that the effects of denervation in the adult animal are pathological. There is physiological (Anderson *et al.*, 1971; Kjerulf *et al.*, 1973; Sharpless, 1975; Stavraky, 1961; Trendelenberg, 1963) and morphological (Bernstein and Bernstein, 1977; Bernstein *et al.*, 1974; Illis, 1973; McCouch *et al.*, 1958) evidence supporting this view.

A second possibility is that the development or recovery of behaviors after damage to the developing CNS is related to a greater reorganization of synaptic connections than is found after adult lesions. According to this hypothesis, increased response depression (spinal shock) does not prevent recovery of function, it simply reflects the functional maturity of descending and local connections at the time of injury. There is a greater propensity for axonal growth after injury in the developing CNS than in the adult (Bernstein and Stelzner, 1983; Bregman and Goldberger, 1983c; Schneider, 1979; Stelzner *et al.*, 1979) supporting this view. Also, cutting afferents to an area during development may result in changes in the synaptic organization of intrinsic connections within a deafferented area (Lenn, 1978). We have found that there is an increased density of dorsal root axons in a zone of the lumbosacral spinal cord denervated by a midthoracic spinal hemisection at the neonatal stage of development that is not found after the same injury is given to weanling rats (Stelzner *et al.*, 1979) further supporting this idea. It may be that during normal encephalization there is competition between descending and local connections leading to the elimination of many local connections that subserve the spared behaviors that are found after neonatal spinal transection.

Further support for this idea comes from studies of spinal hemisection. The effect of spinal hemisection in the developing animal is less clear cut than that seen after spinal transection (Bregman and Goldberger, 1983a,b,c; Prendergast and Shusterman, 1982; Prendergast *et al.*, 1982). Certain responses are spared in the infant operate while other responses are actually worse off than those found after an adult lesion. One hypothesis to explain the deleterious effects of the infant hemisection is that the development and maintenance of local connections on the injured side of the cord, not found after adult injury, and normal, as well as abnormal, supraspinal connections on the intact side, may result in an imbalance in normal functioning. This combination of abnormal local and descending connections on opposite sides of the cord may interfere with the development of responses which either local (after neonatal spinal transection) or descending connections could control.

References

Altman, J., and Sudarshan, K. (1975). *Anim. Behav.* **23**, 896–920.
Anokhin, P. K. (1964). *Brain Res.* **9**, 54–86.
Amassian, V. E., and Ross, R. J. (1978a). *J. Physiol. (Paris)* **74**, 165–184.
Amassian, V. E., and Ross, R. J. (1978b). *J. Physiol. (Paris)* **74**, 185–201.
Anderson, L. S., Black, R. G., Abraham, J., and Ward, A. A. (1971). *J. Neurosurg.* **35**, 444–452.

Barcroft, J., and Barron, D. H. (1937). *J. Physiol. (London)* **91**, 329–351.
Barcroft, J., and Barron, D. H. (1939). *J. Comp. Neurol.* **70**, 477–502.
Barcroft, J., and Barron, D. H. (1942). *J. Comp. Neurol.* **77**, 431–452.
Beck, C. H., and Chambers, W. W. (1970). *J. Comp. Physiol. Psychol.* **70**, 1–22.
Bernstein, D. R., and Stelzner, D. J., (1983). *J. Comp. Neurol.* **221**, 382–400.
Bernstein, D. R., Bechard, D. E., and Stelzner, D. J. (1981). *Neurosci. Lett.* **26**, 55–60.
Bernstein, J. J., Gelderd, J. B., and Bernstein, M. E. (1974). *Exp. Neurol.* **44**, 470–482.
Bernstein, M. E., and Bernstein, J. J. (1977). *J. Neurocytol.* **6**, 85–102.
Bignall, K. E. (1974). *Exp. Neurol.* **42**, 566–573.
Bignall, K. E., and Schramm, L. (1974). *Exp. Neurol.* **42**, 519–531.
Bodian, D. (1966). *Bull. Johns Hopkins Hosp.* **119**, 129–149.
Bodian, D., Melby, E. C., and Taylor, N. (1968). *J. Comp. Neurol.* **133**, 113–165.
Bogen, J. E., and Campbell, B. (1962). *Science* **135**, 309–310.
Bregman, B. S., and Goldberger, M. E. (1983a). *Dev. Brain Res.* **9**, 103–118.
Bregman, B. S., and Goldberger, M. E. (1983b). *Dev. Brain Res.* **9**, 119–136.
Bregman, B. S., and Goldberger, M. E. (1983c). *Dev. Brain Res.* **9**, 137–154.
Brooks, C. M., and Peck, M. E. (1940). *J. Neurophysiol.* **3**, 66–73.
Bruce, I. C., and Tatton, W. G. (1980a). *Exp. Brain Res.* **39**, 411–419.
Bruce, I. C., and Tatton, W. G. (1980b). *Exp. Brain Res.* **40**, 349–353.
Campbell, B. A., Lytle, L. D., and Fibiger, H. C. (1969). *Science* **166**, 635–636.
Carmichael, L. (1933). *In* "A Handbook of Child Psychology." 2nd Ed., pp. 31–159. Clark Univ. Press, Worcester, Massachusetts.
Carmichael, L. (1934). *Genet. Psychol. Monogr.* **16**, 341–491.
Chambers, W. W. (1955). *In* "Regeneration in the Central Nervous System" (W. F. Windle, ed.), pp. 135–147. Thomas, Springfield, Illinois.
Chambers, W. W., Liu, C. N., McCouch, G. P., and D'Aquili, E. (1966). *Brain* **89**, 377–390.
Chambers, W. W., Liu, C. N., and McCouch, G. P. (1973). *Brain Behav. Evol.* **8**, 5–26.
Coghill, G. E. (1929). "Anatomy and the Problem of Behavior." Cambridge Univ. Press, London and New York.
Commissiong, J. W. (1983). *Brain Res.* **264**, 197–208.
Corner, M. A. (1977). *Prog. Neurobiol. (Oxford)* **8**, 279–295.
Coyle, J. T., and Axelrod, J. (1972a). *J. Neurochem.* **19**, 449–459.
Coyle, J. T., and Axelrod, J. (1972b). *J. Neurochem.* **19**, 1117–1123.
Cummings, J. P., Bernstein, D. R., and Stelzner, D. J. (1981). *Exp. Neurol.* **74**, 615–620.
Das, G. D., and Hine, R. J. (1972). *Z. Anat. Entwicklungsgesch.* **36**, 98–114.
Decker, J. D., and Hamburger, V. (1967). *J. Exp. Zool.* **165**, 371–384.
Donatelle, J. M. (1977). *J. Comp. Neurol.* **175**, 207–232.
Drachman, D. B., and Sokoloff, L. (1966). *Dev. Biol.* **14**, 401–420.
Eidelberg, E., Straehley, D., Erspamer, R., and Watkins, C. J. (1977). *Exp. Neurol.* **56**, 312–322.
Fibiger, H. C., Lytle, L. D., and Campbell, B. A. (1970). *J. Comp. Physiol. Psychol.* **72**, 384–389.
Forssberg, H. (1979). *Acta Physiol. Scand. Suppl.* **474**, 1–56.
Forssberg, H., and Grillner, S. (1973). *Brain Res.* **50**, 184–186.
Forssberg, H., Grillner, S., and Sjostrom, A. (1974). *Acta Physiol. Scand.* **92**, 114–120.
Forssberg, H., Grillner, S., and Halbertsma, J. (1980a). *Acta Physiol. Scand.* **108**, 269–281.
Forsberg, H., Grillner, S., Halbertsma, J., and Rossignol, S. (1980b). *Acta Physiol. Scand.* **108**, 283–295.
Fox, M. W. (1966). *Brain Res.* **2**, 3–20.
Fukson, O. I., Berkinblit, M. B., and Feldman, A. G. (1980). *Science* **209**, 1261–1263.
Fulton, J. F. (1949). "Physiology of the Nervous System," 3rd. Ed. Oxford Univ. Press, London and New York.

Gelfan, S. (1963). *Nature (London)* **198**, 162–163.
Gilbert, M., and Stelzner, D. J. (1979). *J. Comp. Neurol.* **184**, 821–838.
Goldberger, M. E. (1972). *Exp. Brain Res.* **15**, 79–96.
Goldberger, M. E. (1974). *In* "Plasticity and Recovery of Function in the Central Nervous System" (D. G. Stein, J. J. Rosen, and N. Butters, eds.), pp. 265–339. Academic Press, New York.
Gottlieb, G., ed. (1976). "Studies on the Development of Behavior and the Nervous System," Vol. 4, Early Influences. Academic Press, New York.
Hamburger, V. (1963). *Q. Rev. Biol.* **38**, 342–365.
Hamburger, V. (1973). *In* "Studies on the Development of Behavior and the Nervous System" (G. Gottlieb, ed.), Vol. 1, Behavioral Embryology, pp. 52–76. Academic Press, New York.
Hamburger, V., Wenger, E., and Oppenheim, R. (1966). *J. Exp. Zool.* **162**, 133–160.
Hart, B. L. (1971). *Physiol Behav.* **6**, 627–628.
Hicks, S. P., and D'Amato, C. J. (1975). *Am. J. Anat.* **143**, 1–42.
Hughlings-Jackson, J. (1886). *Brain* **9**, 1–23.
Humphrey, T. (1964). *Prog. Brain Res.* **4**, 93–135.
Illis, L. S. (1973). *Brain* **96**, 47–60.
Jacobson, M. (1978). "Developmental Neurobiology." Plenum, New York.
Kalil, K., and Reh, T. (1979). *Science* **205**, 1158–1161.
Kennard, M. A. (1936). *Am. J. Physiol.* **115**, 138–146.
Kennard, M. A. (1938). *J. Neurophysiol.* **1**, 477–497.
Kennard, M. A. (1940). *Arch. Neurol. Psychiatry* **44**, 377–397.
Kennard, M. A. (1942). *Arch. Neurol. Psychiatry* **48**, 227–240.
Kjerulf, T. D., O'Neal, J. T., Calvin, W. H. Loesser, J. D., and Westrum, L. E. (1973). *Exp. Neurol.* **39**, 86–102.
Lamprecht, F., and Coyle, J. T. (1972). *Brain Res.* **41**, 503–506.
Langworthy, O. R. (1925). *Am. J. Physiol.* **74**, 1–13.
Langworthy, O. R. (1926). *Contrib. Embryol.* **17**, 125–140.
Langworthy, O. R. (1927a). *Contrib. Embryol.* **19**, 149–175.
Langworthy, O. R. (1927b). *Contrib. Embryol.* **19**, 179–207.
Langworthy, O. R. (1928). *J. Comp. Neurol.* **46**, 201–247.
Langworthy, O. R. (1929). *Contrib. Embryol.* **20**, 127–171.
Langworthy, O. R. (1932). *Arch. Neurol. Psychiatry* **28**, 1365–1382.
Lawrence, D. G., and Kuypers, H. G. J. M. (1968). *Brain* **91**, 1–14.
Lenn, N. J. (1978). *J. Comp. Neurol.* **181**, 93–116,
Leong, S. K., Shieh, J. Y., and Wong, W. C. (1984). *J. Comp. Neurol.* **228**, 18–23.
Loizow, L. A., and Salt, P. (1970). *Brain Res.* **20**, 467–470.
Mabry, P. D., and Campbell, B. A. (1974). *J. Comp. Physiol. Psychol.* **86**, 193–201.
McCouch, G. P., Austin, G. M., Liu, C. N., and Liu, C. Y. (1958). *J. Neurophysiol.* **21**, 205–216.
McCouch, G. P., Liu, C. N., and Chambers, W. W. (1966). *Brain* **89**, 359–376.
Maling, H. M., and Acheson, G. H. (1946). *J. Neurophysiol.* **9**, 379–386.
Martin, G. F., Beals, J. K., Culberson, J. L., Dom, R., Goode, G., and Humbertson, A. O., Jr. (1978). *J. Comp. Neurol.* **181**, 271–290.
Martin, G. F., Cabana, T., Di Tirro, F. J., Ho, R. H., and Humbertson, A. O. (1982). *Prog. Brain Res.* **57**, 131–144.
Meyer, P. M., Horel, J. A., and Meyer, D. R. (1963). *J. Comp. Physiol. Psychol.* **56**, 402–404.
Moorecroft, W. H. (1971). *Brain Res.* **35**, 513–522.
Narayanan, C. H., Fox, M. W., Hamburger, V. (1971). *Behaviour* **40**, 100–134.
Nornes, H. O., Hart, H., and Carry, M. (1980). *J. Comp. Neurol.* **192**, 133–141.
Okado, N., and Oppenheim, R. W. (1985). *J. Comp. Neurol.* **232**, 143–161.

Oppenheim, R. W. (1972). *J. Comp. Neurol.* **146**, 479–506.
Oppenheim, R. W. (1973). *In* "Studies on the Development of Behavior and the Nervous System" (G. Gottlieb, ed.), Vol. 1, Behavioral Embryology, pp. 164–244. Academic Press, New York.
Oppenheim, R. W. (1975). *J. Comp. Neurol.* **160**, 37–50.
Oppenheim, R. W. (1981a). *In* "Maturation and Development: Biological and Psychological Perspectives" (K. J. Connolly and H. F. R. Prechtl, eds.), pp. 73–109. Lippincott, Philadelphia.
Oppenheim, R. W. (1981b). *In* "Studies in Developmental Neurobiology: Essays in Honor of Viktor Hamburger" (W. M. Cowan, ed.), pp. 74–134. Oxford Univ. Press, London and New York.
Oppenheim, R. W. (1982). *Curr. Top. Dev. Biol.* **17**, 257–309.
Oppenheim, R. W., Pittman, R., Gray, M., and Maderdrut, J. L. (1978). *J. Comp. Neurol.* **179**, 619–640.
Pierce, E. T. (1973). *Prog. Brain Res.* **40**, 53–65.
Prendergast, J., and Shusterman, R. (1982). *Exp. Neurol.* **78**, 176–189.
Prendergast, J., and Stelzner, D. J. (1976). *J. Comp. Neurol.* **166**, 163–172.
Prendergast, J., Shusterman, R., and Phillips, T. (1982). *Exp. Neurol.* **78**, 190–204.
Rakic, P. (1977). *Philos. Trans. R. Soc. London Ser. B* **278**, 245–260.
Ramon y Cajal, S. (1911). "Histologie du Systeme Nerveux de l'Homme et des Vertebres." Maloine, Paris.
Rhines, R., and Windle, W. F. (1941). *J. Comp. Neurol.* **75**, 165–189.
Rodier, P. M. (1980). *Dev. Med. Child. Neurol.* **22**, 525–545.
Rubel, E. W. (1978). *Handb. Sens. Physiol.* **9**, 135–237.
Saito, K. (1979). *J. Physiol. (London)* **294**, 581–594.
Scheibel, M., and Scheibel, A. (1964). *Prog. Brain Res.* **9**, 6–25.
Schneider, G. E. (1979). *Neuropsychologia* **17**, 557–585.
Schoenfeld, T. A., Street, C. K., and Leonard, C. M. (1979). *Soc. Neurosci. Abstr.* **5**, 177.
Schreyer, D. J., and Jones, E. G. (1982). *Neuroscience* **7**, 1837–1855.
Sechzer, J. A., Ervin, G. N., and Smith, G. P. (1973). *Exp. Neurol.* **41**, 723–737.
Sedlacek, J. (1978). *Prog. Brain Res.* **48**, 367–384.
Seiger, A., and Olson, L. (1973). *Z. Anat. Entwicklungsgesch.* **140**, 281–318.
Sharpless, S. K. (1975). *In* "The Developmental Neuropsychology of Sensory Deprivation" (A. H. Riesen, ed.), pp. 125–152. Academic Press, New York.
Sherrington, C. S. (1947). "The Integrative Action of the Nervous System," 2nd Ed. Yale Univ. Press, New Haven, Connecticut.
Shieh, J. Y., Leong, S. K., and Wong, W. C. (1983). *J. Comp. Neurol.* **214**, 79–86.
Shurrager, P. S., and Dykman, R. A. (1951). *J. Comp. Physiol. Psychol.* **44**, 252–262.
Singer, H. S., Skoff, R. P., and Price, D. L. (1978). *Brain Res.* **141**, 197–209.
Skoglund, S. (1960). *Acta Physiol. Scand.* **49**, 299–317.
Skoglund, S. (1966). *Nobel Symp.* **1**, 245–261.
Smith, J. L., Smith, L. A., Zernicke, R. F., and Hoy, M. (1982). *Exp. Neurol.* **76**, 393–413.
So, K.-F., Schneider, G. E., and Ayres, S. (1981). *Exp. Neurol.* **72**, 379–400.
Stavraky, G. W. (1961). "Supersensitivity Following Lesions of the Nervous System." Univ. of Toronto Press, Toronto.
Stelzner, D. J. (1971). *Exp. Neurol.* **31**, 337–357.
Stelzner, D. J. (1982a). *In* "Changing Concepts of the Nervous System" (A. R. Morrison and P. L. Strick, eds.), pp. 105–119. Academic Press, New York.
Stelzner, D. J. (1982b). *In* "Brain Stem Control of Spinal Mechanisms" (B. Sjölund and A. Björklund, eds.), pp. 297–321. Elsevier, Amsterdam.
Stelzner, D. J., Ershler, W. B., and Weber, E. D. (1975). *Exp. Neurol.* **46**, 156–177.

Stelzner, D. J., Weber, E. D., and Prendergast, J. (1979). *Brain Res.* **172**, 407–426.
Sugar, O., and Gerard, R. W. (1940). *J. Neurophysiol.* **3**, 1–19.
Tang, Y. (1935). *Chung-kuo Sheng Li Hsueh Tsa Chih* **9**, 339–344.
Tilney, F. (1933). *Bull. Neurol. Inst. N. Y.* **3**, 252–358.
Tracy, H. L. (1926). *J. Comp. Neurol.* **40**, 253–360.
Travis, A. M., and Woolsey, C. N. (1956). *Am. J. Phys. Med.* **35**, 273–310.
Trendelenberg, U. (1963). *Pharmacol. Rev.* **15**, 225–276.
van Mier, P., and ten DonKelaar, H. J. (1985). *Anat. Embryol.* **170**, 295–306.
Vernadakis, A., and Woodbury, D. M. (1969). *Epilepsia* **10**, 163–178.
Vince, M. A. (1973). *In* "Studies on the Development of Behavior and the Nervous System" (G. Gottlieb, ed.), Vol. 1, Behavioral Embryology, pp. 286–323. Academic Press, New York.
Volokhov, A. A. (1961). *Plzen. Lek. Sb. Suppl.* **3**, 141–145.
Weber, E. D., and Stelzner, D. J. (1977). *Brain Res.* **125**, 241–255.
Weber, E. D., and Stelzner, D. J. (1980). *Brain Res.* **185**, 17–37.
Weed, L. H. (1917). *Am. J. Physiol.* **43**, 131–157.
Windle, W. F. (1940). "Physiology of the Fetus." Saunders, Philadelphia.
Windle, W. F. (1970). *Exp. Neurol. Suppl.* **5**, 44–83.
Windle, W. F., O'Donnell, J. E., and Glasshagle, E. E. (1933). *Physiol. Zool.* **4**, 521–541.
Windle, W. F., Fish, M. W., and O'Donnell, J. E. (1934). *J. Comp. Neurol.* **59**, 139–165.
Windle, W. F., Smart, J. O., and Beers, J. J. (1958). *Neurology* **8**, 518–521.
Wise, S. P., and Jones, E. G. (1976). *J. Comp. Neurol.* **168**, 313–344.
Wise, S. P., Hendry, S. H. C., and Jones, E. G. (1977). *Brain Res.* **138**, 538–544.
Wise, S. P., Fleshman, J. W. F., Jr., and Jones, E. G. (1979). *Neuroscience* **4**, 1275–1297.
Woods, J. W. (1964). *J. Neurophysiol.* **27**, 635–644.

9

Neuronal Activity as a Shaping Factor in Postnatal Development of Visual Cortex

WOLF SINGER
Max-Planck-Institut für Hirnforschung
Frankfurt am Main, Federal Republic of Germany

I. Introduction

Analysis of cortical circuitry with morphological methods has revealed a fantastic degree of complexity of intracortical connectivity. For a long time this most complex of all neuronal networks appeared far beyond the reach of physiological analysis. However, we have experienced a very fruitful

Developmental
NeuroPsychobiology

convergence between morphological and physiological approaches and we now possess some insight into the general principles of cortical organization. One clear finding is that intracortical pathways are highly specific and are far from forming a meshwork of random connections. This, in turn, raises the challenging question of how nature solves the problem of specifying the myriads of interneuronal contacts during development. This question applies, of course, to the development of nervous tissue in general but apart from the quantitative differences in complexity there is one feature that distinguishes cortical development from most other developmental processes. The formation and final determination of cortical connections occurs to a considerable extent only postnatally and, hence, at a time when nervous activity is already under the influence of sensory experience. This is important for two reasons. First, because evidence is accumulating that indicates that neuronal activity plays a role in developmental processes; and, second, because it has been established that sensory experience is a necessary prerequisite for the development of certain brain functions. This prompts the speculation that the developing brain might make use of functional criteria for the identification and specification of appropriate interneuronal connections. In the following paragraphs, I will investigate this possibility in some detail, concentrating on the experience-dependent development of visual functions.

From a certain developmental stage onward, neuronal activity becomes an important shaping factor for the specification of neuronal connections. The elimination of excess climbing fibers from Purkinje cells in the cerebellum (Mariani and Changeux, 1981) and of supernumerary motor neuron axons from muscle cells (Jansen and Lomo, 1981), the retraction of ectopic callosal fibers (Innocenti and Frost, 1979), the reorganization of the retinotectal map (Gaze *et al.*, 1979), and the segregation of the afferents from the two eyes in the tectum (Meyer, 1982) and the visual cortex (Stryker, 1981) are all activity-dependent processes. Since Hubel and Wiesel (1965) demonstrated in their pioneering studies that visual experience profoundly affects the development of the mammalian visual system it has become clear that in this system activity-dependent developmental processes extend far into postnatal life. Moreover, when visual experience was manipulated during early development, close correlations could be established between the structural and functional changes at the neuronal level and the alterations of visual functions. Many of these developmental disturbances in animals could further be related to the various forms of amblyopia in humans, emphasizing the crucial role of visual experience in the development of visual functions in animals, including humans.

Because of these manifold implications, the visual system of mammals has become a widely used model for the investigation of activity-dependent developmental processes. Comprehensive reviews of the structural and functional changes following manipulations of early visual experience are now available (e.g., Movshon and Van Sluyters, 1981; Sherman and Spear, 1982; Fregnac and Imbert, 1984). Therefore, only those facts will be repeated here that are relevant for the discussion of neuronal processes that could mediate activity-dependent modifications of neuronal functions. In the first part of this chapter, I shall review work that was aimed at elucidating the algorithms that guide these activity-dependent maturation processes. In the second part, I shall attempt to derive from these developmental mechanisms some predictions on the organization of the mature cortex.

II. Experience Effects on Binocularity

A. Binocular Interactions

By the time kittens or monkeys have opened their eyes, most neurons in the visual cortex respond to stimulation of both eyes (Hubel and Wiesel, 1963a). With normal visual experience, but also with complete deprivation of contour vision, this condition is maintained. However, when visual signals are available but not identical in the two eyes, either because one eye is occluded (Wiesel and Hubel, 1965) or because the images on the two retinae are not in register—as is the case with strabismus (Hubel and Wiesel, 1965), cyclotorsion (Blakemore *et al.*, 1975), or anisometropia (Blakemore and Van Sluyters, 1974)—cortical cells lose their binocular receptive fields. In the first case they stop responding to the deprived eye; in the other cases they segregate into two groups of approximately equal size, one responding exclusively to the ipsilateral eye and the other exclusively to the contralateral eye. These functional changes in eye preference are associated with distortions of the columnar organization. The territories occupied by afferents from the normal eye and by cells responding preferentially to this eye increase at the expense of territories innervated by the deprived eye (Hubel *et al.*, 1977). These effects are obtainable only during a critical period of early development. During this period, but not thereafter, the effects of monocular deprivation can be fully reversed by closing the open eye and at the same time reopening the previously closed eye. This indicates that the efficacy of connections does not only decrease but also can increase as a function of retinal stimulation (Wiesel and Hubel, 1965b). Evidence is available that these modifications are associated with changes of the amplitude of excitatory synaptic currents (Singer, 1977a; Mitzdorf and Singer, 1980). Changes in inhibitory mechanisms appear to be less prominent and may be secondary to modifications of excitatory pathways (Sillito *et al.*, 1981).

B. The Rules Governing Activity-Dependent Modifications

The synaptic and molecular mechanisms underlying these experience-dependent changes of receptive field properties are still largely unknown. There is, however, a solid base of functional phenomena that make it possible to derive the rules according to which the experience-dependent modifications occur. In principle two different possibilities can be conceived of. The criterion for an activity-dependent change of the efficacy of excitatory transmission could simply be the amplitude of activity in either the pre- or the postsynaptic element. Alternatively, and this would have far-reaching consequences, the criterion for modifications could be the statistical correlation of the activity patterns in the interconnected pre- and postsynaptic elements. Available evidence favors the second assumption in that it indicates that the activation of the postsynaptic neuron is one essential prerequisite for a long-term modification of excitatory transmission (Singer, 1976; Rauschecker and Singer, 1979).

If in 4-week-old kittens one eye is closed light-tight and the other exposed to diffuse light whose intensity is continuously modulated, the ocular dominance of cortical neurons does not shift toward the stimulated eye (Singer *et al.*, 1977). In this case the afferents from the stimulated eye are more active than those from the deprived eye, but most cortical cells cannot respond to the activity conveyed by the stimulated eye because their receptive fields are selective for spatial contrast gradients. Another indication of the relevance of postsynaptic activation came from experiments in which we occluded one eye and exposed the other to contours of a single orientation (Singer, 1976; Rauschecker and Singer, 1979). In that case differential gain changes occur only for pathways connecting to those cortical cells whose orientation preference matches the experienced orientations and hence allows them to respond to the signals conveyed by the open eye. For these cells the efficacy of afferents from the stimulated eye increases while that of afferents from the deprived eye decreases. These cells become monocular and excitable only from the open eye. Cells whose orientation preference does not correspond to the orientations seen by the stimulated eye cannot respond to activity from this eye, and the efficacy of afferents to these cells is not modified. These cells remain connected to the deprived eye (Fig. 1). Thus, even though the geniculocortical afferents from the open eye are much more active than those

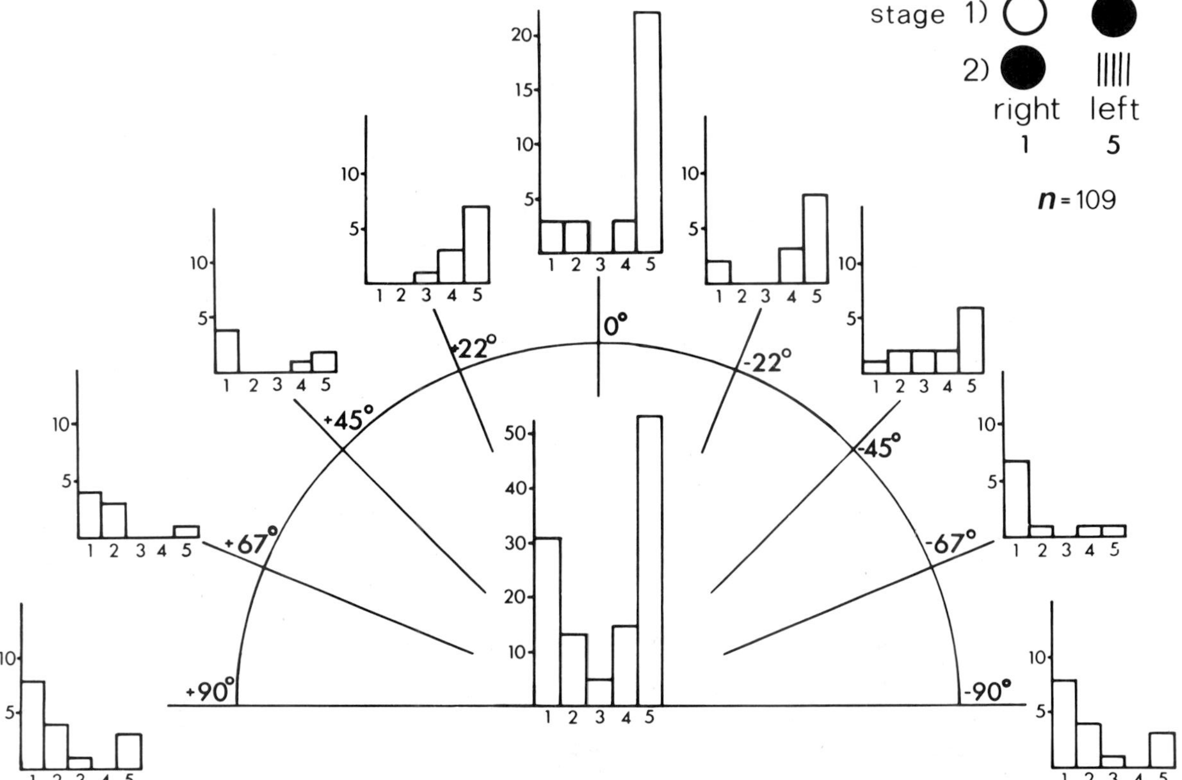

Fig. 1. Ocular dominance distributions from the striate cortices of kittens that had normal monocular experience in one eye prior to monocular experience with contours of a single orientation in the other eye (two kittens horizontal, one kitten vertical). The center graph reflects the ocular dominance distribution of all sampled neurons. Ocular dominance class 1 refers to monocular neurons driven by the previously open eye and ocular dominance class 5 refers to neurons driven exclusively by the newly open eye that was exposed to a selected range of orientations. The ocular dominance distributions around the perimeter have been compiled as a function of the neurons' orientation preference. The degree values indicate the difference between the experienced orientation and the preferred orientation of the unit. From Rauschecker and Singer (1981).

from the deprived eye, the former do not increase their efficacy at the expense of the latter when the postsynaptic cells do not respond.

From a series of related experiments it was concluded that these modifications of excitatory transmission follow rules that closely resemble those postulated by Hebb (1949) and Stent (1973) for adaptive neuronal connections (Rauschecker and Singer, 1981). These rules are summarized in Fig. 2. The main implication of these rules is that the direction of change–increase or decrease of the efficacy of excitatory transmission–does depend on the correlation between the activation of interconnected pre- and postsynaptic elements rather than on the level of activation of the afferent pathways or of the postsynaptic target cells. The experimental results indicate that a modifiable connection increases its efficacy when the probability is high that the afferent pathway is active in temporal contiguity with its postsynaptic target cell. Conversely, a connection becomes less effective when the probability is high that the afferent pathway is silent while the postsynaptic cell is active. Irrespective of the level of activation of the afferent pathway, no differential changes of excitatory coupling seem to occur when the postsynaptic target is inactive.

As indicated in Fig. 3, such activity-dependent modifications have a dual effect when they are extended to conditions in which several subpopulations of afferents converge onto a common target cell. Pathways that convey correlated activity and are capable of driving the follower cell will increase their gain. Hence, the modification algorithm has an associative function in that it enhances interactions between neuronal elements whose activity is correlated. However, the same algorithm also mediates competitive interactions. If converging pathways convey uncorrelated activity, those afferents that drive the postsynaptic cell most effectively will increase their efficacy at the expense of other afferent connections. In more general terms, afferents that are more often active in contiguity with the postsynaptic target increase their efficacy. This will be the case for afferents that convey activity patterns that match best the response preferences of the postsynaptic cell. If this matching criterion is equally fulfilled by all afferent subpopulations, those subpopulations that are activated more often or else that happen to be more efficient right from the beginning will eventually win. They activate the postsynaptic cell more effectively than the other inputs and hence the probability of contiguous pre- and postsynaptic activation is highest for them. As summarized in Fig. 3, such activity-dependent competition can account for the results of a variety of studies in which vision has been manipulated in such a way that correlation between the responses arriving from the two eyes is disrupted. As discussed in the following paragraph it can probably account also for the experience-dependent maturation of orientation-selective receptive fields.

	Rule 1	Rule 2	Rule 3	Rule 4
state of A	+	−	+	−
state of C	+	+	−	−
state of E	↑	↓	↔	↔

Fig. 2. Rules for the modification of excitatory transmission. A, Excitatory afferent; E, efficacy of excitatory transmission; C, postsynaptic neuron. (+) Active; (−) inactive; (↑) increased; (↓) decreased; (⟷) no change.

C. Evidence for Experience-Dependent Development of Orientation Selectivity

In 3-week-old kittens reared without visual experience a substantial fraction of striate cortex cells possess a preference for stimulus orientation (Buisseret and Imbert, 1976; Hubel and Wiesel, 1963a; Sherk and Stryker, 1976). With increasing duration of visual deprivation, the percentage of these cells decreases (Singer and Tretter, 1976); the percentage increases to nearly 100% with normal visual experience (Buisseret and Imbert, 1976; Buisseret *et al.*, 1978). In the latter case cells with preferences for horizontal, vertical, and oblique orientations are about equally frequent. When, however, the kitten experiences only contours of a single orientation the majority of cortical cells adopt preferences for this orientation (Blakemore and Cooper, 1970; Hirsch and Spinelli, 1970; for further citations, see Movshon and Van Sluyters, 1981; Fregnac and Imbert, 1984). Two studies, one based on single cell recording (Rauschecker and Singer, 1981) and the other on deoxyglucose mapping of orientation columns (Singer *et al.*, 1981), both are compatible with the idea that such experience-dependent distortions in the distribution of orientation preferences could result from experience-dependent competition between converging excitatory pathways. The main finding of the deoxyglucose study was that visual experience, when restricted to a single orientation, has relatively little influence on the development of the orientation column system within layer IV but produces massive distortions of the columnar system within nongranular layers. While cells responding to inexperienced orientations were still present within layer IV and were grouped within regularly spaced patches, activity from these cells was no longer relayed to cells in supra- and infragranular layers. By contrast, activity of layer IV cells, whose orientation preference corresponded to the experienced orientations,

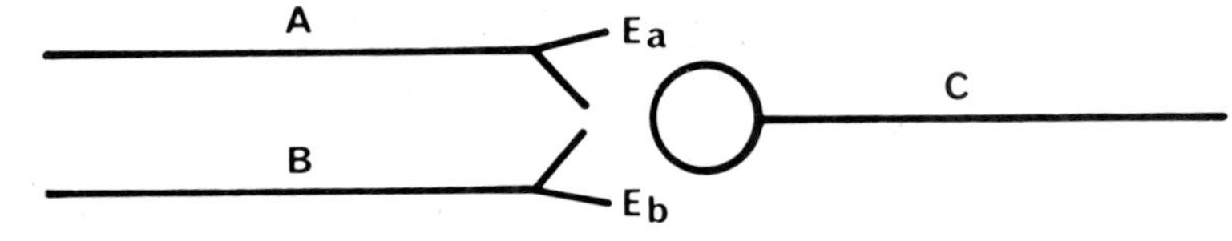

raising conditions	normal	monocular deprivation	strabismus	exposure to restricted range of orientations	binocular deprivation
state of A	+ } in synchrony	+	+ } in asynchrony	+ (vertical contours)	−
state of B	+ } in synchrony	− (deprived)	+ } in asynchrony	+ (horizontal cont.)	−
state of C	+	responds to A	responds better to A(B)	responds only to A (B)	−
change of E_a	↑	↑	↑ (↓)	↑ (↓)	↔
change of E_b	↑	↓	↓ (↑)	↓ (↑)	↔
long-term results	A + B consolidate	A consolidates B disrupts	A consolidates (B disrupts) B disrupts (A consolidates)		nonselective decrease at excitability

Fig. 3. Examples of competititve interactions between converging inputs. A, Excitatory input from right eye; B, excitatory input from left eye; E, efficacy of excitatory transmission from A/B to C; C, postsynaptic neuron. (+) Active; (−) inactive; (↑) increased; (↓) decreased; (⟷) no change.

was now relayed not only to nongranular cells located above and below the active zone in layer IV but spread tangentially to adjacent cells (Fig. 4). In this respect the reorganization of the orientation column system after orientation deprivation closely resembles the reorganization of the ocular dominance columns after monocular deprivation (Hubel *et al.*, 1977). Only the site of competition would have to be assumed to be different. In the case of monocular deprivation, competition occurs where afferents from the two eyes converge onto the common cortical target cells, which are located mainly in layer IV. In the case of orientation deprivation, competition would have to occur at the level where axons from orientation-selective layer IV cells converge onto second order target cells that are located mainly in nongranular layers.

D. The Mechanisms?

The activity-dependent modifications of the efficacy of excitatory pathways that have been described in Section II,D can result from a variety of different neuronal processes. Selective stabilization and removal of particular synaptic connections during synaptogenesis could be one mechanism (see Changeux and Danchin, 1976). Another possibility are selective modifications of the efficacy of already established connections. Here a variety of different processes could intervene: gating of action potential propagation into terminal arborizations, modulation of the amount of transmitter released per action potential, changes of the size of synaptic appositions, modifications of receptor sensitivity, alterations of the electrotonic properties of spines and dendrities, and changes in the threshold of regenerative dendritic responses. In the case of monocular deprivation there are structural changes of the terminal arborizations of geniculate afferents, suggesting selective stabilization and removal of connections (Hubel *et al.*, 1977). However, changes of response properties have been observed to occur within a few hours (Buisseret *et al.*, 1978; Geiger and Singer, 1986), suggesting processes other than selective growth and physical removal of connections. Clearly, much more work is required to determine which of these many possible processes actually mediates the experience-dependent modifications of cortical receptive fields during early development.

III. Evidence for a Central Control of Experience-Dependent Modifications

The modification rules listed in Section II,E imply that the activation of postsynaptic target cells is a necessary prerequisite for the induction of experience-dependent modifications. However, there is now good evidence that the generation of action potentials in the postsynaptic cell may not be

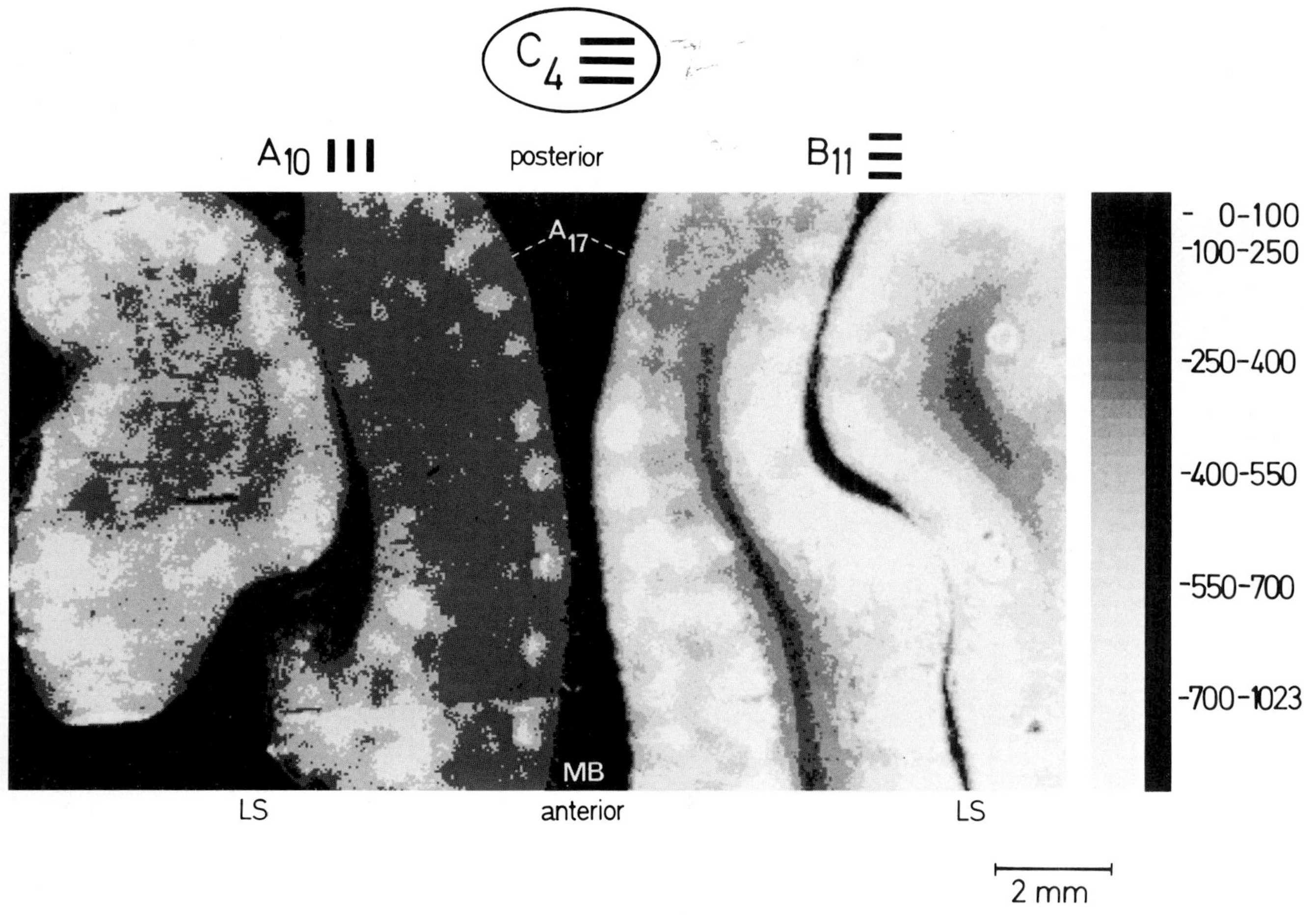

C_4

A_{10}

posterior

B_{11}

A_{17}

MB

LS

anterior

LS

- 0-100
-100-250
-250-400
-400-550
-550-700
-700-1023

2 mm

sufficient to produce a change. Even when contour vision is unrestricted and retinal signals readily elicit responses in the neurons of the visual cortex, vision-dependent modifications of excitatory transmission may fail to occur in a variety of rather different conditions. Thus, cortical cells maintain binocular receptive fields when retinal coordinates are in gross mismatch with other sensory maps, for example, when the eyes are surgically rotated within the orbit by 90° or more (Crewther *et al.*, 1980; Singer *et al.*, 1982) or when large-angle squint is induced in both eyes (Singer *et al.*, 1979). In this case contour vision per se is unimpaired but the abnormal eye position and motility lead to massive disturbances of the kittens visuomotor coordination. Initially the inappropriate retinal signals cause abnormal visuomotor reactions and are effective in influencing cortical ocular dominance. Subsequently, however, the kittens rely less and less on visual cues and develop a near-complete neglect of the visual modality. In this phase, retinal signals no longer modify ocular dominance and they also fail to support the development of orientation-selective receptive fields.

These latter results suggest that retinal signals only influence the development of cortical functions when the animal uses these signals for the control of behavior. This view is compatible with two lines of evidence. First, results from three laboratories indicate that retinal signals never lead to changes of cortical functions when the kittens are paralyzed and/or anesthetized while being exposed to visual patterns. Even though the light stimuli undoubtedly drive cortical cells they fail to bring about changes of ocular dominance (Freeman and Bonds, 1979; Singer, 1979; Singer and Rauschecker, 1982) or to develop orientation selectivity (Buisseret *et al.*, 1978). Second, the very same retinal signals may induce changes in the visual cortex of one hemisphere but not in the other when the latter is "paying less attention" to the visual signals than the former (Singer, 1982). This evidence comes from experiments in which a sensory hemineglect was induced in dark-reared kittens by producing unilateral lesions in the intralaminar nuclear complex of the thalamus. In addition, to induce changes of ocular dominance, these kittens had one eye sutured closed before they were exposed to light. Later electrophysiological analysis revealed that the neurons in the visual cortex of the normal hemisphere had become monocular as is usual with monocular deprivation. However, neurons in the visual cortex of the

---

Fig. 4. Computer processed deoxyglucose autoradiographs of horizontal sections through the occipital cortices of kittens raised with selective exposure to horizontal contours. These sections are from a depth of 2 mm and are orthogonal to the cortical lamination of striate cortex in the medial bank. Section A_{10} is from the hemisphere that, after application of the glucose pulse, had been activated with vertically oriented gratings. Section B_{11} is from the other hemisphere that had been stimulated with a horizontally oriented grating. Six activity ranges were defined and associated with different gray levels (scale on right margin). Note the reduced activity over the left hemisphere, activity over area 17 being restricted mainly to small, equally spaced islands within layer IV. LS, Lateral sulcus; MB, medial bank. From Singer *et al.* (1981).

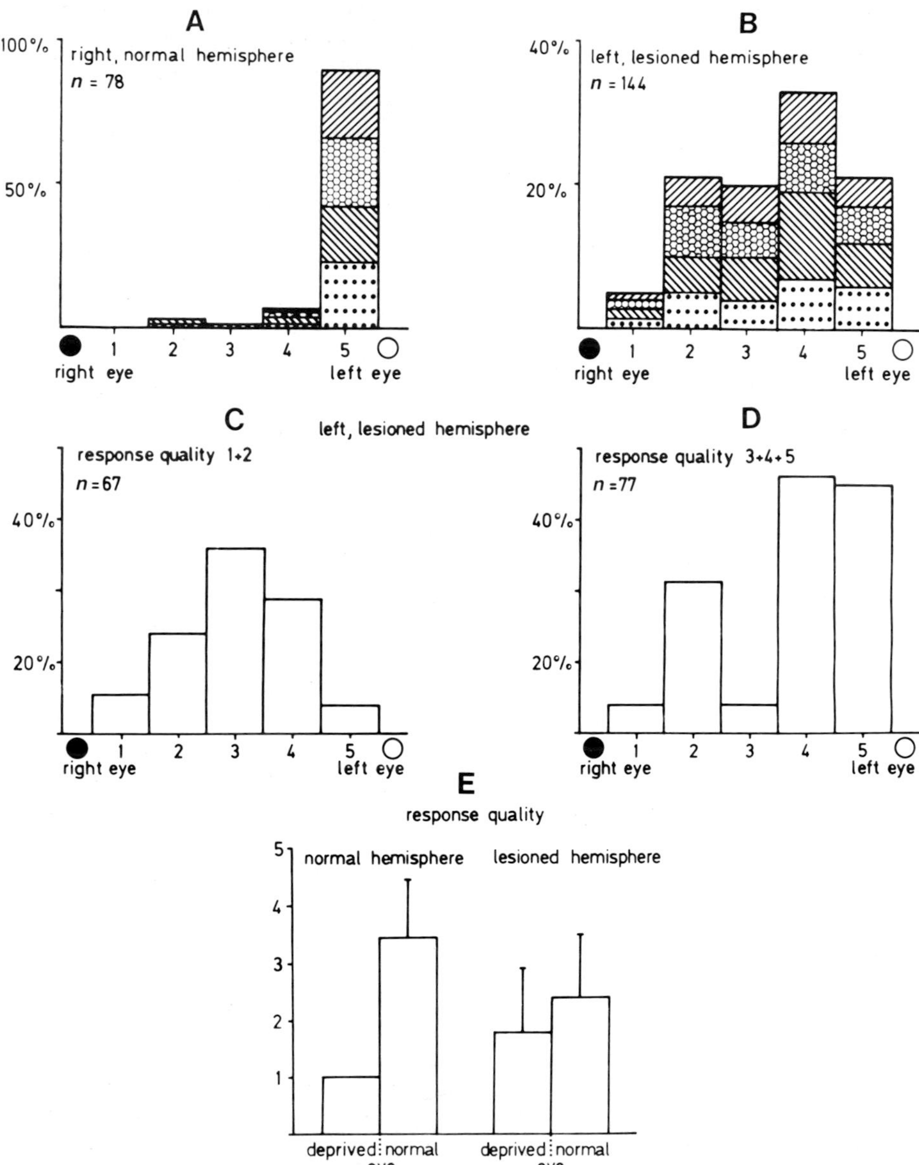

Fig. 5. Ocular dominance (OD) distribution of single cells recorded from striate cortices (area 17) of kittens that were monocularly deprived in the right eye and had a lesion in the medial thalamus of the left hemisphere. (A) Ocular dominance distribution of the neurons recorded in the normal and (B) in the lesioned hemisphere. OD classes 1 and 5 comprise cells excitable exclusively from the right (deprived) or the left (normal) eye; classes 2 and 4 correspond to

hemisphere with the lesion had remained binocular (Fig. 5). In addition, the vigor of the neuronal responses and the selectivity of the receptive fields were nearly as low as they would have been if this hemisphere had been deprived of vision altogether. Thus, although both hemispheres had received identical signals from the open eye, these signals induced modifications only in the normal hemisphere and remained ineffective in the hemisphere which, because of the diencephalic lesion, "attended" less to retinal stimulation. Another finding was that the diencephalic lesion diminished the typical effects of reticular arousal in the lesioned hemisphere. In the visual cortex of the normal hemisphere, stimulation of the mesencephalic reticular formation elicited the characteristic negative field potential, caused a transient desynchronization of the electroencephalogram, and greatly facilitated responses to light stimuli. All of these effects were greatly reduced in the hemisphere with the lesion.

The conclusion that diencephalic central core structures mediated permissive gating signals that are required for the manifestation of vision-dependent modifications in striate cortex and are related to the ascending arousal system receives further support from stimulation experiments. When monocular light stimulation was paired with electrical activation of central core structures, it proved to be possible to induce changes of ocular dominance in kittens that had been anesthetized and paralyzed (Singer and Rauschecker, 1982). Five-week-old, dark-reared kittens were prepared as usual for acute electrophysiological experiments. They were anesthetized with nitrous oxide supplemented by barbiturates and paralyzed with a muscle relaxant. Several hours after surgery we started to stimulate one eye with a slowly moving star pattern while the other eye remained closed. As expected, this procedure did not lead to changes in ocular dominance even when light stimulation was continued over 2 to 3 days. However, in 9 out of 10 kittens in which the light stimulus was paired with brief electrical stimulation of either the reticular formation or medial thalamus, clear changes in ocular dominance toward the open, stimulated eye became apparent after one night of monocular conditioning. These changes were seen at the level of evoked potentials elicited either with phase-reversing gratings or with electrical stimulation of the optic nerves and they were also apparent at the level of single unit receptive fields

---

binocular cells dominated by either of the two eyes; and class 3 comprises binocular cells responding equally well to stimulation of either eye. (A) and (B) show the OD distributions of all responsive cells recorded from the normal hemispheres (A) and the hemispheres with the lesions (B). Data from the four kittens are distinguished by different hatchings. For the histograms (C) and (D), the cells in the hemisphere with the lesion were grouped according to the vigor of responses. The OD distribution in (C) is from cells with abnormally weak responses (vigor indices 1 and 2); and the OD distribution in (D) is from the remaining cells, whose responsiveness was in the normal range (vigor indices 3 to 5). The plot in (E) represents the average vigor indices of responses to the deprived and to the normal eye in the two hemispheres. From Singer (1982).

(Fig. 6). Moreover, there was an indication from both evoked potential and single unit analyses that the gain of excitatory transmission in the pathways from the conditioned eye had increased and that cortical cells had become more selective for contrast gradients and stimulus orientation. These results are in line with the results of the lesion experiments and further corroborate the hypothesis that nonspecific modulatory systems which increase cortical excitability facilitate experience-dependent modifications.

Another manipulation that prevents retinal signals from inducing cortical modifications is the abolition of proprioceptive signals from the extraocular muscles. When this input is disrupted by severing the ophthalmic branch of the third cranial nerve bilaterally, retinal signals neither stimulate the development of orientation selectivity (Trotter *et al.*, 1981) nor do they induce changes of ocular dominance (Buisseret and Singer, 1983). We interrupted proprioception in dark-reared, five-week-old kittens by severing bilaterally the ophthalmic branches of the trigeminal nerves. In addition we either sutured closed one eye or induced strabismus by resecting the lateral rectus muscle of one eye. Subsequently, the kittens were raised in normally lighted colony rooms and were investigated with electrophysiological methods after they had passed the critical period. If modifications had occurred the large majority of cortical cells should have become monocular by that time and should have responded only to the open eye in the monocularly deprived kittens and to either the ipsi- or the contralateral eye in the strabismic kittens. However, in both groups we found the majority of cells still binocular, which suggests that proprioceptive signals from extraocular muscles are involved in gating experience-dependent modifications of cortical connectivity (Fig. 7). In addition to the nonoccurrence of competitive interactions between afferents from the two eyes we also found a number of abnormalities in other receptive field properties. In many cells, responses to optimally aligned light stimuli were sluggish and only a small fraction of the neurons had developed orientation-selective receptive fields. Since eye movements are virtually normal after proprioceptive signals from extraocular muscles are abolished (Fiorentini and Maffei, 1977), and since the uptake of visual signals is not hindered at all, one is led to conclude that the nonretinal information about eye position and/or motility is yet another of the critical parameters that determine the occurrence of long-lasting modifications in response to retinal stimulation.

IV. A Voltage-Dependent Threshold for Hebbian Modifications?

The evidence presented so far indicates that in a variety of conditions retinal signals fail to induce permanent modifications even though they are eliciting responses in cortical cells. Thus, temporal congruence between pre- and postsynaptic activity appears to be only a necessary but not a sufficient condition for the occurrence of adaptive changes. Additional "now print"

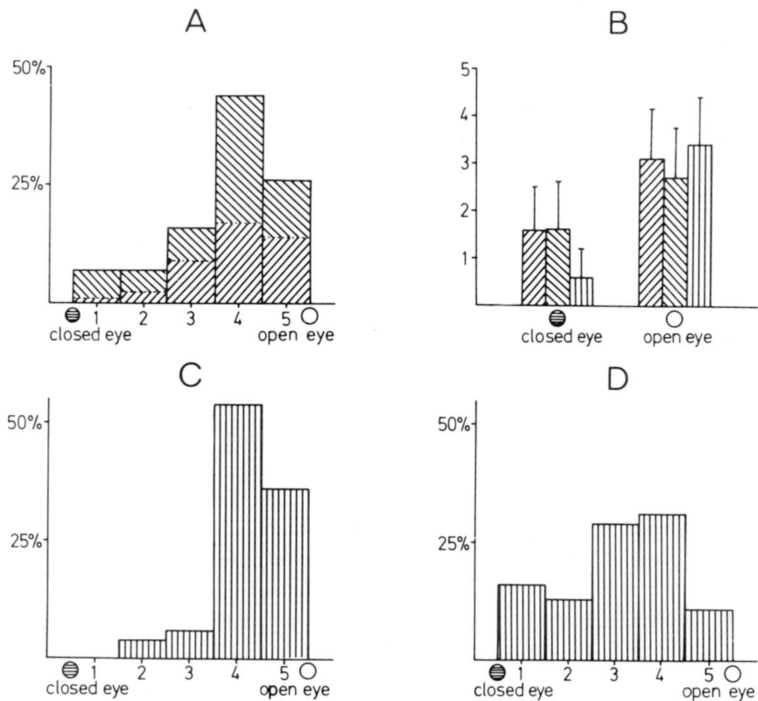

Fig. 6. Ocular dominance distributions (A, C, D) and average indices of response vigor (B) of single cells in area 17. The cells are from two kittens that had 18 h of conditioning monocular light stimulation paired with electric stimulation of medial thalamus. Cells in ocular dominance classes 1 and 5 responded exclusively either to the ipsilateral closed eye or to the contralateral conditioned eye. Cells in class 3 reacted equally well to stimulation of either eye; and cells in classes 2 and 4 responded more vigorously to one of the two eyes. The distribution in (A) (n = 86) summarizes the ocular dominance distribution of all responsive cells in the two kittens Jn4 ( ) and Jn5 ( ). The graph in (B) compares the average vigor of responses from the two eyes with corresponding averages from kittens raised with conventional monocular deprivation (Singer *et al.*, 1982). The average vigor of responses from the conditioned eye has attained nearly the same level as responses from the normal eye in cats raised with conventional monocular deprivation, while the vigor of responses from the deprived eye is abnormally low but not yet as poor as that of responses from the deprived eye in control MDs ( ). The distribution in (C) (n = 48) shows the ocular dominance of mature cells whose response vigor was >3 and whose orientation selectivity was in the normal range, while the distribution in (D) (n = 38) summarizes the remaining, immature cells in which either property was rated abnormal. This comparison reveals that the bias in the total sample of cells (A) is essentially caused by cells that have developed normal response properties. From Singer and Rauschecker (1982).

signals appear to be required and these seem to be available only when retinal signals are attended to and identified as being appropriate in a behavioral context. As the stimulation experiments suggested, the permissive gating signals can be substituted by electrical stimulation of the mesencephalic reticular formation or of the intralaminar nuclear complex of the thalamus.

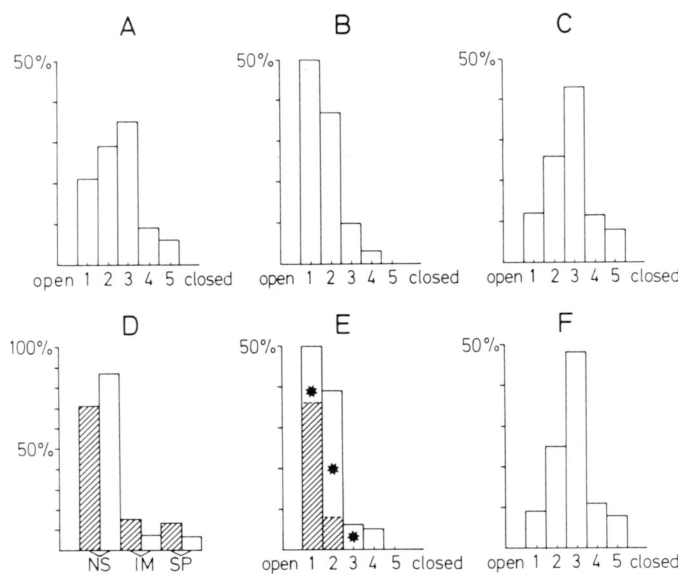

Fig. 7. Ocular dominance (OD) distributions of striate cortex neurons of kittens that had the ophthalmic branches of the fifth cranial nerve severed bilaterally and were raised with one eye closed. (A) All responsive cells ($n = 125$); (B) cells whose response vigor was rated normal (on a scale ranging from 1 to 5, they were rated in classes 3 to 5 ($n = 30$); (C) cells whose response vigor was abnormally low (classes 1 and 2) ($n = 95$). OD classes 1 and 5 correspond to monocular cells excitable exclusively from the normal or the strabismic eye. Class 3 contains binocular cells responding equally well to both eyes, and classes 2 and 4 refer to binocular cells dominated by either the normal or the strabismic eye. (D) Distribution of cells with different degrees of orientation selectivity ($n = 125$). Nonspecific (NS) cells are activated by nonoriented stimuli moving in any direction: immature cells (IM) are activated by orientated stimuli within a large angle ($>150°$); specific cells (SP) are activated by orientated stimuli within a narrow angle ($>60°$). Hatched columns, experimental kittens; blank columns, dark-reared kittens investigated previously (Buisseret *et al.*, 1978). (E) OD distributions of specific (hatched part) plus immature (blank) cells; the stars indicate the relative proportions of simple cells ($n = 37$). (F) OD distribution of nonspecific cells ($n = 88$). From Buisseret and Singer (1983).

We know the effects that stimulation of these structures has on visual processes. It greatly facilitates the transmission of retinal signals through the lateral geniculate nucleus (for review, see Singer, 1977b). Furthermore, stimulation raises cortical excitability, enhancing dramatically the transmission from granular to nongranular layers (for review, see Singer, 1979). A predictable consequence of both effects is that the depolarization of cortical dendrites increases. Thus, it is conceivable that it is critical for the occurrence of adaptive changes that cortical dendrites be sufficiently depolarized. A dendritic process that is voltage dependent and does have a high threshold is that of the activation of dendritic Ca^{2+} channels (Llinas, 1979). We hypothesized, therefore, that the final trigger signal for the occurrence of an adaptive change in response to retinal stimulation might be the influx of Ca^{2+} ions through activated, voltage-dependent Ca^{2+} channels. The prediction of this working hypothesis is that extracellular Ca^{2+} concentrations should transiently decrease with stimulation conditions that induce adaptive changes.

Our results on stimulus-induced changes of extracellular Ca^{2+} concentrations conform with this prediction. When light stimuli arc coincident with electrical activation of the nonspecific activating system—a condition sufficient to induce activity-dependent modifications—sudden decreases of the extracellular Ca^{2+} concentration can be observed. With light stimulation alone or with electrical stimulation of the modulatory projections alone–conditions which do not lead to adaptive changes–no changes of extracellular Ca^{2+} were observed. Likewise, in adult cats in which modifications of striate cortex functions can no longer be induced, even contiguous stimulation of the retina and of the modulatory projections failed to alter extracellular Ca^{2+} concentrations (Geiger and Singer, 1982). Although this covariance between Ca^{2+} changes and the occurrence of adaptive changes is not proof for an involvement of a Ca^{2+} mechanism in experience-dependent modifications of striate cortex functions, it is compatible with such an interpretation.

V. The Chemical Nature of Permissive Gating Systems

Kasamatsu and Pettigrew (1979) showed that neurons of the kitten striate cortex remain binocular despite monocular deprivation when cortical norepinephrine (NE) is depleted by local infusion of the neurotoxin 6-hydroxydopamine (6-OHDA). Since they could demonstrate that microperfusion of the depleted cortical tissue with NE reinstates ocular dominance plasticity (Kasamatsu *et al.*, 1979), these authors proposed that normal NE levels are a necessary prerequisite for ocular dominance plasticity. Subsequently, however, several independent investigations have indicated that ocular dominance changes can be induced despite NE depletion. In these investigations 6-OHDA was either injected prior to monocular deprivation or the noradrenergic input to cortex was blocked by means other than local 6-OHDA application (Adrien *et al.*, 1982; Bear and Daniels, 1983; Daw *et al.*, 1984, 1985a,b). This apparent controversy has now been resolved by the demonstration that ocular dominance plasticity is influenced both by noradrenergic *and* cholinergic mechanisms (Bear and Singer, 1986). Ocular dominance plasticity is abolished only if *both* the noradrenergic pathway from locus ceruleus and the cholinergic projection from the basal forebrain are lesioned. Disruption of either system alone is not sufficient to arrest plasticity. The initial finding that intracortical or intraventricular application of 6-OHDA blocked ocular dominance plasticity could be attributed to an unexpected pharmacological property of 6-OHDA. This drug, in addition to its neurotoxic effect, antagonizes the effects of acetylcholine (Bear and Singer, 1986). These results demonstrate that both neuromodulators acetylcholine and norepinephrine have a permissive function in cortical plasticity, either system being capable of substituting for the other.

At the present stage we can only speculate on the mechanisms by which the two systems could facilitate plasticity. Both modulators close potassium channels (Krnjevic *et al.*, 1971; Madison and Nicoll, 1982; Halliwell and Adams, 1982). This increases the length constant of the dendrites and enhances depolarization in response to EPSPs. Increased length constants would enhance cooperative as well as competitive interactions between pathways converging onto the same dendrite, and enhanced depolarization would increase the probability of Ca^{2+}-channel activation. Another mode of action of the two modulators is via second messenger systems. Norepinephrine stimulates via a β-receptor-coupled cyclase intracellular cAMP production while ACh enhances via the inosital pathway intracellular Ca^{2+}-levels (for review, see Nestler *et al.*, 1984). Synergistic interactions between both second messengers, Ca^{2+} and cAMP, are known to occur and hence NE and ACh could substitute for each other also at this level.

VI. Implications for Normal Development

To conclude the discussion of experience-dependent modifications of striate cortex, we would like to consider briefly the functional role of these processes in normal development. The evidence reviewed above warrants the conclusion that a functional criterion—the degree of correlation of neuronal activity—guides the specification of neuronal connections during an early phase of postnatal development. The fact that this activity-dependent specification process is gated implies further that the developing brain is able to select according to its central state the instances at which sensory activity can assume the role of a shaping factor. Such a selection algorithm is ideally suited to solving specification problems that require that neuronal connections be selected according to functional criteria. In the following paragraphs I shall give examples of such specification problems.

A. Binocular Correspondence

The formation of binocular neurons with precisely corresponding receptive fields in the two eyes poses a specification problem that appears solvable only if the development of connections is guided by functional criteria. The pathways connecting the two eyes with the common binocular target cells in the visual cortex have to be selected such that they come from precisely matching retinal loci. Retinal correspondence depends on parameters such as the size of the eyes, the position of the eyes in the orbit, the interocular distance, and so on, parameters that are obviously under the influence of manifold epigenetic factors. Which retinal loci in the two eyes will actually be matching with each other in the mature system can therefore be anticipated only with moderate accuracy. Pathways originating from corresponding retinal loci can be identified, however, with great precision using functional criteria. These pathways convey highly correlated activity patterns when the animal is fixating a real world object in everyday life. Since the activity-dependent modifications occurring in developing striate cortex selectivity stabilize connections that convey correlated activity, they are ideally suited to optimize the correspondence of binocular connections.

In this context, a crucial role can also be assigned to the gating of adaptive changes by nonretinal afferences. It is obvious that the selection process should take place only when the animal is attentively fixating a target and it should not occur in instances when the eyes are not properly aligned or are moving in an uncoordinated way. In the latter cases the signals from the two eyes are uncorrelated and the activity-dependent modifications would lead to a disruption rather than to an optimization of binocular connectivity. The gating functions mentioned above, in particular those exerted by the proprioceptive afferents from the extraocular muscles and by the ascending reticular activating system, could take care of this additional problem. They would be ideally suited to allow modifications to occur only when the conditions of attentive fixation are fulfilled.

B. Orientation Selectivity in Large-Field Neurons

Another specification problem is raised by cortical cells which possess large, orientation-selective receptive fields. This is particularly the case for the fraction of cortical neurons, located preferentially within layer VI, whose receptive fields are very elongated (Gilbert, 1977; Gilbert and Kelly, 1975). One way to develop such large receptive fields could be to integrate excitatory input from cells whose receptive fields have the same orientation preference and whose position in the cortical sheet is such that their receptive fields line up along lines parallel to the axis of preferred orientation. From the magnification factor of the retinocortical mapping it can be inferred that the large-field neurons would have to sample activity from cells that are distributed over several millimeters within the cortical sheet (Hubel and Wiesel, 1974). Moreover, because of the columnar organization of the striate cortex (Hubel *et al.*, 1978), sampling would have to be discontinuous, sparing orientation slabs whose orientation preference does not correspond to the orientation of the respective large-field neurons. The sampling problem is further complicated by the fact that the axis in the cortex along which sampling has to occur is a function of the orientation preference of the large-field neuron. Cells with extended vertical receptive fields have to sample along axes that are parallel to the representation of the vertical meridian while cells with horizontally oriented fields must sample along axes parallel to the representation of the horizontal meridian and so on. An anatomical substrate for such a discontinuous sampling process could be the far-reaching axon collaterals of cortical pyramidal cells (Lund and Boothe, 1975; Szentagothai, 1975). Particularly relevant in this context is the evidence, first, that these axon collaterals form several discrete turfs whose spacing conforms to the spacing of columnar systems (Gilbert and Wiesel, 1979, 1983); and second, that intracortical projections tend in general to be periodic (Rockland and Lund, 1982, 1983).

It is obvious that the development of such highly specific connections between cortical neurons would require a powerful selection algorithm that is capable of identifying orientation preference and retinal position of each of the selectively interconnected cells. It is difficult to see how such could be accomplished by genetic instructions alone. Again selective stabilization of interconnections according to functional criteria could provide a solution to this specification problem.

Because of the continuity of contours in the outer world it is to be expected that the probability of correlated activation is high for cortical neurons that share the same orientation preference and whose receptive fields are adjacent. This is particularly true if one considers that intracortical inhibitory interactions between adjacent columns have the effect of preventing simultaneous responses to nearby contours with differing orientations (Blakemore and Tobin, 1972). Thus, the very same selection algorithm that we derived from experiments on binocular convergence could serve to solve the specification problem raised by large field, orientation-selective neurons.

These considerations prompt the speculation that we might have to do with a rather general principle of self-organization, a principle which allows the developing system to select and to stabilize neuronal pathways with great precision using functional properties of the neurons as selection criteria.

VII. Concluding Remarks

It follows from this brief survey that we are still at the very beginning of an understanding of the mechanisms that mediate activity-dependent modifications of neuronal transmission in the developing visual cortex. However, even without detailed knowledge of the neuronal mechanisms mediating the changes, we find that the available data have a few interesting implications. The activity-dependent modifications follow rules that allow them to stabilize selectively pathways between neuronal elements that have a high probability of being active at the same time. Hence, this selection process has an associative function, coupling preferentially those cells with each other that share particular functional properties. This selection process could, therefore, serve to develop assemblies of cooperating neurons whose coherent and—if reciprocally coupled—reverberating responses could represent the neuronal code for particular patterns of activity at the sensory surfaces. The fact that these associative processes are not solely dependent on the pattern of activity on the sensory surface but are gated in addition by internally generated signals provides the developing system with the option to form neuronal assemblies not only as a function of the patterns in afferent sensory pathways but also as a function of the central state of the system itself. This implies that the formation of assemblies is a process of active selection whereby the selection criteria emerge first from the genetically determined properties of the system, second from the actual dynamic (behavioral) state which the system maintains while it interacts through its sensory surfaces with the "outer" world, and third from the structure of the sensory stimuli.

References

Adrien, M., Buisseret, P., Fregnac, Y., Gary-Bobo, E., Jubert, M., Tassin, M. P., and Trotterny, (1982). *C. R. Acad. Sci. Paris* **295**, 795–750.

Bear, M. F., Paradiso, M. A., and Daniels, J. D. (1982). *Soc. Neurosci. Abstr.* **8**, 4.

Bear, M. F., and Daniels, J. D. (1983). *J. Neurosci.* **3**, 407–476.

Bear, M. F., and Singer, W. (1986). *Nature (London)* (in press).

Blakemore, C., and Cooper, G. F. (1970). *Nature (London)* **228**, 477–478.

Blakemore, C., and Tobin, E. A. (1972). *Exp. Brain Res.* **15**, 439–440.

Blakemore, C., and Van Sluyters, R. C. (1974). *Br. J. Ophthalmol.* **58**, 176–182.

Blakemore, C., Van Sluyters, R. C., Peck, C. K. and Hein, A. (1975). *Nature (London)* **257**, 584–586.

Buisseret, P., and Imbert, M. (1976). *J. Physiol. (London)* **255**, 511–525.

Buisseret, P, and Singer, W. (1983). *Exp. Brain Res.* **51**, 443–450.

Buisseret, P., Gary-Bobo, E., and Imbert, M. (1978). *Nature (London)* **272**, 816–817.

Changeux, J. P., Danchin, A. (1976). *Nature (London)* **264**, 705–712.
Crewther, S. G., Crewther, D. P., Peck, C. K., and Pettigrew, J. D. (1980). *J. Neurophysiol.* **44**, 97–118.
Daw, N. W., Robertson, T. W., Rader, R. U., Videen, T. O., and Coscia, C. T. (1984). *J. Neurosci.* **4**, 1354–1360.
Daw. N. W., Videen, T. O., Parkinson, D. and Rader, R. U. (1985). *J. Neurosci.* **5**, 1925–1933.
Fiorentini, A., and Maffei, L. (1977). *Nature (London)* **269**, 330–331.
Freeman, R. D., and Bonds, A. B. (1979). *Science* **206**, 1093–1095.
Fregnac, T. and Imbert, M. (1984). *Physiol. Rev.* **64**, 325–434.
Gaze, R. M., Feldman, J. D., Cooke, J., and Chung, S.-H. (1979). *J. Embryol. Exp. Morphol.* **53**, 39–66.
Geiger, H., and Singer, W. (1982). *Int. J. Dev. Neurosci. Suppl.*, R328.
Geiger, H., and Singer, W. (1986). *Exp. Brain Res. Suppl.* (In press).
Gilbert, C. D. (1977). *J. Physiol. (London)* **268**, 391–421.
Gilbert, C. D., and Kelly, J. P. (1975). *J. Comp. Neurol.* **163**, 81–106.
Gilbert, C. D., and Wiesel, T. N. (1979). *Nature (London)* **280**, 120–125.
Gilbert, C. D., and Wiesel, T. N. (1983). *J. Neurosci.* **3**, 1116.
Halliwell, T. V., and Adams, P. R. (1982). *Brain Res.* **250**, 77–78.
Hebb, D. O. (1949). "The Organization of Behavior." Wiley, New York.
Hirsch, H. V. B., and Spinelli, D. N. (1970). *Science* **168**, 869–871.
Hubel, D. H., and Wiesel, T. N. (1963a). *J. Neurophysiol.* **26**, 994–1002.
Hubel, D. H., and Wiesel, T. N. (1963b). *J. Physiol. (London)* **160**, 106–154.
Hubel, D. H., and Wiesel, T. N. (1965). *J. Neurophysiol.* **28**, 1041–1059.
Hubel, D. H., and Wiesel, T. N. (1974). *J. Comp. Neurol.* **158**, 295–306.
Hubel, D. H., Wiesel, T. N., and LeVay, S. (1977). *Philos. Trans. R. Soc. London Ser. B* **278**, 377–409.
Hubel, D. H., Wiesel, T. N., and Stryker, M. P. (1978). *J. Comp. Neurol.* **177**, 361–380.
Innocenti, G. M., and Frost, D. O. (1979). *Nature (London)* **280**, 231–234.
Jansen, J. K. S., and Lomo, T. (1981). *Trends Neurosci.*, July 1981, 178–181.
Kasamatsu, T., and Pettigrew, J. D. (1979). *J. Comp. Neurol.* **185**, 139–162.
Kasamatsu, T., Pettigrew, J. D., and Ary, Mary Louise (1979). *J. Comp. Neurol.* **185**, 163–181.
Krnjevic, K., Pummain, R., and Renaud, L. (1971). *J. Physiol.* **215**, 247–268.
Llinas, R. (1979). *Neurosci. Study Program, 4th 1979*, pp. 555–571.
Lund, J. S., and Boothe, R. G. (1975). *J. Comp. Neurol.* **159**, 305–334.
Madison, D. V., and Nicoll, R. A. (1982). *Nature (London)* **299**, 636–638.
Mariani, J., and Changeux, J.-P. (1981). *J. Neurosci.* **1**, 703–709.
Meyer, R. L. (1982). *Science* **218**, 589–591.
Mitzdorf, U., and Singer, W. (1980). *J. Physiol. (London)* **304**, 203–220.
Movshon, J. A., and Van Sluyters, R. (1981). *Annu. Rev. Psychol.* **32**, 477–522.
Nestler, E. T., Walaas, S. T., and Greengard, P. (1984). *Science* **255**, 1357–1364.
Rauschecker, J. P., and Singer, W. (1979). *Nature (London)* **280**, 58–60.
Rauschecker, J. P., and Singer, W. (1981). *J. Physiol. (London)* **310**, 215–239.
Rockland, K. S., and Lund, J. S. (1982). *Science* **214**, 1532–1543.
Rockland, K. S., and Lund, J. S. (1983). *J. Comp. Neurol.* **216**, 303–318.
Sherk, H., and Stryker, M. P. (1976). *J. Neurophysiol.* **39**, 63–70.
Sherman, S. M., and Spear, P. D. (1982). *Physiol. Rev.* **62**, 738–855.
Sillito, A. M., Kemp, J. A., and Blakemore, C. (1981). *Nature (London)* **291**, 318–322.
Singer, W. (1976). *Brain Res.* **118**, 460–468.

Singer, W. (1977a). *Exp. Brain Res.* **30**, 25–41.
Singer, W. (1977b). *Physiol. Rev.* **57**, 386–420.
Singer, W. (1979). *Neurosci. Study Program, 4th, 1979*, pp. 1093–1109.
Singer, W. (1982). *Exp. Brain Res.* **47**, 209–222.
Singer, W., and Rauschecker, J. (1982). *Exp. Brain Res.* **47**, 223–233.
Singer, W., and Tretter, F. (1976). *J. Neurophysiol.* **39**, 613–630.
Singer, W., Rauschecker, J., and Werth, R. (1977). *Brain Res.* **134**, 568–572.
Singer, W., von Grunau, M., and Rauschecker, J. (1979). *Brain Res.* **171**, 536–540.
Singer, W., Freeman, B., and Rauschecker, J. (1981). *Exp. Brain Res.* **41**, 199–215.
Singer, W., Tretter, F., and Yinon, U. (1982). *J. Physiol. (London)* **324**, 221–237.
Stent, G. S. (1973). *Proc. Natl. Acad. Sci. U.S.A.* **70(4)**, 997–1001.
Stryker, M. P. (1981). *Soc. Neurosci. Abstr.* **7**, 842.
Szentagothai, J. (1975). *Brain Res.* **95**, 475–496.
Trotter, Y., Gary-Bobo, E., and Buisseret, P. (1981). *Dev. Brain Res.* **1**, 450–454.
Wiesel, T. N., and Hubel, D. H. (1965a). *J. Neurosphysiol.* **28**, 1029–1040.
Wiesel, T. N., and Hubel, D. H. (1965b). *J. Neurophysiol.* **28**, 1060–1072.
Wiesel, T. N., and Hubel, D. H. (1974). *J. Comp. Neurol.* **158**, 307–318.

10

Experience and Development in the Visual System: Anatomical Studies

RONALD G. BOOTHE,[1] ELSI VASSDAL, and MARILYN SCHNECK
Department of Psychology, and Department of Ophthalmology
University of Washington
Seattle, Washington

I. Introduction

In this chapter we discuss development of the geniculostriate pathways in the mammalian visual system and the ways that this development can be modified by postnatal visual deprivation. These geniculostriate pathways

[1]Present address: Yerkes Primate Center, Department of Psychology, and Department of Ophthalmology, Emory University, Atlanta, Georgia 30322.

Developmental NeuroPsychobiology

include the connections between retinal ganglion cells and the dorsal lateral geniculate nucleus (LGN), and between the LGN and striate cortex (Broadman's area 17).

Development is a continuous process, but for purposes of exposition we will discuss it in terms of five discrete stages: (1) Generation of neurons, (2) cell migration, (3) formation of pathways between nuclei, (4) cell death, and (5) formation and pruning of connections.

There are billions of neurons and thousands of times this number of connections in the central nervous system. It is obvious that the genome cannot contain sufficient information to specify exactly each connection that is made between neurons. Rather, the genome must specify rules that govern factors such as the types and number of neurons generated and the interactions between neurons. We will consider the literature on anatomical development of the visual system within a framework that deals with the nature of these rules. A very common finding, which may be a general rule, is an initial overproduction of both neurons and connections. Later development involves the selective maintenance of some neurons and connections and elimination of others. This selection process sometimes involves competitive interactions between neurons.

In Section II we provide a brief overview of the five stages that take place during normal visual system development. Next we review the ways in which development can be modified by postnatal visual deprivation, emphasizing studies of monocular deprivation. Finally, we discuss the evidence that monocular deprivation effects are due to a combination of binocularly competitive and noncompetitive mechanisms. The data presented are largely limited to those from the binocular visual systems of cats and monkeys and the primarily crossed visual systems of rodents.

II. Normal Development

A. Generation of Neurons (Cell Birthdays)

The germinal cell lines that give rise to specific populations of neurons might play a critical role in developmental interactions. For example, Jacobson and Hirose (1978) note that, in the frog, each neural retina originates from both sides of the prospective forebrain. This suggests the possibility that retinal ganglion cells that project to the contralateral hemisphere in mammals might be derived from contralateral germ cell lines.

The timing of cell birthdays (Sidman, 1970) may also be an important factor influencing developmental interactions. Rakic (1977b) notes that the primate LGN cells that receive corresponding inputs from the contralateral and ipsilateral eyes are probably formed in succession from proliferative units at the surface of the third ventricle. Therefore, the LGN cells that receive left and right eye inputs are probably derived from the same germinal cell lines, but have their birthdays at different developmental times.

Studies that have examined cell birthdays in the geniculostriate pathways have consistently found that these neurons are born relatively early during development. Johns *et al.* (1979) determined that all retinal ganglion cells are already born at the time of birth in cats. In monkeys the retinal ganglion cells are born during the last half of the first trimester* (Rakic, 1977a). Neurons destined for the LGN have their birthdays within an 8- to 9-day period within the first trimester in monkeys (Rakic, 1977a,b), and during a 2- to 3-day period right after midgestation in rats (Lund and Mustari, 1977). Striate cortex neurons are born in the monkey during a 2-month period falling mostly during the second trimester of gestation (Rakic, 1977a) and during a 1-week period just before birth in rats (Lund and Mustari, 1977).

B. Migration

Neurons are always generated near a ventricle. Once born, the cells must somehow move from their birthplace near the ventricle to their adult positions. Proposed mechanisms of cell migration include extension by the neuroblast of a process to the new region, followed by nuclear translocation within this process (Berry and Rogers, 1965; Morest, 1969, 1970) and movement of the neuroblast along a radial glia cell (Rakic, 1971, 1972). It seems likely that both of these mechanisms operate in different neural populations and at different developmental stages (Rakic, 1975).

Detailed descriptions of cell migration in the retina, LGN, and cortex of the rat have been provided by Raedler and Sievers (1975). In the retina, mitosis takes place near the remnant of the optic vesicle at the external limiting membrane. All neural precursors must then migrate toward the vitreous surface. Cells destined for the LGN are generated in ventricular or subventricular zones near the surface of the third ventricle and migrate laterally after their last cell division to the lateral aspect of the developing diencephalon. Neurons in area 17 proliferate in ventricular and subventricular zones near the surface of the lateral cerebral ventricle. The neuroblasts subsequently migrate to the outermost part of the developing cortical plate.

These cell migration patterns are responsible for the formation of nuclei and lamination within nuclei. Neurons migrate into final position in the rat

*The term *trimester* is used in this chapter to refer to one-third of the total gestation period.

LGN during the last week before birth, and the LGN becomes differentiated from the rest of the thalamus during this same time period (Lund and Mustari, 1977). In primate LGN, the adult laminar pattern is not evident during the first trimester. The laminae are discernible but not completely separated during the second trimester and become gradually more distinct during the third trimester, assuming adult form by birth (Rakic, 1977a,b). If one eye is removed during the first trimester, segregation into laminae does not occur (Rakic, 1981). The striate cortex of the newborn monkey exhibits an adultlike pattern of lamination (O'Kusky and Colonnier, 1982), while in rats the adult pattern of six layers develops postnatally (Lund and Mustari, 1977; Raedler and Sievers, 1975).

In cats and primates there is an additional specialized horizontal migration that takes place in the central region of the retina. This horizontal migration continues into the postnatal period and is associated with the forming of the area centralis in cats (Stone *et al.*, 1982), and the foveal pit in primates (Hendrickson and Kupfer, 1976).

C. Formation of Pathways between Nuclei

1. Retina to Lateral Geniculate Nucleus

In rats the first optic axons arrive at the LGN about 1 week before birth (Lund and Bunt, 1976). In primates, where the inputs from the two eyes are segregated into separate layers in the adult, Rakic (1977a) found that retinal ganglion cell terminals from the left and right eyes intermingle when they first arrive at the LGN and only later segregate out into separate laminae. The retinal inputs segregate at about the same time that the LGN neurons themselves separate into laminae. If one eye is removed during the first trimester, the terminals from the remaining eye appear to form synaptic connections with all geniculate neurons rather than segregating into layers (Rakic, 1981). Thus, the presence of both eyes appears to be necessary for the segregation of afferent inputs within the LGN, as well as for the laminar segregation of the LGN neurons.

2. Lateral Geniculate Nucleus to Striate Cortex

Rakic (1977a) determined that monkey LGN fibers reach the occipital lobe during the second trimester, but do not enter the cortical plate in any substantial number until the third trimester. Anker and Cragg (1974) noted that the connections from the LGN to area 17 are present at birth in cats. In rats, LGN afferents reach the cortical plate before birth but do not distribute their terminals in the adult laminar pattern until postnatal day 8 (Lund and Mustari, 1977).

In adult cats and monkeys, geniculocortical afferents serving the two eyes are segregated into separate regions of cortical layer IV, referred to as ocular dominance columns (Hubel and Wiesel, 1972; LeVay *et al.*, 1975; Shatz *et al.*, 1977; Wiesel *et al.*, 1974). These columns are not present in rats, which have primarily crossed visual pathways. These columns are also not present in cats and monkeys prior to birth. Rather, the prenatal geniculocortical afferents serving the two eyes overlap within layer IV in both the kitten (LeVay *et al.*, 1978) and the monkey (Hubel *et al.*, 1977; LeVay *et al.*, 1980; Rakic, 1976). Segregation into ocular dominance columns does not begin in the kitten until about 3 weeks postnatal, 1 week after eye opening, and is complete by about 6 weeks of age (LeVay *et al.*, 1978). In the fetal monkey the beginnings of segregation are apparent by about 3 weeks before birth (Rakic, 1976), but complete segregation is not established until about 6 weeks postnatal (LeVay *et al.*, 1980). If one eye is removed prior to birth, the ocular dominance columns never develop (Rakic, 1981).

D. Cell Death

In the geniculostriate pathway, as in several other developing neural systems that have been studied, more neurons are born and migrate into position than are present in the adult. Evidence has been obtained for naturally occurring cell death in rat ganglion cells (Cunningham *et al.*, 1981) and in cortex of the cat (Cragg, 1975) and monkey (O'Kusky and Colonnier, 1982). In the monkey, O'Kusky and Colonnier estimate that the total number of neurons in striate cortex of each hemisphere decreases from 192 million at birth to 161 million in the adult, a postnatal decrease of 16%. Cragg (1975) found a 12% decrease in density of neurons in cat cortex.

It has been suggested that this cell death might reflect the operation of a competitive mechanism; neurons compete with each other for a finite number of potential connections, and those that fail to establish an appropriate pattern of connections are eliminated (for a review, see Cunningham, 1982). Seneglaub and Finlay (1981) found evidence consistent with this competition hypothesis in the hamster. If one eye is removed at birth, thereby increasing the terminal field available to the remaining eye, ganglion cell death is reduced in the remaining eye.

E. Formation and Pruning of Connections

A process analogous to cell death of neurons also occurs for connections between neurons. There is often an overproduction of neuropil and synaptic connections during early development, followed by a decrease, or pruning, to the adult level. This overproduction and the subsequent elimination of connections have been proposed as a mechanism for removing inappropriate or redundant connections (for a review, see Purves and Lichtman, 1980; see also Chapter 4, this volume).

Two primary methods have been used to study formation and pruning of neural connections. First, light microscopic studies, primarily using Golgi methods, have looked at growth and elaboration of the neuropil. These studies make indirect inferences about neural connections by examining developmental changes in availability of potential postsynaptic sites, that is, by looking at growth of dendrites and changes in spine frequency. Second, electron microscopic studies have looked directly at synaptogenesis.

1. Lateral Geniculate Nucleus Connections

Golgi studies have been conducted with rats (Parnavelas *et al.*, 1977), cats (Morest, 1969), and monkeys (Garey and Saini, 1981). There are several features of developing LGN cells that are characteristic of immature neurons: growth-cone-like enlargements and varicosities along dendritic shafts; excessive numbers of dendritic and somatic spines; and fine, long, hairlike processes, or filopodia, on all neurons including populations that bear few or no spines in their mature form. In monkeys, enlargements resembling growth cones begin to disappear during the late prenatal period, and few are seen postnatally (Garey and Saini, 1981). Growth cones continue to be observed into the postnatal period in rat and cat LGN (Morest, 1969; Parnavelas *et al.*, 1977).

Even though growth cones are seldom seen postnatally in the monkey, some LGN neurons continue to show postnatal increases in the numbers of secondary and tertiary branches (Garey and Saini, 1981). However, the most prominent postnatal changes occur in the numbers of spines and filopodia. Spine frequencies increase to levels that are higher at birth than in the adult. This is followed by a decrease to adult levels after birth. All main neuronal types follow a similar sequence, but neurons in magnocellular layers tend to mature before those in parvocellular layers. Fully mature dendritic trees seem to be established in most LGN neurons by 4 weeks after birth, although some cells in parvocellular laminae may continue to develop longer.

Differentiation of synaptic profiles occurs largely prenatally in the monkey. Synaptogenesis begins at the caudal pole of the nucleus, which receives input from the central retina, and proceeds toward the rostral pole, which receives input mostly from the peripheral retina (Hendrickson and Rakic, 1977).

Quantitative measurements of synaptogenesis in cat LGN reveal a fourfold increase in synaptic density during the first postnatal month (Cragg, 1975). This is followed by a more gradual increase of about 10% to the adult level. The rat LGN shows a tenfold increase in synaptic density during the first 2 weeks after birth (Karlsson, 1967). By 2 weeks the synaptic density appeared to have reached adult levels, but the results at older ages were too variable to allow an exact specification of the time course of development.

2. *Striate Cortex Connections*

Elaboration of dendrites in the striate cortex has been found to go through a similar sequence to that in the LGN. This sequence has been described in the rat (Juraska, 1982; Juraska and Fifkova, 1979a; Miller and Peters, 1981; Parnavelas *et al.*, 1978) and the monkey (Lund *et al.*, 1977). Initially there are only a few large dendrites with very little branching. The apical dendrites of pyramidal cells develop before the basilar or oblique dendrites. Dendrites at early stages of development are short and varicose with abrupt changes in diameter along their length, and they often exhibit characteristic growth cones. By the time of birth (in the monkey) or the time of eye opening (in the rat), the dendritic branching pattern appears similar to that in the adult (Boothe *et al.*, 1979; Juraska, 1982; Juraska and Fifkova, 1979a).

During the first week after birth in the rat, fine, long, hairlike processes, or filopodia, as well as immature spines of various shapes, are found extended from dendritic tips and shafts and also from cell bodies (Juraska and Fifkova, 1979a; Miller and Peters, 1981). As the dendrites grow and branch the filopodia eventually disappear and the spines slowly change into their mature adult shape. Some spines have a mature adult shape even though electron microscopy reveals that the spinal apparatus is not developed (Juraska and Fifkova, 1979b). Miller and Peters (1981) suggest that spines pass through a specific sequence of shape changes during maturation. Dendritic spines begin as low, broad protrusions, gradually become taller stumps, and finally acquire their mature mushroom shape (see Chapter 13, this volume).

A dense population of spines on dendrites is found at early postnatal ages, regardless of whether or not a heavy population is present in the adult form of the cells (Lund *et al.*, 1977); Parnavelas *et al.*, 1978). In rats, spine frequency increases until the third postnatal week and then decreases to adult levels by the fourth week on nonpyramidal neurons (Parnavelas *et al.*, 1978). On pyramidal neurons the pattern of spine growth and loss varies with the type of dendritic branch (e.g., bifurcating, terminal) and the cell layer (Juraska,

1982). Neurons in monkey striate cortex continue to increase their spine frequency until 8 weeks after birth (Boothe *et al.*, 1979). After 8 weeks, monkey striate neurons show a decrease in spine number. As in the rat, this pruning follows different time courses for different neuronal populations. Cells in cortical layers that receive primarily a parvocellular geniculate input (layer IVC β) decrease to adult levels sooner than cells in layers receiving primarily a magnocellular input (layer IVC α). For some neuronal populations the spine frequency is still above adult levels at 9 months after birth (Boothe *et al.*, 1979).

Dendritic spines are sites of synaptic contacts (Colonnier, 1968; Gray, 1959; LeVay, 1973), and developmental changes in spine number reflect changes in the number of type 1 synapses (Mates and Lund, 1983a,b). Type 2 synapses onto cell somas follow a similar pattern, first increasing and then decreasing during postnatal development (Mates and Lund, 1983c). O'Kusky and Colonnier (1982) made quantitative measurements of synapses. They also found an overshoot of synapses in monkey striate cortex with an excess of neural connections made between birth and 6 months of age, followed by a decrease to the adult value. They estimate that total numbers of synapses increase from approximately 389 billion at birth to 741 billion at 6 months and then decrease to 381 billion in the adult.

Cragg (1972) reported that only 1.5% of the adult number of cortical synapses is present in kittens 8 days after birth. There is a period of rapid development from 8 days to 5 weeks after birth. The number of synapses per neuron also remains very low until about day 8, then rises rapidly (Cragg, 1975). Synaptic numbers overshoot the adult level and these excess synapses are maintained until after day 108. Winfield (1981) examined kittens of several postnatal ages to specify the time course of the overshoot. A maximum synaptic density was found at 70 days of age. After this age the density diminished, falling off 17% by 110 days and 30% by adulthood. Winfield (1981) concluded that most, if not all, of the reduction is due to loss of asymmetric axospinous synapses.

Lund and Mustari (1977) found prenatal synapses only in layer I of rat cortex. At birth there were synapses in layers I and II. By 6 days postnatal, large numbers of synapses were found in all cortical layers. Juraska and Fifkova (1979b) also reported finding very few synapses at birth and many of those that were present looked immature. Axosomatic synapses were first observed at day 3. By the second postnatal week many synapses appeared to be mature in form and the majority could be classified as asymmetrical (Juraska and Fifkova, 1979b).

Miller and Peters (1981) studied Golgi-impregnated pyramidal neurons in layer V of rat cortex. Somatic spines on these neurons, which form a single synapse with a symmetric junction, are present during the first week, but are rarely seen after the second week. Dendrites form synapses with symmetric densities as early as day 3 and asymmetric ones by day 9. Flat, stubby dendritic spines are seen to make symmetric synapses, whereas longer thin and mushroom-shaped spines form the characteristic adult asymmetric junctions.

III. Effects of Visual Deprivation on Development

Visual deprivation studies, primarily involving monocular suturing of the eyelids, have been conducted that address the role of functional activity on the developmental process.

A. Effects of Deprivation on the Lateral Geniculate Nucleus

1. Synaptology

Wilson and Hendrickson (1981) examined terminal profiles and synapses in the parvocellular, magnocellular, and interlaminar zones of monocularly visually deprived monkeys. They failed to find any qualitative or quantitative differences between synaptology of zones receiving input from the deprived retina compared to the open eye retina, or compared to the LGN in normal monkeys.

2. Cell Size

Monocular visual deprivation can produce permanent effects on LGN cell soma size. In cats, LGN cells in layers receiving input from a lid-sutured eye are typically 20–40% smaller in cross-sectional area than those in nondeprived laminae (Chow and Stewart, 1972; Garey *et al.*, 1973; Guillery, 1972; Guillery and Stelzner, 1970; Hickey, 1980; Hickey *et al.*, 1977; Kalil, 1980; Wan and Cragg, 1976; Wiesel and Hubel, 1963). Similar changes in geniculate cell size following monocular deprivation have also been reported in monkeys (Headon and Powell, 1973, 1978; Hubel *et al.*, 1977; Levay *et al.*, 1980; Vital-Durand *et al.*, 1978; von Noorden, 1973).

Hickey *et al.* (1977) compared LGN cell sizes in deprived and normal cats and concluded that the interlaminar differences are caused by a combination of a 20–25% reduction of cell size in the deprived laminae and a 10–15% hypertrophy of cells in the nondeprived layers. Hickey (1980) compared the time courses of normal and visually deprived development of LGN cell size. Normal LGN cells grew rapidly during the first postnatal month and then gradually approached adult levels. In deprived animals, the cells from all layers also grew rapidly during the first month. After 28 days, however, the cells in the deprived layers stopped growing and then underwent slow shrinkage. Cells in the nondeprived layers showed a small amount of hypertrophy.

Kalil (1980) concluded similarly that deprivation alters the growth of LGN neurons. He found that in kittens deprived prior to eye opening, affected cells continue to grow, although more slowly than normal, for 3 to 4 weeks postoperatively, at which time all further growth stops. Subsequently, there is marked cellular atrophy and shrinkage. Kalil (1980) noted that cells in the normally innervated binocular segment grow somewhat faster than normal, but in contrast to Hickey *et al.* (1977; Hickey, 1980), found no evidence for hypertrophy in these cells. Wan and Cragg (1976) also found that cells in the nondeprived laminae appear to grow faster than normal. They also noted that cell size was more variable in the nondeprived laminae. This increased variability could account for some of the discrepancies in the literature.

3. *Subpopulations of Lateral Geniculate Nucleus Neurons*

Sherman *et al.* (1972) suggested, on the basis of physiological studies, that the population of large Y cells are most affected by monocular deprivation. A number of anatomical findings are consistent with this hypothesis. Garey and Blakemore (1977) used horseradish peroxidase to retrogradely label cat geniculate cells projecting to either area 17 or 18. Monocular deprivation had a greater effect on large LGN cells that project to area 18 (presumed Y cells) than on small LGN cells that project to area 17 (presumed X cells). Lin and Sherman (1978) found even more dramatic results following longer deprivation periods. Few labeled cells were found in deprived layers following injections into area 18 and those that were labeled were faint. Lin and Sherman (1978) speculate that severe deprivation may disrupt HRP labeling among Y cells projecting to area 18. Using the anatomical criterion of the presence (X cells) or absence (Y cells) of a cytoplasmic laminar body, LeVay and Ferster (1977) reported that Y cells are fewer in number and smaller in size in deprived laminae.

Friedlander *et al.* (1982) recorded activity of single neurons in the deprived cat LGN and then injected horseradish peroxidase into the same neurons. They found that physiological X cells were within the normal soma size range, whereas physiological Y cells were abnormally small. In addition, some Y cells had both abnormal physiology and morphology, while other Y cells had abnormal physiology but normal morphology. Finally, some cells had normal

X cell physiology and normal Y cell morphology. Friedlander *et al.* (1982) speculate that morphological Y cells initially accept X- and Y-cell retinal inputs, which then compete for postsynaptic space during development. By some (as yet unknown) mechanism, Y inputs takeover during the normal pruning process, but during deprivation conditions, X inputs are favored.

These reported differences between X-like and Y-like cells in susceptibility to deprivation might be related to cell size. Hoffman and Hollander (1978) reported that large LGN cells shrink more than smaller cells following deprivation. On the other hand, Hickey *et al.* (1977) made between-animal comparisons of cells from within the same laminae and found that the percentage change in cell size was the same for the subpopulations of the 10 largest and 10 smallest cells.

There is some evidence that X-like and Y-like cells in primates are segregated into the parvocellular and magnocellular layers, respectively (Marrocco, 1976; Sherman *et al.*, 1976). However, neither Vital-Durand *et al.* (1978) nor von Noorden and Crawford (1978) found clear differences in overall deprivation effects between parvocellular and magnocellular layers. Von Noorden and Crawford did report a timing difference, with parvocellular layers reaching a maximum deprivation effect by 4 weeks after lid suture, while magnocellular layers continued to show additional shrinkage for longer periods.

Other laminar differences have also been noted but their significance is not clear. Both Wan and Cragg (1976) and Movshon and Dursteller (1977) reported that short-term closure has more pronounced affects on layer-A cells than on layer-A1 cells of the cat LGN. These differences are not apparent following longer deprivation periods. Hickey (1980) reported that the small-cell C laminae of the cat are affected less than the A laminae.

B. Effects of Deprivation on Cortex

1. *Synaptogenesis*

Monocular deprivation results in a decrease in numbers of dendritic spines in the hemisphere contralateral to the deprived eye in rats (Fifkova, 1970a; Rothblat and Schwartz, 1979). Fifkova (1970b) used electron microscopy to show that the mean density of synapses in rat cortex receiving input from a monocularly deprived eye was 20% less than on the control side. One caveat is that Fifkova's work does not include correction of density measures for size differences or calculations of synapses per neuron, both of which are necessary to describe synaptic changes accurately (cf. Cragg, 1975). For example, using such methods, Winfield (1981) found that unilateral eyelid closure in the cat did not result in any significant changes in synaptic density for the cortex as a whole. However, the average number of synapses per neuron declined by 30% from normal. O'Kusky and Colonnier (1982) studied monocular deprivation in the monkey and found that eye closure for 6 months' duration did not alter the number of neurons or synapses in striate cortex.

2. *Ocular Dominance Columns*

In the cat, the segregation of geniculate terminals into ocular dominance columns in the cortex appears to be dependent on visual experience since these columns are absent in dark-reared cats (Swindale, 1981). It has been reported that the normal pattern of ocular dominance columns developed in a single, dark-reared monkey (LeVay *et al.* 1980), suggesting that, for this species, visual experience is not necessary for the segregation process.

The process of segregation into ocular dominance columns may be redirected in both cats and monkeys if a period of monocular deprivation is imposed. Following monocular deprivation, the pattern of termination of geniculocortical afferents in layer IV is markedly different from that seen in normally reared animals. For both the cat (Shatz and Stryker, 1978; Thorpe and Blakemore, 1975) and the monkey (Hubel *et al.*, 1977; LeVay *et al.*, 1980), the territory representing the deprived eye is severely decreased, with a corresponding expansion of territory serving the visually experienced eye.

C. Sensitive Periods

Monocular deprivation has no lasting anatomical effects on adult animals. Rather, the effects of monocular deprivation are all associated with sensitive periods during early development. There are sensitive periods during which deprivation can induce the specific deficits and also sensitive periods during which these deficits can be reversed.

1. *Rats*

The period during which cortical spine density can be altered by monocular deprivation in rats begins about the time of eye opening and extends approximately 30 days (Rothblat and Schwartz, 1979). Reverse suture at 45 days of age and lasting for 30 days does not lead to recovery (Schwartz and Rothblat, 1980).

2. *Cats*

The sensitive period for anatomical changes in the cat LGN as a result of monocular deprivation begins in the first month and probably ends following the third month (Garey *et al.*, 1973; Hubel and Wiesel, 1970). The peak sensitivity appears to occur at about 4 weeks after birth, when significant differences in cell size can be produced by lid closures as short as 12 h (Movshon and Dursteller, 1977).

Reverse suture at 5 or 6, but not at 8 weeks of age reverses the effects of monocular deprivation on cat LGN cell size (Dursteller *et al.*, 1976; Garey and Dursteller, 1975). A number of studies found that reverse suture after about 3 months of age does not reverse these cell size changes (Cragg *et al.*, 1976; Dursteller *et al.*, 1976; Garey and Dursteller, 1975; Hoffman and Hollander, 1978; Hubel and Wiesel, 1970; Wiesel and Hubel, 1965). An exception to this finding was reported by Chow and Stewart (1972), who found some evidence for reversal effects instituted as late as 20 months after birth. Their finding is intriguing because of the fact that the cats showing these late reversal effects were ones forced to use their formerly deprived eyes in behavioral discrimination tasks following reverse suture. However, Hoffman and Hollander (1978) failed to get reversal in three cats similarly trained on a pattern discrimination task following reverse suture.

The effects of monocular deprivation on cat LGN cell size can also be reversed by opening the deprived eye and at the same time enucleating the nondeprived eye (Garey and Dursteller, 1975) or crushing its optic nerve (Cragg *et al.*, 1976). These procedures can lead to reversal even after 3 months of age. Furthermore, Spear and Hickey (1979) demonstrated that enucleation or optic nerve crush will produce reversal of LGN cell size deficits even if the deprived eye remains closed. The reversal does not occur immediatcly. Rather, the deprived laminae neurons resume their growth and attain near normal sizes within about 3 months.

Neither the sensitive periods for producing anatomical changes in cortical ocular dominance nor for reversing these effects have been determined in cats.

3. *Monkeys*

In the monkey LGN, monocular closure during the first 3 months after birth produces differences in cell size between deprived and nondeprived layers (Headon and Powell, 1973, 1978; von Noorden and Crawford, 1978; von Noorden and Middleditch, 1975). Slight cell size differences have been found following later closures up until about 7 months after birth. Vital-Durand *et al.* (1978) found no differences in LGN cell size following 6 months of monocular deprivation, beginning after 1 year of age. However, Headon *et al.* (1981) used between-animal comparisons to demonstrate that later closures can cause a shrinkage of monkey LGN cells in both the deprived and nondeprived parvocellular layers. This late period of susceptibility appears to begin in the fourth month, with maximal sensitivity during the sixth to ninth months and declining in the second year. These later changes had probably not been noted by previous investigators because the magnitude of shrinkage is similar in deprived and nondeprived layers, and for this reason is not apparent in within-animal comparisons.

LeVay *et al.* (1980) conducted an extensive study of the sensitive period in monkey cortex. They found that layer IV ocular dominance columns are susceptible to monocular deprivation during a relatively short period spanning the first 10 weeks, with greater sensitivity during the first 6 weeks. Later closures produced no change in the width of the layer IV columns. It is noteworthy that physiological ocular dominance in upper cortical layers can still be shifted by closure at these later ages. Thus, it appears the period of sensitivity to monocular deprivation may be different for the ocular dominance columns seen anatomically in layer IV than for physiological ocular dominance columns seen in other cortical layers.

Reverse suture curing the first 2 postnatal months can reverse the effects of monocular deprivation on monkey LGN cell size (Garey and Vital-Durand, 1981; Vital-Durand *et al.*, 1978). Ocular dominance column shifts produced in striate cortex during the first few weeks after birth can also be altered by reverse suture (LeVay *et al.*, 1980; Swindale *et al.*, 1981). Reverse suturing at 3 weeks can cause true reversal of column sizes such that the originally deprived eye columns become larger than did columns serving the later deprived eye. However this anatomical reversal is limited to layers receiving input from the parvocellular laminae of the LGN. Reverse suture at 6 weeks of age will produce a recovery of the bands in the parvocellular receiving laminae to normal widths but will not lead to a complete reversal. Reverse suture after 1 year of age does not lead to recovery (LeVay *et al.*, 1980).

IV. Binocularly Competitive and Noncompetitive Mechanisms

A. Competitive Mechanisms

Wiesel and Hubel (1965) first proposed an interaction between the inputs from the two eyes, a binocularly competitive mechanism, to account for the physiological effects of lid suture on cortical neurons. Since then a number of investigators have extended this general notion of binocular competition to try to account for the various physiological and anatomical effects of monocular deprivation.

1. *Lateral Geniculate Nucleus Cell Size*

There are now a number of lines of evidence consistent with the hypothesis that changes in LGN cell size result in part from some type of binocular competition. Most of the initial studies found that binocular deprivation had relatively little effect upon LGN cell growth (Chow and Stewart, 1972; Guillery, 1973; but see also Wiesel and Hubel, 1965), and that the monocular segment, a region where binocular interactions are precluded, does not show any significant changes in cell size following short periods of monocular deprivation (Guillery and Stelzner, 1970). Guilllery (1972) created an artificial monocular segment in the LGN by producing a small lesion in the retina of the nondeprived eye. Neurons in this critical segment, which had no cells to compete with from a corresponding region of the normal eye, were of approximately normal size following monocular deprivation while cells in neighboring regions showed the characteristic shrinkage. Similar critical segment results have also been obtained with monkeys (von Noorden *et al.*, 1976).

Guillery (1972) suggested two specific ways that LGN cells might compete. First, there might be direct interlaminar interactions within the geniculate. Second, LGN axon terminals might interact where they terminate in the cortex. LGN neurons that compete successfully and form large cortical arborizations would then grow larger cell bodies to support their larger terminal fields. These two mechanisms are not mutually exclusive, and there is evidence in favor of both.

Many reports have demonstrated a correlation between relative geniculate cell sizes in the deprived versus nondeprived layers and the relative sizes of ocular dominance columns (Dursteller *et al.*, 1976; Garey and Vital-Durand, 1981; Huber *et al.*, 1977; Movshon and Dursteller, 1977; Shatz and Stryker, 1978; Sherman *et al.*, 1974; Vital-Durand *et al.*, 1978). But exceptions to this general rule have also been reported (LeVay *et al.*, 1980); von Noorden and Crawford, 1978). Deprivation or reversal instituted at ages older than 6 weeks can cause ocular dominance changes but have no effects on LGN cell size.

Casagrande and colleagues conducted comparative studies that bear on the issue of where the competitive interactions occur. In the tree shrew, LGN layers 2, 3, and 4 all receive input from the ipsilateral eye (Casagrande *et al.*, 1978). Similarly, in the cortex of the tree shrew the terminal arborizations of layer 3 lie in laminae adjacent to those of 2 and 4. If competition involves interactions between adjacent LGN layers or between adjacent terminal fields in the cortex, then layer 3 should be spared the effects of monocular deprivation. Consistent with these predictions, Casagrande *et al.* (1978) report that monocular deprivation has no effect on LGN cell size in layer 3 of the tree shrew.

Like the tree shrew, *Galago* (bush baby) LGN layers 2, 3, and 4 all receive input from the ipsilateral eye, and the remaining layers (1, 5, 6) from the contralateral eye (Casagrande and Joseph, 1980). On the other hand, *Galago* has ocular dominance columns arranged in the cortex such that terminals from all layers are in close proximity and potentially capable of competing with one another. If competition is between adjacent layers in the LGN, then one might expect cells in layers 3 and 6 to be spared from monocular deprivation as in the tree shrew. On the other hand, if competition is between LGN terminals in cortex, then these layers should exhibit the characteristic deprivation shrinkage. In fact, in *Galago*, layers 3 and 6 show some effects of monocular deprivation but not to the degree seen in the other layers (Casagrande and Joseph, 1980). These results suggest that deprivation shrinkage effects in LGN cells are due to a combination of interactions at both loci.

2. *Cortical Ocular Dominance Columns*

Hubel *et al.* (1977) noted that it is only within the regions of cortex receiving binocular input that the deprived eye's input is reduced. They proposed a potential explanation based on the observation that LGN inputs to layer IV from the two eyes overlap at birth and retract during normal development until they no longer overlap. Hubel *et al.* (1977) proposed that deprivation renders the LGN afferents from the deprived eye unable to compete successfully against afferents from the normal eye for postsynaptic sites during this retraction process. Thus, the deprived eye afferents end up with smaller columns. It is now known that this explanation is inadequate because deprivation effects can still be produced after retraction is complete (LeVay *et al.*, 1980).

Swindale (1980) elaborated on the kinds of competitive interactions between LGN afferents that might be sufficient to account for the formation of ocular dominance columns in layer IV. He proposed a model in which right and left eye geniculate terminals are initially randomly intermixed. Interactions between like synapses (same eye terminals) are assumed to be stimulating over short distances but inhibitory for longer distances. The reverse is true for interactions between synapses of opposite eye type. Computer simulations of this model reproduce many of the morphological features of ocular dominance stripes in normal and monocularly deprived animals. For this model successfully to produce a segregation of terminals according to eye type, LGN terminals carrying signals from the two eyes must somehow be distinguished, that is, labeled and recognized as belonging to different populations.

Ocular dominance outside of the geniculate recipient zone cannot be directly due to this same mechanism because cortical cells are not prespecified as being associated with the left or right eye. The evidence reviewed above demonstrates that the ocular preference of a cortical cell can be modified by deprivation and even reversed by later reverse deprivation. Thus, labeling of cortical cells according to ocular dominance must be determined at least partially by visual experience.

Neural activity has been suggested as one means whereby visual experience could influence the labeling of cortical cells. Several investigators have proposed mechanisms whereby axons having highly correlated neural activity could successfully compete for postsynaptic sites at the expense of less correlated inputs (e.g., Changeux and Danchin, 1976). Such a mechanism might account for some of the deprivation effects on ocular dominance. If monocular deprivation simply decreased the correlated neural activity coming from cells in the deprived eye, then these cells might be put to a competitive disadvantage relative to the more highly correlated inputs from the normal eye (see Chapter 9, this volume).

In summary, it seems that a two-stage process of binocular competition is needed to account for the current deprivation literature. Labeled LGN afferents to the cortex compete for cortical cells in the geniculate recipient zone. Axons from these first-order cortical cells then compete for synaptic connections to higher order cortical cells.

B. Mechanisms That Are Not Binocularly Competitive

In addition to binocularly competitive mechanisms, there is now strong evidence for additional effects of deprivation on LGN size. Hickey *et al.* (1977) reviewed three lines of evidence. First, cats show deprivation effects when reared with one eye deprived after having had the other eye enucleated. Second, small but significant changes in cell size can be found in both the binocular and monocular segments following long-term binocular deprivation. Third, small but significant changes in cell size in the monocular segment of deprived layers of the LGN can be found following long-term monocular deprivation. However, with regard to this last point, Kalil (1980) demonstrated that the magnitude of cell size changes diminishes as measurements are made farther from the border of the binocular segment, suggesting that part of this effect may involve interactions between the binocular and monocular segments.

The evidence for a deprivation effect that does not involve binocular competition is even stronger in monkeys. Monkeys show less difference between binocular and monocular segments (von Noorden and Middleditch, 1975; von Noorden *et al.*, 1976) and long-term deprivation may obliterate the difference. Binocular deprivation also has marked effects on geniculate cell growth in monkeys (Headon and Powell, 1978; Vital-Durand *et al.*, 1978).

Finally, the fact that rats, which have primarily a crossed visual pathway, show effects from monocular deprivation also indicates that deprivation can act through mechanisms other than binocular competition.

V. Summary and Conclusions

We have discussed development as consisting of five stages: (1) generation of neurons, (2) cell migration, (3) formation of connections between nuclei, (4) cell death, and (5) formation and pruning of connections.

The result of these five stages of development is to produce an adult mammalian visual system with a high degree of specificity of neural connections. The specificity of connections is not directly controlled by the genome at any stage. Rather, the genome confers some rules of interaction and the specificity reflects the outcomes of these interactions between developing neurons. The developmental interactions are characterized by initial overproduction and subsequent selective loss of both neurons and connections. Competition is an example of a rule that might be used to govern which neurons and connections are maintained and which are lost.

The nature of developmental competitive interactions has been widely studied by conducting monocular deprivation experiments in which binocular competitive interactions can be manipulated experimentally. The competitive interactions that have been widely studied under the exaggerated conditions of monocular deprivation probably also play a large role during normal development.

Acknowlegments

Preparation of this chapter was supported in part by National Institutes of Health research grant EY03956.

References

Anker, R. L., and Cragg, B. G. (1974). *J. Comp. Neurol.* **154,** 29–42.

Berry, M., and Rogers, A. W. (1965). *J. Anat.* **99**, 691–709.

Boothe, R. G., Greenough, W. T., Lund, J. S., and Wrege, K. (1979). *J. Comp. Neurol.* **186**, 473–490.

Casagrande, V. A., and Joseph, R. (1980). *J. Comp. Neurol.* **194**, 413–426.

Casagrande, V. A., Guillery, R. W., and Harting, J. K. (1978). *J. Comp. Neurol.* **179**, 469–486.
Changeux, J.-P., and Danchin, A. (1976). *Nature (London)* **264**, 705–712.
Chow, D. L., and Stewart, D. L. (1972). *Exp. Neurol.* **34**, 409–433.
Colonnier, M. (1968). *Brain Res.* **9**, 268–287.
Cragg, B. G. (1972). *Invest. Ophthalmol.* **11**, 377–385.
Cragg, B. G. (1975). *J. Comp. Neurol.* **160**, 147–166.
Cragg, B., Anker, R., and Wan, Y. K. (1976). *J. Comp. Neurol.* **168**, 345–354.
Cunningham, T. J. (1982). *Int. Rev. Cytol.* **74**, 163–186.
Cunningham, T. J., Mohler, I. M., and Giordano, D. L. (1981). *Dev. Brain Res.* **2**, 203–215.
Dursteller, M. R., Garey, L. J., and Movshon, J. A. (1976). *J. Physiol. (London)* **261**, 189–210.
Fifkova, E. (1970a). *J. Comp. Neurol.* **140**, 431–438.
Fifkova, E. (1970b). *J. Neurobiol.* **1**, 285–294.
Friedlander, M. J., Stanford, L. R., and Sherman, S. M. (1982). *J. Neurosci.* **2**, 321–330.
Garey, L. J., and Blakemore, C. (1977). *Science* **195**, 414–416.
Garey, L. J., and Dursteller, M. R. (1975). *Neurosci. Lett.* **1**, 19–23.
Garey, L. J., and Saini, K. D. (1981). *Exp. Brain Res.* **44**, 117–128.
Garey, L. J., and Vital-Durand, F. (1981). *Proc. R. Soc. London Ser. B* **213**, 425–434.
Garey, L. J., Fisken, R. A., and Powell, T. P. S. (1973). *Brain Res.* **52**, 363–369.
Gray, E. G. (1959). *Nature (London)* **183**, 1592–1593.
Guillery, R. W. (1972). *J. Comp. Neurol.* **144**, 117–130.
Guillery, R. W. (1973). *J. Comp. Neurol.* **148**, 417–422.
Guillery, R. W., and Stelzner, D. J. (1970). *J. Comp. Neurol.* **139**, 413–422.
Headon, M. P., and Powell, T. P. S. (1973). *J. Anat.* **116**, 135–145.
Headon, M. P., and Powell, T. P. S. (1978). *Brain Res.* **143**, 147–154.
Headon, M. P., Sloper, J. J., Hiorns, R. W., and Powell, T. P. S. (1981). *Brain Res.* **229**, 187–192.
Hendrickson, A. E., and Kupfer, C. (1976). *Invest. Ophthalmol. and Visual Sci.* **15**, 746–756.
Hendrickson, A., and Rakic, P. (1977). *Anat. Rec.* **187**, 602.
Hickey, T. L. (1980). *J. Comp. Neurol.* **189**, 467–481.
Hickey, T. L., Spear, P. D., and Kratz, K. E. (1977). *J. Comp. Neurol.* **172**, 265–282.
Hoffman, K.-P., and Hollander, H. (1978). *J. Comp. Neurol.* **177**, 145–158.
Hubel, D. H., and Wiesel, T. N. (1970). *J. Physiol. (London)* **206**, 419–436.
Hubel, D. H., and Wiesel, T. N. (1972). *J. Comp. Neurol.* **146**, 421–450.
Hubel, D. H., Wiesel, T. N., and LeVay, S. (1977). *Philos. Trans. R. Soc. London Ser. B* **278**, 377–409.
Jacobson, M., and Hirose, G. (1978). *Science* **202**, 637–639.
Johns, P. R., Rusoff, A. C., and Dubin, M. W. (1979). *J. Comp. Neurol.* **187**, 545–556.
Juraska, J. M. (1982). *J. Comp. Neurol.* **212**, 208–213.
Juraska, J. M., and Fifkova, E. (1979a). *J. Comp. Neurol.* **183**, 247–256.
Juraska, J. M., and Fifkova, E. (1979b). *J. Comp. Neurol.* **183**, 257–268.
Kalil, R. (1980). *J. Comp. Neurol.* **189**, 483–524.
Karlsson, U. (1967). *J. Ultrastruct. Res.* **17**, 158–175.
LeVay, S. (1973). *J. Comp. Neurol.* **150**, 53–86.
LeVay, S., and Ferster, D. (1977). *J. Comp. Neurol.* **172**, 563–584.
LeVay, S., Hubel, D. H., and Wiesel, T. N. (1975). *J. Comp. Neurol.* **159**, 559–576.
LeVay, S., Stryker, M. P., and Shatz, C. J. (1978). *J. Comp. Neurol.* **179**, 223–244.
LeVay, S., Wiesel, T. N., and Hubel, D. H. (1980). *J. Comp. Neurol.* **191**, 1–51.
Lin, C. S., and Sherman, M. (1978). *J. Comp. Neurol.* **181**, 809–832.
Lund, J. S., Boothe, R. G., and Lund, R. D. (1977). *J. Comp. Neurol.* **176**, 149–188.

Lund, R. D., and Bunt, A. H. (1976). *J. Comp. Neurol.* **165**, 247–264.
Lund, R. D., and Mustari, M. J. (1977). *J. Comp. Neurol.* **173**, 289–306.
Marrocco, R. T. (1976). *J. Neurophysiol.* **39**, 340–353.
Mates, S., and Lund, J. (1983a). *J. Comp. Neurol.* **221**, 60–90.
Mates, S., and Lund, J. (1983b). *J. Comp. Neurol.* **221**, 91–97.
Mates, S., and Lund, J. (1983c). *J. Comp. Neurol.* **221**, 98–105.
Miller, M., and Peters, A. (1981). *J. Comp. Neurol.* **203**, 555–573.
Morest, D. K. (1969). *Z. Anat. Entwicklungsgesch.* **128**, 290–317.
Morest, D. K. (1970). *Z. Anat. Entwicklungsgesch.* **130**, 265–305.
Movshon, J. A., and Dursteller, M. R. (1977). *J. Neurophysiol.* **40**, 1255–1265.
O'Kusky, J., and Colonnier, M. (1982). *J. Comp. Neurol.* **210**, 291–306.
Parnavelas, J. G., Mounty, E. J., Bradford, R., and Lieberman, A. R. (1977). *J. Comp. Neurol.* **171**, 481–500.
Parnavelas, J. G., Bradford, R., Mounty, E. J., and Lieberman, A. R. (1978). *Anat. Embryol.* **155**, 1–14.
Purves, D., and Lichtman, J. W. (1980). *Science* **210**, 153–157.
Raedler, A., and Sievers, J. (1975). *Adv. Anat. Embryol. Cell Biol.* **50**, 5–88.
Rakic, P. (1971). *Brain Res.* **33**, 471–476.
Rakic, P. (1972). *J. Comp. Neurol.* **145**, 61–84.
Rakic, P. (1974). *Science* **183**, 425–427.
Rakic, P. (1975). *UCLA Forum Med. Sci.* **n18**, 3–39.
Rakic, P. (1976). *Nature (London)* **261**, 467–471.
Rakic, P. (1977a). *J. Comp. Neurol.* **176**, 23–52.
Rakic, P. (1977b). *Philos. Trans. R. Soc. London Ser. B* **278**, 245–260.
Rakic, P. (1981). *Science* **214**, 928–930.
Rothblat, L. A., and Schwartz, M. L. (1979). *Brain Res.* **161**, 156–161.
Schwartz, M. L., and Rothblat, L. A. (1980). *Exp. Neurol.* **68**, 136–146.
Seneglaub, D. R., and Finlay, B. L. (1981). *Science* **213**, 573–574.
Shatz, C. J., and Stryker, M. P. (1978). *J. Physiol (London)* **281**, 267–283.
Shatz, C. J., Lindstrom, S., and Wiesel, T. N. (1977). *Brain Res.* **131**, 103–116.
Sherman, S. M., Hoffman, K.-P., and Stone, J. (1972). *J. Neurophysiol.* **35**, 532–541.
Sherman, S. M., Guillery, R. W., Kaas, J. H., and Sanderson, K. J. (1974). *J. Comp. Neurol.* **158**, 1–18.
Sherman, S. M., Wilson, J. R., Kaas, J. H., and Webb, S. V. (1976). *Science* **192**, 475–477.
Sidman, R. L. (1970). *In* "Contemporary Research Methods in Neuroanatomy" (W. Nauta and S. Ebbesson, eds.), pp. 252-274.
Spear, P., and Hickey, T. L. (1979). *J. Comp. Neurol.* **185**, 317–328.
Stone, J., Rapaport, D. H., Williams, R. W., and Chalupa, L. (1982). *Dev. Brain Res.* **2**, 231–242.
Swindale, N. V. (1980). *Proc. R. Soc. London Ser. B* **208**, 243–264.
Swindale, N. V. (1981). *Nature (London)* **290**, 332–333.
Swindale, N. V., Vital-Durand, F., and Blakemore, C. (1981). *Proc. R. Soc. London Ser. B* **213**, 435–450.
Thorpe, P. A., and Blakemore, C. (1975). *Neurosci. Lett.* **1**, 271–276.
Vital-Durand, F., Garey, L. J., and Blakemore, C. (1978). *Brain Res.* **158**, 45–64.
von Noorden, G. K. (1973). *Invest. Ophthalmol. Visual Sci.* **12**, 727–738.
von Noorden, G. K., and Crawford, M. L. J. (1978). *Invest. Ophthalmol. Visual Sci.* **17**, 762–768.
von Noorden, G. K., and Middleditch, P. R. (1975). *Invest. Ophthalmol. Visual Sci.* **14**, 674–683.
von Noorden, G. K., Crawford, M. L. J., and Middleditch, P. R. (1976). *Brain Res.* **111**, 277–285.

Wan, Y. K., and Cragg, B. (1976). *J. Comp. Neurol.* **166**, 365–372.
Wiesel, T. N., and Hubel, D. H. (1963). *J. Neurophysiol.* **26**, 978–993.
Wiesel, T. N., and Hubel, D. H. (1965). *J. Neurophysiol.* **28**, 1029–1040.
Wiesel, T. N., Hubel, D. H., and Lam, D. M. K. (1974). *Brain Res.* **79**, 273–279.
Wilson, J. R., and Hendrickson, A. E. (1981). *J. Comp. Neurol.* **197**, 517–539.
Winfield, D. A. (1981). *Brain Res.* **206**, 166–171.

11

Experience and Visual Development: Behavioral Evidence

RICHARD C. TEES
Department of Psychology
The University of British Columbia
Vancouver, British Columbia, Canada

I. Introduction: General Perspectives and Methodology

A. Nativism versus Empiricism

Although the classical controversy between nativism and empiricism has profoundly influenced neuropsychological theory, in contemporary research it has become largely a matter of emphasis. What had originally been a philosophical issue, then a matter of psychological dogma, gradually was recognized as a set of empirical problems. It became clear that neither those who believed that the perceptual capacities of the organism were exclusively inherited, nor those who maintained that these capacities were almost all

Developmental
NeuroPsychobiology

acquired, had a respectable collection of relevant facts. The fundamental questions that needed to be answered were What are the perceptual capabilities of a particular species at birth, prior to sensory experience? To what degree did subsequent experiential and maturational factors collaborate to influence further perceptual development (Tees, 1976)?

With the publication of Hebb's *Organization of Behavior* (1949), not only was the significance of early sensory preconditioning as a potential antecedent of mammalian visual development recognized, but also the importance of the technique of deprivation (and enrichment) in helping to answer these fundamental questions. Deprivation of visual input from birth allows investigation of visual competence in an older, visually naive organism whose response capabilities permitted evaluation. The differences (or lack of differences) found in comparison of the deprived animal's behavior with that of an animal reared under normal environmental conditions have been used (rightly or wrongly) to establish the relative contribution normally made by experiential factors in the development of the specific discriminative behavior in question (Riesen, 1961). With respect to those few perceptual abilities that can be measured shortly after birth, this technique could be easily adapted to help establish whether subsequent sensory experience plays any role and, if so, what kind of role. For example, do certain aspects of the visual environment *maintain* or sustain maturing or stable innate competences? Does sensory experience *facilitate*, sharpen, or "fine tune" rudimentary perceptual ability? Finally, do changes in early stimulation history *induce* qualitative and significant shifts in the ontogeny of certain aspects of visual behavior (Ganz, 1978; Gottlieb, 1976; Tees, 1976)?

In addition to the question of initial competences and the general role (if any) subsequent experience might play, specific issues such as the timing, crucial stimulus factor(s), size, generality, and potential reversibility of environmental influence are also subject to further analysis. The paradigm has one other important purpose. In looking at brain and behavior during development, we can use the growth of the brain to provide neural alterations that can be monitored and correlated with behavioral changes (Globus, 1975). Obviously, there are vast numbers of neural and behavioral measures that show similar rates of changes over time. The significance of these correlations is, of course, unclear. Sometimes the relative independence of behavioral and neural variables can be revealed through manipulations such as controlled rearing.

B. General Theoretical Positions

In terms of theorizing about perceptual development and its neural substrate, Hebb (1949) provided an alternative to the early nativist (and empiricist) position by suggesting that mammals at birth possessed basic neural circuitry underlying certain fundamental discriminative abilities, while circuitry underlying others (e.g., perception of shape and form) required further elaboration, gradually established through learning and experience. There were other models of behavioral development with strong biological or neuropsychological orientations (e.g., Gottlieb, 1971, 1976; Kuo, 1967; Schneirla, 1965; Scott, 1962; Thompson and Grusec, 1970) and, beginning in 1959, physiologists began to provide information about the selective responsiveness of neurons within the visual system to specific spatiotemporal arrays of visual stimuli (e.g., Hubel and Wiesel, 1959). A conception of sensory neurons as pattern or feature detectors emerged, and the role played by visual experience and the genome in the development of neural circuitry began to be outlined (e.g., Grobstein and Chow, 1976). Some of those involved in this research argued that the neural connections underlying visual abilities are genetically determined (Wiesel and Hubel, 1974). Such individuals were making predictions about behavioral development that were consistent with the hypothetical outcomes to Figure 1A,B,C. Others (Pettigrew, 1974) argued that genes code only on rough outline and complexities of the final connections are established in response to environmental stimulation (see Fig. 1D) which could involve competition in which certain synaptic connections are selectively stabilized by experience, and others are lost through attrition (see Fig. 1E) when no appropriate stimulation is provided (Lund, 1978; Walk, 1978; Chapters 4, 9, and 13, this volume). Variations on these themes included the idea of two classes of neurons underlying visual competence—one modifiable, one not (Jacobson, 1974)—and the notion that the breadth of any genetic constraints on the influence of experiences (representing the range of acceptable and hence influential environments) might differ, depending on the neural and behavioral characteristics under study (Aslin, 1981; Chapter 9, this volume).

Most of these theories of visual development incorporate the concept of a critical period (e.g., Hubel and Wiesel, 1970). While many of the problems associated with using an invariant maturational schedule as an explanatory concept have been widely discussed previously (e.g., Gottlieb, 1976; Zolman, 1976), one simple point that needs to be made is that a variety of mechanisms can be postulated which would lead to the same observed outcome, for example, a change in visual development resulting from early, but not later, sensory deprivation or restriction. Building on an analogy offered by Bateson (1979), the developing animal could be pictured as a train traveling past an environmental setting. All windows would be closed for the first part of the trip; then, at particular stages in the journey, the windows in certain cars would open, exposing the occupants (representing neural circuitry underlying visual behavior) to the outside world. One version of the process involved is that a little while later those windows would close permanently. This version has been readily accepted by a number of researchers (e.g., Hubel and Wiesel,

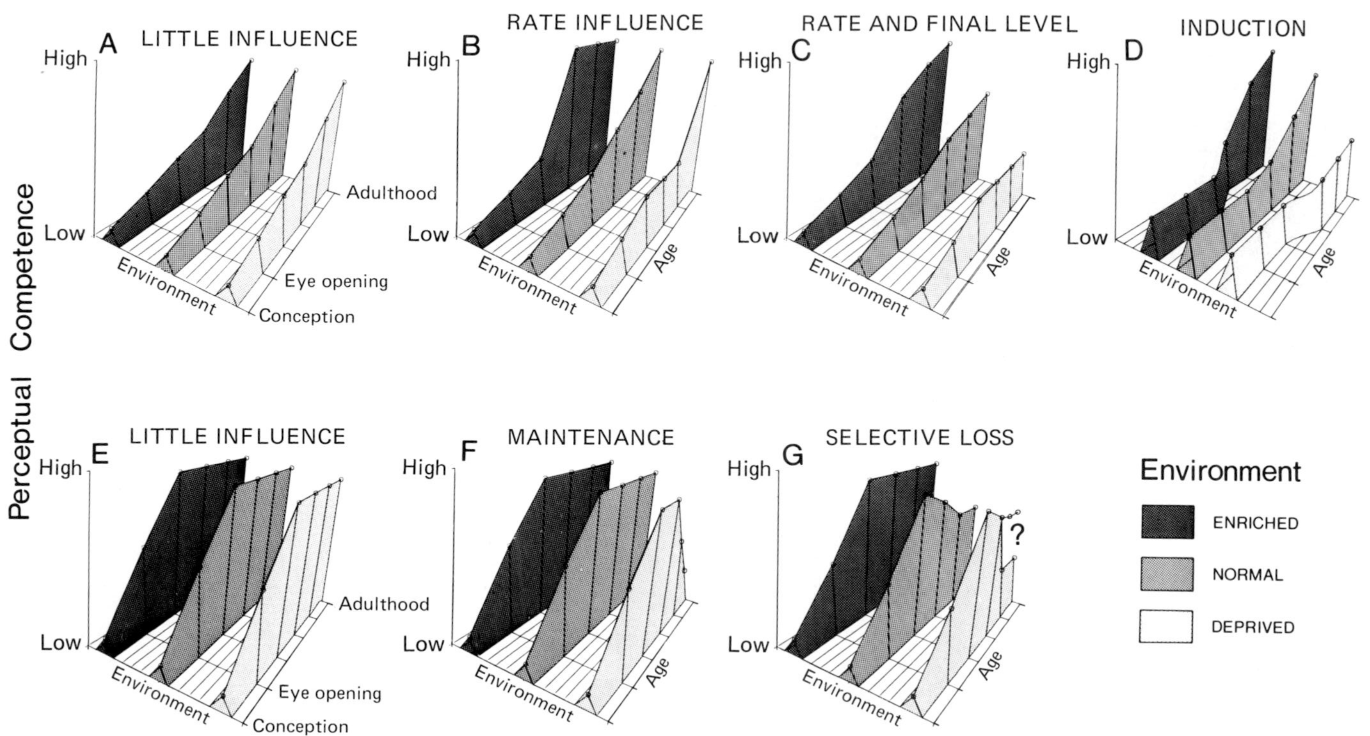

Fig. 1. The role of stimulation history in the development of a specific visual competence (and its possibly related neutral substitute) in the case of an ability either partially developed at eye opening (A,B,C,D) or fully developed at eye opening (E,F,G)—some hypothetical outcomes. Experience might have little effect (A + E), maintain (F), accelerate the rate of development (B), facilitate the final level of competence (C), induce a qualitative change in a particular competence (D), or attune an ability selectively (G), with losses in all but enriched environmental circumstances.

1970) looking at the effect of monocular deprivation (MD) on cortical physiology and visual competence in the cat. However, it is clear now (Timney *et al.*, 1980) that the age at which a cat is susceptible to the effects of MD depends on the animal's history of previous exposure and that period can be extended indefinitely (see Section IIA). The original assumption about the nature of the period was incorrect in this instance. One could argue that, at a particular stage of development, the windows opened and never did shut. At least in part, it was the changes in the neural circuitry and environmental factors, not the timing per se, that gave rise to the observed outcome. Perhaps a more neutral or refined version of the process needs to be entertained until definite evidence is available.

At a more fundamental level, the permanence of changes due to deprivation or restriction that have been attributed to the passing of a critical period should be adequately evaluated and the significance of the results taken into account. A true critical-period hypothesis demands irreversibility of the effects, but many examples of developmental arrest being overcome by later enriched environments have already been reported (Chow and Stewart, 1972; Riesen, 1966), seemingly without much effect on the frequency with which the term *critical* is used. Evidence of environmentally induced plasticity in the visual system of adult animals also is available (Creutzfeldt and Heggelund, 1975; Salinger *et al.*, 1977). Yet Rosenzweig and Bennett (1978) have suggested that the brain changes resulting from interaction with an enriched environment are not comparable to those effected by visual environmental manipulations because the effects of visual deprivation are restricted to critical periods early in life, while the effects of enrichment are not. Obviously, the species and the behavior being examined are important, but for mammals a case can be made for "optimum" periods for development of a number of categories of visual behavior in which timing is important, but only as important as the sequence of stimulation.

If we attempt to examine available evidence as to potential support or rejection of any of the theoretical ideas cited earlier, we see that many investigations are conducted without clear reference to a particular theoretical position and several (e.g., Aslin, 1981; Gottlieb, 1976) have decried the lack of contextual or theoretical framework that could help to guide experimentation and interpret empirical results. However, the theories themselves lack detail and are often ambiguous as to what specific predictions they make regarding the outcome of any particular piece of research (Aslin, 1981; Bond, 1972). In fact, they are not mutually exclusive and are often complementary. However, part of the explanation for their ambiguity lies in the difficulty in specifying what constitutes evidence of a particular competence (see Tees, 1976, for a more complete discussion).

While awareness of the potential complexity of discriminative behavior is a desirable goal in biopsychological research effort, it is particularly important in the case of developmental questions. Careful behavioral analysis has often revealed that earliest visual reactions are different in kind from those of the reactions of older organisms and that these differences are not simply a matter of relative efficiency (Tees, 1976). Moreover, there are significant differences between the encoding processes underlying feature detection and those underlying recognition of form, configuration, or structure (Dodwell, 1970; Hebb, 1972; Sutherland, 1968). We need adequate evidence on each of these perceptual abilities as well as discussion of the kinds of circuitry that would have to be incorporated into any (theoretical) neural networks designed to account for these abilities (Allman *et al.*, 1985; Barlow, 1979; Cavanagh, 1978; Milner, 1974). Statements about the ontogeny of, for example, pattern perception or visual attention and supposedly related brain–behavior relationships may be unwarranted when they are based on faulty assumptions about what is actually being measured by particular techniques.

C. Methodological Considerations

Manipulation of early sensory experience has involved a variety of procedures, each with its own goals, advantages, and disadvantages (reviewed by Tees, 1976; Riesen and Zilbert, 1975). These attempts to control visual stimulation can be roughly subsumed under three headings: deprivation, biased rearing, and enrichment. Most frequently employed has been deprivation in which the objective is to eliminate or substantially reduce light-produced activity (or change) in the visual system. Atypical or biased rearing has been employed often in conjunction with an overall deprivation regime. During a limited daily period of exposure, deprived animals are presented with particular and limited visual arrays (e.g., stationary lines in a single orientation, or lines moving in one direction) or exposure (e.g., prismatic displacement or stroboscopic) conditions.

Enrichment represents a third experimental strategy to test propositions about perceptual and neural ontogeny. If early restrictive manipulation is hypothesized to impede perceptual development, and biased rearing to alter it in a specific way, then enrichment should be examined as to the possibility of accelerating development. While enrichment has been less frequently used in studies of visual development, exposure to regularly changed stimulus objects and shapes has been incorporated into experimental designs (Rosenzweig and Bennett, 1978).

Interpretations of the effects of any of these manipulations are sometimes inconsistent or even contradictory. Differences in the relative maturity of particular species at birth, in life span, in duration of controlled rearing, as well as disparity in measures of behavioral consequences, all contribute to this apparently confusing picture. Many of these problems and considerations have been discussed in detail previously (e.g., Chow, 1973; Tees, 1976; Chapter 10, this volume). I would like to focus on just two examples of potential sources of confusion. The most frequently employed deprivation techniques have been rearing in total darkness from birth (DR) and diffuse light rearing, usually effected by means of eyelid suture. Eyelid suturing has been viewed as depriving the developing visual system of normal amounts of light, all pattern stimuli, and of causing anatomical, physiological, and behavioral effects that have been interpreted as being very similar to those caused by DR. As a matter of fact, several studies have combined the results from DR and sutured animals (e.g., Cynader *et al.*, 1976). Suturing was thought to provide the diffuse light necessary to protect the visual system from peripheral deterioration associated with prolonged DR (Riesen, 1966) and to be more convenient to use than opaque contact lenses or goggles. Recent estimates, however, have suggested the reduction in the amount of light is only one log unit, due to factors such as the high transmittance of young animals' eyelids (Loop and Sherman, 1977). Of more significance is the reported presence of small holes along the suture scars of deprived eyes, which are only apparent upon close scrutiny. These small lid openings turn out to provide sufficient spatial vision to enable a cat to perform effectively on a line orientation discrimination task (Loop and Sherman, 1977). Obviously small eyelid holes might occur in one eye and contribute to a lack of comparable visual input during binocular eye suturing. Even without evidence of these holes, sutured cats are able rapidly to learn a flicker discrimination (Loop *et al.*, 1980) and their striate cortex neurons are responsive to spatial aspects of stimuli during development (Spear *et al.*, 1978). The relative temporal asynchronies in the afferent discharges coming from the two sutured eyes could easily be expected to produce abnormal binocular interactions at the cortex that would not occur under conditions of DR. There is already evidence of abnormal ocular dominance histograms for cells of sutured cats (Kratz and Spear, 1976). Further direct comparisons of prolonged DR versus binocular suture confirm that the two preparations yield striking differences in their initial effects on cortical physiology, behavior, and the recovery possible following subsequent experience (Mower *et al.*, 1981, 1982). Two points are clear. The contradictory results from different laboratories on the recoverability following binocular deprivation are, in large part, explainable in terms of these differences. In spite of the results of these direct comparisons, investigators continue to view the two as yielding equivalent effects.

While most of the speculation regarding the difficulties researchers have had relating the ontogeny of structure to function have focused on the impreciseness of behavioral testing techniques and the resultant poor quality picture of visual functioning (e.g., Blake, 1981), recent evidence has pointed to some methodological problems associated with the collection of physiological evidence about function. Sampling biases arising from differences in soma size of certain classes of cells, differences in their responsiveness, fatigability, and receptive field size can easily result in underestimation of the presence of important types of cells in visual structures of neonates and DR animals. More recent work (Eysel *et al.*, 1979; Leventhal and Hirsch, 1978; Singer, 1978) has shown that sampling bias has led to a somewhat distorted picture of the effect of sensory restriction and biased rearing on the physiology of the visual system (see Section IIA). There is another kind of "Heisenbergian" problem associated with physiological recording. It is possible that young, visually inexperienced cortical neurons mirror in responsiveness whatever stimuli each experimenter presents during an experiment (Pettigrew, 1978). There is certainly evidence that at least temporary and significant changes can be induced during recording (Tsumoto and Freeman, 1981). Uncertainty as to whether the proportion of cells sampled with microelectrodes is representative and what properties result from the highly selective and arbitrary stimulation during the experiment obviously makes it difficult to discuss the relevance of recorded neuronal subsets to morphological changes or impairments of behavior.

II. Specific Visual Capacities and Sensory Experience

In light of these initial considerations and cautions regarding methodology and theory, the following section examines investigations relating to the role played by experience in the development of specific and representative visual functions in mammals, notably those functions mediating acuity, depth, visual guidance, and form discrimination. Since this chapter is not intended to be an all-inclusive review or to provide historical perspective, some topics will get only a cursory examination. The emphasis will be on experiential contributions since the studies reviewed usually involve manipulations of environmental, not of genetic, variables. An attempt will be made to relate physiological and behavioral effects and/or to suggest some direction for future research on these important questions. A number of reviews (e.g., Greenough, 1976) have attempted to assess the available information on the effects of controlled rearing involving social or general mutlimodal restriction, stimulation, and enrichment on the ontogeny of emotionality, learning, and problem-solving abilities and their underlying neural mechanisms; while others (Ganz, 1978; Mitchell and Timney, 1984; Tees, 1976) have provided more exhaustive reviews of the behavioral effect of controlled rearing.

A. Visual Acuity

The role of maturational and experiential factors in the development of visual acuity has been examined quite extensively, particularly in the case of the visually deprived rabbit, cat, and monkey. One of the rationales offered for focusing on the assessment of visual acuity as opposed to other abilities has been the assumption that acuity is more likely to be correlated with response characteristics of single neurons than with other measures of visual functioning (e.g., Timney and Mitchell, 1979). This, as well as other assumptions (e.g, about the relative sensitivity of various behavioral techniques) has been the subject of considerable discussion. For example, assessment of acuity using optokinetic nystagmus (OKN) in the binocularly deprived rabbit (e.g., Collewijn, 1977), cat (Van Hof-Van Duin, 1976a), and monkey (Riesen, *et al.*, 1964) has revealed that up to 7 months of DR cause only slight deficits in visual acuity during initial testing. Even these appear to be eliminated following exposure to normal patterned light for 30 days or less. However, the significance of these reports has been questioned. The OKN response can be reliably elicited only if the striated stimuli fill most of the animal's visual field. Such displays are difficult to arrange without introducing low spatial frequency artifacts. Obviously, any statements about a close correlation between these changes in visual acuity and in response characteristics of single units in the cortex also induced by DR would have to be made cautiously in view of the fact that visual acuity is not adversely affected by removal of visual cortex when it is measured using OKN (Smith, 1937). Regal *et al.* (1976) have also looked at visual acuity using a forced-choice, preferential looking technique. Acuity of 3- to 6-month-old DR macaques (7.5–5.5 cycles per degree) was also found to be close to or within the range of same-age LR controls (12.0–7.5 cycles per degree) after 30 days of patterned light stimulation. Still, the preferential looking method (like OKN) also depends on primitive innate orientation responses that likely depend on different neural pathways (e.g., Y cell/retinotectal) from those we are interested in (e.g., x cell/geniculostriate). Fortunately, there is other evidence on the binocularly deprived mammal's visual ability using assessment techniques with which significant acuity deficits are found following striate cortex ablation (Berkeley and Sprague, 1979). Using a two-choice, food-

reinforced, jumping stand procedure, Timney and Mitchell (1979) report that, for cats DR up to the age of 6 months, the time course of recovery after an initial period of "blindness" very much parallels that observed in maturing controls kept under normal light conditions from birth (see Fig. 1c). The "final acuities" of 4- to 6-month-old DR cats (7.0–5.7 cycles per degree) were also well within the range of normal controls (Fig. 2). [Interestingly, a comparable period of deprivation by binocular eyelid suture leaves the cat less able to resolve detail (3.0–3.7 cycles per degree), underlining the point made earlier about significant differences between the effects of the two deprivation regimes (Mower *et al.*, 1981; Smith, 1981a).] The extent of recovery for cats DR for 12 months or longer is not yet clear. As in the case of an early study of the effect of deprivation of two chimpanzees deprived of patterned light for 18 months, there is evidence that irreversible and substantial loss of function may result from prolonged deprivation (Riesen, 1966). Cats reared in darkness for 1 year were not able to achieve acuities greater than 2 cycles per degree, even after more than 4 months in a normally illuminated environment (Timney and Mitchell, 1979). This is in agreement with recent physiological findings (Chapter 9, this volume) that time-dependent changes (i.e., degradation) do take place in the visual system in the absence of visual input over a prolonged period. Boothe (1981), using a variety of methods, has tracked the course of improving acuity in normal monkeys. Acuity development seems to improve steadily over the first 9 months of age and be highly correlated with gestational rather than postnatal age. This seems to indicate that, within a range of relatively normal, mildly restrictive visual environments, acuity may largely be determined by physical maturation with no particular period being "critical."

The fact that 6-month-old DR cats and monkeys show evidence of reasonably good acuity within 2 months of entering a normally illuminated environment raises several important questions. Perhaps the possibility that the more serious consequences of DR are in respect to an animal's responsiveness to visual input, and not to its ability to resolve detail (see Section IIC). Alternatively, if we view this as representing substantial recovery of the competence itself, it suggests that the plasticity of the part of the sensory system in mammals underlying acuity judgments is not lost after the first 3–4 months of life, at least not in visually naive DR animals. Finally, if experience (or rather lack of it) had modified the connections of neurons in those parts of the CNS that are necessary for basic, local feature analysis of the environment and if this modification was substantial and permanent, these basic deficits would have important implications for any further evaluation of other visual competences.

With respect to the cat, there is some evidence of physiological recovery

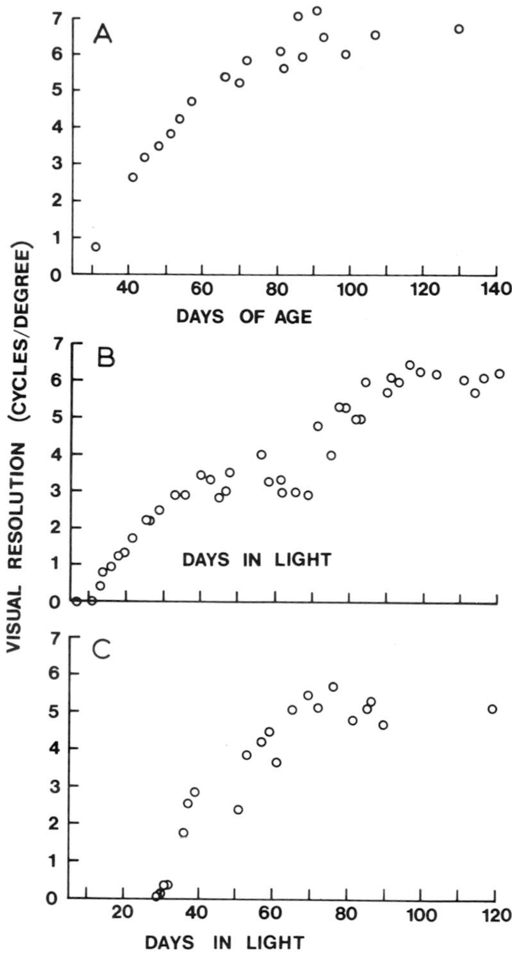

Fig. 2. The development of visual acuity (as measured on the jumping stand) in cats raised under different conditions of visual deprivation. (A) Normal kittens raised in a lighted environment. (B) Cats dark-reared until 4 months of age. (C) Cats dark-reared until 6 months of age. Adapted from Timney and Mitchell (1979).

to parallel its behavioral recovery (Cynader *et al.*, 1976). Physiologically, the visual cortex of a 4- to 6-month-old DR cat initially resembles that of a naive kitten with many poorly responsive cells as well as a few cells with normal response characteristics, including those of normal, simple, as well as perhaps complex and hypercomplex cells (Cynader *et al.*, 1976; Mower *et al.*, 1982; Wiesel and Hubel 1965a). After a 6-month period of normal patterned light, the number of cells exhibiting orientation selectivity increases substantially. While the ocular motility during the period in the light is necessary for recovery of orientation sensitivity, active body movement is not (Buisseret *et al.*, 1978). Overall, the evidence that is available suggests that both the physiological (Mitchell, 1981) and behavioral consequences (as far as acuity is concerned) of 6 months of DR have been somewhat overestimated and for degree of recovery over a subsequent period of 1 to 6 months in the light underestimated (e.g., Wiesel and Hubel, 1965b). There are several reasons why this may be the case. First, binocular suture was usually the technique employed to deprive the animals. Second, the deprived animal initially failed to respond to visual input (i.e., it appeared to be blind) and continued to do poorly, if inappropriate measures of its ability were taken. What was needed were methods that motivated LR and DR animals to perform at the limits of their ability. For example, Blake and Di Gianfilippo (1980) reported that DR cats after 6 months in the light were still less able than LR cats to resolve detail (with cutoff frequencies of 2 versus 4 cycles per degree). However, using a conditioned suppression technique, their LR cats were resolving detail at a full octave below that reported by Timney *et al.* (1980) for normally reared cats (6 cycles per degree). Strangely enough, Blake and Di Gianfilippo report that their DR cats ignored the stimuli and were even observed to lick for reinforcement with both eyes closed. They seemed impressed that they were able (despite these problems) to obtain reliable threshold estimates (and virtually the same data points) for the two DR cats on repeated testing.

The consequences of monocular patterned light deprivation (MD) for the cat and monkey appear more severe and persistent than binocular deprivation (BD) with respect to measures of acuity through the deprived eye (e.g., Ganz and Fitch, 1968; von Noorden *et al.*, 1970). This is consistent with the general proposition that asymmetrical exposure to significant visual input produces a *competitive disadvantage* for the deprived eye in terms of its developing neural substrate. This leaves the eye in a poorer position in terms of visual capacity than if the light deprivation had been more complete but symmetrical (e.g., Chow and Stewart, 1972). In effect, recovery is slower in the case of MD simply because it involves the animal's overcoming the effects of this earlier competitive neuronal interaction as well as the consequences of disuse. However recent evidence has served to highlight the idea that the severity and permanence of the consequences of early MD (Wiesel and Hubel, 1965b) on basic discriminative ability may have been exaggerated. While several investigators (e.g., Hubel and Wiesel, 1970) have reported that even brief periods (e.g., 10 days) of MD during the "most sensitive" period of cat's visual development (30–60 days of age) results in substantial deficits in acuity and changes in cortical physiology with respect to the briefly deprived eye, this finding hinges on the history of the animal and its overall stimulation. For example, 4 h daily of MD through the entire period has little or no impact, provided the animal is being raised with normal patterned light during the remainder of each day (Olson and Freeman, 1980). Recovery from brief periods of MD does occur even in the dark, albeit more slowly (Freeman, 1979).

The effects of continuous, early MD on acuity has been more extensively and carefully studied in the case of the cat (than the monkey or rodent). Kittens, initially deprived monocularly for 45 or 60 days, attained nearly normal grating resolution (4–6 c/deg) through their deprived eye over a 60-day recovery period (Mitchell *et al.*, 1977). While this recovery is facilitated by reverse suturing in which the initially deprived eye is artificially placed at a competitive advantage, recovery is also observed in the case of kittens whose deprived eyes are opened without such reversal (Fig. 3).

This behavioral recovery is also paralleled by marked physiological recovery. For example, a substantial proportion (50%) of neurons recorded from these animals could be influenced through the deprived eye after the period of light exposure. Longer periods of MD result in slower (Fig. 3) behavioral and neural recovery (Giffin and Mitchell, 1978). Removal of the nondeprived eye (rather than reverse suture) in adult cats and monkeys that have been MD earlier also results in an immediate and sizable additional increase in the percentage of responsive cells and parallel improvements in both the rate of behavioral recovery and the final activities achieved (Hendrickson *et al.*, 1977; Smith, 1981b). While this is reassuring to those investigators who are interested in making a simple connection between neural and behavioral indices, several observations need be made about this connection. First, there is no particular reason to suspect that the total *number* of visually responsive neurons (or even the number of orientation sensitive units) "connected" to an eye should provide the physiological basis for visibility or acuity measured through the eye. It clearly does not in the case of the hamster (Emerson, 1980), whose acuity is not related to the number of orientation-selective cells in the visual cortex responsive to a particular line orientation. It is more likely that the spatial resolving properties of the most sensitive units would be more closely related to the animal's ability. Unfortunately, most studies have not attempted to measure the spatial resolution of these cortical units. Moreover, if one proposes that sustained X retinal ganglion cells in the area centralis, resulting in activation of small receptive field cortical cells, represent the basis for the upper limits of an animal's acuity, then these cells exhibit a relatively low degree of binocular activation (Leventhal and Hirsch, 1980). Even a small difference in eye

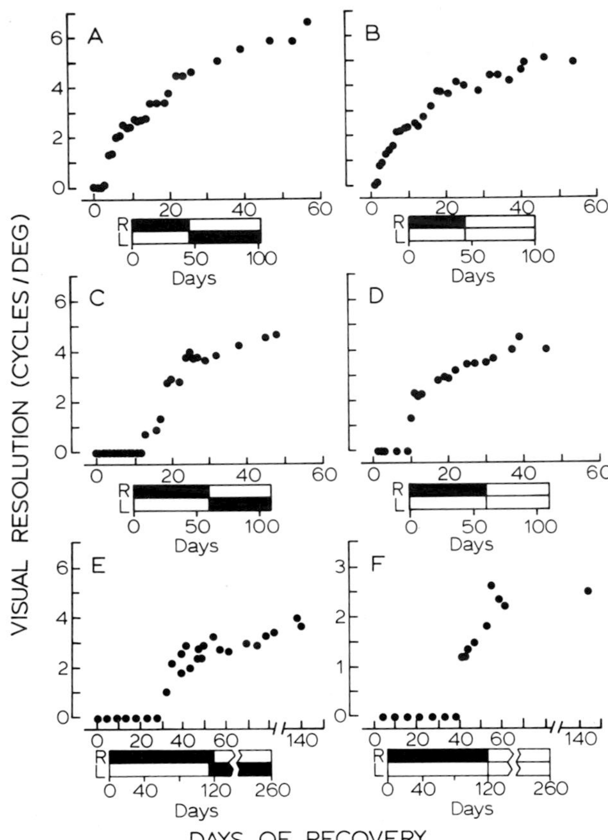

Fig. 3. Comparison of the rate of recovery of vision in the deprived eyes of six animals whose early visual history is depicted schematically below each graph. (A) and (B) show recovery of acuity in the deprived eyes of kittens monocularly deprived to 45 days of age; (C) and (D) to 60 days of age; and (E) and (F) depict recovery following monocular deprivation to 120 days of age. At the termination of the deprivation one member of each pair (A, C, and E) was reverse-sutured while the other (B, D, and F) received binocular visual input. Adapted from Giffin and Mitchell (1978).

alignment, possibly within the range commonly observed for normal animals, may be sufficient to limit the degree of binocular integration displayed by extremely selective units with small receptive fields. Until recently, physiological investigation has underestimated (and underexamined) the population of small receptive field, X-cell projection systems (Stryker and Shatz, 1976). The best of the investigations (e.g., Hoffman and Sherman, 1975; Leventhal and Hirsch, 1980) seems to suggest that MD has less impact on these units than on other classes of cells. As a consequence, if the physiological basis of acuity close to threshold reflects the operation of this system of units, then improvement in acuity following MD seems a reasonable prospect, since the competitive disadvantage is less than it might be. Mower and Christen (1982) observed that 12 months of MD in cat does result in a loss of sensitivity in X cells of the deprived laminae of LGN but only at higher spatial frequencies. Interestingly, unlike several studies of the resolving abilities of LGN cells, the spatial resolution of the best of the LGN cells they recorded from (5–6 cycles per degree) was similar to that a normal cat exhibits behaviorally at that age. Moreover, while the distribution for the deprived eye was skewed in comparison to that of the nondeprived eye, one-third of the deprived eye's cells showed spatial resolving capacities between 2.5–5.5 cycles per degree.

While it seems reasonable to argue that 6 months of BD might "freeze" or retard the development of some cortical functions to the stage reached early in development, the magnitude of the initial deficit in behavioral performance following MD is greater than if its effect were simply to arrest development of acuity at its level at the time of occlusion. Giffin and Mitchell (1978) have shown the acuities of kittens after 23 days of MD were always less than those achieved prior to this period of occlusion (if the MD period took place within 2 months of age). This decline may reflect an impairment best related to visually guided behavior (see Section IIC for further discussion) rather than to acuity itself.

The studies on the neural consequences of MD in animals have suggested that expansion of the nondeprived eye's neural representation accompanies the atrophy of that of the deprived eye (see Chapter 9, this volume). The only behavioral support for the idea that this expansion might be accompanied by supernormal resolving ability is that of Freeman and Bradley (1980), who reported the nondeprived eye of monocularly deprived humans had a supernormal vernier acuity.* It could be that the difference reflects a change in visual responsiveness (see Section IIC) rather than basic resolving ability. It is also plausible that vernier offsets are detected by noisy neural arrays in which additional neurons (as opposed to neurons with high spatial frequency cutoffs) would result in increased sensitivity.

One of the central concepts to emerge from early studies of MD has been that of the critical period. Persuasive evidence was offered that indicated

*Their evidence would be more persuasive if they had utilized highly practiced subjects and obtained the level of "hyperacuity" that can be observed in this situation (e.g., Levi and Klein, 1982).

that the critical period for susceptibility to MD began when a kitten was 3 weeks of age and continued until the animal was 3 months old (Blakemore and Van Sluyters, 1974; Hubel and Wiesel, 1970). In contrast to the profound effects of deprivation instituted during this period, MD initiated in adulthood was shown to have little effect; other environmental manipulations (e.g., surgically induced strabismus) seemed to cause marked changes only if the manipulation occurred before 3 months. As mentioned earlier, this interpretation has been weakened by a number of findings. For example, the evidence suggests that the effects of early MD on acuity and cortical units are not permanent (e.g., Cynader *et al.*, 1976). More recently, results of Cynader and co-workers suggest that the critical period for the effects of MD may be prolonged, apparently indefinitely, by rearing cats in the dark from birth (Cynader and Mitchell, 1980; Timney *et al.*, 1980). The behavioral and physiological effects of a 6-week period of monocular occlusion imposed on cats DR from birth to 6 months of age or longer are much more pronounced than those observed after similar periods of occlusion imposed on LR animals or binocularly sutured animals (Mower *et al.*, 1981). Effects were also observed in the case of 4-month-old DR animals who had undergone periods of monocular suture after a second period of 1 or 2 months of binocular vision. During this period of binocular vision, acuity had developed to a level of 3–4 cycles per degree. When tested after 6 weeks of subsequent MD, it was much poorer initially (0.1–1.0 cycles per degree) than prior to monocular occlusion. Two months of further binocular experience resulted in its acuity (4.5–5.0 cycles per degree) being still somewhat lower than the other eye. In this instance, both an animal's age and its entire history of exposure play a significant and complex role in its sensitivity to environmental manipulation.

Recently Mitchell and his co-workers (Mitchell, 1985; Mitchell and Murphy, 1984) have reopened the question of the permanence of the physiological and behavioral indices of visual recovery resulting from reverse occlusion. Surprisingly, careful analysis revealed that this recovery was extremely labile. When kittens were given normal binocular experience after 4 to 19 weeks of reverse occlusion, the gains of the initially deprived eye were lost within a few weeks. Moreover, the acuity of the originally nondeprived eye never actually returned to the level that it had achieved prior to its occlusion. While this evidence is somewhat disconcerting for advocates of several positions about the role (or lack of it) of sensory experience and for clinicians using conventional patching in the case of human amblyopics, it does offer a paradigm in which different regimes might be evaluated. As it has turned out, distributed rather than massed reverse occlusion has proven to be extremely effective in terms of cortical and behavioral recovery. Brief daily periods of forced visual exposure of initially deprived eye embedded in periods of normal binocular vision proved to promote permanent recovery with respect to the initially deprived eye while not compromising the visual ability of the nondeprived eye. This work suggests several ideas about how one might determine the nature of the physiological changes occurring during MD and reverse occlusion. For example, if the loss of binocularity in the visual cortex in the MD cats is actually more severe in area 19 than in areas 17 or 18 (Leventhal, 1984), and if MD affects intracortical connections more than geniculocortical ones to area 19, one could directly test this by studying binocularity and acuity in area 19 of normal and MD cats in which area 17 has been inactivated. Most importantly, Mitchell's studies (e.g., 1985) reveal the importance of clever and systematic studies of environmental conditions that might be optimum to generate the recovery. Too often researchers have measured the effects of MD under a limited set of circumstances and drawn strong conclusions about the general issue of the malleability of neurons and functions in respect to any environmental manipulation.

A number of investigators have looked at acuity after biased rearing. Their purpose was to test notions concerning the modifiability of neurons in the visual cortex (see Chapter 9, this volume) and potentially to develop animal models of clinical disturbances in human spatial vision. For example, early alternating monocular occlusion reduces the proportion of binocular neurons in the cat without affecting the numbers of neurons driven by each eye, but the perceptual deficit that follows is not a reduction in acuity (Blake and Hirsch, 1975). Induced uniocular exotropia (divergent squint) also does not yield deficits in acuity in cats or monkeys (Hubel and Wiesel, 1965; von Noorden, 1977) unless the squint angle is extreme, i.e., 80° (Singer, 1978).* On the other hand, exotropia (uniocular convergent squint) has been shown to result in apparent amblyopia in both the cat (Hubel and Wiesel, 1965; Jacobson and Ikeda, 1979) and monkey (Boothe, 1981) if the squint is induced within the first 2 to 3 months.

A theory of developmental arrest (rather than degeneration) appears to

*Focusing on the effects of surgical eye rotation in cats, Singer (Chapter 9, this volume) has offered physiological evidence that experience-dependent changes in neuronal connections during early development require the target cells to respond to the retinal signals, something that does not occur under conditions of extreme environmental alteration. Interestingly, in spite of the lack of neural reorganization under these extreme conditions, behavioral adaptation takes place in terms of a kitten's ability to detect differences in the orientation of lines seen through the rotated eyes (Mitchell *et al.*, 1976). Timney and Peck (1981) have reported that while large unilateral and bilateral rotation of eyes do yield acuity deficits, the magnitude of these deficits seems to be much less than that that would be predicted by Singer's observations on cortical physiology.

find support from the results of the studies of the development of spatial resolution by normal kittens (Mitchell *et al.*, 1976) and by kittens with convergent squint initiated at different ages throughout this early period of susceptibility. There is a clear relationship between the age of onset of squint and the relative degree of loss of spatial resolution in the central 10° of the visual field (Jacobson and Ikeda, 1979). The absolute timing of the occurrence of this period of susceptibility is, of course, dependent on previous stimulation history but, given the same environmental circumstances, the period of effectiveness for strabismus may end earlier than that of MD.

The developmental plot of spatial resolving properties of the X-cell system of the cat's squinting eye against age-of-squint initiation is also highly consistent (see Fig. 4). Kiorpes and Boothe (1980) have examined the time course of the development of acuity following surgically induced estropia in monkeys. If the surgery is performed early (6 days postnatally) the acuity of the deviating eye develops normally during the first 4 weeks after surgery, then either improves at a slower rate (one animal) or not at all (one animal). Their data do not reveal whether the timing of the delay or the appearance is invariant (or not). In cat, Von Grunau (1979) has also reported a continued period of normal development after onset of estropia before acuity differences emerge and speculated that it was triggered by the changes produced in neural circuitry rather than an invariant schedule. Jacobson and Ikeda (1979) have suggested that the poor acuity is the result of the habitual exposure to blurred images leading to nondevelopment of the X cells of the squinting eye, and there are both physiological and behavioral data supporting this proposition. For example, Bennett *et al.* (1980) attempted to examine this question directly by inducing squint optically (by means of prism goggles). Ocular motility and lack of fusion would occur under these conditions and yet the images, though conflicting, would be of equal quality. Most of the prism-reared subjects, while developing mild convergent misalignment, showed equal and normal acuity in the eyes. The behavioral data on monocular contrast sensitivity in cats with conventionally caused squint suggest gain that the disturbances of vision involves more than acuity (see Section IIC). Contrast sensitivity for a wide range of spatial frequencies is impaired (Jacobson and Ikeda, 1979). We do not as yet have systematic behavioral data on the potential for recovery following early squint in infrahumans. However, in infants with squint, acuities of the deviating eyes can be improved with reverse patching through the first 2 years (Held, 1981).

Many of the early studies of biased rearing employed cats whose eyes were selectively exposed to a field of either vertical or horizontal striations for daily periods of several hours during the first few months or weeks of life (e.g., Blakemore and Cooper, 1970). Cats reared during the first few weeks

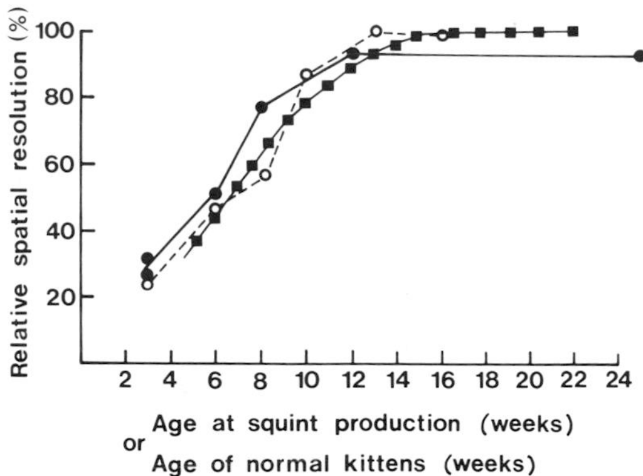

Fig. 4. Comparison of the visual acuity of adult cats with convergent squint at different ages with that of normal kittens of different ages. The spatial resolution of the squinting eye (relative to the nonsquinting eye) is presented as a fraction of age of squint initiation. The neurophysiological data of the relative spatial resolution of X cells in the LGN driven from the area centralis of squinting eyes of similarly reared adult cats are also displayed. (●—●) Squinting cats; (○---○) squint LGN cells; (■—■) normal kittens. Adapted from Mitchell *et al.* (1976); Ikeda *et al.* (1978); and Jacobson and Ikeda (1979).

of life in environments in which only contours in a single orientation were presented showed diminished ability in later testing to resolve striations of the orthogonal orientation (Hirsch, 1972; Muir and Mitchell, 1975). Several points should be made about these results. As yet no study has attempted to test recovery of function with deliberate reversal of contour orientation or other special exposure conditions. Although the perceptual deficits are persistent in the sense that they remained after a year of exposure to normally lighted colony environments, they are considerably smaller than one would expect from the neurophysiological evidence. Biased exposure seemed to result in an absence of cortical neurons having a preferred orientation perpendicular to that experienced. Yet the consequence of early selective rearing was not blindness for lines perpendicular to the contours presented in the initial environment, but only a slight deficit in acuity comparable to that of an optically corrected human astigmatism.

Reasonable interpretation of these findings (Tees, 1976) was that the orientation-sensitive units, whose receptive fields were dramatically altered, had little to do with the function (acuity) in question. Another is that the units that were recorded from in the initial studies were not representative of the total population. Clearly, considerable support has accumulated for both possibilities. Leventhal and Hirsch (1977) reported that such biased rearing has less impact on the (earlier overlooked) small receptive field, X-cell system and layer IV neurons in the striate cortex (see Chapter 9, this volume).

Finally, the literature on the limited comparable human data suggests that scattering of light and obscuring of the retinal image by congenital cataracts do not prevent the development of foveal fixation and reasonable levels of acuity after removal of cataracts later in life. The extent and time course of the reduced visual acuity associated with visual anomalies such as strabismus and the recovery resulting from prolonged occlusion of the dominant eye is comparable to that obtained in MD cats and monkeys (Flom, 1970; Mitchell and Timney, 1984). Preliminary results of a treatment program in which exposure to high-contrast gratings rotating in front of the amblyopic eye was used reveal that 21 min of such exposure yields large, extremely rapid improvement of the formerly amblyopic eye's acuity (Campbell *et al.*, 1978) in children as old as 10 years of age. Such results, if verified by further work, suggest that the acuity deficits caused by squint are less severe (in terms of subsequent modifiability) if appropriate visual input is provided. According to Campbell *et al.*; the optimum input at this stage of an individual's stimulation history does not represent a normal environment (see Section iii).

Evidence of modification of human visual systems by early experience due to meridional amblyopia caused by astigmatism is also available. Optically corrected astigmatic subjects whose early visual environment was one in which contours in a particular orientation were imaged considerably less sharply than others were found to show slight but significant acuity deficits similar to those found in cases of biased reared cats (Mitchell *et al.*, 1973). While the parallel between the effect of selective (stripe) rearing in kittens and that occurring in humans with meridional amblyopia seems a reasonable one (Mitchell, 1978), it is interesting that the majority of all infants up to the age of 1 year show astigmatism of one or more dioptics. The incidence of astigmatism does not decline to adult levels until after the age of 2 years. Obviously, throughout this period, these infants are being exposed to poorly focused images, yet meridional amblyopia does not develop in the majority of cases. At the very least, we must conclude that this form of amblyopia "develops" after 1 or 2 years in humans and it is aberrant input after this time, and not before, that is significant as far as acuity is concerned. Work in this same laboratory (Held, 1981) indicates that the development of amblyopia resulting from binocular anomalies follows a different timetable. Strabismic amblyopia appears to be present during the first year. Other evidence (e.g., Levi and Klein, 1982) also points to differences in the timing and nature of the visual deficits (amblyopia) caused by symmetrical and asymmetrical aberrant visual input (see Chapter 9, this volume).

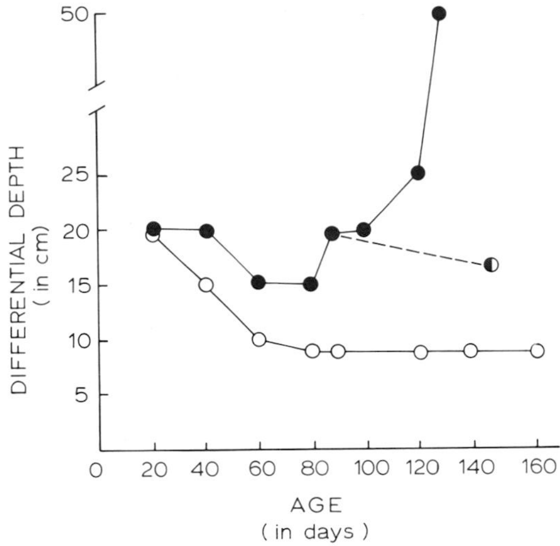

Fig. 5. Differential visual depths at which at least 75% of the light-reared (○–○), dark-reared (●–●), and dark-then-light-reared (◐–◐) rats descended to the optimally shallow sides.

B. Depth Perception

A great deal of research has been carried out especially using the visual cliff apparatus (reviewed by Walk, 1978); on the development of depth perception in mammals and the role played by experience. Early experiments by Walk and Gibson (1961) indicated that the rat, at eye opening and after DR until 90 days, could discriminate depth on the cliff. They initially proposed that depth perception in the rat (unlike other mammals) was innate, requiring neither experience nor maturation beyond eye opening.

None of these early investigations attempted systematically to study performance on the cliff as a function of age and all used a deep side in which the deep-side pattern was 50–135 cm below the glass surface, not a very sensitive measure of ability. Systematically testing naive DR and LR rats on a modification of the cliff in which the depth of the deep side could be varied, we have demonstrated that the ontogeny of the rat's depth perception was also a result of a collaboration of a number of sources of development (Tees, 1974). The influence of an innate component is reflected in the ability of the 20-day-old rats to discriminate depth on the cliff at 20 cm of differential depth (see Fig. 5). The contribution of maturation and nonvisual experience can be seen in the improvement by older DR rats. The influence of visual experience perhaps can be said to take two forms: the first (facilitative or sharpening), as evidenced in the decreasing differential thresholds obtained by the maturing LR rat; the second, a maintenance role, the lack of which is seen in the deterioration in performance after prolonged DR. In the case of the rat DR until 90 days of age, we found that 60 days of subsequent exposure to patterned light ameliorated the serious decline in performance on the visual cliff that is normally observed after continuous DR (Tees and Midgley, 1978). The visual experience not only protected the performance from decline but also yielded some evidence of improvement or partial recovery. Such a period of visual experience yielded recovery at most differential depths, but not at the 9-cm level, close to the threshold in this particular situation (see Fig. 5). This difference might not reflect a difference in perceptual ability per se. Performance on the cliff depends on the ability to see the difference between the two sides and to "care" about the difference (see Section IIC).

In other mammalian species, recovery (or occurrence) of depth discrimination has been shown to require several weeks of patterned light experience after initial testing in the case of DR young and adolescent rabbits, cats, and monkeys (see Walk, 1978). In terms of functional recovery after an extensive period of early DR, Van Hof-Van Duin (1976a) reported that binocularly DR cats were indistinguishable from normal cats on the cliff after 10 weeks of visual experience following an initial period of 7 months of DR. Van Hof-Van Duin's cats were given a somewhat enriched postdeprivation period that may have accelerated recovery, and other findings also suggest that the quality as well as the amount of timing of experiential therapy has an influence on recovery (Baxter, 1966; Eichengreen *et al.*, 1966). However, several investigations (Kaye, 1982; Timney, 1981) have found that the monocular depth perception of 4-month-old DR cats to be very much comparable to those of normal animals after they have spent 3–4 months in normal light.

It was also unfortunate from the point of view of understanding recovery that the earliest study of DR cats that looked at near-threshold depth discriminative ability employed 6 months of (alternating) MD, and persistent deficits were found in binocular (but not monocular) depth discrimination even after 2 years of patterned-light experience (Blake and Hirsch, 1975). Since the only evident (then) physiological consequences of alternating occlusion was a permanent and substantial loss of cortical cells responsive to either eye, it was reasonable for Blake and Hirsch to conclude that these binocular neurons were crucially involved in stereoscopic depth perception. However, some element of doubt lingered about this conclusion (see Packwood and Gordon, 1975). The slight interocular misalignment itself resulting from alternating occlusion could have been sufficient to prevent stereopsis on these tests themselves and hence this could have been the factor influencing performance. More importantly, the proportion of binocular cells in a binocularly DR cat was actually not very different from that encountered in LR controls and such cats also seem to suffer a permanent loss in binocular depth perception (Kaye *et al.*, 1982).

The best evidence on the potential recovery (or lack of it) of the electrophysiological consequences of DR is that from the cat (see Chapter 9, this volume). Several studies have indicated partial recovery in DR cats with 6–12 months in normal light. Hoffman and Cynader (1977) reported that the behavioral recovery of visual cliff behavior reported by Van Hof-Van Duin (1976a) seemed to be attended by a reversal of the physiological changes in LGn cells found after 4 or 7 months of early deprivation. Cynader *et al.* (1976) also reported that cortical units showed partial recovery of orientation selectivity even after 1 year of visual deprivation, but directional selectivity and binocularity remained permanently impaired. On the basis of their findings, Cynader and his co-workers suggested that they have some evidence of subsequent gain in specificity with "therapeutic" visual experience to correlate with evidence of behavioral recovery. Perhaps, but I do not believe that it is the right evidence. We would expect animals exhibiting initial deficits (and recovery) in depth perception to lack cortical or collicular neurons that respond selectively to increasing or decreasing size or to some other depth-related cue. Regan *et al.* (1979) outlined a set of neural mechanisms underlying discrimination of objects in depth and tried (successfully in the LR cat) to discover single units in the cortex with appropriate receptive field properties. Changes in these kinds of neurons in DR animals should be the focus of any physiological study that attempts to relate neural and behavioral mechanisms in this particular instance. One possibility worth considering is that the binocularly DR and MD animal has the (monocular) neural circuitry to detect objects at depth, and the apparent period of recovery reflects changes in the substrate underlying visual responsiveness in general.

The story on brain–behavior relationships with respect to stereoscopic depth perception has cleared up recently. The emergence of binocular depth perception in kittens (measured behaviorally) coincides precisely with the maturation of cortical disparity-tuned neurons (Timney, 1981). The DR cats who fail to develop stereoscopic depth perception even after 4 months in the light also fail to develop precise tuning for retinal disparity in the case of their relatively large number of binocularly excitable cortical cells (Kaye *et al.*, 1982). Both BD and MD (Timney, 1984) during the first 4 months seem to affect retinal disparity and stereoscopic depth perception permanently. The sudden onset and rapid development of the same capacity was found in a recent study of infants with anomalous visual experience (Held, 1981) whose abilities also seemed to be highly susceptible to disruption between 2 to 5 months of age. The only paradox remaining may be that this does not fit well with the idea (Mitchell, 1981) that stereoscopic vision should continue to remain plastic while the young organism's head and body continues to change in size.

Held and Hein (1963) showed that depth perception in the cat not only depends on patterned visual experience, but that this exposure must be concurrent with an systematically dependent upon self-produced movement. It is not clear with respect to performance on the visual cliff whether or not a distorted relationship between self-produced movements and visual stimulation has an especially disruptive effect. The restricted (i.e., passive) cat is learning during its daily exposure to light that its own movements are unrelated to perceived motion, lines, and so on, in the environment. It is conceivable that performance differences may be directly related to this aberrant association (and neglect) rather than to a lack of ability to distinguish depth cues. Consequently, we would expect to find little change due to the exposure history in the physiological mechanisms described above (Regan *et al.*, 1979). The adjustment would not be in the visual system itself but would involve visuomotor changes or recalibrations of position sense (i.e., map) of various parts of the body.

C. Visually Guided Behavior

One of the recent trends in analysis of the function of the visual system has been the partial dissociation of those neural mechanisms underlying visual information processing, "perception" or discrimination, and those underlying visually elicited orientation or locomotion (Schneider, 1969), the geniculocortical pathway being identified with perception and the retinotectal pathway with orientation.

Normal adult vision involves both orientation and perception simultaneously or in close interaction. The capacity to orient to parts of the environment may have its own developmental history, which intuitively might precede (with some overlap) the development of certain perceptual abilities. It is the presence of this orienting behavior to stimulus change that could lead directly to the acquisition of the representative processes that function in a perceiving, normal, adult organism. However, ideally, orienting behavior itself should be analyzed as a complex response with a number of component elements, each of which might have its own developmental history and be affected by lack of visual experience or lesion differentially (Midgley and Tees, 1983). For example, orientation involves interruption of ongoing behavior, location of the stimulus with appropriate head and postural adjustment, tracking or following the stimulus, and, potentially, further visually guided responses such as approach or avoidance. The characteristics of the stimulus itself, such as its pattern, motion, distance, and size, would be expected to influence the appearance (or lack of it) of such visually related, investigative sequences of behavior. Film or video tape would seem essential in such an analysis (Goodale and Murison, 1975). Unfortunately the ontogeny of visually guided behavior has seldom been examined in this way.

Investigation of the consequences of experiential manipulations on the ontogeny of these visuomotor capacities have primarily focused on easily and quickly administered tasks such as perimetry, visual tracking, visual placing, or reaching. Differential responsiveness to light and dark or to high-contrast targets appears in the rat before it has opened its eyes (Crozier and Pincus, 1937). It is also observed in rats reared in darkness, either until their eyes are open or until maturity, then tested for visual orientation (Hebb, 1937). Dark- or strobe-reared hamsters, when tested in adults, have been found to orient to sunflower seeds presented in all portions of their visual field as effectively as to LR controls (Rhoades and Chalupa, 1978). The failure to find differences due to lack of experience may reflect the sensitivity of behavioral measures used. Midgley and Tees (1983) have discovered deficits in DR rodents even in the simplest of visuomotor responses—the interruption of ongoing behavior. Dark-reared rats, beginning at 90 days of age, exhibit some neglect (lack of orientation) over a wide part of their visual field of stimuli of various degrees of complexity. We do not know yet how permanent these changes are. Interestingly, there is also evidence that DR rats may be more responsive to changes in visual displays than are LR rats at earlier stages of development. DR rats begin to orient preferentially to different kinds of visual displays later than do their LR controls as their overall visual responsiveness declines (Tees *et al.*, 1980). In any event, evidence of behavioral changes related to orientation may develop that will parallel the neural changes (Lund, 1978) reported to appear with continuous DR and with strobe rearing.

In the case of the cat and primate, BD for the first few weeks or months delays visual placing and other visually guided responses to obstacles or moving objects, but not simple head turn or "interruption" to these same stimuli. Riesen (1966) reports that recovery involves as little as 5 h of patterned light experience for BD cats and 19 h for deprived monkeys. More recent data on the cat from several laboratories suggest that complete recovery of the standard set of visuomotor behavior (tracking, placing, reaching, obstacle avoidance) is obtained within 1–2 months postdeprivation, even after 10 months of DR (Crabtree and Riesen, 1979; Van Hof-Van Duin, 1976a). While rearrangement of the normal relationship between self-produced movements and patterned light experience prevents the initial appearance of visual placing in the cat (Held and Hein, 1963; Hein *et al.*, 1979), it quickly returns with such experience. Dependence of the development of one component of visually coordinated guided reaching on another component has been demonstrated. Kittens acquired guided reaching when provided with visual feedback from limbs in an otherwise dark surrounding after—but not before—they had acquired visually guided locomotion (Hein and Diamond, 1972). Kittens DR at birth for 1 to 3 months and then exposed to a luminously painted limb for 10 days failed to exhibit appropriate visual reaching responses. Only after 30 h of visual stimulation, systematically associated with free locomotion and, hence, after the development of guided locomotion, did a second short test experience with a luminous limb result in appropriate visual guided reaching. Clearly this particularly elegant experiment revealed a sequential developmental scheme in terms of acquisition of guided locomotion and guided reaching in which the kitten had to acquire a body-centered spatial framework first, so that the later experience with the limbs might be incorporated into its behavior.

There is other evidence that recovery of such functions after long periods of deprivation may be either much slower or absent. For example, Sherman (1974) reported that, after 6 months of diffuse light-rearing (involving eye suturing), persistent monocular visual field deficits were evident in which cats failed to show visually guided behavior in the ipsilateral hemifield. However, I would speculate that the *persistent* visual field deficits in BD animals are a secondary consequence of change in interocular alignment, for example, and the special risks associated with eyelid suturing as a method of restricting vision equally in both eyes. Van Hof-Van Duin (1976a) did not observe such neglect in DR cats. While it would be helpful to have more systematic and detailed analyses of visuomotor deficits and neglect in BD- and biased-reared animals, these deficiencies in orientation abilities after periods as long as 6 to 8 months of DR appear to be not as severe or permanent as those associated with pattern vision. As far as attempts to relate neurophysiological consequences of deprivation (usually in the retinotectal part of the visual system) to these deficiencies, there are some disconcerting pieces of information. First, different kinds of restriction (strobe versus dark rearing) that yield quite different patterns of electrophysiological changes may have very similar outcomes as far as behavior is concerned (Rhoades and Chalupa, 1978). Moreover, while the potential for neural recovery is not uninfluenced by postdeprivation environmental conditions, such recovery appears to proceed at a substantially slower rate than the rate of behavioral recovery of visual orientation abilities (Blake, 1979). I think that these apparent different rates of recovery are a reflection of several factors. For one thing, the focus of a physiological recovery study is usually on the proportion of cells that appear to show normal responsiveness. If there is any relationship between the lack of receptive field characteristics being examined and a visual competence, the behavior is likely to appear or be present if and when a small subset of the total population is functional. Sometimes the dependence between the neurons and the behavior can be made more clear if the measurements of visual ability are systematic and close to the limits but, inevitably, the quantitative measures of neural and behavioral recovery are unlikely to match up very well.

One oversimplification that is, however, consistent with the findings on visuomotor competence, the brain, and experience, is that binocularly deprived mammals fail to develop cortical mechanisms (Midgley and Tees, 1983) underlying visual orientation while the retinotectal mechanisms also underlying similar competence are available to some extent to support relatively rapid recovery of these visual behaviors. The evidence on the effects of early and late lesions of the SC and cortex in both rodents and cats is certainly consistent with this idea (Finley *et al.*, 1980; Sherman, 1977). It is the combination of DR and SC lesions that has serious consequences for the development of visuomotor orientational skills.

A second observation about these relationships has to do with competition. While the focus of observations on cortical mechanisms underlying orientational behaviors has been area 17 (e.g., Sherman, 1977), deficits due to lack of visual experience are more likely to be the results of changes central to area 17. For example, the orientational deficits of monkeys who were binocularly deprived (by eyelid suture) for 7 to 10 months that persisted for several months after normal colony experience were much more consistent with the lack of physiological recovery in association cortex (e.g., area 7). Since area 7 receives input from several systems (e.g., vision and somesthesis), the lack of significant input in one leads to very few cells being visually responsive (i.e., a decrease in its neural representation) as a result of this early age intermodal competition that has a significant impact on the use of orientational skills (and the use of vision) of these long-term binocularly sutured monkeys (Hyvarinen *et al.*, 1981).

With respect to this category of visual behavior and the ideas about competitive disadvantages, we feel that a review of the consequences of MD is particularly relevant. While it is clear that MD results in more serious and longer lasting deficits in visual acuity, pattern recognition, and soon, than does binocular DR (Sections IIA and IID), neglect of visual arrays is probably at the heart of the visual problem that MD animals have to overcome during postdeprivation testing.

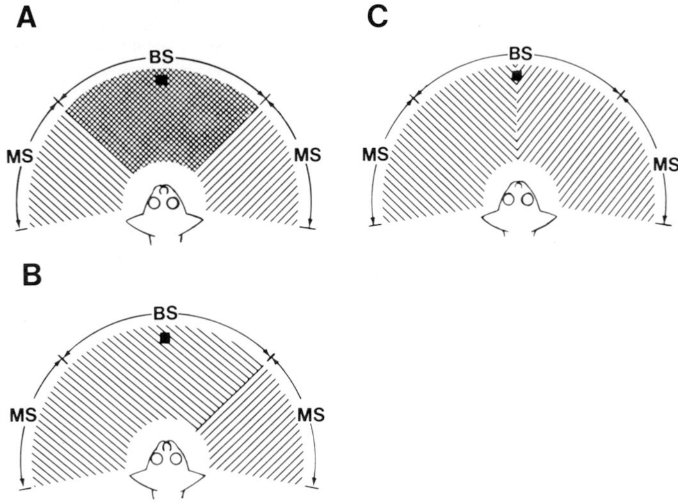

Fig. 6. Summary of idealized visual field perimetry testing of cats after "correction" for eccentric fixation. (A) Normal cat; (B) monocularly deprived cat; (C) binocularly deprived cat. (■) Fixation object; (▧) left eye field; (▨) right eye field; BS, binocular segment; MS, monocular segment. Adapted from Sherman (1973).

While testing reveals that cats reared with monocular closure for 6 months exhibit visual placing, these same animals fail to respond to novel visual stimuli in the binocular (central) part of the visual field (Fig. 6) when given perimetry testing with the deprived eye, and this failure persists (Hein and Diamond, 1971; Sherman, 1973). Other visuomotor behaviors in such cats (e.g., visually guided reaching, obstacle avoidance, tracking) also still show impairments 2 years after reverse closure of their formerly deprived eye (Van Hof-Van Duin, 1977). We do not have as good a behavioral picture of the orientation skills of the MD rodent or primate (including man) yet but what we do have is consistent with the above (Joseph and Casagrande, 1980; Moran and Gordon, 1982; Rothblat *et al.*, 1978).

In all mammals examined, however, MD results in a dramatic loss of input via the binocular segment of the Y pathway associated with the deprived eye (to cortex and then SC) (Hoffman and Sherman, 1975). According to Sherman (1979), the visual neglect in MD is a direct result of suppression of the retinotectal pathways (and input) by the imbalanced, abnormally developed corticotectal pathway. Results of physiological studies employing surgical removal of suppressing corticotectal projection to SC or injection of bicuculline have also supported the idea of suppression (= neglect); both techniques have been interpreted as yielding removal of a basic inhibitory influence of the experienced eye and its projections on the deprived connections (e.g., Duffy *et al.*, 1976). In the recovery of visual orientational skills, abolition of the activity of the nondeprived eye (through reverse suture, eye or visual cortex removal, bicuculline injection) may be initially more important than visual experience per se (see Mitchell, 1981; Mitchell and Timney, 1984). This has been well documented and discussed. However, further improvement (physiologically or behaviorally) may still depend on cumulative visual experience. In kittens reared so that each eye receives normal input on alternate days, visual fields and orientation skills are normal (Tumosa *et al.*, 1980). If one eye receives eight times (8 h versus 1 h) as much experience as the other, the field of the less experienced eye is restricted to the temporal hemifield (but includes the binocular segment). We need to know more about the ability of the cat with its less experienced eye; we do know that the visual field changes resulting from unequal alternating MD are different from those produced by MD itself. More importantly, daily alternating (and equal) monocular exposure results in cortical and collicular units whose properties are normal with the exception of their lack of binocularity (Blake, 1979). The lack of perceptual import of this change in cortical connectivity is such that the alternators not only do not show neglect *but can even readily attend to and combine* information from both eyes simultaneously (see Fig. 7) to solve a pattern discrimination in which such an ability is required (Von Grunau and Singer, 1979). The presence of a normal population of binocular cells in area 17 or SC is not a prerequisite for the use of binocular signals. The proposition that other cortical areas may differ in their susceptibility to manipulations that interfere with binocularity is certainly worth further physiological investigation (e.g., Leventhal, 1984).

D. Form Perception

Visually deprived mammals, including the rat, cat, chimpanzee, and monkey, require two to three times as many trials as normally reared animals to learn discriminations such as *N* versus *X* or + versus 0 (reviewed by Ganz, 1978; Riesen and Zilbert, 1975). Such retarded performance has been used to indicate the important contribution that visual experience plays in the development of the mechanism underlying form perception in mammals (e.g., Hebb, 1949). In the past, some have argued (Melzack, 1962) that DR animals are intellectually and/or emotionally inferior to their LR counterparts and, hence, the more difficult the discrimination (or task) of whatever kind, the more deleterious the effect of deprivation on performance. However, evidence has accumulated over the years that suggests that this alternative interpretation

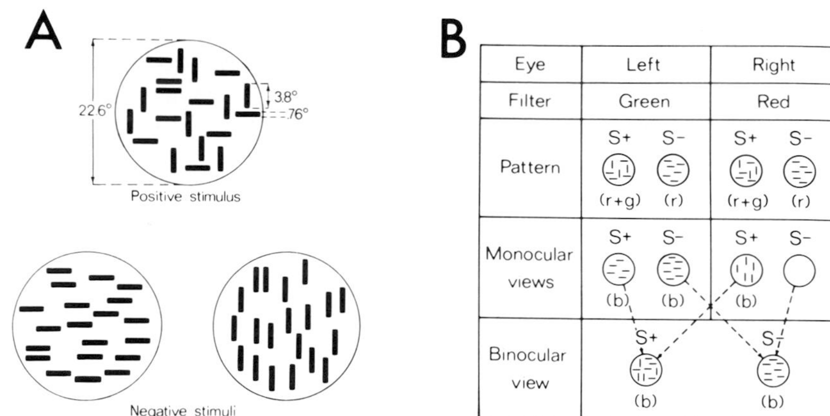

Fig. 7. (A) Example of randomized positive (S +) and negative (S-) stimuli. The small pattern elements were red (r) or mixed red and green (r + g) but were always seen as black (b) through the lenses. (B) Schematic diagram of the test shown are the positive stimulus and one of the negative stimuli along with their monocular and binocular views. Adapted from Von Grünau and Singer (1979).

is not correct. Problem difficulty per se is not critical. Visual experience plays no more of a role when the stimuli employed are light intensities or single rectangles (see Fig. 8) and when near-threshold stimulus differences are used so that such tasks are made more difficult for the normal rat than an *N* versus *X* discrimination (e.g., Tees, 1968, 1972, 1979).

Such results illustrate the most frequently employed behavioral technique on the effect of early experience. To learn about the mechanism underlying the way normal and restricted animals classify shapes and process visual information, one looks at the relative ease or difficulty with which these animals are able to distinguish two representative stimulus arrays. Less frequently employed have been transfer or stimulus equivalence tests in which that which has been learned from an initial exposure to particular training shapes or forms is examined. The results of this kind of testing are important; the classical question associated with pattern recognition can be identified as a problem of stimulus equivalence (Lashley, 1938; Sutherland, 1968). It is the neural and psychological processes underlying the ability to recognize a pattern or object in spite of variations in its size, orientation, color, and contrast that are at issue.

Some evidence has accumulated in relation to experience and generalization after edge and contour training in mammals. Ganz and Wilson (1967) found that after training on a horizontal versus vertical line discrimination, the visually naive monkey was able to generalize as well as the LR monkey to

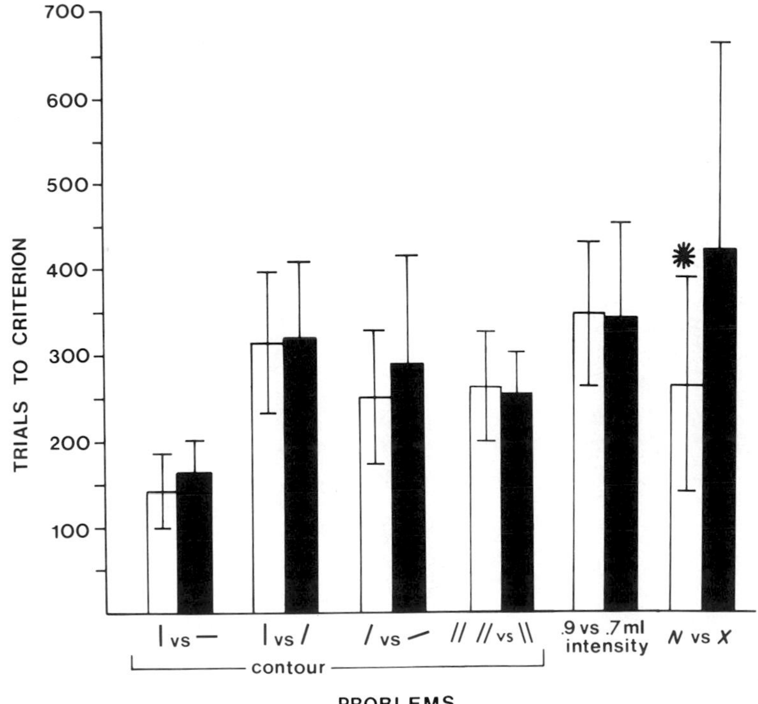

Fig. 8. Performance by separate groups of light-reared (□) and dark-reared (■) rats in learning a complex pattern (N versus X), a near-threshold intensity discrimination, and a four-line orientation discrimination of varying degrees of difficulty in either two- or three-choice (oddity) test situations. Only performance on the complex pattern yielded significant (*) differences due to stimulation history.

stimuli that had undergone fragmentation, reversal of figure-ground brightness relations, and shifts to the contralateral retina. These investigators suggested the operation of an "innate" line-recognition mechanism based on findings concerning the receptive fields of simple cortical cells in the visual system whose properties paralleled their generalization data. Ganz and his co-workers (Ganz and Haffner, 1974; Ganz *et al.*, 1972) provided similar evidence with respect to the ability of the DR cat to generalize to new versions of line stimuli.

We have somewhat more data on the rat. For example, after learning to discriminate a single rectangle from an equally bright target containing no rectangle, both LR and DR rats showed peaked generalization gradients when faced with transfer stimuli in different angular orientations (Tees, 1972).

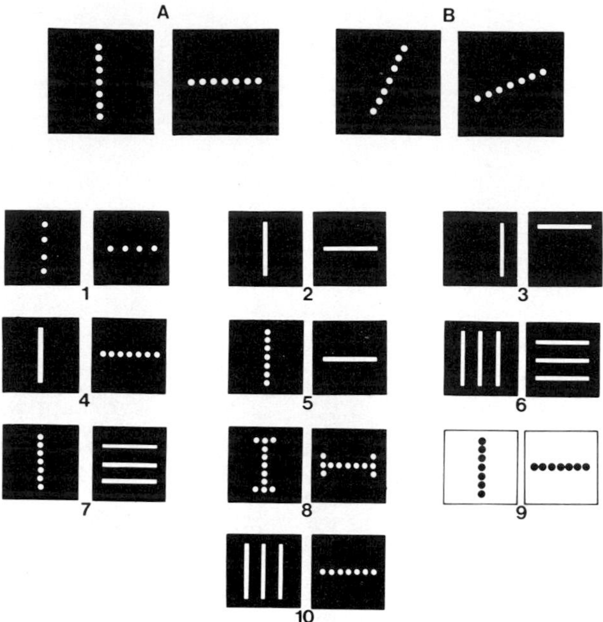

Fig. 9. Original training stimulus arrays. (A) 90°–0°; (B) 22.5°–67.5°. 1–10, Sample transfer tests for training stimuli (A) to cats originally trained with horizontal and vertical (A) dot stimulus pairs.

While a discrimination involving 45° of angular orientation separation in dot patterns proved to be more difficult than one involving 90°, DR rats performed as effectively as LR rats on both (Tees *et al.*, 1976). After initial acquisition, rats transferred well to new stimulus patterns containing fewer dots, stripes, and so on, provided a relative difference in orientation was maintained (see Fig. 9). We have even raised rats in a planetarium-like environment (Bruinsma and Tees, 1977) for 2 h daily from age 15 to 45 days, and that experience had little impact on how these dot-reared rats perceived similar arrangements of dot or continuous lines (as revealed in transfer tests).

Visually naive (and biased reared) rats seem to possess not only the predisposition to isolate a figure from its ground (Hebb, 1937), and the capacity to respond differentially to visual, oblique, and horizontal stimuli, but also to generalize from that experience even to lines in different angular orientations. On the other hand, with identifications involving forms differing only in their elements' relationship to one another (e.g., N versus X), DR animals learn more slowly and learn less (as evidenced in stimulus equivalence tests) about these concatenations of lines in different orientations than do

LR animals, at least insofar as mammals such as rat, cat, and monkey are concerned (see Riesen and Zilbert, 1975; Tees, 1976).

Differences between shapes such as N versus X involve relational properties between sets of linear elements. Until recently, these have not been examined very systematically. We have looked at the role played by visual experience in the discrimination of two of the most elemental relationships—contour separation per se and angle or junction. Visually naive DR rats learned a task in which spatial distance (i.e., the horizontal separation between vertical contours) was a critical feature more slowly than did their LR counterparts and performed less effectively on transfer tests even after 150 trails of overtraining (Tees, 1979). The same pattern of results (Tees and Midgley, 1982) was observed with respect to the rats' abilities to discriminate simple types of intersections between lines (e.g., T versus ⊥). In short, the capacity to perceive even basic relationships between linear elements as contour separation or angle appears to depend on visual experience for its development (see Fig. 10).

The evidence obtained from testing BD rats, cats, and monkeys (see Ganz, 1978 or Tees, 1976, for a review) who are exposed to normal environments after deprivation is consistent with the above conclusion relating to the relative importance of early experience. It is deficits in the ability to recognize complex patterns and their variations during transfer testing that seem the most resistent to improvement during a recovery period (e.g., Chow and Stewart, 1972; Tess and Midgley, 1978; Van Hof-Van Duin, 1976a,b).

Most of the research on behavioral recovery has tended to focus on what I argue is a special case, the lack of recovery following MD (Cynader *et al.*, 1976). While there are physiological consequences due to disuse in the deprived eye's monocular cortical segment in addition to those well-established effects in the binocular segment (Wilson and Sherman, 1977), and potentially permanent deficits in pattern discriminative abilities certainly have been reported in MD cat (Dews and Wiesel, 1970; Ganz and Haffner, 1974) and rat (Rothblat *et al.*, 1978); however, these consequences may be made more complicated by the general deficits related to visual attention (i.e., neglect; see Section IIC). While Van Hof-Van Duin (1976b) found cats MD for 8 to 10 months still exhibiting seriously impaired visuomotor skills 3 months after reverse suture, she did not find sizeable pattern discrimination deficits in these cats when she used large stimuli and a simple response that required less visuomotor coordination and body movement in relationship to the targets. The success of her cats could reflect less demand placed either (1) on the ability to recognize configurational patterns and their transformations, or (2) on visual attentional mechanisms. Results of work on visual pattern discrimination in cats with moderate cylodeviation of one or both eyes would support the idea that monocular input restriction results in significant attentional losses (neglect) later due to the competitive disadvantages introduced during development (Peck and Wark, 1981). Others (e.g., Spear and Ganz, 1975) have also reported some recovery of pattern discriminative ability in MD cats, but this ability is lost when the visual cortex is removed. The visual cortex is functional in these cats and can be influenced by the deprived eye in spite of some evidence based on single unit recordings of a lack of neuronal function in the visual cortex (e.g., Hoffman and Cynader, 1977).

The stimuli employed in testing pattern recognition in LR and DR mammals so far have been relatively simple. We wanted to examine an organism's ability to perceive similarities and differences between stimuli in which the invariant-from-stimulus array to array was more complex and had to be computed over a wide area. We have confirmed an earlier tentative report (Sutherland and Williams, 1969) that rats, having learned to discriminate the checkerboard pattern from a pattern containing an irregularity, treat as equivalent to the original (negative) stimulus a pattern containing a variety of irregularities, including a change in the shape of the checkers in both patterns (Tees, 1979), and they do this at least after 150 trials of overtraining. Almost half of our DR rats (10 of 22) failed to learn the original task within 1000 trials, and those 12 that did learned it very slowly and performed very poorly on most of the transfer stimuli, even after overtraining (see Fig. 11). From the results of the transfer tests, it appears that the LR rat is able to use an abstract description of the difference between the patterns and could not have been encoding a difference in spatial frequency, an exact representation of the pattern, or even a list of features.

While at this stage I am not in a position to state precisely what properties or rules were used by our rats to "describe" the regular and irregular patterns, the results serve to remind us of the level of perceptual processing that must be accounted for in theories of shape recognition (e.g., Milner, 1974), even with respect to the rat and the special significance of visual experience in its development.

Ganz (1978) and Pettigrew (1974) have focused on the special role visual experience might play in the development of inhibitory synaptic connections critical to the development of subordinate neurons, having the "veto" properties that are necessary to mediate configurational discriminations. Visual experience might be viewed as changing the gating action of these neurons from OR gates or AND gates into AND–NOT gates in which suboptimal features exert considerable inhibition (see Chapter 9, this volume, for further ideas about the development of the underlying mechanism). While such a speculation represents an excellent step toward breaking out of this earlier simple view of pattern discrimination and its neural substrate, I think the main site of environmental stimulation influence on the system is probably

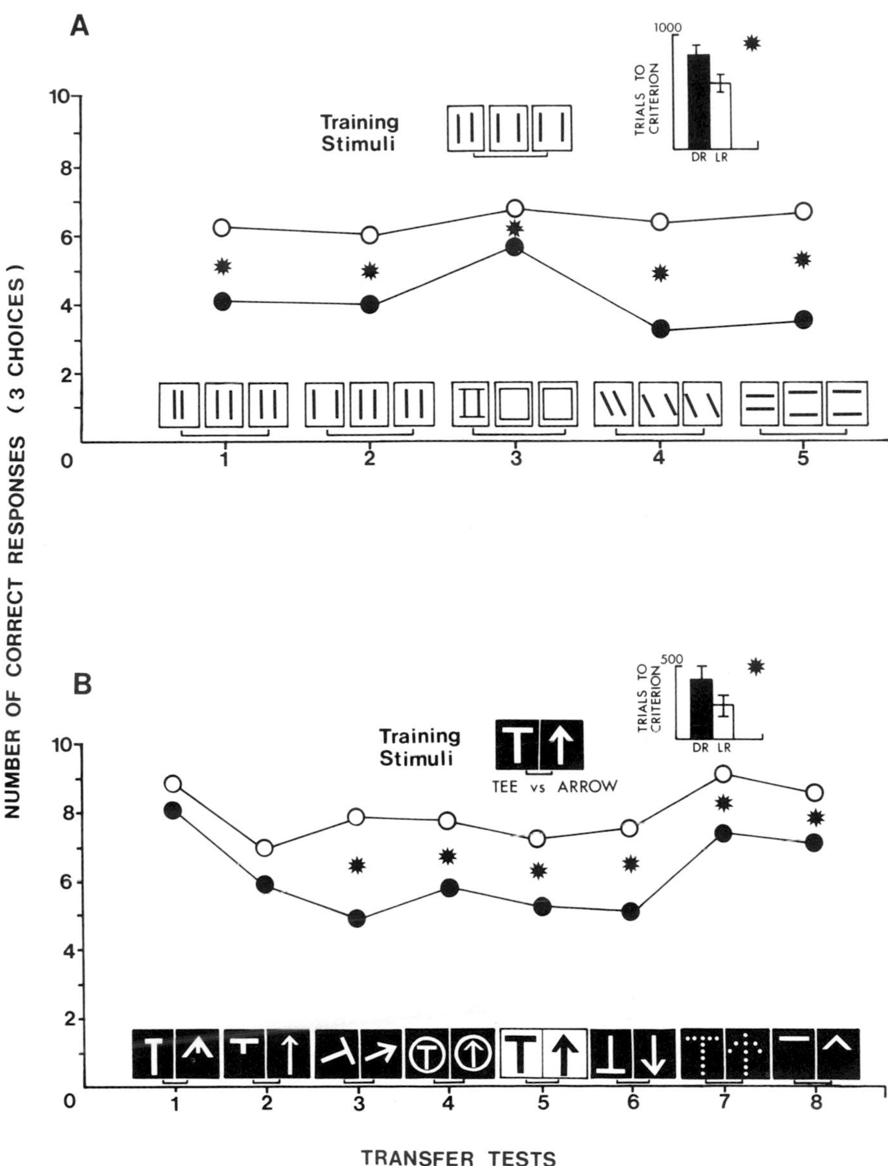

Fig. 10. Mean number of correct responses made by 90-day-old DR (○—○) and LR (●—●) rats to some test shapes employed during transfer testing. In each case LR rats transferred significantly (*) more successfully than did DR rats even after 150 overtraining trials after initial acquisition with the original training stimulus shown. (A) Contour separation; (B) contour intersection.

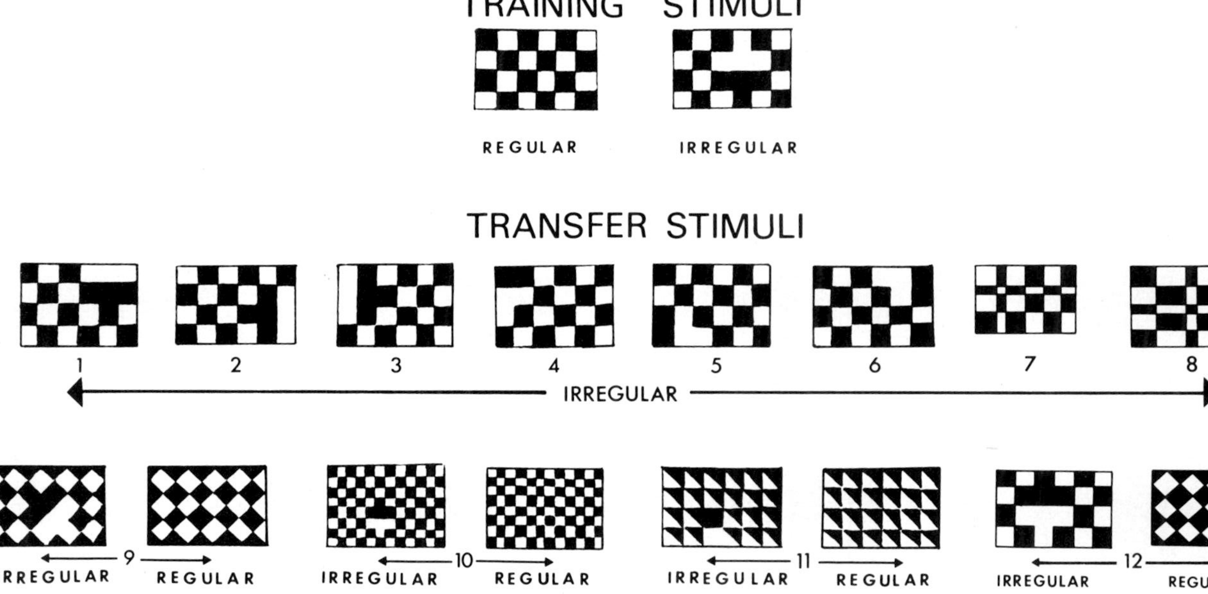

Fig. 11. Initial checkerboard training stimuli and 12 of the transfer test patterns.

not at the feature-detecting mechanisms of the striate cortex (e.g., see Leventhal, 1984). The site of additional processing of visual pattern information must certainly involve other structures such as the inferotemporal cortex, where physiological experiments at least hint at substantial integration in which responses to stimuli appear to be independent of size, position, orientation, and even spatial frequency to some extent, and whose responsiveness emerges gradually during a later period postnatally (Desimone and Gross, 1979; Humphrey and Weiskrantz, 1967; Mayers *et al.*, 1971). I am not arguing that such a site would be the final representation of the visual pattern. All levels of analysis would be accessible to the normal system. This analysis would be in parallel with other descriptions of the stimuli (in which information about position, color, movement, and line orientation is coded). These other descriptions are apparently available to the visually inexperienced animal. The inferotemporal (and extrastriate) cortex would either be the potential locus of form-specific encoding developed from the line segment and/or spatial frequency information of striate cortex, or would provide a somewhat more independent transformation of the visual input in which size and position invariance is achieved.

While results of single-unit recordings have revealed something about the coding of features or pattern elements, we know little about the structure or pattern code (e.g., see Garner, 1974). Even with respect to the sensory elements of the code, there has been controversy over such matters as to whether we should focus on the properties of visual neurons as analyzers of specific spatial frequencies or of line segments (see Cavanagh, 1978). Unfortunately, pattern synthesis could result from the building up of neural representatives through cumulative experience (Hebb, 1949), automatic processing involving synthesis of Fourier components (Blakemore and Campbell, 1969), the action of specific mathematical operators such as Lie transformations (Dodwell, 1983), or in some other fashion, and there is little agreement on the specific characteristics of an adequate model of pattern recognition. As a result there has been little to guide neurophysiological investigation. However, researchers have begun to focus on stimulus-specific properties of extrastriate neurons that seem better suited to perform the (e.g.) local–global comparisons made evident by study of normal perceptual functioning (e.g., Treisman and Gelade, 1980; Julesz, 1981) that could serve as a basis for discrimination of figures from ground, preattentive vision, and so on (Allman *et al.*, 1985).

III. Concluding Observations

While the deprivation (and enrichment) paradigm has indeed furnished us with a strategy for trying to relate the ontogeny of underlying neural mechanisms to visual capabilities, the results of such attempts have not been overwhelming (see also Blake, 1979, 1981; Greenough, 1976). One basic problem seems to be that we have underestimated the difficulty of the task. As Riesen (1971) pointed out, behavior is overdetermined. There is evidence that in many instances many antecedural conditions are necessary but are not sufficient with regard to particular kinds of visual behavior. Physiological and anatomical substrates are interchangeable to some extent and, moreover, shift during development so that changes that are important at one change may not be at another. A few other observations are in order.

1. While there have been both physiological (Berman and Daw, 1977) and behavioral (Dodwell *et al.*, 1976) studies that suggest that particular synapses, receptive field characteristics, and aspects of behavior follow different time courses during development, many investigations have taken a very simple-minded view about the nature of sensory experience and its effects. The sensory input of a normal environment is not a field of vertical or horizontal lines. A few hours of such input embedded in a regime of total patterned light deprivation may have a dramatic impact on cortical physiology if delivered between days 28–35 of a kitten's life; however, the ever-changing stimulation available normally simply does not (Peck and Blakemore, 1975). A typical restrictive environment thus results in atypical competitive advantages and disadvantages between potential visual stimuli that compress and alter the timing and nature of subsequent visual development. While the neural circuitry underlying discrimination of contour orientation may normally be primarily influenced during the first few months, the neural circuitry underlying more complex discriminations of configuration and shape is likely influenced during a later period by different visual stimuli, but only if other necessary parts of the neural mechanism have been developed.

For me, a kind of stage hypotheses would represent a first step in trying to conceptualize what is going on. Particular kinds of sensory stimulation would be optimal at different ages in terms of their effect on the ontogeny of pattern-related abilities. This sequence of stimulation would be as important as absolute age. Thus, the characteristics of selectivity/sensitivity "filters" for those experiential inputs that can modify behavior during any particular age range reflect the neural circuitry established, in part, by previous stimulation history. Within this framework, the influence of normal experience in visual development cannot be viewed as either passive (or maintenance) or active (or facilitative), but its role must be seen to be dependent on the context, stage, function, and so on in question. There are several kinds of experiments I would use to illustrate this notion, including the behavioral and physiological experiment of Timney *et al.* (1980) in which the timing of the sensitive period in the kitten depended on previous visual experience (or lack of it) and those of Dodwell and Tees on the emergence of visual pattern-seeking behavior in the deprived cat and rat, which develops well after a "most sensitive" period as far as striate neurons are concerned (Dodwell *et al.*, 1976; Tees *et al.*, 1980). Turkewitz and Kenny (1982) outlined evidence that the effect that changes in the nature of experience within a particular modality has on perceptual abilities is a function of the stage of organization of other modalities at the same time.

2. I raise this issue along with a plea against premature oversimplification of what will doubtless turn out to be a very complex physiological mechanism. In any event, we often find ourselves trying to relate changes resulting from sensory restriction with respect to specific visual abilities (e.g., acuity) with changes in cortical physiology that are not likely to be related to the behavior (e.g., number of orientation-sensitive cortical units). Sometimes we do not have the right kinds of physiological data because it is difficult to obtain them, but sometimes it is because we have not bothered to collect them.

3. Recent evidence raises the possibility that there may be synapses that are anatomically normal but physiologically ineffective (Marks, 1978). For some physiologists that has meant that functional connections are best reflected in the response properties of the cell in question. One can make an equally chauvinistic remark about the functional validity of behavioral processes. Though, by definition, there are more direct measures of detailed connectivity among neurons, if one is interested in the functional significance of changes in either physiology or anatomy, then clearly behavior must be a most powerful and sensitive indicator. At this stage I would like to argue that the least one can do is to appreciate behavioral evidence when it exists. The kinds of data accumulated on the visual acuity of DR and biased-reared cats since the early 1970s (e.g., Hirsch, 1972; Muir and Mitchell, 1975) should have prompted a swifter reassessment of the dramatic physiological consequences of these changes in early stimulation history.

4. One point that emerges clearly from this survey of the evidence on visual behavior and early experience is that we are in desperate need of more and better data. Kittens have been raised in a variety of atypical environments containing contours and patches moving in just on direction or in stroboscopic or prismatic viewing conditions (e.g., Berman and Kaw, 1977; Chapter 9, this volume). These special arrangements appear to produce striking changes in receptive field characteristics of cortical neurons consistent with these early conditions—provided that the atypical conditions are not extreme (Singer, 1979). What usually is not available are the results of careful behavioral testing of these animals' visual competence using tasks selected to measure the significance of the environmentally induced physiological abnormalities. The behavioral "facts" (if there are any) are usually so general, irrelevant, limited,

and/or poorly collected as to make attempts to draw inferences about potential brain–behavior relationships sheer speculation. The conclusion that needs to be drawn is not that one must be distrustful of behavioral data (Blake, 1981); rather, it is that extensive testing is needed in which several paradigms are used that are designed to offer the developing or potentially handicapped animal maximum opportunity to reveal what visual abilities it has. Procedures that allow separate components of behavioral sequences to be individually assessed, near-threshold stimulus values to be tested, and stimulus equivalence trials to be incorporated to permit evaluation of the use of alternative cue or perceptual strategies—any or all might be needed in order to provide relevant information about the behavioral significance of physiological changes induced by special rearing conditions. Monitoring of these behavioral and neural indices over a reasonable period of further exposure to a specially designed therapeutic or normal visual environment is also likely to provide significant information about potential relationships.

5. One promising strategy that allows us to test ideas about experientially induced plasticity and the role that particular neural sites play in development of particular visual competences is the use of surgical intervention in conjunction with manipulations of early visual experience. The time course of visuomotor behavioral development in LR and DR animals with neonatal SC lesions (e.g., Finlay *et al.*, 1980) can provide additional insight into the site and nature of deficit caused by early DR. The effect of neonatal section of the corpus callosum on the development of depth perception indicates that it plays a significant role in LR cats (Elberger, 1980; 1984). Increasingly effective use of sensitive anatomical tracing techniques (Lund, 1978) has revealed, for example, an expanded ipsilateral retinotectal pathway resulting from neonatal eye removal; the behavioral role of this pathway is beginning to be examined (Midgley and Tees, 1982). The effects of visual experience on these projections and on visuomotor development are also subject to assessment and should further facilitate analysis of the roles these effects normally play in visual, behavioral, and neural development.

Acknowledgments

This chapter was written in connection with research activities supported by National Sciences and Engineering Research Council of Canada (research grant APA 0179). The assistance of K. F. Tees, E. McCririck, K. Searcy, C. Bawden, and S. Lee in preparing the manuscript is gratefully acknowledged.

References

Allman, J., Miezin, F., and McGuinness, E. (1985). *Ann. Rev. Neurosci.* **8**, 407–430.

Aslin, R. N. (1981). *In* "Development of Perception" (R. Aslin, J. Alberts, and M. Peterson, eds.). Academic Press, New York.

Barlow, H. B. (1979). *In* "Developmental Neurobiology of Vision" (R. D. Freeman, ed.), pp. 1-16. Plenum, New York.
Bateson, P. (1979). *Anim. Behav.* **27**, 470-486.
Baxter, B. L. (1966). *Exp. Neurol.* **14**, 224-237.
Bennett, M. J., Smith, E. L., Harwerth, R. S., and Crawford, M. L. (1980). *Brain Res.* **193**, 33-45.
Berkeley, M. A., and Sprague, J. M. (1979). *J. Comp. Neurol.* **187**, 679-702.
Berman, N., and Daw, N. W. (1977). *J. Physiol. (London)* **265**, 249-259.
Blake, R. (1979). *Percept. Psychophysics* **26**, 423-448.
Blake, R. (1981). *In* "The Development of Perception" (R. Aslin, J. R. Alberts, and M. R. Peterson, eds.), pp. 95-110. Academic Press, New York.
Blake, R., and Di Granfilippo, A. (1980). *J. Neurophysiol.* **43**, 1197-1205.
Blake, R., and Hirsch, H. V. B. (1975). *Science* **190**, 1114-1116.
Blakemore, C., and Campbell, F. W. (1969). *J. Physiol. (London)* **203**, 237-260.
Blakemore, C., and Cooper, G. F. (1970). *Nature (London)* **228**, 477-478.
Blakemore, C., and Van Sluyters, R. C. (1974). *J. Physiol. (London)* **237**, 195-216.
Bond, E. K. (1972). *Psychol. Bull.* **72**, 225-245.
Boothe, R. G. (1981). *In* "Development of Perception" (R. N. Aslin, J. R. Alberts, and M. R. Peterson, eds.), Vol. 2, pp. 218-242. Academic Press, New York.
Bruinsma, Y., and Tees, R. C. (1977). *Bull. Psychon. Soc.* **10**, 433-435.
Buisseret, P., Gary-Bobo, E., and Imbert, M. (1978). *Nature (London)* **272**, 816-817.
Campbell, F. W., Hess, R. F., Watson, P. G., and Banks, R. (1978). *Br. J. Ophthalmol.* **11**, 748-755.
Cavanagh, P. (1978). *Perception* **7**, 168-177.
Chow, K. L. (1973). *Handb. Sens. Physiol.* **7**, 599-630.
Chow, K. L., and Stewart, D. L. (1972). *Exp. Neurol.* **34**, 409-433.
Collewijn, H. (1977). *Exp. Brain Res.* **27**, 287-300.
Crabtree, J. W., and Riesen, A. H. (1979). *Dev. Psychobiol.* **12**, 291-303.
Creutzfeldt, O. D., and Heggelund, P. (1975). *Science* **188**, 1025-1027.
Crozier, W., and Pincus, G. (1937). *J. Genet. Psychol.* **17**, 105-111.
Cynader, M., and Mitchell, D. E. (1980). *J. Neurophysiol.* **43**, 1026-1040.
Cynader, M., and Berman, N., and Hein, A. (1976). *Exp. Brain Res.* **25**, 139-156.
Desimone, R., and Gross, C. G. (1979). *Brain Res.* **178**, 363-380.
Dews, P. B., and Wiesel, T. N. (1970). *J. Physiol. (London)* **206**, 437-455.
Dodwell, P. C. (1970). "Visual Pattern Recognition." Holt, New York.
Dodwell, P. C. (1983). *Percept. Psychophysics* **34**, 1-16.
Dodwell, P. C., Timney, B. N., and Emerson, U. F. (1976). *Nature (London)* **260**, 777-778.
Duffy, F. H., Snodgrass, S. R., Burchfiel, J. L., and Conway, J. L. (1976). *Nature (London)* **260**, 256-257.
Eichengreen, J. M., Coren, S., and Nachmias, J. (1966). *Science* **151**, 830-831.
Elberger, A. J. (1980). *Vision Res.* **20**, 177-187.
Elberger, A. J. (1984). *Behav. Brain Res.* **11**, 223-231.
Emerson, V. F. (1980). *Exp. Brain Res.* **38**, 43-52.
Eysel, U. T., Grusser, O. J., and Hoffman, K. P. (1979). *Exp. Brain Res.* **34**, 521-539.
Finlay, B. L., Marder, K., and Cordon, D. (1980). *J. Comp. Physiol. Psychol.* **94**, 506-518.
Flom, M. C. (1970). *In* "Early Experience and Visual Information Processing in Perceptual and Reading Disorders" (F. A. Young and D. B. Lindsley, eds.), pp. 291-301. Nat. Acad. Sci., Washington, D.C.
Freeman, R. D. (1979). *In* "Developmental Neurobiology of Vision" (R. D. Freeman, ed.), pp. 305-315. Plenum, New York.
Freeman, R. D., and Bradley, A. (1980). *J. Neurophysiol.* **43**, 1645-1653.
Ganz, L. (1978). *Handb. Sens. Physiol.* **3**, 437-488.

Ganz, L., and Fitch, M. (1968). *Exp. Neurol.* **22**, 638-660.
Ganz, L., and Haffner, M. E. (1974). *Brain Res.* **20**, 67-87.
Ganz, L., and Wilson, P. D. (1967). *J. Comp. Physiol. Psychol.* **63**, 258-269.
Ganz, L., Hirsch, N. V. B., and Tieman, S. B. (1972). *Brain Res.* **44**, 547-568.
Garner, W. (1974). "The Processing of Information and Structure." Erlbaum, Hillsdale, New Jersey.
Giffin, F., and Mitchell, D. E. (1978). *J. Physiol. (London)* **274**, 511-537.
Globus, A. (1975). *In* "The Developmental Neuropsychology of Sensory Deprivation" (A. H. Riesen, ed.), pp. 9-91. Academic Press, New York.
Goodale, M. A., and Murison, R. C. (1975). *Brain Res.* **88**, 243-261.
Gottlieb, G. (1971). *In* "The Biopsychology of Development" (E. Tobach, L. R. Aronson, and E. Shaw, eds.), pp. 67-128. Academic Press, New York.
Gottlieb, G. (1976). *In* "Studies on the Development of Behavior and the Nervous System" (G. Gottlieb, ed.), Vol. 3, pp. 25-54. Academic Press, New York.
Greenough, W. T. (1976). *In* "Neural Mechanisms of Learning and Memory" (M. R. Rosenzweig and E. L. Bennett, eds.), pp. 255-278. MIT Press, Cambridge, Massachusetts.
Grobstein, P., and Chow, K. L. (1976). *In* "Studies on the Development of Behavior and the Nervous System" (G. Gottlieb, ed.), Vol. 3, pp. 155-193. Academic Press, New York.
Hebb, D. O. (1937). *J. Genet. Psychol.* **51**, 101-126.
Hebb, D. O. (1949). "The Organization of Behavior." Wiley, New York.
Hebb, D. O. (1972). "Textbook of Psychology." Saunders, Philadelphia.
Hein, A., and Diamond, R. M. (1971). *J. Comp. Physiol. Psychol.* **76**, 219-224.
Hein, A., and Diamond, R. M. (1972). *J. Comp. Physiol. Psychol.* **81**, 394-398.
Hein, A., Vital-Durand, F., Salinger, W., and Diamond, R. (1979). *Science* **204**, 1321-1322.
Held, R. (1981). *In* "Development of Perception" (R. N. Aslin, J. R. Alberts, and M. R. Peterson, eds.), Vol. 2, pp. 279-296. Academic Press, New York.
Held, R., and Hein, A. (1963). *J. Comp. Physiol. Psychol.* **56**, 872-876.
Hendrickson, A., Boles, J., and McLean, E. B. (1977). *Invest. Ophthalmol. Visual Sci.* **16**, 469-473.
Hirsch, H. V. B. (1972). *Exp. Brain Res.* **15**, 405-423.
Hoffman, K.-P., and Cynader, M. (1977). *Philos. Trans. R. Soc. London Ser. B* **278**, 411-424.
Hoffman, K. P., and Sherman, S. M. (1975). *J. Neurophysiol.* **38**, 1049-1059.
Hubel, D., H., and Wiesel, T. N. (1959). *J. Physiol. (London)* **148**, 574-591.
Hubel, D. H., and Wiesel, T. N. (1965). *J. Neurophysiol.* **28**, 1041-1059.
Hubel, D. H., and Wiesel, T. N. (1970). *J. Physiol. (London)* **206**, 419-436.
Humphrey, N. K., and Weiskrantz, L. (1967). *Nature (London)* **215**, 595-597.
Hyvarinen, J., Hyvarinen, L., and Linnankoski, I. (1981). *Exp. Brain Res.* **42**, 1-8.
Ikeda, H., Tremain, K. E., and Ennon, G. (1978). *Exp. Brain Res.* **31**, 207-220.
Jacobson, M. (1974). *In* "Studies on the Development of Behavior and the Nervous System" (G. Gottlieb, ed.), Vol. 2, pp. 151-166. Academic Press, New York.
Jacobson, S. G., and Ikeda, H. (1979). *Exp. Brain Res.* **34**, 11-26.
Joseph, R., and Casagrande, V. A. (1980). *Behav. Brain Res.* **1**, 165-186.
Julesz, B. (1981). *Nature (London)* **290**, 91-97.
Kaye, M., Mitchell, D. E., and Cynader, M. (1982). *Dev. Brain Res.* **2**, 37-53.
Kiorpes, L., and Boothe, R. G. (1980). *Invest. Ophthalmol. Visual Sci.* **19**, 841-845.
Kratz, K., and Spear, P. (1976). *J. Comp. Neurol.* **170**, 141-152.
Kuo, Z. Y. (1967). "The Dynamics of Behavioral Development." Random House, New York.
Lashley, K. S. (1938). *J. Genet. Psychol.* **18**, 123-193.
Leventhal, A. (1984). *In* "Development of Visual Pathways in Mammals" (J. Stone, B. Dreher, and D. Rapaport, eds.) pp. 347-361. Liss, New York.
Leventhal, A. G., and Hirsch, H. V. B. (1977). *Proc. Natl. Acad. Sci. U.S.A.* **74**, 1272-1276.

Leventhal, A. G., and Hirsch, H. B. V. (1978). *J. Neurophysiol.* **41**, 948–962.
Leventhal, A. G., and Hirsch, H. V. B. (1980). *J. Neurophysiol.* **43**, 1111–1132.
Levi, D. M., and Klein, S. (1982). *Nature (London)* **298**, 268–270.
Loop, M. S., and Sherman, S. M. (1977). *Brain Res.* **128**, 329–339.
Loop, M. S., Petuchowski, S., and Smith, D. C. (1980). *Visual Res.* **20**, 49–57.
Lund, R. D. (1978). "Development and Plasticity of the Brain." Oxford Univ. Press, London and New York.
Marks, R. F. (1978). *Trends Biochem. Sci* **3**, 9–12.
Mayers, K. S., Robertson, R. T., Rubel, E. W., and Thompson, R. T. (1971). *Science* **171**, 1037–1038.
Melzack, R. (1962). *Science* **137**, 978–979.
Midgley, G. C., and Tees, R. C. (1982). Paper presented at the Canadian Psychological Association meetings, Montreal, Quebec, June.
Midgley, G. C., and Tees, R. C. (1983) *Behav. Neurosci.* **97**, 624–638.
Milner, P. M. (1974). *Psychol. Rev.* **81**, 521–534.
Mitchell, D. E. (1978). *In* "Perception and Experience" (R. D. Walk and H. L. Pick, Jr., eds.), pp. 37–75. Plenum, New York.
Mitchell, D. E. (1981). *In* "Development of Perception" (R. N. Alsin, J. R. Alberts, and M. R. Peterson, eds.), Vol. 2, pp. 3–43. Academic Press, New York.
Mitchell, D. E. (1985). Paper presented at the Canadian Psychological Association meetings, Halifax, Nova Scotia, June.
Mitchell, D. E., and Murphy, K. M. (1984). In "Development of Visual Pathways in Mammals" (J. Stone, B. Dreher, and D. Rapaport, eds.), pp. 381–392. Liss, New York.
Mitchell, D. E., and Timney, B. (1984). In "Handbook of Physiology" (I. Darian-Smith, ed.), Vol. 2, pp. 507–555. American Physiological Society, Bethesda, Maryland.
Mitchell, D. E., Freeman, R. D., Millodot, M., and Haegerstram, G. (1973). *Visual Res.* **13**, 535–558.
Mitchell, D. E., Giffin, F., and Muir, D. (1976). *Exp. Brain Res.* **25**, 109–113.
Mitchell, D. E., Cynader, M., and Movshon, J. A. (1977). *J. Comp. Neurol.* **176**, 53–64.
Moran, J., and Cordon, B. (1982). *Visual Res.* **22**, 27–36.
Mower, G. D., and Christen, W. G. (1982). *Dev. Brain Res.* **3**, 475–460.
Mower, G. D., Berry, D., Burchfiel, J. L., and Duffy, F. H. (1981). *Brain Res.* **220**, 255–267.
Mower, G. D., Caplan, C. J., and Letson, G. (1982). *Behav. Brain Res.* **4**, 209–215.
Muir, D. W., and Mitchell, D. E. (1975). *Brain Res.* **85**, 459–477.
Olson, C. R., and Freeman, R. D. (1980). *Exp. Brain Res.* **38**, 53–56.
Packwood, J., and Gordon, B. (1975). *J. Neurophysiol.* **38**, 1485–1499.
Peck, C. K., and Blackmore, C. (1975). *Exp. Brain Res.* **22**, 57–68.
Peck, C. K., and Wark, R. C. (1981). *Exp. Brain Res.* **44**, 317–324.
Pettigrew, J. D. (1974). *J. Physiol. (London)* **237**, 49–74.
Pettigrew, J. D. (1978). *In* "Neuronal Plasticity" (C. W. Cotman, ed.), pp. 311–330. Raven, New York.
Regal, D. M., Boothe, R., Teller, D. Y., and Sackett, G. P. (1976). *Visual Res.* **16**, 523–530.
Regan, D., Beverley, K. I., and Cynader, M. (1979). *Proc. R. Soc. (London) Ser. B* **204**, 485–501.
Rhoades, R. W., and Chalupa, L. M. (1978). *J. Comp. Neurol.* **177**, 17–32.
Riesen, A. H. (1961). *In* "Functions of Varied Experience" (D. W. Fiske and S. K. Maddi, eds.), pp. 57–80. Dorsey, Homewood, Illinois.
Riesen, A. H. (1966). *Prog. Physiol. Psychol.* **1**, 116–147.
Riesen, A. H. (1971). *In* "Brain Development and Behavior" (M. B. Sterman, D. J. McGinty, and A. M. Adinolf, eds.), pp. 59–70. Academic Press, New York.
Riesen, A. H., and Zilbert, D. E. (1975). *In* "The Developmental Neuropsychology of Sensory Deprivation" (A. H. Riesen, ed.), pp. 211–246. Academic Press, New York.

Riesen, A. H., Ramsey, R. L., and Wilson, P. D. (1964). *Psychon. Sci.* **1**, 33–34.
Rosenzweig, M. R., and Bennett, E. L. (1978). *In* "Studies on the Development of Behavior and the Nervous System" (G. Gottlieb, ed.), Vol. 4, pp. 289–327. Academic Press, New York.
Rothblat, L. A., Schwartz, M. L., and Kasdan, P. M. (1978). *Brain Res.* **158**, 456–460.
Salinger, W. L., Schwartz, M. A., and Wilkinson, P. R. (1977). *Brain Res.* **125**, 257–263.
Schneider, G. E. (1969). *Science* **163**, 895–902.
Schneirla, T. C. (1965). *Adv. Study Behav.* **1**, 1–74.
Scott, J. P. (1962). *Science* **138**, 949–958.
Sherman, S. M. (1973). *Brain Res.* **49**, 25–45.
Sherman, S. M. (1974). *Brain Res.* **73**, 491–501.
Sherman, S. M. (1977). *J. Comp. Neurol.* **172**, 231–246.
Sherman, S. M. (1979). *In* "Developmental Neurobiology of Vision" (R. D. Freeman, ed.), pp. 79–97. Plenum, New York.
Singer, W. (1978). *Brain Res.* **157**, 351–355.
Singer, W. (1979). *In* "Developmental Neurobiology of Vision" (R. D. Freeman, ed.), pp. 135–148. Plenum, New York.
Smith, D. C. (1981a). *J. Comp. Neurol.* **198**, 667–676.
Smith, D. C. (1981b). *Science* **213**, 1137–1139.
Smith, K. U. (1937). *J. Genet. Psychol.* **50**, 137–156.
Spear, P. D., and Ganz, L. (1975). *Exp. Brain Res.* **23**, 181–201.
Spear, P. D., Tong, L., and Langsetno, A. (1978). *Brain Res.* **155**, 141–146.
Stryker, M. P., and Shatz, C. J. (1976). *Soc. Neurosci. Abstr.* **2**, 1137.
Sutherland, N. S. (1968). *Proc. R. Soc. Edinburgh Sect. B Biol.* **71**, 296–317.
Sutherland, N. S., and Williams, C. (1969). *J. Exp. Psychol.* **21**, 77–84.
Tees, R. C. (1968). *Can. J. Psychol.* **22**, 294–298.
Tees, R. C. (1972). *J. Comp. Physiol. Psychol.* **3**, 494–502.
Tees, R. C. (1974). *J. Comp. Physiol. Psychol.* **86**, 300–308.
Tees, R. C. (1976). *In* "Studies on the Development of Behavior and the Nervous System" (G. Gottlieb, ed.), Vol. 3, pp. 281–326. Academic Press, New York.
Tees, R. C. (1979). *Dev. Psychobiol.* **12**, 485–497.
Tees, R. C. (1984). *Behav. Neurosci.* **98**, 969–978.
Tees, R. C., and Midgley, G. (1978). *J. Comp. Physiol. Psychol.* **92**, 742–751.
Tees, R. C., and Midgley, G. (1982). *Can. J. Psychol.* **36**, 488–496.
Tees, R. C., Bruinsma, Y., and Midgley, G. (1976). *Dev. Psychobiol.* **11**, 31–49.
Tees, R. C., Midgley, G., and Bruinsma, Y. (1980). *J. Comp. Physiol. Psychol.* **94**, 1003–1018.
Thompson, W. R., and Grusec, J. E. (1970). *In* "Carmichael's Manual of Child Psychology" (P. H. Mussen, ed.), Vol. 1, pp. 565–654. Wiley, New York.
Timney, B. (1981). *Invest. Ophthalmol. Visual Sci.* **21**, 493–496.
Timney, B. (1984). *In* "Development of Visual Pathways in Mammals" (J. Stone, B. Dreher, and D. Rapaport, eds.), pp. 405–423. Liss, New York.
Timney, B., and Mitchell, D. E. (1979). *In* "Developmental Neurobiology of Vision" (R. D. Freeman, ed.), pp. 149–160. Plenum, New York.
Timney, B., and Peck, C. K. (1981). *Behav. Brain Res.* **3**, 289–302.
Timney, B., Mitchell, D. E., and Cynader, M. (1980). *J. Neurophysiol.* **43**, 1041–1054.
Tsumoto, T., and Freeman, R. D. (1981). *Exp. Brain Res.* **44**, 347–351.
Tumosa, N., Tieman, S. B., and Hirsch, H. V. B. (1980). *Science* **208**, 421–423.
Turkewitz, G., and Kenny, P. A. (1982). *Dev. Psychobiol.* **15**, 357–368.
Van Hof-Van Duin, J. (1976a). *Brain Res.* **104**, 233–241.

Van Hof-Van Duin, J. (1976b). *Brain Res.* **111**, 261–276.
Van Hof-Van Duin, J. (1977). *Exp. Brain Res.* **30**, 353–368.
von Grünau, M. W. (1979). Paper presented at the European conference on Visual Perception, Noordwijkerhante, The Netherlands.
von Grünau, M. W., and Singer, W. (1979). *Exp. Brain Res.* **34**, 133–142.
von Noorden, G. K. (1977). *Adv. Ophthalmol.* **34**, 93–114.
von Noorden, G. K., Dowling, J. E., and Ferguson, D. C. (1970). *Arch. Ophthalmol.* **84**, 206–214.
Walk, R. D. (1978). *In* "Perception and Perceptual Development" (R. D. Walk and H. L. Pick, Jr., eds.), Vol. 1, pp. 77–103. Plenum, New York.
Walk, R. D., and Gibson, E. J. (1961). *Psychol. Monogr.* **75**, 1–44.
Wiesel, T. N., and Hubel, D. H. (1965a). *J. Neurophysiol.* **28**, 1029–1040.
Wiesel, T. N., and Hubel, D. H. (1965b). *J. Neurophysiol.* **28**, 1060–1072.
Wiesel, T. N., and Hubel, D. H. (1974). *J. Comp. Neurol.* **158**, 307–318.
Wilson, J. R., and Sherman, S. M. (1977). *J. Neurophysiol.* **40**, 891–903.
Zolman, J. F. (1976). *In* "Knowing, Thinking, and Believing" (L. Petrinovich and J. L. McGaugh, eds.), pp. 85–114. Plenum, New York.

12

Neural Correlates of Development and Plasticity in the Auditory Somatosensory, and Olfactory Systems

BEN M. CLOPTON
Kresge Hearing Research Institute
The University of Michigan School of Medicine
Ann Arbor, Michigan

I. Introduction

In higher vertebrates the uniqueness of an individual's sensory systems appears to be partially determined by the sensory input to that system during development. Descriptions of experiential effects must, of necessity, involve normative parameters of response for a sensory system, and this has been most elaborated for the visual system. Conceptually, the designation of visual relational entities such as center–surround, edges, lines, and orientation has promoted the analysis of experiential influences. Experiments in the auditory, somatosensory, and olfactory systems have also demonstrated developmental sequences that illustrate or are directly relevant to experiential influences on the development of these systems, but they have been less oriented toward stimulus processing because their central representations of stimulus parameters are, in general, less well known. Our knowledge of these other sensory systems emphasizes the anatomical correlates of normal development and the consequences of experimental intervention on structure and function.

It is difficult to identify the encoding schemes for higher order sensory attributes in the nonvisual senses in the adult. Developmental processes may clarify how the extraction of sensory information in these modalities is accomplished. Most superficially this would rest on correlative relationships between progressive anatomical and physiological changes and behavioral capacities that suggest sensory and perceptual mechanisms. More convincing evidence would be the selective abolishment or enhancement of perceptual processing through experimental intervention known to affect specific neural mechanisms. As the sophistication of intervention increases, the dissection of perceptual processes becomes practical, and the plasticity of sensory processing may promote this approach.

This chapter will selectively summarize the development of neural mechanisms in the auditory, somatosensory, and olfactory systems. The auditory system has seen an especially high volume of work on its development and plasticity recently. In all of these modalities the common approach to the study of development has been either to characterize the "normal" progression of events or to restrict or interrupt afferent input and "deprive" the system. In the somatosensory system the distortion of receptotopic projection patterns has revealed plasticity in central pathways. Few studies have used early exposure to selected, salient stimulus subsets to bias development as have studies in the visual system.

The auditory, somatosensory, and olfactory systems have some distinct advantages for the study of developmental processes. Characteristics of the peripheral sensory organs and their central projections include regenerative capabilities, receptotopic mapping, central interactions between stimuli (e.g., binaural acoustic inputs) that can be independently controlled so as to reveal central mechanisms and precise embryological and postnatal sequences of development. In addition, species-specific characteristics such as extreme variations in developmental histories provide experimental opportunities in these sensory systems.

Development in any sensory system, indeed for any neuron, can be delineated as a sequence of cell birth; migration to a permanent site; growth and differentiation of the soma, dendrites, and axons; maturation of internal biochemical processes; efferent axonal growth; contact and interaction with other neurons; establishment of efferent and afferent synapses; and other morphological and functional events including cell death. Cell death may, in fact, come quite early as it appears to play a central role in the regulation of the size of neuronal populations at initial stages of maturation. In part of the olfactory system, cell death and regeneration occur throughout a lifetime. If we can extrapolate from available experimental evidence, often based on admittedly crude intervention in the lives of neurons, many of the developmental processes in the nonvisual sensory systems can be influenced by "experiential" conditions in the broad sense. A brief survey of the normative development in the somatosensory, auditory, and olfactory systems will provide a reference against which plasticity in the systems will be discussed.

II. The Auditory System

A. Maturation Times

The following summary of events characterizing development in auditory pathways is derived from studies in mice, albino rats, gerbils, chickens, and, to a lesser extent, cats. Rats, mice, and chickens have been the most commonly chosen species for studies of auditory development because of their well-known maturational sequences, their reproductive rates, and their convenience for experimentation in relation to larger species. The chick embryo has an especially well-defined embryonic life (Hamburger and Hamilton, 1951) and is easily studied at early stages of growth in the egg (Levi-Montalcini, 1949).

Since pre- and postnatal time references are necessary, gestational days for the mammals will be designated, for example, as G18, designating the eighteenth day of gestation, and embryonic days for the birds will be noted by the letter E. The letter P will indicate postnatal and posthatching days for both classes. Mice and rats have gestational times of 19 and 21 days, respectively, that is, G19 = P0 for mice, and G21 = P0 for rats. Embryonic times for chickens run to about 21 days at which time hatching occurs, that is, E21 = P0.

B. Onset of Peripheral Transduction

The development of the auditory system is highly variable among species. Extensive prenatal development in some mammals is associated with reactivity to sounds in the uterine environment. The human fetus responds to intense sounds up to 16 weeks before birth (Ando and Hattori, 1970; Johansson *et al.,* 1964; Murphy and Smyth, 1962; Sakabe *et al.,* 1969). At the other extreme, some mammals such as the mink and opossum, do not respond to sounds for a month or more after birth (Foss and Flottorp, 1974; McCrady *et al.,* 1937). Birds have functional auditory systems long before hatching so that airborne sounds can activate their developing auditory pathways by conduction across the shell. In the chick embryo this begins at about E14 (Jackson and Rubel, 1978).

Mammals commonly used for the study of auditory development have the onset of hearing in the first few weeks after birth, the cat at about P5, and myomorph rodents such as mice and rats at about P10. Limitations on hearing in these animals are a conductive block of sound transmission from the eardrum to the inner ear and maturational processes in the cochlea. Mesenchyme, which has filled the middle ear to this time, is resorbed to allow ossicular movement and vibrations of the eardrum and round window. Mechanical changes in the cochlea appear to remove obstructions to vibration of the basilar membrane and associated structures (Hinojosa, 1977).

Physiological indices of peripheral auditory maturation are the appearance of the endocochlear potential (EP) and the cochlear microphonic (CM). The EP is a large positive voltage maintained in the fluid of the scala media of the cochlea by metabolic activity of the stria vascularis, which lies on one wall of this compartment (Bosher and Warren, 1971). This potential rises from essentially zero to 80–90 mV over just a few days in the rat pup as peripheral transduction of sounds into neural activity begins (Bosher and Warren, 1971). The rise of the EP in the kitten is much more prolonged, lasting most of the first month of life (Fernandez and Hinojosa, 1974). The CM appears at this time, resulting from the modulation of currents promoted by the EP due to the changing resistance of hair cells induced by mechanical displacement of their stereocilia (Davis, 1965). These physiological events correlate with the appearance of Preyer's reflex to intense sounds (Foss and Flottorp, 1974). The onset of peripheral function thus appears to be the limiting factor for response to sounds.

Cochlear maturation, according to a number of anatomical features, proceeds from base to apex in humans (Bast and Anson, 1949), rodents (Hinojosa, 1977; Ruben, 1967), and kittens (Pujol and Marty, 1970). In the mouse Ruben (1967) demonstrated that cells of the embryonic spiral ganglion destined for the basal turn undergo their final division on about G12 while those of the apical turn appear on G15–G16. The axons of the eighth nerve in cat are myelinated in this sequence (Romand and Romand, 1982), and in the hamster the dorsal cochlear nucleus (DCN) appears to receive its initial ingrowth of cochlear afferents from basal fibers (Sweitzer and Cant, 1984). This receptotopic gradient of maturation would seem to imply that initial sensitivity would be greatest to high-frequency sounds, but this has not been observed (Pujol and Marty, 1970). Mechanical changes in the middle ear and cochlea appear to limit high-frequency sensitivity (Relkin and Saunders, 1980; Relkin *et al.,* 1979). Recordings of CM from the developing gerbil cochlea point to maturational changes in the mechanics underlying tuning. Changes in frequency response indicate that the traveling waves for lower frequencies are initially maximal nearer the base of the cochlea than in the adult, with responses to higher frequencies coming with age (Harris and Dallos, 1984). Evidence from the developing chick indicates that the frequency map of the cochlea changes with development; low frequencies initially induce membrane displacement more basalward, subsequently shifting toward apical sites (Rubel and Ryals, 1983). High frequencies progressively activate basal sites as the frequency map shifts. As will be seen, a similar gradient of maturation for central structures is seen (Lippe and Rubel, 1983).

C. Central Pathways

Auditory pathways in mammals consist of the cochlear hair cells, neurons of the spiral ganglion, the cochlear nuclei (CN), the nuclei of the superior olivary complex (SOC), the nuclei of the lateral lemniscus, the inferior colliculus, the medial geniculate of the thalamus, and numerous cortical areas including some primarily associated with auditory processing. Obligatory synapses occur at the bases of inner hair cells, in the cochlear nuclei, the inferior colliculus, and the thalamus with many less direct connections. In birds the homologs of the CN are the nucleus magnocellularis (NM) and nucleus angularis (NA). The NM appears homologous to the anteroventral CN (AVCN), and the NA to the posteroventral and dorsal CN (PVCN and DCN). The nucleus magnocellularis projects to both the ipsilateral and contralateral nucleus laminaris (NL), which is homologous to the medial superior olive (MSO) in mammals (Campbell and Boord, 1974).

Many investigations on the development and plasticity of the auditory system have involved brainstem auditory nuclei because their anatomy and physiology are most extensively documented. Tonotopic mapping due to organized cochleotopic projections is well established in these nuclei and can serve to distinguish neuronal subsets in the afferent input to a nucleus. Some secondary brainstem nuclei, the MSO of mammals and NL of birds, receive afferents bilaterally from the primary nuclei. Thus, in addition to providing a spatial representation of frequency, they compare binaural inputs. In birds and mammals these comparisons take place at neurons with bipolar dendritic trees, one side receiving input from ipsilateral and the other from contralateral stimuli. Such anatomical connections invite experimental investigation of relational interactions between independent stimuli from the two ears. They represent a simple form of integration in the brainstem that is easily studied and often involved in plasticities of the system.

D. Central Development

The birth of neurons in the cochlear nuclei of mice begins at about G10 and continues until after birth (Taber-Pierce, 1967; Mlonyeni, 1967; Martin and Rickets, 1981). These neurons undergo their final mitosis on the rhombic lip of the developing brainstem and migrate to their adult positions. Neurons that will eventually be the largest are the first to appear, and successively smaller ones are subsequently born until birth. Nine morphological types of neurons can be identified in the Nissl-stained mouse CN, which has three major divisions, the AVCN, PVCN, and the DCN (Martin, 1981). Each of these divisions, in turn, has characteristic distributions of its neuronal morphological types which indicate subdivisions.

Figure 1 summarizes the "birth" or final mitosis sequences for neurons in subdivisions of the mouse CN (Martin and Rickets, 1981). Larger neurons of the DCN appear on G10–G10.5. In the VCN, larger neurons have their terminal mitosis on G12–G13. Two successive waves of neuronal birth for DCN neurons occur. Smaller granular neurons of the VCN have a peak birth period at G15. Glial cells are born from P0 to at least P14. In chick embryos, neurons that constitute the NM arise entirely at E2.5, whereas those going to make up the NL are born about 1 day later on E3.5–E4 (Rubel *et al.,* 1976).

After undergoing terminal mitosis, these neurons must then migrate to their adult loci and develop mature axonal and dendritic processes. In the mouse the birth of cochlear hair cells and neurons of the spiral ganglion occurs after P0 (Ruben, 1967). This sequence is in agreement with peripheral development, which limits initial responses to sounds in mice at P10 or later. The chick is precocial in this respect since experimentally induced sounds evoke

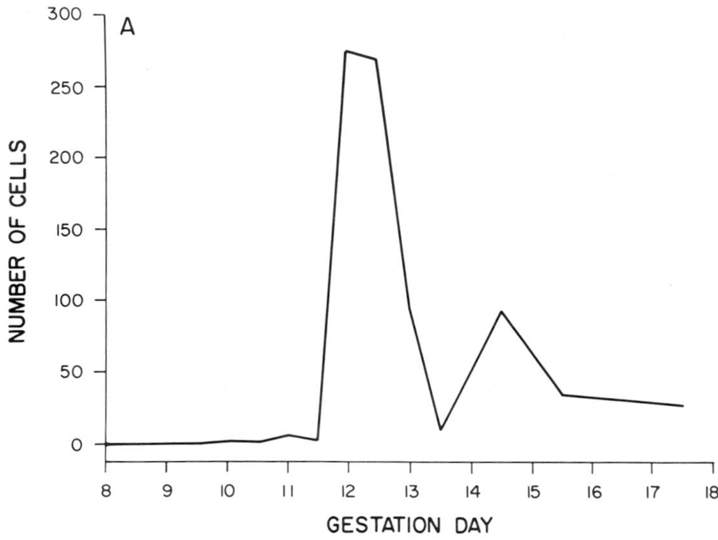

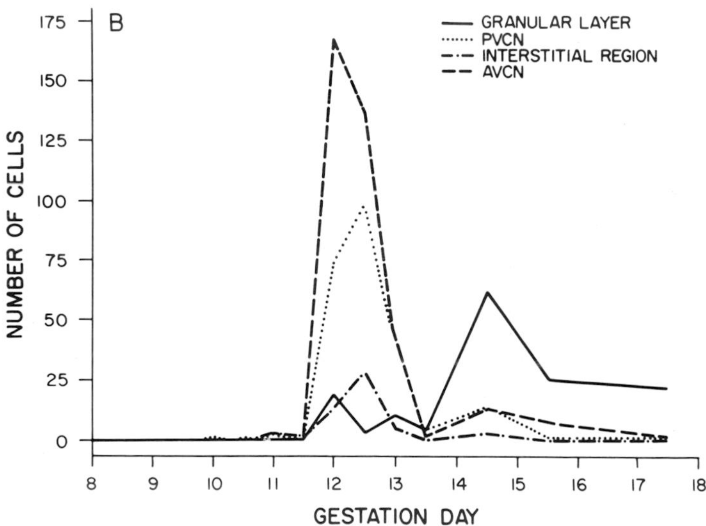

Fig. 1. Dates for the final mitosis or "birth" of different classes of auditory neurons in the VCN (A,B) and the DCN (C,D). The appearance of total neuron populations for the ventral and dorsal nuclei is shown by the single lines in (A) and (C), and subpopulations are given in (B) and (D). From Martin and Rickets, (1981) *(continued).*

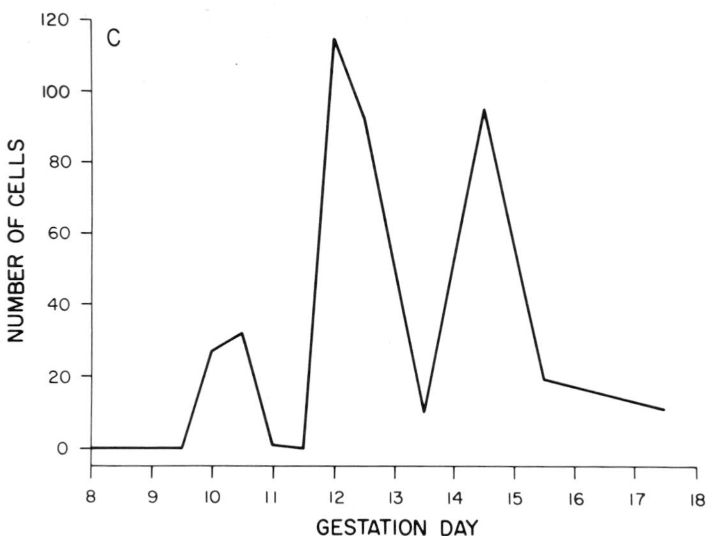

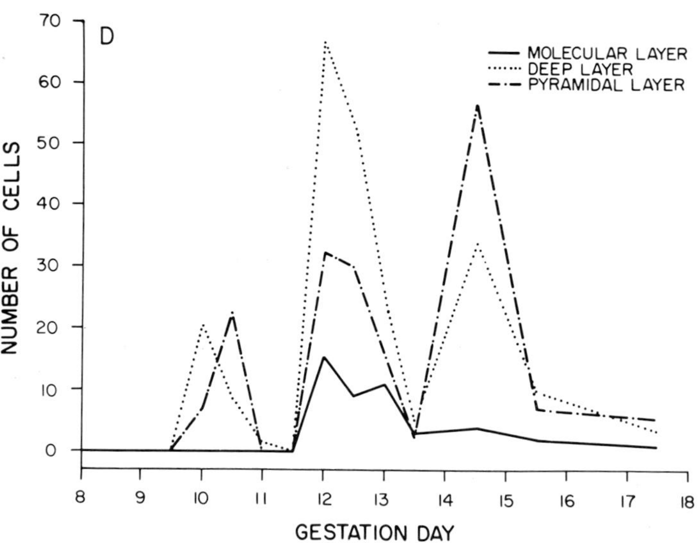

Fig. 1. *(Continued.)*

responses at brainstem nuclei as early as E11 (Saunders *et al.,* 1973), about 1 week after the final mitosis of NM and NL neurons and following development of the cochlea (Knowlton, 1967). The processes of birth, migration, and initial differentiation in central auditory neurons are not influenced by the presence or absence of nearby cochlear tissue. However, the influence of afferents arriving later from the cochlea is central to the development of CN neurons and higher nuclei.

E. Somatic, Dendritic, and Axonal Stabilization

Following the birth, migration, and arrival of a neuron at its destination, a sequence of temporally overlapping events takes place that determines the connectivity and the survival of the neuron. These events include somatic and nuclear growth, the ingrowth of extrinsic afferents, synaptogenesis, the dynamics of dendritic elaboration, pruning, and stablization, and, in many instances, cellular death. This period thus encompasses the establishment of functional connections and the determination of whether or not a neuron will survive.

In the cochlear nucleus of the 1- or 2-day-old kitten, as many as 10 cellular types can be identified in Nissl-stained material (Larsen, 1984). During the first postnatal month the cytoplasmic and nuclear volumes increase rapidly, then a slower growth is observed for as long as 3 months after birth. Some larger neurons, such as the large spherical cells of the AVCN, are the last to reach their adult sizes even though they are the first to be born. Smaller neurons of the CN, possibly medium-sized cells with short axons that serve as interneurons, have shorter growth periods. A relationship many exist in the AVCN between the maturational gradient of the cochlea and celluar growth. Smaller spherical cells purported to respond to high frequencies (Rose *et al.*, 1959) mature more rapidly than the larger spherical cells that receive more inputs from apical regions of the cochlea (Osen, 1969).

A general sequence can be roughly given for dendritic maturation. As afferent axons approach the developing neurons, fine dendritic processes grow from the soma. These dendritic trees extend for about the same distances from the soma as they do in the adult but have more branches and are much smaller in diameter. A quantitative analysis of developing motoneurons in the spinal cord (Tatton *et al.*, 1983) indicates that this proliferation of "netlike" trees can greatly increase an approaching axon's chance of encountering a dendrite; the chance of an axonal–dendritic contact is 3 to 12 times greater in this early period. The dendritic trees then undergo a pruning process, with a reduction in the number of branches and an increase in the thickness of the dendrites. Although variations on this process are seen, it is remarkably common in sensory and motor pathways.

Specific descriptions for the establishment of mature axonal or dendritic structure in the auditory brainstem have been given for the NM (Jhaveri and Morest, 1982), the trapezoid body (Morest, 1968, 1969b), medial superior olive (Rogowski and Feng, 1981), NL (Smith, 1981), and inferior colliculus (Meininger and Baudrimont, 1981). Morest (1969a) presented an overview of dendritic development for many areas of the brain by noting the appearance of growth cones at the tips of dendrites with filopodia, both of which disappear when the neuron reaches maturity. The persistence of many of these growth cones indicates that modification of the anatomical structure of dendrites may extend beyond initial periods of development. Specific neurons may have restricted periods during development when their dendrites undergo these changes, and there may be regional gradients across which a progression of dendritic development can be observed at any one time. In general, the dendrites of large neurons tend to stabilize first, while the dendrites of smaller, Golgi type II (short axon) neurons differentiate later. These statements apply to thalamic neurons with exceptions notable in other areas. It is the case in the auditory system, as in other sensory systems, that neuronal maturation tends to occur in a temporal sequence from the periphery to cortex (Hendrickson and Rakic, 1977; Woolsey *et al.*, 1981).

Rogowski and Feng (1981) examined the stabilization of neurons in the MSO of rat during the postnatal period using Nissl and Golgi techniques. They characterized the development of this nucleus of the SOC as having an initial stage of rapid development followed by a stabilization phase. The first phase covered P0 to about P14, and it was in this period that a large increase in the diameter of the soma, a mean of 6.8 μm at P0 to 10 μm at P8, was followed by a decrease to about 8 μm by the end of the stage of rapid development. These observations, drawn from Nissl material, are probably correlated with dendritic proliferation during the initial period. Near the end of this period the external meatus opens, and the central pathways are subject to increased sound-induced afferent activity. Golgi material indicated that dendrites were numerous at about P16, the beginning of the stabilization phase. They were relatively thick at this time, and smaller processes arose from the branches. From about P14 to P18 a significant decrease in the number of dendritic branches occurred. Figure 2 compares MSO neurons at P12 and P30, illustrating the results of the stabilization process.

F. Plasticity of Neuronal Stabilization

The mechanisms of early plasticity in the auditory system center around processes that stabilize neuronal anatomy and physiology. Due to technical and interpretive difficulties with electrophysiological approaches, anatomical

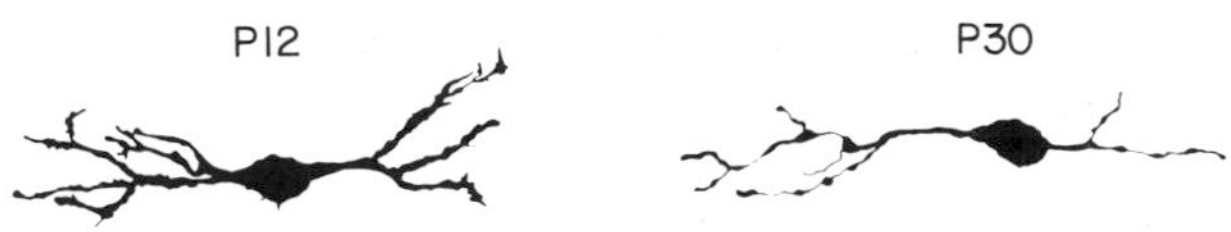

Fig. 2. Bipolar neurons in the MSO of albino rats for P12 and P30. Stabilization for the dendritic trees is evident as a thinning of dendrites with a reduction in the number of branches. From Rogowski and Feng (1981).

observations are more abundant. A brief description of dendritic and somatic stabilization in nuclei concerned with binaural processing will be given because such nuclei can best be related to the available physiological evidence.

The MSO and NL are the homologous nuclei that process binaural input. In both cases anatomical plasticity has been demonstrated under conditions involving the manipulation of early sound experience. These nuclei have populations of neurons with bipolar dendritic trees, input to one tree coming from the ipsilateral CN and input to the other coming from the contralateral CN. this anatomical arrangement provides the opportunity for the integration of sound information from the two ears. The existence of such neurons was predicted by Jeffress (1948) from psychophysical considerations, but the precise manner in which time and intensity cues at the ears are combined is not yet completely understood. It is clear, however, that such neurons provide insight into an integrative process whose inputs are independent and controllable with significant experiential plasticity during development.

A study by Smith (1981) in the chick embryo examined the maturation of dendrites and somas in NL neurons. At E8.5 to E9 the neurons in NL have reached their characteristic places and have axonal projections, but afferent synapses are either not yet present or at least not yet functional. Soon afterward up to 20 to 30 fine dendritic sprouts arise, and a gradient for dendritic tree size in the rostromedial-to-ventrolateral direction begins to develop. This gradient corresponds to the frequency axis of the nucleus with high-frequency afferents entering the rostromedial portion and low-frequency afferents reaching the ventrolateral part of the nucleus. An extensive proliferation of dendrites occurs during the E9 to E16 period, with significant pruning following toward the end of this period, ultimately resulting in fewer branches for the dendritic tree. Differential proliferation and pruning leads to 3 to 12 times as many dendritic trees in the ventrolateral portion of the nucleus. Also, during this period, a high rate of cell death is observed. Synaptic formation is extensive and widespread at this time, suggesting that afferent synapses play an important role in determining not only dendritic extent but also overall neuronal survival.

Several investigators have provided evidence that afferent sensory activity influences these events. Smith (1981) demonstrated that the intensity and duration of noise in a given frequency range activating a segment of the NL

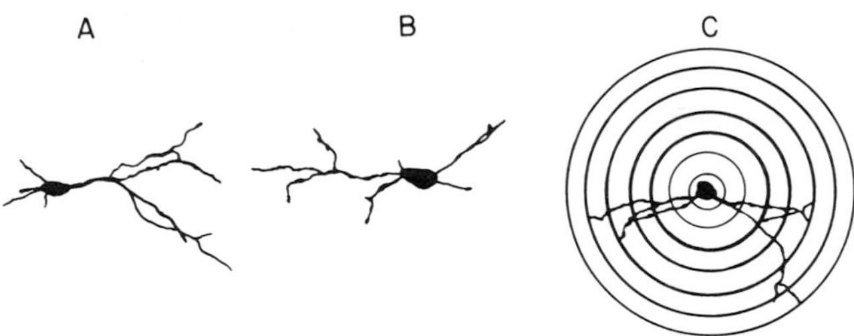

Fig. 3. Three classes of dendritic morphology in the MSO of rats include right dominance (A), left dominance (B), and relatively balanced (C) dendritic trees. Monaural deprivation during development significantly changes the numbers of neurons in these categories, dominance of dendritic trees on the side of the undeprived ear becoming more common. Binaural deprivation does not produce significant changes in dominance patterns. From Feng and Rogowski (1980).

correlated with measures of dendritic growth. This implies that dendritic development is under the control of sensory input in this instance. These results are supported by observations in the albino rat from Feng and Rogowski (1980). In their experiment, monaural and binaural sound deprivation was induced from P12 to P60 by occluding the external ear canal of one or both ears. At the end of this period, as shown in Fig. 3, neurons in the MSO showed significant changes in the probability of dendritic asymmetry relative to undeprived normals. Dendritic trees in monaurally deprived animals tended to be much less developed in extent and branching on the side innervated by the deprived ear but normally developed on the other side, which received normal input. Binaural deprivation does not produce such asymmetries. As with the chick, the relative level of afferent activity in the rat appears to determine the size and branching of dendritic trees at this site of binaural processing.

G. Functional Correlates of Plasticity

Our understanding of the relationship between anatomical asymmetries induced in the brainstem auditory system by developmental deprivation and functional deficits is not complete, but converging evidence from physiological and behavioral studies complements the structural findings. Single-unit recording at one of the two bilaterally paired inferior colliculi of rats indicates that early monaural deprivation causes an almost total loss of the integration of afferent activity from the two ears. In the adult rat with normal, binaural experience, this consists of the suppression of responses to clicks presented to the opposite (contralateral) ear by more intense clicks presented to the ear on the same side (ipsilateral) as the recording. After monaural deprivation during the first weeks of hearing, ipsilateral suppression evoked by sounds to the deprived ear is absent, but these sounds continue to evoke activity in the contralateral colliculus, indicating that the ear and some of the central pathways remain functional (Silverman and Clopton, 1977). Binaural deprivation during the same period does not strongly affect binaural interaction. These results are obtained only if the monaural deprivation falls within a sensitive period from about P10 to P30 (Clopton and Silverman, 1977). As illustrated in Fig. 4, these observations imply a period of functional plasticity in brainstem auditory processing that agrees well with the anatomical findings previously discussed. A direct association between changes in the structure of MSO neurons and the IC physiological observations has not been made, but it is likely that these changes reflect the same binaural comparators located in the SOC with projections to the IC. Monaural deprivation during early development has been shown to disrupt localization behaviors in guinea pigs (Clements and Kelly, 1978).

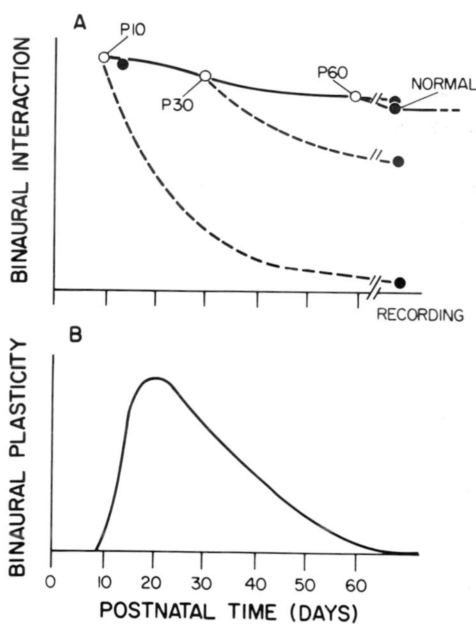

Fig. 4. Binaural interaction in neurons of the inferior colliculus, as indicated by the suppression of responses to contralateral clicks by ispilateral clicks, is shown in (A). Interaction is present soon after the onset of hearing in the rat, but is is abolished by monaural deprivation from P10. Deprivation from P60 has no effect, and an intermediate effect is produced by deprivation from P30. A period of plasticity can be inferred as in (B). (●) Actual observations (○) inferred data points; (—) presumed course of binaural interaction in normal development; (---) changes in interaction for deprivation from P10, P30, and P60. From Silverman and Clopton (1977); Clopton and Silverman (1977).

III. The Somatosensory System

A. Vibrissal Receptors

The somatosensory system has diverse inputs coming from an array of unique receptor structures. In a sense it can be considered to have multiple systems; thus, a discussion of its development and plasticity must be addressed to a restricted set of receptors and their central projections. Numerous studies pertaining to plasticity have involved somatosensory spinal mechanisms (e.g., Liu and Chambers, 1958; Mendell *et al.*, 1978), but more recent investigations have focused on the mystacial vibrissae of rodents (Woolsey *et al.*, 1975; Woolsey *et al.*, 1981). These hair-follicle receptors project through the trigeminal ganglion to the contralateral somatosensory cortex.

B. Central Receptotopic Organization

All stations of the projections, except for the ganglion (Gregg and Dixon, 1973), demonstrate an isomorphic receptotopic organization (Nord, 1967; Belford and Killackey, 1979; Woolsey, 1967). The stimulus dimensions encoded are not well understood, but it is likely that the surface texture and shapes of objects encountered are analyzed by the vibrissal system (Shipley, 1974). This conclusion is reached from observations of neural responses in the trigeminal nuclei when vibrissae are stimulated with varying time-displacement parameters. These temporal correlates of stimulus properties are difficult to relate to anatomical correlates of plasticity, but the receptotopic organization provides a robust reference against which to gauge experimental effects. Specifically, the effect on central representation in the more mature animal of disrupting the peripheral receptotopic organization soon after birth is determined.

The projections of the vibrissal system include the trigeminal ganglion (first-order neurons), the principal nucleus of layer V and spinal trigeminal complex (second-order neurons), somatomsensory portions of the contralateral thalamus (third-order neurons), and the sensorimotor cortex. Staining for the metabolic enzyme succinic dehydrogenase reveals the details of segmentation in the brainstem and thalamic stations that represent the row arrangements of the whiskers (Belford and Killackey, 1979). Subcortical organization appears as "barreloids," while the cortical organization, evident especially in layer IV of sensorimotor cortex, has evoked the appellation of "barrels" because of the three-dimensional arrangement of the cells (Woolsey and Van der Loos, 1970; Welker and Woolsey, 1974). The vibrissal row arrangement that is maintained throughout the central pathways is shown in Fig. 5. It consists of four to eight whiskers in each row across five rows. The rows are labeled A through E in the dorsal-to-ventral direction (Rice and Van der Loos, 1977).

C. Development of Vibrissal Pathways

As with the time course of development in the visual and auditory systems, that in the vibrissal pathways is from peripheral receptors to cortical projection areas. As summarized by Woolsey *et al.*, (1981), the final mitosis of cells for the peripheral receptors occurs at about E12; trigeminal neurons are born on E11 to E13; the neurons of the brainstem nuclei arise at E10 to E11; the thalamic neurons, on E10 to E14; and neurons of the cortical barrels divide at E15 to E17. The cells of the peripheral receptors reach adult form almost immediately, but the neurons of the central nuclei and cortex take from P0 to P4 to attain their adult appearance. The development of the barrel field in the cortex has been described in some detail (Killackey and Belford, 1979; Rice, 1974; Rice and Van der Loos, 1977). The overall picture is of a gradient with cortical maturation occurring last. This sequence suggests that organizational "instructions" are passed from the periphery to central stations during the developmental sequence (Woolsey, 1978).

D. Plasticity of Cortical Barrels

When the pattern of vibrissae is disrupted by cauterization at the periphery in the newborn animal, the corresponding pattern of barrels in the cortex is also distorted when observed later (Van der Loos and Woolsey, 1973). This is not due to the degeneration of an established pattern at the cortex but to the induction of an aberrant pattern from the periphery onto a developing cortex (Jeanmonod *et al.*, 1977). The destruction at the periphery prevents the development of barrels at the cortex; the disappearance of barrels that are already present is not seen. Neighboring barrels expand into the cortical area normally devoted to the damaged barrels. Similar distortions in the thalamic barreloids can be seen. In both cases a sensitive period for the effect can be demonstrated with cortical plasticity for peripheral changes up to P7 while thalamic plasticity ends on about P5 (Woolsey *et al.*, 1979). As with initial maturational processes, plasticity appears to decline in a peripheral-to-cortical sequence.

The changes in the gross mystacial pattern seen in the cortex after peripheral

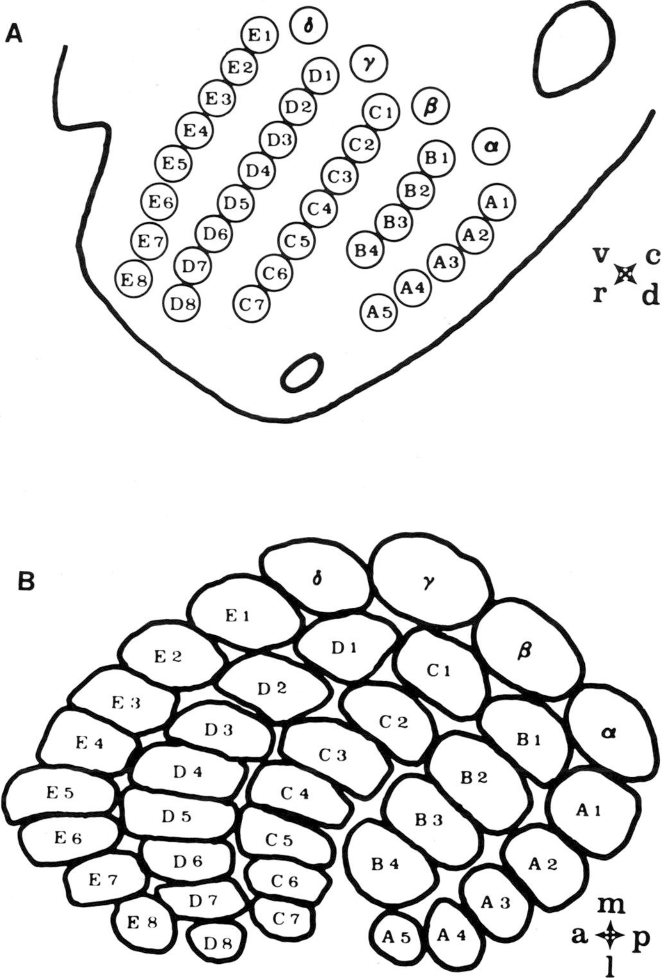

Fig. 5. Pattern of mystacial vibrassae in the rat. This pattern is carried throughout the trigeminal pathway. In (A) the arrangement and nomenclature for the whiskers are indicated for the face, and in (B) the topology for the corresponding pattern in the cortex is shown. From Woolsey *et al.* (1981).

cauterization are also observed in specific details of dendritic development in the area. Harris and Woolsey (1981) used electrocautery of row C of the pattern in mice to determine how anatomical parameters in two types of layer-IV cortical neurons lying in the row-C area were changed. No differences in soma size or measures of dendritic tree branching or extent were observed in Golgi preparations, but the dendritic territories of neurons in the row-C area were significantly different. Although the extent of the territory did not change, the orientation of the tree was such that it entered the adjacent row B and row D areas. The sensitive period during which the dendritic orientations could be affected included only P1. Only 32% of the row-C dendritic-tree territories fell within their barrel for a P1 destruction of row-C whiskers, but 57 to 69% fell in the appropriate barrel for P2 and P4 manipulations or for undisturbed normals.

Harris and Woolsey (1981) reported in more detail on dendritic territories in this preparation and observed that cortical effects from peripheral destruction of row C held until P3. Neurons in cortical areas designated as row B or D were determined to be row-C neurons whose dendritic trees were reoriented toward these neighboring rows. Competition for active thalamic afferents thus appeared as shifts in tree morphology and "expansions" of the undamaged rows at the expense of the damaged one. The absence of changes in soma size and the extent of dendritic territories suggests genetic control of these neuronal parameters. Spatial orientation of the territories, however, remains plastic during the immediate postnatal period and is influenced by afferent activity relayed through the thalamus.

IV. The Olfactory System

A. Organization and Stimulus Encoding

Although the olfactory system is phylogenetically ancient and pervasive in terrestrial vertebrates, knowledge of how it is stimulated and olfactory qualities encoded is not well developed. No support for receptotopic organization that is representative of stimulus properties or other conveniently manipulable receptor parameters exists for this sytem. However, it offers great opportunity for observing plasticity because of high levels of cellular death and renewal throughout development and adulthood (Graziadei and Monti Graziadei, 1978a,b). The neuroepithelium of the nasal cavity contains bipolar, primary neurons that encode olfactory stimuli. Each neuron has a distal, ciliated dendrite extending to near the surface of the neuroepithelium where it transduces olfactory stimuli, possibly responding to a variety of olfactory stimuli (Van Drongelen, 1978). The topography of the neuroepithelium, although functionally obscure, projects in an orderly manner through unmyelinated axons collected into discrete bundles passing through the lamina cribrosa of the ethmoidal bone to the nearby olfactory bulb of the brain. The primary neurons undergo continuous death and renewal, so the connections of new neurons to the olfactory bulb must be

Fig. 6. Neurons of the olfactory bulb and their connections as shown by Ramon y Cajal (1911) from Golgi-stained material. The major cell types, discussed in the text, are the mitral cell (M), the granule cell (G), and the tufted cell (T). Axons from basal cells of the neuroepithelium enter (above) to form glomeruli, and efferents from mitral and tufted cells project through the lateral olfactory tract (below). From Shepherd (1970).

remade continuously (Graziadei and Monti Graziadei, 1978a,b). Basal cells in the neuroepithelium provide the reservoir of undifferentiated primary neurons. Activation of basal cell differentiation is tied to the number of degenerating primary neurons, and axotomy at the lamina cribosa stimulates this process by producing massive loss of epithelial neurons.

The olfactory bulb has a layered structure with distinctive neuronal types. Ramon y Cajal's (1911) representation of the layered structure of the bulb is shown in Fig. 6. The mitral cell is the principal neuron. Its apical dendrite forms a glomerulus in the most external (glomerular) layer with afferents from the primary neurons of the olfactory epithelium, and its axon projects centrally through the lateral olfactory tract as the major output of the bulb. It is in the glomeruli that reordering of connections must be continual, as portions of the converging afferents degenerate and new axons arrive. Just below the glomerular layer, the basal dendrites of the mitral cells spread laterally within the plexiform layer that lies between the glomerular layer and the layer of mitral cell bodies. Tufted cells have cell bodies lying in the plexiform layer and are similar to mitral cells although their projections are less well defined. Only mitral and tufted cells have efferent projections from the bulb. These go to make up the lateral olfactory tracts that project widely to phylogenetically older portions of the cortex and underlying structures. The granule cell is a special neuron that does not have an axon but has only basal and apical dendrites. It appears to provide reciprocal inhibition to mitral cells and has been compared to the amacrine cell of the retina (Shepherd, 1970). Its cell body lies more centrally than that of the mitral cell, and its apical dendrites project into the plexiform layer. The granule cell population continues to increase long after birth in the rat (Kaplan and Hinds, 1977). Periglomerular cells lie near to and connect neighboring glomeruli with their short axons.

B. Early Development

The primary neurons of the neuroepithelium in rodents are present and have initiated axonal projections to the bulb on E10 (Cuschieri and Bannister, 1975; Graziadei and Monti Graziadei, 1978a. Their axons reach the bulb on E11, but thc cilia on their distal dendrite, which are necessary for sensory transduction, are not seen until E13. They continue to mature into the postnatal period (Leonard, 1981). The mitral cells of the bulb are born between E10 and E16, and the arrival of axons from the neuroepithelium apparently initiates a prolonged interaction between the primary neurons and mitral cells, which supports maturational processes in both (Hinds, 1968a,b). Axons projecting from the olfactory bulb through the lateral olfactory tract are observed in the pyriform cortex at birth, and their numbers grow for weeks afterward (Westrum, 1975; Leonard, 1975).

C. Plasticity

Investigations of plasticity in the olfactory system have involved experimentally produced anosmia by blockage of nares (Meisami, 1976), destruction of the neuroepithelium with a bath of zinc sulfate (Smith, 1938), and surgical interruption of olfactory pathways either at the lamina cribosa or lateral olfactory tracts (Westrum, 1975). The unique regenerative properties of olfactory projections makes transections a viable approach to determining what mechanisms regulate connectivities in this system.

Transection of primary axons going to the bulb causes the degeneration of their synaptic contacts in the glomeruli in only a few days (Graziadei and Monte Graziadei, 1978a). In 20 days the axons of regenerating primary neurons have entered the bulb again, and their contacts with the glomeruli are well established by 30 days. Even without disruption of the afferent axons, the bulb normally undergoes lesser degrees of afferent degeneration and reconnection as primary neurons die and are replaced. Allison and Warwick (1949) determined that the rabbit has 50 million primary neurons converging onto less than 2000 glomeruli that affect 45,000 mitral cells. The maintenance of a functional sensory system under conditions of continual renewal and great convergence is remarkable and not at all understood.

Use of the relatively mild deprivation caused by cauterization and subsequent blockage of one nostril leads to a readily apparent underdevelopment of the olfactory bulb on that side (Meisami, 1976; Meisami and Manochehri, 1977). When this is done near birth, the deprived bulb weighs 25% less than the nondeprived bulb by P60, and its external dimensions are correspondingly reduced. Deafferentiation of the bulb by transection at the lamina cribosa causes a greater effect, a 41% reduction in weight at P25 being seen in one study. The effects of deafferentiation are complicated by nutritional deprivation common with anosmic animals. Both unilateral naris blockage and bulb deafferentiation cause a number of measurable changes in the deprived bulb including enzyme activity associated with membrane pumps' maintaining intracellular potentials and synaptic activity (Meisami, 1978).

The disruption of the lateral olfactory tracts early in life leads to changes in projections of mitral cells to the cortex (Devor, 1976; Westrum, 1975). Such disruption has been utilized to investigate principles of synaptic formation. It has been observed that mitral cells will form alternative synaptic contacts, but their total dendritic arborizaton tends to be conserved (Devor, 1976). Loss of olfactory afferents in the pyriform cortex leads to the reinnervation of cortical neurons by nearby tract axons (Westrum, 1975). In summary, the plasticity of the olfactory system provides a useful model not yet fully exploited for the manipulation of synaptic degeneration and formation.

V. Dimensions of Plasticity

The correlates of plasticity in the auditory, somatosensory, and olfactory systems have been documented primarily with anatomical examples. Observations in the auditory system suggest orderly sequences for the maturation of individual neurons and formation of synaptic connectivity.

Incoming afferent axons appear to regulate the maturational process, and beyond that, the activity in those afferents affects that regulation. The elaboration of dendritic trees has been identified as a major correlate of plasticity having an initial stage largely independent of afferent input and a period of stabilization subject to extrinsic influences. At the same time, the volumetric domain of dendritic trees seems to be largely predetermined, as evinced by studies of plasticity in the somatosensory cortex, (in possible contrast to some visual cortex studies; see Chapter 10, this volume). Evidence from these modalities, as from elsewhere in the nervous system, implicates the presence and activation of synapses as being critical to the full realization of adult neuronal morphology and survival in at least some populations.

Except for a few auditory studies, little relationship has been established in these nonvisual systems between stimulus parameters and either anatomical or physiological indices of plasticity. The investigations on early monaural deprivation have successfully associated physiological and anatomical changes in brainstem centers. The lack of afferent activity from the deprived side restricts the extent of dendritic trees that receive afferents from that side. The physiological effects of reduced dendritic trees can be predicted to some extent since temporal and spatial integration are governed partially by the structure of dendritic trees (e.g., Rall, 1969; Rall and Rinzel, 1973). Such considerations make a possible quantitative characterization of changes in binaural integration due to early monaural deprivation in terms of physiology of single neurons and their processing of time and intensity differences of sounds at the two ears.

VI. Implications for Behavior

The transition from changes in neuronal anatomy, physiology, and biochemistry to their behavioral consequences is difficult. At one level one must take deficits in behavioral capacities after developmental stimulus deprivation or disruption of pathways as clues to underlying mechanisms. However, the literature on behavior is replete with behavioral deficits subsequently identified as deficits in the testing paradigm. The nature of animal behavior requires caution in the interpretation of behavioral correlates of structural and functional changes in brain tissue. To infer that modified dendritic fields for neurons in a nucleus or in one of its subpopulations are responsible for a significant behavioral change requires evidence that does not yet exist in most cases. Correlative relationships clearly hold between some gross behavioral competences and neuronal measures in the auditory system. They have been demonstrated primarily through effects on sound thresholds and binaural interactions.

As suggested at the beginning of this chapter, a major potential of developmental observations is that variation of normal or modified development will expose relationships between neuronal mechanisms and behavior. This promise has begun to be realized in the nonvisual senses, and as parametric studies relate stimulus and response measures to normal or induced formative neuronal changes, our understanding of both the developing and adult brain will increase.

Acknowlegments

Preparation of this manuscript was partially supported by grant no. NS18692 from the National Institutes of Health. The author would like to thank Dr. Susan Shore and Ms. Yolande Au for their suggestions.

References

Allison, A. C., and Warwick, R. T. (1949). *Brain* **72**, 186–197.
Ando, Y., and Hattori, H. (1970). *J. Acoust. Soc. Am.* **47**, 1128–1130.
Bast, T. H., and Anson, B. J. *The temporal bone and the ear.* (1949). Thomas, Springfield, Illinois.
Belford, G. R., and Killackey, H. P. (1979). *J. Comp. Neurol.* **183**, 305–322.
Bosher, S. K., and Warren, R. L. (1971). *J. Physiol.* (*London*) **212**, 739–761.
Campbell, C. G. B., and Boord, R. L. (1974). *Handb. Sens. Physiol.* **5**, 337–362.
Clements, M., and Kelly, J. B. (1978). *J. Comp. Physiol. Psychol.* **92**, 34–44.
Clopton, B. M., and Silverman, M. S. (1977). *J. Neurophysiol.* **40**, 1275–1280.
Cuschieri, A., and Bannister, L. H. (1975). *J. Anat.* **119**, 471–498.
Davis, H. (1965). *Cold Spring Harbor Symp. Quant. Biol.* **30**, 181–190.
Devor, M. (1976). *J. Comp. Neurol.* **166**, 49–72.
Feng, A. S., and Rogowski, B. A. (1980). *Brain Res.* **189**, 530–534.
Fernandez, C., and Hinojosa, R. (1974). *Acta Oto-laryngol.* **78**, 173–186.
Foss, I., and Flottorp, G. (1974). *Acta Oto-laryngol.* **77**, 202–214.
Graziadei, P. P. C., and Monti Graziadei, G. A. (1978a). *Handb. Sens. Physiol.* **9**, 55–83.
Graziadei, P. P. C., and Monti Graziadei, G. A. (1978b). *In* "Neuronal Plasticity" (C. W. Cotman, ed.), pp. 131–153. Raven, New York.
Gregg, J. M., and Dixon, A. D. (1973). *Arch. Oral Biol* **18**, 487–498.
Hamburger, V., and Hamilton, H. L. (1951). *J. Morphol.* **88**, 49–92.
Harris, D. M., and Dallos, P. (1984). *Science* **225**, 741–743.
Harris, R. M., and Woolsey, T. A. (1981). *J. Comp. Neurol.* **196**, 357–376.
Hendrickson, A. E., and Rakic, P. (1977). *Anat. Rec.* **187**, 602.
Hinds, J. W. (1968a). *J. Comp. Neurol.* **134**, 287–304.
Hinds, J. W. (1968b). *J. Comp. Neurol.* **134**, 305–322.
Hinojosa, R. (1977). *Acta Oto-laryngol.* **84**, 238–251.
Jackson, H., and Rubel, E. W. (1978). *J. Comp. Physiol Psychol.* **92**, 682–696.
Jeanmonod, D., Rice, F. L. and Van der Loos, H. (1977). *Neurosci. Lett.* **6**, 151–156.
Jeffress, L. A. (1948). *J. Comp. Physiol. Psychol.* **41**, 35–39.
Jhaveri, S., and Morest, D. K. (1982). *Neuroscience* **7**, 837–853.
Johansson, B., Wedenberg, E., and Westin, B. (1964). *Acta Oto-laryngol.* **57**, 188–192.

Kaplan, M. D., and Hinds, J. W. (1977). *Science* **197**, 1092–1094.
Killackey, H. P., and Belford, G. R. (1979). *J. Comp. Neurol.* **183**, 285–304.
Knowlton, V. Y. (1967). *J. Morphol.* **121**, 179–208.
Larsen, S. A. (1984). *Acta Oto-laryngol. Suppl.* 417, 1–43.
Leonard, C. M. (1975). *J. Comp. Neurol.* **162**, 467–486.
Leonard, C. M. (1981). *In* "Development of Perception" (R. N. Alsin, J. R. Alberts, and M. R. Peterson, eds.), pp. 383–410. Academic Press, New York.
Levi-Montalcini, R. (1949). *J. Comp. Neurol.* **91**, 209–242.
Lippe, W., and Rubel, E. W. (1983). *Science* **219**, 514–516.
Liu, C. N., and Chambers, W. W. (1958). *Ama Arch. Neurol. Psychiatry* **79**, 46–61.
McCrady, E., Jr., Wever, E. G., and Bray, C. W. (1937). *J. Exp. Zool.* **75**, 503–517.
Martin, M. R. (1981). *J. Comp. Neurol.* **197**, 141–152.
Martin, M. R., and Rickets, C. (1981). *J. Comp. Neurol.* **197**, 169–184.
Meininger, V., and Baudrimont, M. (1981). *J. Comp. Neurol.* **200**, 339–355.
Meisami, E. (1976). *Brain Res.* **107**, 437–444.
Meisami, E. (1978). *Prog. Brain Res.* **48**, 211–230.
Meisami, E., and Manochehri, S. (1977). *Brain Res.* **128**, 170–175.
Mendell, L. M., Sassoon, E. M., and Wall, P. D. (1978). *J. Physiol. (London)* **285**, 299–310.
Mlonyeni, M. (1967). *Brain Res.* **4**, 324–344.
Morest, D. K. (1968). *Z. Anat. Entwicklungs gesch.* **127**, 201–220.
Morest, D. K. (1969a). *Z. Anat. Entwicklungs gesch.* **128**, 290–317.
Morest, D. K. (1969b). *Z. Anat. Entwicklungs gesch.* **128**, 271–289.
Murphy, K. P., and Smyth, C. N. (1962). *Lancet* **5**, 972–973.
Nord, S. G. (1967). *J. Comp. Neurol.* **130**, 343–356.
Osen, K. (1969). *Acta Oto-laryngol.* **67**, 352–359.
Pujol, R., and Marty, R., (1970). *J. Comp. Neurol.* **139**, 115–126.
Rall, W. (1969). *Biophys. J.* **9**, 1483–1508.
Rall, W., and Rinzel, J. (1973). *Biophys. J.* **13**, 648–688.
Ramon y Cajal, S. (1911). "Histologie du System Nerveux de l'Homme et des Vertebres." Maloine, Paris.
Relkin, E. M., and Saunders, J. C. (1980). *Acta Oto-laryngol.* **90**, 6–15.
Relkin, E. W., Saunders, J. C., and Konkle, D. F. (1979). *J. Acoust. Soc. Am.* **66**, 133–139.
Rice, F. L. (1974). *Anat. Rec.* **178**, 447.
Rice, F. L., and Van der Loos, H. (1977). *J. Comp. Neurol.* **171**, 545–560.
Rogowski, B. A., and Feng, A. S. (1981). *J. Comp. Neurol.* **196**, 85–97.
Romand, R., and Romand, M. R. (1982). *J. Comp Neurol.* **204**, 1–5.
Rose, J., Galambos, R., and Hughes, J. (1959). *Bull. Johns Hopkins Hosp.* **104**, 211–251.
Rubel, E. W., and Ryals, B. M. (1983). *Science* **219**, 512–514.
Rubel, E. W., Smith, Z. D. J., and Miller, L. C. (1976). *J. Comp. Neurol.* **166**, 469–490.
Ruben, R. J. (1967). *Acta Oto-laryngol. Suppl.* **220**, 1–44.
Sakabe, N., Arayama, T., and Suzuki, T. (1969). *Acta Oto-laryngol. Suppl.* **252**, 29–36.
Saunders, J. C., Coles, R. B., and Gates, G. R. (1973). *Brain Res.* **63**, 59–74.
Shepherd, G. M. (1970). *Neurosci. Second Study Program* pp. 539–552.
Shipley, M. T. (1974). *J. Neurophysiol.* **37**, 73–90.
Silverman, M. S., and Clopton, B. M. (1977). *J. Neurophysiol.* **40**, 1266–1274.
Smith, C. G. (1938). *Can. Med. Assoc. J.* **39**, 138–149.
Smith, Z. D. J. (1981). *J. Comp. Neurol.* **203**, 309–333.
Sweitzer, L., and Cant, N. B. (1984). *J. Comp. Neurol.* **225**, 228–243.
Taber-Pierce, E. (1967). *J. Comp. Neurol.* **131**, 27–54.
Tatton, W. G., Hay, M., McCrea, D., and Bruce, I. C. (1983). *Exp. Brain Res.* **52**, 461–465.

Van der Loos, H., and Woolsey, T. A. (1973). *Science.* **179**, 395–398.
Van Drongelen, W. (1978). *J. Physiol. (London).* **277**, 423–435.
Welker, C., and Woolsey, T. A. (1974). *J. Comp. Neurol.* **158**, 437–454.
Westrum, L. E. (1975). *Exp. Neurol.* **47**, 442–447.
Woolsey, T. A. (1967). *Johns Hopkins Med. J.* **121**, 91–112.
Woolsey, T. A. (1978). *Brain Behav. Evol.* **15**, 325–371.
Woolsey, T. A., and Van der Loos, H. (1970). *Brain Res.* **17**, 205–242.
Woolsey, T. A., Welker, C., and Schwartz, R. (1975). *J. Comp. Neurol.* **164**, 79–94.
Woolsey, T. A., Anderson, J. R., Wann, J. R., and Stanfield, B. B. (1979). *J. Comp. Neurol.* **184**, 363–380.
Woolsey, T. A., Durham, D., Harris, R. M., Simons, D. J., and Valentino, K. (1981). *In* ''Development of Perception'' (R. N. Aslin, J. R. Alberts, and M. R. Petersen, eds.), pp. 259–292. Academic Press, New York.

13

What's Special about Development? Thoughts on the Bases of Experience-Sensitive Synaptic Plasticity

WILLIAM T. GREENOUGH
Department of Psychology, Department of Anatomical Sciences, and Neural and Behavioral Biology Program
University of Illinois
Champaign, Illinois

I. Introduction

A. The Range of Brain Information Storage

The premise of this book is that developmental neurobiology and developmental psychobiology represent a continuum, rather than independent fields. This chapter extends that continuum one step farther, to suggest that development, adulthood, and aging represent a continuum, with many processes continuing throughout life, particularly those involved with the storage of extrinsically (to the brain) originating information. The term "information storage" is used in a very broad sense here to include events ranging from the effects of early developmental organizational processes, such as steroid hormone modification of hypothalamic organization, through later organization of sensory system function by use, to learning and memory of the very specific nature characteristic of the adult mammalian nervous system. Recent work has increasingly indicated that some, if not many, of the mechanisms underlying these seemingly diverse categories of brain information storage may be similar and, in particular, that altered patterns of synaptic organization may be a common process whereby extrinsic events alter brain function on a more or less permanent basis.

In this chapter, I focus primarily upon information acquired as a result of behavioral interaction with the environment—"experience." The work reviewed indicates that (1) experience alters the synaptic organization of the brain, (2) this occurs in adulthood as well as during development, and (3) experiences effective in altering synaptic patterns include standard learning paradigms. The work has additionally suggested, however, that there may be an important difference between the manner in which the pattern of connectivity is influenced by extrinsic events in some cases in early development and the manner in which similar effects become manifest in later life. Specifically, I propose that, when a highly predictable, or expected, aspect of experience is incorporated during development, the neural process may involve anticipatory overproduction of connections, whereas storage of information from an unexpected event involves the formation of synapses in response to that event.

B. Efficacy and Pattern Hypotheses of Synaptic Plasticity

For the approximately 100 years since the concept of the synapse was developed, it has been the focus of theories of neural information storage, and the two mechanisms first proposed for modification of neural organization, alteration of the strength of preexisting synapses (Tanzi, 1893) and alteration, through formation or loss, of the pattern of synapses (Ramon y Cajal, 1893), have remained dominant. Evidence exists for both sorts of change in development. Changes in the sizes of synaptic components and in most other aspects of morphology that might indicate efficacy change appear, in general, to be larger following early postnatal sensory experience manipulations (e.g., Cragg, 1967; Fifkova, 1970; Freire, 1978; Tieman, 1984; See Greenough and Chang, 1985, for review) than after later developmental experience manipulations such as differential environmental complexity (Diamond *et al.*, 1975; Sirevaag and Greenough, 1985; Wesa *et al.*, 1982; West and Greenough, 1972) if one report, later repudiated by the secondary authors (Møllgaard *et al.*, 1971), is discounted. The sole exception to this pattern is an increase in the frequency of synapses with performations in the paramembrane densities, which appeared in visual cortex following postweaning rearing of rats in complex environments (Greenough *et al.*, 1978) and after visual discrimination training in adult rabbits (Vrensen and Nunes Cardozo, 1981). This structural feature remains to be understood in terms of a functional role, although it has been proposed that perforated synapses may reflect "splitting" or creation of new synapses (Dyson and Jones, 1984; Carlin and Siekevitz, 1983; Greenough *et al.*, 1978; Nieto-Sampedro *et al.*, 1982). Because other proposed changes have not been described throughout the life span, the remainder of this review is confined to a consideration of two mechanisms, experience-expectant and experience-dependent synaptogenesis, whereby experience may affect the pattern of connections.

The proposal that these two types of synaptogenesis should be considered separately arises both from consideration of the types of information that the nervous system must store in early development and in later life and from consideration of the neural mechanisms that seem to be operative in altering synaptic connection patterns at these two times. It is likely that the two processes would not be clearly separable in some cases in development, but the fact that they seem separable in other cases suggests we should consider them to be different. Specifically, experience-expectant information storage appears to involve (1) a nervous system "waiting for" an experiential event that has occurred reliably in the history of the species, (2) critical or sensitive periods, such that if the expected information is not provided the system will organize itself around whatever information is present, and (3) anticipatory (or experience-independent) overproduction of connections prior to (and/or during) the experiential event, with retention of a subset of those connections determined by the character of the experience. In contrast, experience-dependent information storage appears to involve (1) unpredictable events unique to the experience of the individual, (2) no sensitive periods, but rather lifelong (or nearly so) capacity to operate normally, and (3) synaptogenesis that occurs in response to the event.

II. Overproduction of Synapses and Expected Information

Much of the basic organization of the nervous system appears to be laid out with minimal influence of the activity of the nerve cells, as several of the preceding chapters (e.g., Chapters 3, 4, and 7) indicate. A primary role of experience-associated neural activity during early stages of postnatal development, at least in mammals where it has been best studied, appears to be to provide information of a detailed nature, such as pattern and form in the visual system, that cannot (or need not) be supplied by intrinsic sources. Obviously, this information must be "correct" relative to the needs of the developing system if it is to be useful. As experiments providing "incorrect" sensory information indicate (see Chapters 9, 10, 11, and 12, this volume), incorrect information can irreversibly damage the organizational pattern. Incorrect information is unlikely to be provided by the organism's natural environment, of course, because that environment has been reasonably constant over the organism's evolutionary history, or at least the elements that the organism's evolution has prepared it to use have been consistent. Thus mammals have evolved developmental schemes that take advantage of certain types of information that the environment can be expected to provide (Piaget, 1980).

Aslin (1981) has recently proposed the term "attunement" to refer to the experience-based shaping or fine tuning of intrinsically highly organized systems, such as the mammalian visual system, in development. Aslin proposed that Gottlieb's (1976) term "induction" be reserved for phenomena that appear only after appropriate experience, and has also suggested a number of subcategories, based upon the breadth and timing of sensitivity of a developmental process to experience. The terms used here, "experience expectant" and "experience dependent" are not meant to oversimplify the points made in his elegant analysis, but rather to reflect more accurately what seems to be going on in the nervous system in two somewhat different types of information storage processes.

One mechanism that appears to be quite common in neural development, the overproduction of synapses followed by selective retention of a subset of them, appears to have evolved to allow incorporation of expected information, or perhaps this mechanism has been adapted to this use from an original involvement in trial-and-error sampling of intrinsic organizational signals. Overproduction of nerve cells may be associated with synapse or afferent overproduction in some cases, but probably not all, as noted by Lund and Chang (Chapter 4, this volume). Examples of the overproduction and subsequent loss of synapses, often guided by neural activity, are abundant in both the peripheral (Purves and Lichtman, 1980; Hopkins and Brown, 1985) and central (LeVay *et al.*, 1980; Mariani and Changeux, 1981; see Chapter 10, this volume) nervous system. It remains unclear whether the mechanism of loss is active elimination of certain afferents or their passive withdrawal due to the absence of sufficient levels of neural activity and/or a (possibly activity dependent) trophic factor (see Chapter 9, this volume).

In either case, the functional consequence is that connections appropriate to the past experiences of and future demands upon the system are preserved.

When experience is involved in the selection of which afferents survive, it is important that appropriate experience be available at the proper time, since overproduction processes appear to be quite time-locked. If appropriate experience does not occur within a restricted period, the system appears to organize itself in accord with whatever activity patterns are present. In addition to possible neuromodulatory structures or substances (e.g., Kasamatsu and Pettigrew, 1979; see Chapters 5 and 9, this volume), this overproduction–selection mechanism appears to be an important aspect, in at least several cases, of what have been termed critical or sensitive periods.

III. Synaptogenesis and Unexpected Information

A. Experience and Continuing Synaptogenesis

In recent years, earlier indications that synapses continue to form throughout life (Ramon y Cajal, 1909–1911; Sotelo and Palay, 1971) have been extensively supported, and increasing evidence indicates that the experience of a mammal may govern both the number and the pattern of connections in a wide array of neocortical and subcortical regions. This line of research had its origins in the pioneering work of Rosenzweig, Bennett, and Diamond (e.g., Bennett *et al.*, 1964; Rosenzweig *et al.*, 1972) and their collaborators who first showed that the postweaning environment of rodents influenced the size and cellular composition of various regions of the neocortex. Their reports that rats reared in, or, in adulthood, placed in complex (or enriched, relative to the laboratory norm) environments (EC) had substantially thicker and heavier cortices than did littermates placed individually (IC) or in small groups (SC) in laboratory cages indicated a previously unsuspected plasticity of brain structure. Although the effects of sensory deprivation procedures on central neural structure and function could be profound, these effects were largely or exclusively limited to comparatively brief periods of neonatal sensitivity (Fifkova, 1968; Hubel and Wiesel, 1970; LeVay *et al.*, 1980; Rothblat and Schwartz, 1979). In contrast, significant sensitivity to the physical and social complexity of the housing environment was retained well beyond traditionally conceived periods of development. Later reports from this laboratory indicated that neuronal somata were larger and glial cells more frequent in the EC rats in the occipital cortex, the most affected area (Diamond, 1967; Diamond *et al.*, 1966).

Evidence that these differential rearing procedures might affect synaptic connectivity came from quantitative Golgi stain studies, in which measures of occipital cortical neuronal dendritic fields revealed dramatic size differences favoring EC rats (Greenough and Volkmar, 1973; Holloway, 1966; Volkmar and Greenough, 1972). In general, SC littermates had values intermediate between EC and IC counterparts but tended to be closer to the ICs. A study of the frequency of postsynaptic spines on deep occipital cortex pyramidal neuronal dendrites additionally indicated that the larger dendritic fields of EC rats were also somewhat more densely innervated than those of IC littermates (Globus *et al.*, 1973). Similar, though smaller, differences in the extent of dendritic fields were found in some other cortical regions, suggesting that the effects were not due merely to visual stimulation, whereas still other cortical regions appeared unaffected, suggesting that the effects were not due to generally acting hormonal or metabolic influences (Greenough *et al.*, 1973). However, sex differences in the response of brain structure to experience suggest important modulatory influences, probably involving prior or current hormonal conditions (Chapter 14, this volume). In addition, there is some indication of noradrenergic modulation of some complex environment effects (Mirmiran and Uylings, 1983; O'Shea *et al.*, 1983; Pearlman, 1983), although reports that antiadrenergic drug treatments reduce exploratory behavior (Benloucif et al., 1984; Coyle and Singer, 1975) have called the interpretation of these experiments into question. On the other hand, evidence for central nonadrenergic modulation of adult memory storage processes (Gold, 1984) keeps the modulatory concept viable. The occurrence of similar experience effects on dendritic fields in nonneocortical brain regions (e.g., Floeter and Greenough, 1979; Juraska *et al.*, 1985; Pysh and Weiss, 1979) suggests that this sort of brain plasticity may be widespread if not characteristic of central nervous system neurons.

There has been concern that the capricious nature of the Golgi technique may not provide impregnation of a representative sample of neurons and little information, other than the Golgi stain based report of Globus *et al.* (1973) noted above, is available regarding the density of innervation of dendritic fields in differentially reared rats. This led us to conduct studies that have confirmed the interpretation of the original quantitative Golgi studies as indicating relatively increased numbers of synaptic connections in animals reared in more complex surroundings. If innervation density of dendritic fields is equivalent in groups reared in environments varying in physical and social complexity (or, as the Globus *et al.*, 1973, report cited above suggests, innervation density in higher in EC rats), then the number of synapses per neuron should be highest in EC rats, intermediate in SC rats, and lowest in IC rats. Turner and Greenough (1983, 1985) have tested this, calculating the density of synapses in layers I–IV of occipital cortex from

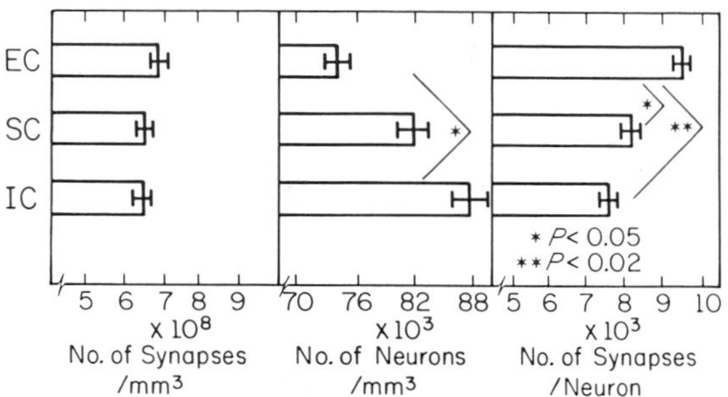

Fig. 1. Synaptic and neuronal density and synapses per neuron, corrected for group differences in size according to Coupland (1968) for synapses and Weibel (1979) for neuronal nuclei, in upper visual cortex of rats reared for 30 days after weaning in environmental complexity (EC), social cages (SC), or individual cages (IC). The lower density of neurons in EC and SC rats is assumed to reflect increases in neuropil and associated tissue elements that move neurons farther apart. The greater number of synapses per neuron in the EC group confirms predictions from Golgi stain studies showing greater dendritic length per neuron. After Turner and Greenough (1985). Copyright 1984, Elsevier Biomedical Press.

electron micrographs and the density of neuronal nuclei in these layers from light microscope sections. In both cases, stereological corrections were made for treatment group differences in the sizes of tissue elements. The results, shown in Fig. 1, indicated nearly equivalent synaptic density in EC, SC, and IC rats, but relatively lower neuronal density in relatively more experienced groups (presumably due to the interpolation of additional amounts of neuronal cytoplasm, glial tissues, etc.), such that the ratio of synapses per neuron was about 20% greater in EC rats than in ICs, with SCs intermediate, as in the light microscopic studies. Similar results have recently been presented by Bhide and Bedi (1984). [A prior study of synapse density, uncorrected for size differences, indicated relatively higher values in IC than in EC rats (Diamond *et al.*, 1975). We have no immediate explanation for this discrepancy.] Thus, at least for this brain region and experience manipulation paradigm, data from quantitative measurement of Golgi-impregnated tissue appear to reflect, at least qualitatively, differences in numbers of synapses per neuron. A similar study of dark-reared kittens by Cragg (1975) indicated lower synapse-to-neuron ratios than found in light-reared controls, paralleling an earlier report (Coleman and Riesen, 1968) of reduced dendritic field dimensions in dark-reared cats. Together, these studies give us some confidence that quantitative measurement of Golgi-impregnated dendritic fields reflects the directions of changes in synaptic numbers.

We have also confirmed that quantitative Golgi measures reflect, reasonably accurately, the amount of dendrite per neuron. Using the same

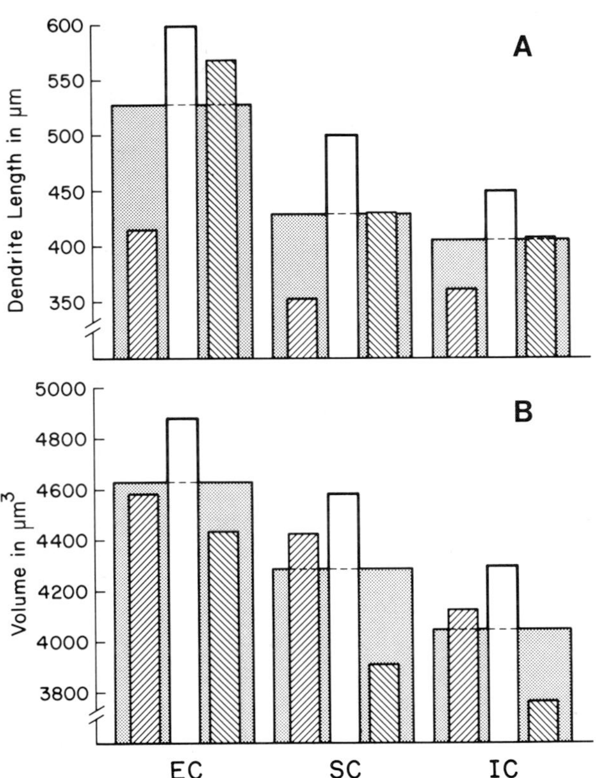

Fig. 2. (A) Comparison of length of dendrites of three upper occipital cortex neuron types using quantitative light microscopic Golgi techniques. Shaded bars represent mean dendrites per neuron. (▨) Layers II–III pyramidals; (□) layer IV stellates; (▧) layer IV pyramidals. (B) Volume of dendrite for the same cortical region per neuronal nucleus in a combined electron and light microscopic study, using stereological measures. Shaded bars represent averages. (▨)Layer I; (□) layers II–III; (▧) layer IV. Six littermate triplet sets of animals were studied with Golgi measures (data from Greenough and Volkmar, 1973), and eight similar triplet sets were used in the stereological study. Data from Sirevaag and Greenough (1984).

micrographs, Sirevaag and Greenough (1984) quantified dendritic volume per neuronal nucleus. Their data are presented in Fig. 2, along with those of Greenough and Volkmar (1973) for mean total dendritic length per neuron in upper visual cortex. Although we do not know precisely what volumes to expect from the Golgi data, the similarity of the differences in dendritic volume per neuron certainly suggests no great bias in estimates from quantitative Golgi measures.

The Rosenzweig *et al.* (1972) studies of cortical weight and thickness indicated that qualitatively comparable effects of differential housing

Table I

Mean number of Branches at Each Order Away from the Cell Body in Layer IV Stellate Neurons in Occipital Cortex in 450-day-old Rats Housed for 45 Days in EC or IC Environments[a]

Order of branch	Number	Number
1	7.9	7.5
2	9.2	7.3*
3	7.6	5.3
4	4.1	2.5
5	1.9	0.9**

[a]From Green, *et al*, 1983.
*$p < .05$.
**$p < .01$.

occurred in adult rats. Quantitative Golgi studies, not yet confirmed by light/electron microscopic techniques, have similarly indicated increased amounts of dendrite per neuron in occipital cortex of EC relative to IC or SC baseline young adults (Uylings *et al.*, 1978; Juraska *et al.*, 1980) or middle-aged (Green *et al.*, 1983) rats. Results for middle-aged rats are shown in Table I. These results indicate that the effects of differential experience upon synaptic connectivity are not restricted to periods early in development and further suggest that modification of synaptic connections may be a lifelong information storage mechanism.

B. Effects of Training on Adult Brain Structure

The finding that the structure of the adult mammalian brain is modulated by experience led to studies of the types of experience that could affect it. As noted above, adult mammals are generally resistant to effects of simple sensory deprivation, suggesting that this phase of brain development cannot easily be reversed. Brain organization, perhaps, could go forward but not backward, as one might expect with a process involving selective stabilization of overproduced connections. The additional dendritic branching in the adult EC brain presumably was contributing to further brain organization or information storage, but it seemed likely that the information was of a different type from the expected information of early development. If the new connections formed as a result of experience were involved in the storage of information arising from that experience, then altered synaptic connectivity might also be involved in the storage of memory of a more traditionally viewed sort. One (not critical) test of this hypothesis was to determine whether similar changes occurred when rats learned new behaviors. Greenough *et al.*

(1979) examined occipital cortex of young adult rats that had been trained for about a month on a changing series of daily maze problems. Compared to littermate controls, that were handled at the time that the trained animals were put in the maze, the maze-trained rats had larger upper apical dendritic fields in both layer IV and V pyramidal neurons, while no differences were seen in adjacent layer IV stellate neurons. Similar findings for gross brain weight and size measures have been reported in both postweaning (Bennett *et al.*, 1979) and middle-aged (Cummins *et al.*, 1973) rats.

These studies are compatible with the hypothesis that altered synaptic connectivity is involved in memory storage, but they do not control for the possibility that the training procedure itself induces dendritic changes via more general causes, such as stress, motor activity, and so forth. No simple experiment could, of course, rule out all possible alternative causes, but experiments using within-subject designs can examine possible contributions of general factors, such as hormonal or metabolic consequences of the training procedure. We have performed two such experiments. In the first of these (Chang and Greenough, 1982) a surgical split-brain procedure (corpus callosum sectioning) was used to isolate the hemispheres. Since the visual system of the rat is largely crossed, occluding one eye largely limits visual input to the ipsilateral hemisphere. We constructed opaque contact occluders, which were inserted prior to a daily maze training period similar to that of the prior experiment. Examination of nontrained split-brain rats indicated that mere insertion of the occluder in one eye for 3 to 4 h on a daily basis had no effect on the structure of layer V pyramidal neurons in the contralateral occipital cortex, relative to those opposite the nonoccluded eye. In the trained rats, however, occipital cortex layer V pyramidal upper apical dendritic fields were larger opposite the open eye, as shown in Fig. 3, indicating that the increased dendritic branching was a specific result of visual input from the training experience, rather than a consequence of general hormonal or metabolic effects of the training procedure.

Another adult training experiment used a within-animal reach training procedure and involved quantification of layer V pyramidal neurons in the forelimb region of somatosensory–somatomotor cortex (Greenough *et al.*, 1985b). Most rats show a preference for use of one forepaw in a task such as reaching into a tube for food although, unlike humans, there is no general species bias toward left or right. We trained rats to reach with either the preferred forepaw, the nonpreferred forepaw, both (in alternate sessions), or neither (eating food from a dish in the training apparatus), allowing about 1600 reaches over about 2 weeks. Apical dendrites of layer V pyramidal neurons were examined. Results are shown in Table II. Significant effects of training were seen between the bilaterally trained and the nontrained

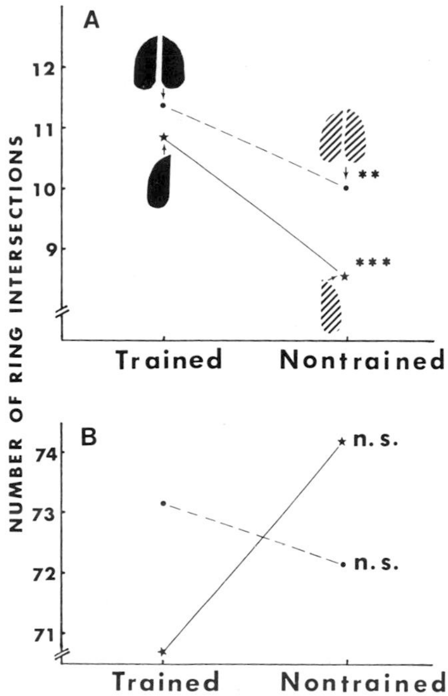

Fig. 3. Effects of maze training on ring intersection measure of length of oblique branches from apical dendrites of layer V visual cortex pyramidal neurons. Unilaterally trained split-brain animals show greater branching in the upper region in the hemisphere opposite "trained" eye. Bilaterally trained split-brain animals had more dendrite in the upper region than did nontrained split-brain controls. The smaller difference in the bilaterally trained split-brain animals may reflect their having been trained with alternating daily eye occlusion, such that each hemisphere received only half as much training. While these localized differences are quite large, they represent differences of only 2.5–5%, relative to the entire apical dendrite. Nonetheless, they clearly indicate, as do the results in Table II, that these effects do not result from general hormonal or metabolic effects of the training procedure. (A) Distal region; (B) proximal region. ns, Not significant. Chang and Greenough (1982). Copyright 1982, Elsevier Biomedical Press.

control group and within the reverse trained group on both total apical dendritic length and the number of oblique branches from the apical dendritic shaft. While these overall differences were not significant for the group trained on the preferred forepaw ("Practice" in Table II), there were significantly more second-order branches opposite the trained forelimb in this group. Again, these results indicate an effect specific to a part of the brain involved in governing the learned behavior, as opposed to a more general hormonal or metabolic effect of the training procedure upon the brain.

Collectively, these experiments provide substantial support for the

Table II

Effects of Unilateral or Bilateral Training on a Reaching Task on Branching of Apical Dendrites of Layer V Pyramidal Neurons in Rat Motor-Sensory Forelimb Cortex

	Control	Reversal	Practice	Alternation
Number of oblique branches per neuron, summed across orders				
Trained/Preferred[a]	10.92	13.19**	13.28	13.55
Nontrained	12.14	11.30	12.63	12.43
Total apical dendritic length per neuron				
Trained/Preferred[a]	546.5	658.4*	773.3	773.6
Nontrained	610.9	566.6	712.9	743.0

[a]Trained for reversal and practice, preferred for control and alternation. For combined control versus combined alternation hemispheres, both number of branches and total dendritic length are significant at $p < .0001$. Data from Greenough *et al* (1985b).

*$p < .05$.

**$p < .01$.

hypothesis that altered patterns of synaptic connectivity (or at least numbers of synapses) are involved in adult information storage processes including those associated with relatively conventional forms of learning. Obviously, a great deal of additional work will be necessary before we can even begin to understand how these synapses function in the information storage process, much less what other mechanisms are involved. Nonetheless, we can begin to speculate as to how increased numbers or altered patterns of synaptic connections might arise as a result of neural activity associated with training or other types of experience.

IV. Mechanisms of Synapse Pattern Change in Later Development and Adulthood

A. Modulated Turnover versus Experience-Dependent Synaptogenesis in the Storage of Unexpected Information

It was noted above that certain features of the environment, such as patterned light, can be expected to be present, based upon the features that have been present during the evolution of the species, and can thus be used to provide information incorporated in the developmental organization of the brain. In contrast, the genome could not possibly anticipate the presence of a Hebb-Williams maze pattern or the miniature garbage dumps sometimes called enriched environments. Thus, while the brain must be prepared to store important information, it is probably not possible (or metabolically very efficient, if it were possible) to put in place the anticipatory wiring necessary to encode by its selective stabilization all possible environmental contingencies.

Thus the incorporation of new information into the synaptic wiring of the brain, if this occurs, must involve synapses which have not been present throughout the life (or adult life) of the organism.

One proposal as to how new synapses might arise involves an extension of the synapse overproduction model of development sometimes termed synapse turnover (e.g., Cotman and Nieto-Sampedro, 1984; Dyson and Jones, 1984). One formulation of this model (Greenough, 1978) proposed that transient or potential connections are formed between proximate neural elements on a constitutive (experience-independent) basis and that their long-term survival as synapses required confirmation by the same (or a similar) mechanism to that operative in the developmental overproduction–loss process. Thus temporary or potential synapses would constantly form more or less at random, and the neural activity associated with a to-be-encoded event would selectively stabilize those available transient synapses most appropriate to the encoding of the information. At this point, however, there is little support for the view that synapses constantly form without direction or without some type of activating mechanism in the brain (in fact, this would be rather difficult to demonstrate), although there is some evidence for instability or turnover of synaptic connections in some regions of the intact adult brain (e.g., Sotelo and Palay, 1971) and considerable evidence for axonal sprouting or reactive synaptogenesis in deafferented brain regions (reviewed by Cotman and Nieto-Sampedro, 1984).

An alternative to this view is that synapses can form "on demand" in response to experience-associated neural activity, either as a direct result of the neural activity involved with the processing of information from experience or as a result of some neuromodulatory event triggered by an experience from which information is to be stored. Evidence that putative neuromodulators affect both developmental organizational processes (e.g., Kasamatsu and Pettigrew, 1979) and adult memory processes (e.g., Gold, 1984) provides some support for the latter hypothesis. Historically, the idea that specific connections can form in response to neural activity has received little support since the initial proposal of electrically directed "neurobiotaxis" of Kappers (1917), largely because plausible mechanisms for the expression of neural activity in specific growth patterns have not been presented.

B. Synapses on Demand?

Recent evidence that synapses can form rapidly in adult brain tissue in response to neural activity may force us to reconsider the possibility of synaptogenesis on demand. Lee *et al.* (1980, 1981) reported that new synapses form in hippocampal subfield CA1 *in vitro* and *in vivo* following patterns of electrical stimulation that induce a long-lasting increase in postsynaptic responsiveness termed long-term potentiation (LTP). We have followed up this finding (Chang and Greenough, 1984) and have noted that two types of synapses, those on dendritic shafts and those on stubby, headless "sessile" spines (Jones and Powell, 1969), form in relatively large numbers within minutes of the eliciting stimulation and persist for at least 8 h thereafter in the hippocampal slice *in vitro* (see Fig. 4). It seems unlikely that this rapid synapse formation is based upon modulation of a synapse turnover process. Interestingly, the source of at least some putative extrinsic neuromodulatory inputs such as brain stem noradrenergic projections is removed from these *in vitro* preparations, although axons within the slice may remain viable, and there is evidence for adrenergic modulation of LTP in the intact hippocampal formation (Gold, 1984).

Given that significant numbers of synapses form in response to neural activity, it seems reasonable to consider such a process a possible mechanism underlying the appearance of increased numbers of connections (or, at least, increased postsynaptic surface) following postweaning and adult exposure to complex environments and learning tasks. It seems quite reasonable that, if the nervous system encodes some types of extrinsically originating information in its pattern of connections, it would have the ability to generate connections in those positions most appropriate to the incorporation of that information, and those positions might well be ones activated by the processing of the information.

The hypothesis that synapses form at accelerated rates in response to experience is testable, provided that we have a way of recognizing synaptogenesis or newly formed synapses. Studies of synaptogenesis in early development have suggested potential morphological indicants of synaptic birthdays, although we should not assume that identical morphological changes will mark synaptogenesis in later development or adulthood. In the developing visual cortex and some other telencephalic regions, a morphological sequence has been described that provides a basis for an initial search for synaptogenic markers at later stages of life. Several studies (Adinolfi, 1972; DeGroot and Vrensen, 1978; Dyson and Jones, 1980, 1984; Freire, 1978; Hwang and Greenough, 1984; Juraska and Fifkova, 1979; Miller and Peters, 1981) suggest that immature synapses have fewer vesicles, less evident postsynaptic densities, and in the case of spine synapses, more small, headless "sessile," postsynaptic components (similar to those described following LTP), relative to their mature counterparts. In addition, studies of reactive (to deafferentation) synaptogenesis in the hippocampal dentate gyrus (Steward, 1983) have indicated a higher frequency of appearance of polyribosomal aggregates (PRA) at the base of spines and, especially, within the spine itself during periods of rapid synaptogenesis.

The appearance of PRA in spines, in particular, may indicate synaptogenesis or newly formed synapses. We (Hwang and Greenough, 1984)

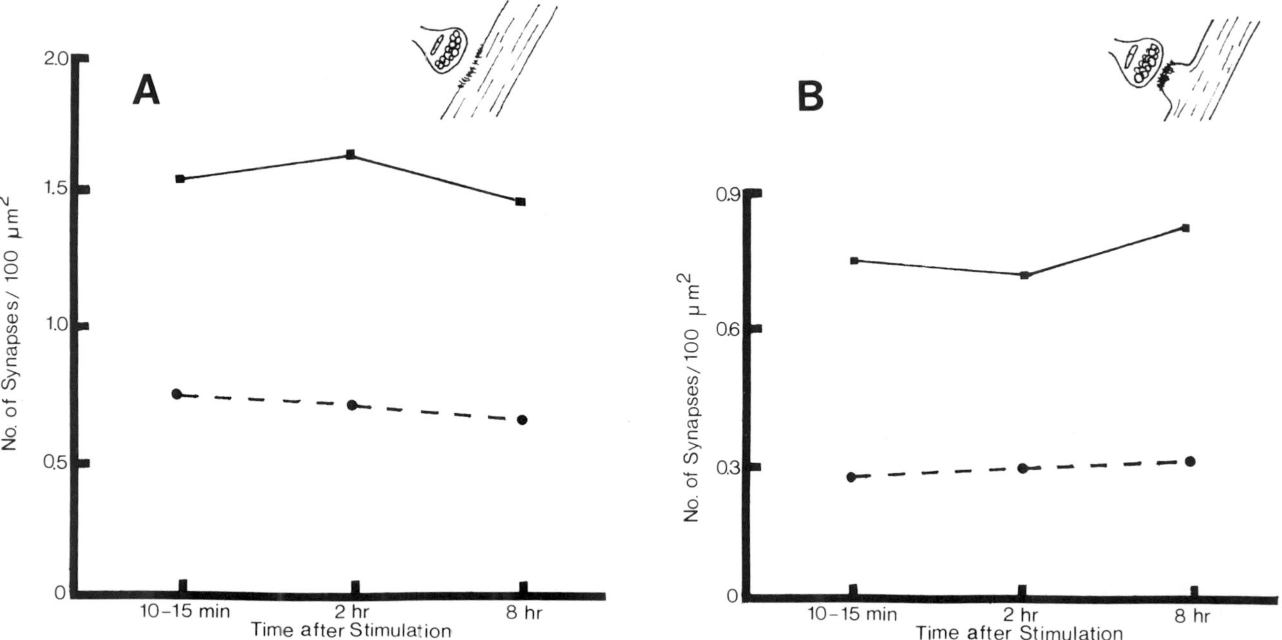

Fig. 4. Increases in areal density of (A) shaft and (B) sessile spine synapses following long-term potentiation-producing stimulation (100 Hz for 1 sec or 200 Hz for 0.5 sec) of Schaffer collaterals in region CA1 of the *in vitro* rat hippocampal slice. Density of the more common headed spine synapses was not affected. Equivalent stimulation at lower frequency or more prolonged stimulation at equivalent frequencies, which did not produce long-term potentiation, did not affect synapse density. (■) LTP; (●) control. From Chang and Greenough (1984). Copyright 1984, Elsevier Biomedical Press.

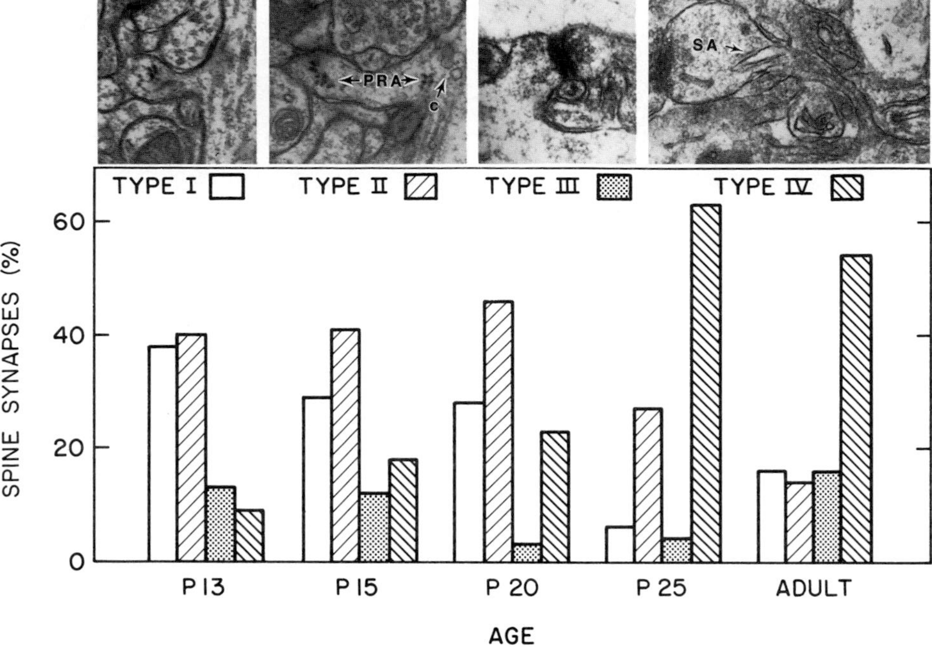

Fig. 5. (Top) Categorization of spine shape (from serial reconstruction). (□) Type I, short sessile spines; (▨) type II, longer sessile spines; (▩) type III, spines with a small head (usually short and/or thin-stemmed); (▧) type IV, large-headed "mushroom" spines. Types I and II, identified by the presence of cisternal material (c), were often without apparent presynaptic elements at the early ages (P13–20). SA, Spine apparatus; PRA, polyribosomal aggregate. (Bottom) Distribution of spine types as a function of age along large vertical dendrites in lower layer IV of rat occipital cortex. Types I, II, and III appear to mature to type IV. Data from Hwang and Greenough (1984).

have recently studied this and other characteristics of the development of spine synapses on large, vertically oriented dendrites in the lower one-third of layer IV of rat visual cortex (presumed to be apical dendrites of layer V pyramids) using reconstruction from serial thin electron microscope sections. These dendrites were selected so that we could examine the same synapse population across ages, in contrast to prior studies, which probably examined a mixed synapse population. We found strong confirmation, across postnatal ages of 13, 15, 20, and 25 days, and adulthood, for the transition from stubby or longer, sessile spines (types I and II in Fig. 5) to thin, headed (type III) spines to mushroom-shaped (type IV) spines with maturation. Moreover, there was clear pattern in the location of PRA, which were much more commonly

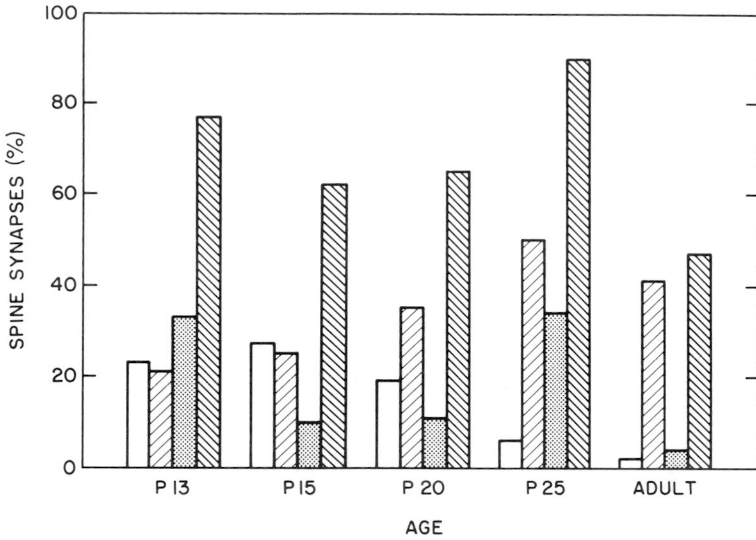

Fig. 6. Percentage of spines along large vertical dendrites in lower layer IV with PRA in the locations indicated at the ages examined (P, postnatal day). PRA were much more likely to appear in the head-stem region of spines (including those apparently noninnervated) during younger ages and appeared to move to the base of spines when innervation occurred and the spines assumed more mature forms with well-developed spine apparati. (□) Spines with PRA in head and stem; (▨) spines with PRA at base; (▩) spines with PRA at both locations; (▧) total spines with PRA. Data from Hwang and Greenough (1984).

seen in the head and stem region during development than in adulthood, as shown in Fig. 6. These findings support those of Steward (1983) and suggest that PRA location may serve as an indicator of synaptogenesis. Thus, if synapses form in response to experience, we might expect to see a higher frequency of spine synapses with PRA in the head and stem in appropriate brain regions of animals in more stimulating surroundings.

Recently, Greenough *et al.* (1985a) studied the location of polyribosomal aggregates in animals taken from EC, SC, and IC environments after a 30-day period of postweaning differential housing. The results, averaged across the upper four layers of visual cortex, are presented in Fig. 7. It is clear that, if PRA location in dendritic spine heads and stems is an indicator of synaptogenesis, more new synapses are forming in the EC animals. This suggests, although converging methods of newly forming synapse identification are still to be applied, that synapses are forming *in response* to experiential demand for information storage.

It should be noted that if new connections are formed they need not be,

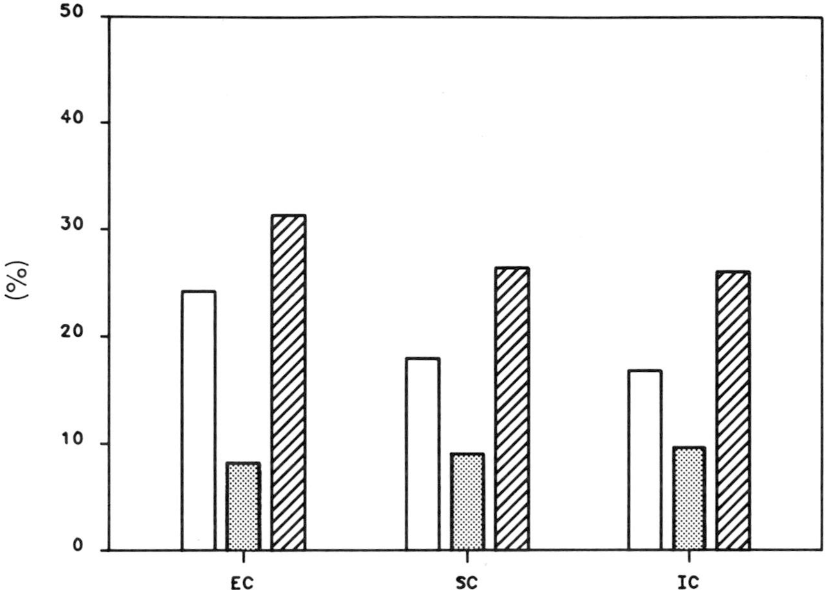

Fig. 7. Location of polyribosomal aggregates in spines in layer IV of occipital cortex of rats from environmental complexity (EC), social cage (SC), or individual cage (IC) rearing environments. The first two bars do not sum to the third bar in each condition because some spines have PRA In both head or stem and base. (□) Spines with PRA in head and stem; (▩) spines with PRA at base; (▨) total spines with PRA. Statistics (χ^2): for both proportion of spines associated with PRA (in head, stem, and/or base) and proportion with PRA in head and stem: EC versus IC, $p < .001$; EC versus SC, $p < .01$ Data from Greenough *et al.* (1985a). Copyright 1985, Guilford Press.

at the outset, sufficient for or universally appropriate to the information to be encoded. It seems quite likely that the mechanism used throughout functional subsystems in development would also be used in the more restricted regions of synaptogenesis resulting from later information storage demands—relatively nonspecific generation of potential connections among proximate elements, followed by selection of connections most appropriate to the encoding of information from the events that triggered the synaptogenesis. (The basis for survival might well be temporally contiguous pre and postsynaptic activation of the Hebb synapse type described in more detail by Singer, Chapter 9 of this volume.) One cannot entirely rule out the alternative, however, that the central nervous system "knows," from its organization, history, and current activity pattern, which connections to form.

V. Conclusions

These data, taken with others indicating greater numbers of synapses in animals exposed to a greater variety of experience or to training, provide reasonable support for the view that synaptogenesis is triggered by neural activity, or by some other neural signal, in the late developmental and adult information storage process. Thus, while a basic mechanism of developmental brain organization appears to be carried forward for use in information storage in adulthood, the mechanisms governing its occurrence appear to be different. It is proposed that, in the case of expected experience in the developing organism, synaptogenesis is triggered largely independently of the event providing information to be incorporated, occurs on a system-wide basis (e.g., primary visual cortex), and is restricted to a brief period, during or following which selection mechanisms operate to determine the pattern of organization that will survive. In the later experience-dependent process, since both the timing and character of the to-be-stored information is unpredictable, synaptogenesis is proposed to be triggered by the information-providing event, to occur locally among activated neurons (and, perhaps, among their neighbors), and to occur throughout much, if not all, of the life of the organism.

Acknowledgment

Work not yet published and preparation of this manuscript was supported by NSF, NIMH, ONR, and the Retirement Research Foundation.

References

Adinolfi, A. M. (1972). *Exp. Neurol.* **34**, 383–393.

Aslin, R. N. (1981). *In* "Development of Perception" R. N. Aslin, J. R. Alberts, and M. R. Peterson, eds.), Vol 2. Academic Press, New York.

Benloucif, S., Rosenzweig, M. R., and Bennett, E. L. (1984). *Soc. Neurosci. Abstr.* **10**, 1174.

Bennett, E. L., Diamond, M. C., Krech, D., and Rosenzweig, M. R. (1964). *Science* **146**, 610–619.

Bennett, E. L., Rosenzweig, M. R., Morimoto, H., and Hebert, M. (1979). *Behav. Neurol. Biol.* **26**, 1–22.

Bhide, P. G., and Bedi, K. S. (1984). *J. Comp. Neurol.* **227**, 305–310.

Carlin, R. K., and Siekevitz, P. (1983). *Proc. Natl. Acad. Sci. U. S. A.* **80**, 3517–3521.

Chang, F.-L. F., and Greenough, W. T. (1982). *Brain Res.* **232**, 283–292.

Chang, F.-L. F., and Greenough, W. T. (1984). *Brain Res.* **309**, 35–46.

Coleman, P. D., and Riesen, A. H. (1968). *J. Anat.* **102**, 363–374.

Cotman, C. W., and Nieto-Sampedro, M. (1984). *Science* **225**, 1287–1294.
Coupland, R. E. (1968). *Nature (London)* **217**, 384–388.
Coyle, I. R., and Singer G. (1975). *Psychopharmacologia* **44**, 253–256.
Cragg, B. G. (1967). *Nature (London)* **215**, 251–253.
Cragg, B. G. (1975). *Exp. Neurol.* **46**, 445–451.
Cummins, R. A., Walsh, R. Budtz-Olsen, O. E., Konstantinos, T., and Horsfall, C. R. (1973). *Nature (London)* **243**, 516–518.
DeGroot, D., and Vrensen, G. (1978). *Brain Res.* **147**, 362–369.
Diamond, M. C. (1967). *J. Comp. Neurol.* **131**, 357–364.
Diamond, M. C., Law, F., Rhodes, H., Lindner, B., Rosenzweig, M. R., Krech, D., and Bennett, E. L. (1966). *J. Comp. Neurol.* **125**, 117–125.
Diamond, M. C., Lindner, B., Johnson, R., Bennett, E. L., and Rosenzweig, M. R., (1975). *J. Neurosci. Res.* **1**, 109–119.
Dyson, S. E., and Jones, D. G. (1980). *Brain Res.* **183**, 43–59.
Dyson, S. E., and Jones, D. G. (1984). *Dev. Brain Res.* **13**, 125–137.
Fifkova, E. (1968). *Nature (London)* **220**, 379–381.
Fifkova, E. (1970). *J. Neurobiol.* **1**, 285–294.
Floeter, M. K., and Greenough, W. T. (1979). *Science* **206**, 227–229.
Freire, M. (1978). *J. Anat.* **126**, 193–201.
Globus, A., Rosenzweig, M. R., Bennett, E. L., and Diamond, M. C. (1973). *J. Comp. Physiol. Psychol.* **82**, 175–181.
Gold, P. E. (1984). *In* "Neurobiology of Learning and Memory" (G. Lynch, J. L. McGaugh, and N. M. Weinberger, eds.), pp. 374–382. Guilford, New York.
Gottlieb, G. (1976). *In* "Studies on the Development of Behavior and the Nervous System" (G. Gottlieb, ed), Vol 3, Neural and Behavioral Specificity, pp. 25–54. Academic Press, New York.
Green, E. J., Greenough, W. T., and Schlumpf, B. E., (1983). *Brain Res.* **264**, 233–240.
Greenough, W. T. (1978). *In* "Brain and Learning" (T. Teyler, ed.), pp. 127–145. Greylock, Stamford, Connecticut.
Greenough, W. T., and Chang, F.- L. F. (1985). *In* "Synaptic Plasticity and Remodeling" (C. Cotman, ed.), pp. 335–372. Guilford, New York.
Greenough, W. T., and Volkmar, F. R. (1973). *Exp. Neurol.* **40**, 491–504.
Greenough, W. T., Volkmar, F. R., and Juraska, J. M. (1973). *Exp. Neurol.* **41**, 371–378.
Greenough, W. T., West, R. W., and DeVoogd, T. J. (1978). *Science* **202**, 1096–1098.
Greenough, W. T., Juraska, J. M., and Volkmar, F. R. (1979). *Behav. Neural Biol.* **26**, 287–297
Greenough, W. T., Hwang, H.-M., and Gorman, C. (1985a). *Proc. Natl. Acad. Sci. U. S. A.* **82**, 4549–4552.
Greenough, W. T., Larson, J. R., and Withers, G. S. (1985b). *Behav. Neural Biol.* **44**, 301–314.
Holloway, R. L. (1966). *Brain Res.* **2**, 393–396.
Hopkins, W. G., and Brown, M. C. (1985). *In* "Synaptic Plasticity and Remodeling" (C. Cotman, ed.) pp. 169–199. Guilford, New York.
Hubel, D. H., and Wiesel, T. N. (1970). *J. Physiol. (London)* **206**, 419–436.
Hwang, H. M., and Greenough, W. T., (1984). *Soc. Neurosci. Abstr.* **10**, 579.
Jones, E. G., and Powell, T. P. S. (1969). *J. Cell Sci.* **5**, 509–529.
Juraska, J. M., and Fifkova, E. (1979). *J. Comp. Neurol.* **183**, 257–268.
Juraska, J. M., Greenough. W. T., Elliott, C., Mack, K., and Berkowitz, R. (1980). *Behav. Neural Biol.* **29**, 157–167.
Juraska, J. M., Fitch, J., Henderson, C., and Rivers, N. (1985). *Brain Res.* **333**, 73–80.
Kappers. C. U. A. (1917). *J. Comp. Neurol.* **27**, 261–298.
Kasamatsu, T., and Pettigrew, J. D. (1979). *J. Comp. Neurol.* **185**, 139–162.

Lee, K. S., Schottler, F., Oliver, M., and Lynch, G. (1980). *J. Neurophysiol* **44**, 247–258.

Lee, K., Oliver, M., Schottler, F., and Lynch, G. (1981) *In* "Electrophysiology of Isolated Mammalian CNS Preparations" (G. A. Kerkut and H. V. Wheal, eds.), pp. 189-211. Academic Press, New York.

LeVay, S., Wiesel, T. N., and Hubel, D. H. (1980). *J. Comp. Neurol.* **191**, 1-51.

Mariani, J., and Changeux, J.-P. (1981). *J. Neurosci.* **1**, 696–702.

Miller, M., and Peters, A. (1981). *J. Comp. Neurol.* **203**, 555–573.

Mirmiran, M., and Uylings, H. B. M. (1983). *Brain Res.* **261**, 331-334.

Møllgaard, K., Diamond, M. C., Bennett, E. L., Rosenzweig, M. R., and Lindner, B. (1971). *Int. J. Neurosci.* **2**, 113-128.

Nieto-Sampedro, M., Hoff, S. F., and Cotman, C. W. (1982). *Proc. Natl. Acad. Sci. U. S. A.* **79**, 5718-5722.

O'Shea, L., Saari, M., Pappas, B. A., Ings, R., and Stange, K. (1983). *Eur. J. Pharmacol.* **92**, 43–47.

Pearlman, C. (1983). *Physiol. Behav.* **30**, 161-163.

Piaget, J. (1980). "Adaptation and Intelligence: Organic Selection and Phenocopy" (S. S. Eames, transl.) University of Chicago Press, Chicago.

Purves, D., and Lichtman, J. W. (1980). *Science* **210**, 153-157.

Pysh, J. J., and Weiss, M. (1979). *Science* **206**, 230–232.

Ramon y Cajal, S. (1893). *Arch. Anat. Physiol., Anat. Abt.* **17**, 319–428.

Ramon y Cajal, S. (1909–1911). "Histologie du système Nerveux de l'Homme et des Vertebres" (L. Azoulay, transl.) Vol. 2. CSIC, Madrid (1952).

Rosenzweig, M. R., Bennett, E. L., and Diamond, M. C. (1972). *In* "Macromolecules and Behavior" (J. Gaito, ed.), 2nd Ed., pp. 205-277. Appleton, New York.

Rothblat, L. A., and Schwartz, M. L. (1979). *Brain Res.* **161**, 156–161.

Sirevaag, A. M., and Greenough, W. T. (1984). *Soc. Neurosci. Abstr.* **10**, 579.

Sirevaag, A. M., and Greenough, W. T. (1985). *Develop. Brain Res.* **19**, 215-226.

Sotelo, C., and Palay, S. L. (1971). *Lab. Invest.* **25**, 653-671.

Steward, O. (1983). *Cold Spring Harbor Symp. Quant. Biol.* **48**, 745–759.

Tanzi, E. (1893). *Riv. Sper. Freniatr. Med. Alienazioni Ment.* **19**, 419-472.

Tieman, S. B. (1984). *J. Comp. Neurol.* **222**, 166–176.

Turner, A. M., and Greenough, W. T. (1983). *Acta Stereol.* **2** *Suppl.* **1**, 239-244.

Turner, A. M., and Greenough, W. T. (1985). *Brain Res.* **329**, 195-203.

Uylings, H. B. M., Kuypers, K., and Veltman, W. A. M. (1978). *Prog. Brain Res.* **48**, 261-274.

Volkmar, F. R., and Greenough, W. T. (1972). *Science* **176**, 1445-1447.

Vrensen, G., and Nunes Cardozo, J. (1981). *Brain Res.* **218**, 79-97.

Weibel, E. R. (1979). "Stereological Methods. Vol. 1. Practical Methods for Biological Morphometry." Academic Press, New York.

Wesa, J. M., Chang, F.-L. F., Greenough, W. T., and West, R. W. (1982). *Dev. Brain Res.* **4**, 253-257.

West, R. W., and Greenough, W. T. (1972). *Behav. Biol.* **7**, 279-284.

14

Sex Differences in Developmental Plasticity of Behavior and the Brain

JANICE M. JURASKA[1]
Department of Psychology
Indiana University
Bloomington, Indiana

I. Introduction: Human Development

Considerable evidence indicates that among humans, males are more vulnerable during development than are females. Human males are more prone to developmental disorders than are females. Males have a higher incidence of autism, dyslexia, stuttering , childhood seizures, cerebral palsy, and hyperactivity (Taylor and Ounsted, 1972). Males also show greater variability than females in IQ scores and even in mathematical abilities where they are superior on average but are overrepresented on both ends of the

[1]Present address: Department of Psychology, University of Illinois, Champaign, Illinois 61820.

scale (Maccoby and Jacklin, 1974). Although more males appear to be conceived, male fetuses are more likely to miscarry or be stillborn than female fetuses. This results in only a slight excess of live male births (Taylor and Ounsted, 1972). Furthermore, in times of high stress, proportionally more females are born, presumably due to fewer male conceptions or to more male miscarriages (Wilson, 1975). Males are less likely to cope successfully with major stressors during childhood such as separation from the primary caretaker, parental illness or discord, or absence of the father (Rutter, 1970; Werner and Smith, 1982). Males thus appear to be the more vulnerable and variable sex during development, while females appear to be more buffered against extremes.

Currently the causes for male variability are not known. The developmental surge in androgens (see Chapter 6, this volume) represents the major difference between the sexes during development and is probably involved at least indirectly. However, the contribution of the androgen surge to nonreproductive functions is not well understood, although it has been hypothesized that dyslexia and stuttering are linked to early androgen levels (Geschwind and Behan, 1982). One factor that may contribute to male variability is the males' slower developmental rate, which places them in each developmental stage longer and consequently more at risk to both advantage and disadvantage (Taylor and Ounsted, 1972; Werner and Smith, 1982). On a neural level, it is notable that brain weight is reported to stabilize at a later age in the development of males and also to decline later (Ordy *et al.*, 1975). Although the human work is important, the animal literature is essential in any discussion of male vulnerability or variability, not only because we can better investigate neural mechanisms in animals but also because human development is influenced by cultural expectations of the male and female roles. In the remainder of the chapter I will review the animal literature on sex differences in behavioral vulnerability and discuss possible mechanisms for this phenomenon, including differences in neural responses to the environment. Lastly, I will take a brief look at human sex differences in light of the animal findings.

II. Sex Differences in Animals in Response to the Environment

There is an extensive, although somewhat fragmented, literature indicating that males (usually rats) show greater vulnerability or variability than females in response to various environmental manipulations. This literature can be arbitrarily divided into studies that deal with malnutrition and those that vary levels of stimulation.

A. Malnutrition

While sex differences are certainly not inevitable in malnutrition studies, when they do occur male rats typically are more affected than female rats on a variety of measures. Female rats showed brain weight changes, especially in the forebrain, in response to an enriched environment, regardless of their nutritional history, while malnourished male rats did not exhibit brain changes in response to the environment (McConnell *et al.*, 1981). Decreases in cerebellar cross-sectional area due to early malnutrition were more marked in male than in female rats (Jordan and Howells, 1978). The body growth of males was more sensitive than that of females to present and past nutritional status (Crutchfield and Dratman, 1980; Jones and Friedman, 1982). Puberty was advanced in undernourished males that were environmentally stimulated (preweaning handling and postweaning enrichment) while the date of female puberty was unaffected (Hansen *et al.*, 1978). Males were also more behaviorally affected by malnutrition than females in tasks such as nest finding (Fleischer and Turkewitz, 1979b), visual discriminations (Barnes *et al.*, 1966; Fleischer and Turkewitz, 1979a; Galler and Manes, 1980; Galler, 1981) and a go/no-go avoidance (Mathura and Harper, 1979). Conversely, female rats were more affected in activity measures and active avoidance when fed a high-carbohydrate diet (Morley *et al.*, 1977). It is possible that alterations in the subcomponents of a diet can interact with sex differences in transmitter levels, such as those described by Dohanich *et al.* (1982), Orensanz *et al.* (1982), and Fischette *et al.* (1983), and are not related to the effects of general malnutrition.

The above sex differences in the effects of malnutrition as well as an extensive literature on sex differences in obesity and heat production have been viewed by Hoyenga and Hoyenga (1982) as part of the same phenomenon. The Hoyengas claim that mammalian females evolved under somewhat different selection pressures than males due to their different sex roles. Consequently, females have evolved a metabolic system that allows for better survival of famine and this affects the developing as well as the reproductively mature female. Thus sex differences in response to malnutrition would be expected. Sex differences in response to other environmental events are less easily explained.

B. Isolation and Stimulation

1. *Behavior*

There is considerable behavioral and some neural evidence that indicate that many types of stimulation during development affect males more than females. Males and females appear to be differentially sensitive to manipulations affecting the adrenal–pituitary axis. Adrenalectomy of pregnant female rats affected the adult corticoid response to stress of male but not female offspring (Levine, 1974). Also, the sexual behavior of male rats was disrupted if their mothers were stressed during pregnancy, presumably due to interference with the prenatal testosterone surge (Ward, 1974). In contrast, prenatally stressed females showed normal sexual and maternal behavior (Beckhardt and Ward, 1983) and once again appeared to be the more buffered sex. The simple paradigm of experimenter-handling of rats at 1–10 days of age had more profound effects on males than females when they were later tested for the response of ACTH to a shock-induced fighting paradigm (Erskine *et al.*, 1975), and behaviorally in arousing tasks such as avoidance learning (Weinberg and Levine, 1977) and reactions to stimulus variability (Weinberg *et al.*, 1978). Female mice and rats have been reported to have higher basal levels of corticosteroids than males (Kitay, 1961; Hennessy *et al.*, 1977; Weinberg *et al.*, 1982), implying that neural control of the pituitary–adrenal axis may be sexually dimorphic and possibly differentially sensitive to environmental influences. Interestingly, the presence of a sex difference in human corticosteroids is dependent on the phase of the menstrual cycle in which comparisons are made, with females secreting comparatively more during the follicular phase (Schoneshofer and Wagner, 1977), thus studying sex by environment interactions in adrenal responsivity in humans may also entail a complicating menstrual cycle factor.

Male rats and monkeys are more sensitive to being reared in isolation than females. The sexual behavior of males of both these species is disrupted by isolated rearing. In rats, rearing alone from either 2 weeks of age or from birth resulted in deficits that were not motivational but rather appeared to stem from a lack of practice in appropriately climbing or mounting another animal (Gerall *et al.*, 1967; Gruendel and Arnold, 1969). With practice this deficit was overcome in some rats (Gerall *et al.*, 1967; Hard and Larsson, 1968). There were no apparent deficits in lordosis behavior of isolated female rats but the more active components of female sexual behavior, such as darting, have apparently not been investigated (Moore, 1985).

Isolation of male rats can also impair recovery from medial preoptic area (MPOA) lesions. In adult males, lesions of the MPOA resulted in deficits in copulatory behavior, regardless of housing conditions. These deficits also occurred in MPOA-lesioned prepuberal rats if the rats were isolated. However, significant recovery of sexual behavior occurred if the rats were housed in a mixed sex group (Twiggs *et al.*, 1978). This difference in recovery between the age groups was not influenced by puberal testosterone (Meisel, 1983); rather the play that accompanies group housing in young rats (both lesioned and intact) may have facilitated recovery (Leedy *et al.*, 1980).

Recovery from lesion-induced deficits in female sexual behavior has not been investigated. This work indicates that social experience is essential to the development of male sexual behavior in rats.

In rhesus monkeys, isolation during development was more debilitating for the sexual behavior of both males and females than was true for rats, but again the effects of rearing in isolation were more severe for the males and not readily reversible while females showed some recovery with experience (Goy and Goldfoot, 1974; Sackett, 1974). Isolated male monkeys, like male rats, appeared to lack appropriate motor skills, in particular the double foot clasp mount, for successful copulation. In addition, they lacked the social skills needed to elicit female cooperation (Goy and Goldfoot, 1974; Moore, 1985).

Isolation and the converse—stimulating, complex environments—often are found to affect nonsexual behaviors more in males than females. In rats, isolated and group housing conditions imposed on adults differentially influenced the rate at which males, but not females, extinguished a taste aversion (Chambers and Sengstake, 1978). The authors also found that isolation lowered male testosterone levels and hypothesized that lower testosterone could account for the isolation effect. Sex by environment (isolated versus enriched) statistical interactions were found for exploratory behavior (Joseph, 1979) and active avoidance (Freeman and Ray, 1972) with male rats being more affected by the rearing environment than females. In contrast, we tested rats of both sexes reared in a complex or isolated environment in a 17-arm radial maze and found that complex-reared rats of both sexes were superior to isolated rats in their performance (Juraska *et al.*, 1984). Obviously, greater male vulnerability to environmental rearing conditions is task specific.

Many nonsexual behaviors are more affected in male than female rhesus monkeys by the conditions of rearing. Mitchell *et al.* (1966) found that the parity of the mother affected male but not female play behavior. Isolated male monkeys were more affected than their female counterparts in self-aggression, in withdrawn behaviors (clutching, rocking), in exploration of novel environments and objects, and in appropriate social behavior (Sackett, 1974). Sackett also noted that isolation-reared females learned to inhibit inappropriate behaviors in social situations more readily than did isolation-reared males. Based upon the wide array of behavioral differences between isolated male and female monkeys (sex differences in group-reared monkeys were trivial by comparison), Sackett termed females "the buffered sex."

Thus, there is evidence from several different rearing paradigms in both rats and rhesus monkeys that males are often more affected than females by the rearing environment, to their advantage and disadvantage.

2. Brain

The behavioral differences between the sexes in vulnerability to the environment imply that there must be neural differences at well. However, examination of this phenomenon at the neural level has been far less extensive; most studies of neural plasticity have used only male animals. One exception is the work of Diamond and her colleagues who placed young adult male and female rats in complex and isolated environments. Their results indicated that male rats may have greater differences due to housing conditions in the depth of the occipital cortex, while female rats may show greater differences in the somesthetic cortex (Diamond *et al.*, 1971; Hamilton *et al.*, 1977). The major problems with this conclusion is that statistical comparisons were made only within each sex and not between sexes, so the results can only be taken as suggestive.

I predicted that areas of the brain that are influenced by the environment such as the cerebral cortex and hippocampus would mirror the behavioral trends and show greater environmental susceptibility in neural structure in males than in females. Therefore, I performed a series of experiments in which male and female littermate sets of hooded rats were reared in either a complex environment (EC), which consisted of 12 same-sex rats housed with a variety of objects that were changed daily or in an isolated environment (IC) in which each rat was housed in a standard laboratory cage. These differential environments began at weaning and continued for 1 month. Thus all rats were postpuberal at the time their brains were examined. EC rats of both sexes weighed significantly less than IC rats at the time their brains were examined. All brains were Golgi-Cox stained and neurons were sampled from three cell populations in the visual cortex and the granule cells of the hippocampal dentate gyrus.

In the visual cortex the pattern of change was dependent on the cell population examined (Juraska, 1984). Layer V pyramidal cells showed no sex differences in either their apical or basilar dendritic fields but differences due to the environments were evident in both sexes (see Fig. 1). In layer IV stellate neurons there were both environmental and sex effects. While it appeared that a sex by environment interaction could be present in the data with males being the more variable sex, the interaction just missed significance at $p = .059$. However, the layer III pyramidal cells showed very striking sex by environmental interactions (both apical and basilar dendrites), with males being more affected by the rearing environments. Thus more pronounced male vulnerability to the environment during development at the neural level is not a general rule but does occur in some neuronal populations.

On the other hand, dendritic plasticity in the granule cells of the dentate gyrus was much greater in females following rearing in the same differential environments (Juraska *et al.*, 1985). These results, illustrated in Fig. 2, were so surprising to us that we replicated them in a separate group of animals

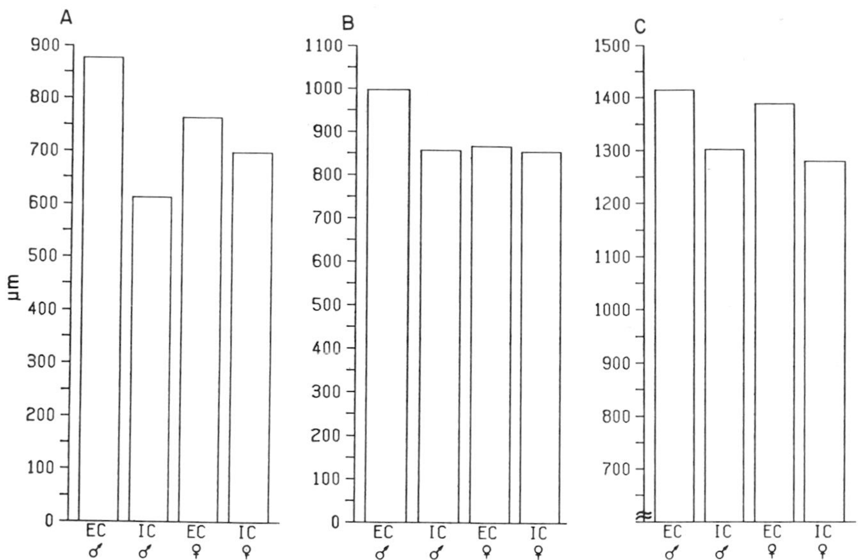

Fig. 1. The average total dendritic length per neuron for (A) layer III pyramidal basilar, (B) layer IV stellate, and (C) layer V pyramidal basilar dendrites in the visual cortex. (A) EC $>$ IC; $p < .00001$; interaction, $p < .01$. (B) EC $>$ IC; $p < .05$; ♂ $>$ ♀; $p < .05$. (C) EC $>$ IC; $p < .006$. From Juraska (1984).

(Fig. 3). The greatest differences between female EC and IC groups occurred primarily in the outer two-thirds of the dendritic tree in both replications—the site of termination of the perforant pathway from the entorhinal area (Storm-Mathisen, 1977). The dendritic differences found in females in this portion of the dendritic tree indicate that there may be sex differences in the plasticity of the perforant pathway. The results from both the visual cortex and dentate gyrus suggest that male and female brains are wired somewhat differently in regions other than the reproductively related hypothalamus (see Chapter 6, this volume, for a discussion of hypothalamic sex differences).

Additional evidence for greater female plasticity in the hippocampus comes from the work of Milner and Loy (1980). They found greater axonal sprouting of sympathetic fibers in the hippocampal dentate and CA3 areas of female rats following fimbria/fornix cuts. The smaller male sprouting response was dependent on the presence of testosterone during development but not postpuberally (Milner and Loy, 1982). Apparently the sex of the animal and the region of the brain are very important factors in neural plasticity.

It may be overly simplistic to expect that the brain will always reflect the behavioral trend toward male vulnerability in a direct way (i.e., behavioral variability equals dendritic variability). However, sex differences in response to the environment are a way of expanding the range of variability in the

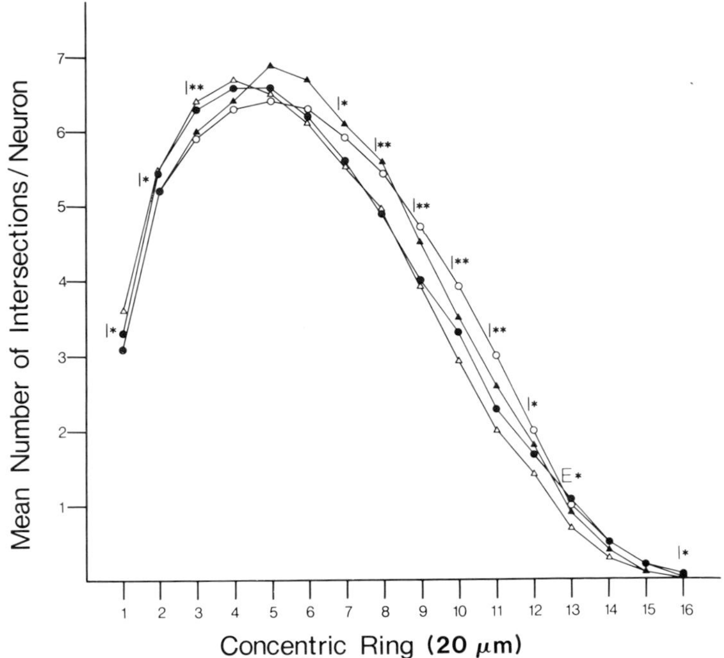

Fig. 2. The average number of intersections between the dendritic tree of dentate granule cells and a series of concentric rings at 20-μm intervals in replication 1. (E) A significant environment effect; (I) a significant interaction between the environment and sex factors. (●) EC ♂; (○) EC ♀; (▲) IC ♂; (Δ) IC ♀. (*) Significant at $p < .05$; (**) Significant at $p < .01$. From Juraska *et al.* (1985).

brain and behavior that may further illuminate the complicated relationship between the two.

One interesting aside that came from my work on sex differences in dendritic plasticity is that sex differences in these brain regions between males and females from the same environment are dependent on the type of environment. In previous work, the overall size of the cortex and hippocampus were measured in rats that presumably had been housed in isolated conditions. Results have been inconsistent, either with males showing larger areas of cortex and hippocampus than females (Pfaff, 1966; Yanai, 1979b) or with no differences between the sexes in these areas (Yanai, 1979a). In my work in the visual cortex, no sex differences could be found in neurons from rats in IC (Fig. 1); however, sex differences in favor of the males were evident in two out of three cell populations in EC rats. Thus the environment can encourage or discourage sex differences in the visual cortex. In dentate granule cells, male ICs had more dendritic material than female ICs in both

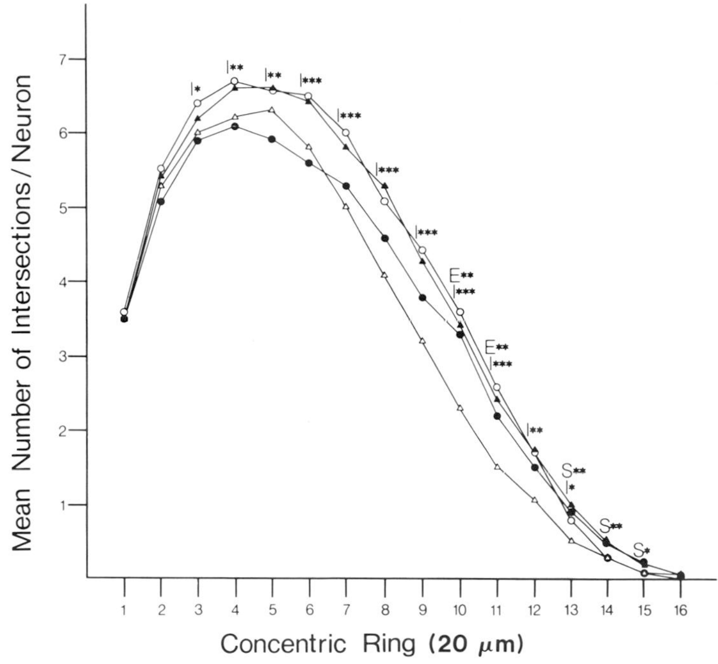

Fig. 3. The concentric ring analysis of granule cells as in Fig. 2 for replication 2. (E) A significant environment effect; (S) a significant sex effect; (I) a significant interaction between the environment and sex factors. (●) EC♂; (○) EC ♀; (▲) IC♂; (Δ) IC ♀. (*) Significant at $p < .05$; (**) significant at $p < .01$; (***) significant at $p < .001$. From Juraska *et al.* (1985).

replications while this pattern was reversed in the EC rats (female > male) (Figs. 2 and 3). The data from both of these brain areas show that sex differences in the brain are not an absolute phenomenon as was implied by previous work, rather their existence and direction are influenced by the environment.

C. Maternal Influences

While human sex differences are an interaction of biological and cultural influences, we often view sex differences in animals as relatively pure manifestations of biological factors. However, Moore and Morelli (1979) have found more maternal licking of the anogenital regions of male than of female rat pups. This differential treatment of male and female offspring may be considered an analog to sex differences in parental expectations in humans. This effect appeared in both natural and foster mothers (Moore and Morelli,

1979; Richmond and Sachs, 1984) and was dependent on odor cues, especially those present in the pup urine (Moore, 1981). Injections of estrogen, testosterone, and dihydrotestosterone into female pups all elicited care of the type given male pups from mothers (Moore, 1982). An important question is how this differential maternal care influences later behavior. Moore (1985) has reported that artificial anogenital stimulation of female pups later increased the amount of male sexual behavior (i.e., more of the motor pattern of intromission) when these females were ovariectomized and given testosterone as adults. On the other hand, males raised with mothers that licked them less due to olfactory deficits had longer latencies for every phase of male sexual behavior as adults (Moore, 1984). Females that were licked less from the same litters when given testosterone following ovariectomy in adulthood also showed longer latencies and less male sexual behavior (Moore, 1984). Thus, while male pups are licked more than female pups, maternal licking facilitates male sexual behavior in both sexes. It is possible that other behaviors are influenced as well. Not only are there sex differences in response to the environment, but Moore's work also indicates that the sexes experience somewhat different environments as early as their first 2 weeks of life.

D. Mechanisms

No unitary mechanism will easily account for all of the behavioral and neural data discussed above—it is too diverse; and while many of the data support the view of the male as the sex that is more susceptible to the environment, this is not invariable. To discuss the mechanisms for sex differences in responsiveness to the environment is to discuss sex differences in general. Any time an organism is different it interacts with environmental events differently, and these differences can then be further augmented by later environmental events in a cascading manner. During development males are exposed to circulating androgens that influence the sexual morphology of the brain (see Chapter 6, this volume) and body, and this makes them different from females in more than just the reproductive realm. Certainly the androgen surge is an extra step in male development. Variations in this step still leave a viable, albeit often more femalelike, organism. This means that there may be more variability in males due to factors that influence the androgen surge. Since females do not have this extra developmental step, they tend to vary less. Circulating androgens (and estrogens, as discussed in Chapter 6) can influence parts of the brain that do not have a direct role in sexual behavior, such as the cortex and hippocampus, through conversion and binding to estrogen receptors that are present in these areas during development (Sheridan, 1979). In addition, dimorphism in structure and function could be induced by dimorphic afferent input into a region, such that areas without steroid receptors may also be affected. Thus, although little work has been focused on this question, early gonadal steroids are probably responsible for many of the sex differences in response to the environment.

Early androgen effects can occur at several different levels. First, the androgen surge can (and apparently does in the hypothalamus) directly influence brain growth, which translates into structural and functional neural differences and behavioral differences. Second, if the hypothalamus is altered, the androgen surge can influence hormones apart from the gonadal steroids. One such candidate is the corticosteroids which, as discussed above, appear to be at higher levels in females than males. These hormones can alter behavior, for example, through corticosteroid receptors in the hippocampus (Stumpf and Sar, 1978), and they can act on metabolic functions. Third, the androgen surge appears to alter several aspects of metabolism including metabolic rate, heat production, and fat and muscle levels, as reviewed by Hoyenga and Hoyenga (1982). Lastly, through pheromones and other means (such as sexually distinct morphology), the androgen surge influences how other animals respond to the individual, as exemplified by the sex differences in maternal licking (Moore and Morelli, 1979). Thus, through many different avenues the sexes act on and are acted upon by the environment differently, potentially resulting in differences between the sexes in plasticity of the brain and behavior.

III. Implications for Human Sex Differences

In addition to sex differences in human behavioral capabilities (Maccoby and Jacklin, 1974), there recently has been extensive documentation of sex differences in cerebral lateralization, with females appearing to have more bilateral representation of language capabilities and, to a lesser extent, of performance IQ than males (McGlone, 1980; Inglis and Lawson, 1981). From these data it appears that males are more vulnerable to brain injuries, in that a particular capability can be wiped out with a localized lesion. Another view of these results has been put forth by Kimura (1983), who reexamined the studies of sex differences in verbal lateralization. She found that females only appear to be relatively spared in verbal abilities following lesions since more vascular accidents occur in the posterior portion of the left (verbal) hemisphere while female verbal abilities are more readily affected by anterior left hemisphere lesions. This suggests a sex difference in intrahemispheric organization of language rather than a lateralization difference. These studies of sex differences in functional organization of the human cerebral cortex, as well as recent evidence indicating that human females have a larger splenium (posterior portion) of the corpus collosum than males (de Lacoste-Utamsing and Holloway, 1982), demonstrate that sex differences in the cerebral cortex are not unique to rats. These studies also indicate that the nature of sex differences in the human brain are not simple (e.g., heavier cortex in males) but are organizational with differences in both brain morphology and function.

One obvious question is that of the extent to which these neural and behavioral sex differences in humans are the result of environmental interactions. My work in the visual cortex and hippocampal dentate gyrus demonstrates that the presence and direction of neural sex differences are dependent on the rearing environment (Juraska, 1984; Juraska *et al.*, 1985). This view of neural sex differences is shared by Peterson (1982), who points out that the biological system interacts with the psychosocial environment to produce a functioning organism. It is a mistake to think of sex differences as strictly intrinsic or environmental in origin. It is the interaction of these factors that is important and such interaction can lead to a good deal of variability that we have not yet explored.

Acknowledgments

The author's work was supported by National Institutes of Health grant HD14949 and the MacArthur Foundation.

References

Barnes, R. H., Cunnold, S. R., Zimmerman, R. R., Simmons, H., MacLeod, R. B., and Krook, L. (1966). *J. Nutr.* **89**, 399–410.
Beckhardt, S., and Ward, I. L. (1983). *Dev. Psychobiol.* **16**, 111–118.
Chambers, K. C., and Sengstake, C. B. (1978). *Physiol. Behav.* **21**, 29–32.
Crutchfield, F. L., and Dratman, M. B. (1980). *Biol. Neonate* **38**, 203–209.
de Lacoste-Utamsing, C., and Holloway, R. L. (1982). *Science* **216**, 1431–1432.
Diamond, M. C., Johnson, R. E., and Ingham, C. (1971). *Int. J. Neurosci.* **2**, 171–178.
Dohanich, G. P., Witcher, J. A., Weaver, D. R., and Clemens, L. G. (1982). *Brain Res.* **241**, 347–350.
Erskine, M. S., Stern, J. M., and Levine, S. (1975). *Physiol. Behav.* **14**, 413–420.
Fischette, C. T., Biegon, A., and McEwen, B. S. (1983). *Science* **222**, 333–335.
Fleischer, S. F., and Turkewitz, G. (1979a). *Dev. Psychobiol.* **12**, 137–149.
Fleischer, S. F., and Turkewitz, G. (1979b). *Dev. Psychobiol.* **12**, 245–254.
Freeman, B. J., and Ray, O. S. (1972). *Dev. Psychobiol.* **5**, 101–109.
Galler, J. R. (1981). *Dev. Psychobiol.* **14**, 229–236.
Galler, J. R., and Manes, M. (1980). *Dev. Psychobiol.* **13**, 409–416.
Gerall, H. D., Ward, I. L., and Gerall, A. A. (1967). *Anim. Behav.* **15**, 54–58.
Geschwind, N., and Behan, P. (1982). *Proc. Natl. Acad. Sci. U. S. A.* **79**, 5097–5100.

Goy, R. W., and Goldfoot, D. A. (1974). *Neurosci. Study Program 3rd,* pp. 571–581.
Gruendel, A. D., and Arnold, W. J. (1969). *J. Comp. Physiol. Psychol.* **67**, 123–128.
Hamilton, W. L., Diamond, M. C., Johnson, R. E., and Ingham, C. A. (1977). *Behav. Biol.* **19**, 333–340.
Hansen, S. Larsson, K., Carlsson, S. G., and Sourander, P. (1978). *Dev. Psychobiol.* **11**, 51–61.
Hard, E., and Larsson, K. (1968). *Brain Behav. Evol.* **1**, 405–419.
Hennessy, J. W., Levin, R., and Levine, S. (1977). *J. Comp. Physiol. Psychol.* **91**, 770–777.
Hoyenga, K. B., and Hoyenga, K. T. (1982). *Physiol. Behav.* **28**, 545–563.
Inglis, J., and Lawson, J. S. (1981). *Behav. Brain. Sci.* **5**, 307.
Jones, A. P., and Friedman, M. I. (1982). *Science* **215**, 1518–1519.
Jordan, T. C., and Howells, K. F. (1978). *Brain Res.* **157**, 202–205.
Joseph, R. (1979). *J. Psychol.* **101**, 37–43.
Juraska, J. M. (1984). *Brain Res.* **295**, 27–34.
Juraska, J. M., Henderson, C., and Müller, J. (1984). *Dev. Psychobiol.* **17**, 209–215.
Juraska, J. M., Fitch, J., Henderson, C., and Rivers, N. (1985). *Brain Res.* **333**, 73–80.
Kimura, D. (1983). *Can. J. Psychol.* **37**, 19–35.
Kitay, J. I. (1961). *Endocrinology (Baltimore)* **68**, 818–824.
Leedy, M. G., Vela, E. A., Popolow, H. B., and Gerall, A. A. (1980). *Physiol. Behav.* **24**, 341–346.
Levine, S. (1974). *In* "Sex Differences in Behavior" (R. C. Freidman, R. M. Richart, and R. L. Vande Wiele, eds.), pp. 87–98. Wiley, New York.
Maccoby, E. E., and Jacklin, C. N. (1974). "The Psychology of Sex Differences." Stanford Univ. Press, Stanford, California.
McConnell, P., Uylings, H. B. M., Swanson, H. H., and Verwer, R. W. H. (1981). *Behav. Brain Res.* **3**, 411–415.
McGlone, J. (1980). *Behav. Brain Sci.* **3**, 215–263.
Mathura, C. B., and Harper, A. H. (1979). *Physiol. Psychol.* **7**, 444–446.
Meisel, R. L. (1983). *Behav. Neurosci.* **97**, 785–793.
Milner, T. A., and Loy, R. (1980). *Anat. Embryol.* **161**, 159–168.
Milner, T. A., and Loy, R. (1982). *Brain Res.* **243**, 180–185.
Mitchell, G. D., Ruppenthal, G. C., Raymond, E. J., and Harlow, H. F. (1966). *Child Dev.* **37**, 781–791.
Moore, C. L. (1981). *Anim. Behav.* **29**, 383–386.
Moore, C. L. (1982). *J. Comp. Physiol. Psychol.* **96**, 123–129.
Moore, C. L. (1984). *Dev. Psychobiol.* **17**, 347–356.
Moore, C. L. (1985). *In* "The Comparative Development of Adaptive Skills: Evolutionary Implications" (E. S. Gollin, ed.), pp. 19–56. Erlbaum, Hillsdale, New Jersey.
Moore, C. L., and Morelli, G. A. (1979). *J. Comp. Physiol. Psychol.* **93**, 677–684.
Morley, B. J., Cohen, S. L., and Poplawsky, A. (1977). *Neurosci. Lett.* **7**, 347–352.
Ordy, J. M., Kaack, B., and Brizzee, K. R. (1975). *In* "Aging"(H. Brody, D. Harman, and J. M. Ordy, eds.), Vol. 1: Clinical, Morphologic, and Neurochemical Aspects in the Aging Central Nervous System. Raven, New York.
Orensanz, L. M. Guillamon, A., Ambrosio, E., Segovia, S., and Azvara, M. C. (1982). *Neurosci. Lett.* **30**, 275–278.
Peterson, A. C. (1982). *Behav. Brain Sci.* **5**, 312.
Pfaff, D. W. (1966). *J. Endocrinol.* **36**, 415–416.
Richmond, G., and Sachs, B. D. (1984). *Dev. Psychobiol.* **17**, 87–89.
Rutter, M. (1970). *In* "The Child in His Family" (E. J. Anthony and C. Koupernik, eds.), Vol. 1, pp. 165–196. Wiley, New York.
Sackett, G. P. (1974). *In* "Sex Differences in Behavior" (R. C. Friedman, R. M. Richart, and R. L. Vande Wiele, eds.), pp. 99–122. Wiley, New York.
Schoneshofer, M., and Wagner, G. G. (1977). *J. Clin. Endocrinol. Metab.* **45**, 814–817.
Sheridan, P. J. (1979). *Brain Res.* **178**, 201–206.

Storm-Mathisen, J. (1977). *Prog. Neurobiol. (Oxford)* **8**, 119–181.
Stumpf, W. E., and Sar, M. (1978). *Am. Zool.* **18**, 435–445.
Taylor, D. C., and Ounsted, C. (1972). *In* "Gender Differences: Their Ontogeny and Significance" (C. Ounsted and D. C. Taylor, eds.), pp. 215–240. Churchill, London.
Twiggs, D. G., Popolow, H. B., and Gerall, A. A. (1978). *Science* **200**, 1414–1415.
Ward, I. L. (1974). *In* "Sex Differences in Behavior" (R. C. Friedman, R. M. Richart, and R. L. Vande Wiele, eds.), pp. 3–17. Wiley, New York.
Weinberg, J., and Levine, S. (1977). *Dev. Psychobiol.* **10**, 161–169.
Weinberg, J., Krahn, E. A., and Levine, S. (1978). *Dev. Psychobiol.* **11**, 252–259.
Weinberg, J., Gunnar, M. R., Brett, L. P., Gonzalez, C. A., and Levine, S. (1982). *Physiol. Behav.* **29**, 201–210.
Werner, E. E., and Smith, R. S. (1982). "Vulnerable but Invincible: A Longitudinal Study of Resilient Children and Youth." McGraw-Hill, New York.
Wilson, E. O. (1975). "Sociobiology: The New Synthesis." Belknap, Cambridge, Massachusetts.
Yanai, J. (1979a). *Acta Anat.* **103**, 150–158.
Yanai, J. (1979b). *Acta Anat.* **104**, 335–339.

15

The Development of Olfactory Control over Behavior

ELLIOTT M. BLASS
Department of Psychology
The Johns Hopkins University
Baltimore, Maryland

I. Introduction

This chapter is concerned with the ontogeny of mammalian olfaction, particularly as it relates to infant–mother interactions in the rat and other animals during the nesting period. It focuses on three vital social behaviors, orientation (Section II), huddling (Section III), and suckling (Section IV), whose expression is strongly influenced by olfactory stimuli. These behaviors, though differing in their motor properties and time of onset, share an important feature concerning the development of olfactory control. At first,

Developmental
NeuroPsychobiology

a broad rage of odors can influence their expression. The range narrows with time. Ultimately, only a particular odor helps either to elicit a behavior or to determine where it will occur.

To better understand the mechanisms that allow olfactory conditioning to occur, I will also review studies that have identified and isolated some of the features of the mother and her interactions with her pups that are critical for conditioning in the nest (Section V).

Section VI seeks correlations between behavioral and neural development. The chapter concludes by offering a behavioral model of how olfactory stimuli might gain control over behavior.

This chapter is written from the perspective of a developmental psychobiologist. Identifying proximal causation of behavioral function in developing organisms is a clear and important goal of psychobiology. Of equal importance is the position that developmental psychobiology occupies relative to sociobiology on the one hand and functional neurology and neurochemistry on the other. To the extent that evolutionary pressures are shaped phenotypically, it behooves us to understand the developmental events that help prepare the adult to meet the demands of its station (e.g., kin recognition, mate selection, parental behavior). Yet infantile behavior should not be viewed primarily as a precursor of the adult form and function. As articulated by Blass (1980) and especially by Galef (1981) and Oppenheim (1981), the functions of behavioral and morphological modifications during early development must also be appreciated during the neonatal period itself and not necessarily seen only as a foundation for an adaptive adult trait. These ideas as they pertain to individual behavior patterns must be evaluated empirically.

The links with developmental neurobiology are at least as strong. Techniques for establishing birth dates of neural processes; detecting the presence of neurotransmitter release, reuptake, and metabolic mechanisms; the recording of unit activity; and assessing metabolically active loci have allowed for a careful demarcation of neural development. Behavioral studies provide the potential for assessing neurological function and how these systems themselves might be modified as a result of particular and identifiable behavioral events (cf. Gottlieb, 1981; Jacobson, 1974; Sperry, 1971).

The Nest Setting

Opportunities for olfactory conditioning are present at birth. Indeed, specific experiences during or surrounding the birth event have been shown to influence olfactory determination of certain behaviors (Pedersen and Blass, 1982; Smotherman, 1982; Stickrod *et al.*, 1982). The nest setting has two characters, the dam and her offspring, who interact reciprocally (Alberts and Gubernick, 1983; Rosenblatt, 1970, 1976). The features and behavior of one player determine the behavior of the other. For example, increased body temperature helps terminate nursing by rat dams (Leon, 1978), while cooling, among other processes, elicits ultrasonic calling in infant rodents (Okon, 1971, 1972). Contact per se quiets seemingly distressed kittens (Freeman and Rosenblatt, 1978a,b). In rats contact between mother and infants elicits licking by the mother, although olfactory cues are also utilized (Moore, 1981; Moore and Morelli, 1979). Ulceration in rats caused by premature separation is eased by milk injected into the gastrointestinal tract but not by other stimuli (Ackerman *et al.*, 1978). In turn, the mother recaptures some of her fluid lost in milk by licking the pup's anogenital region (Friedman *et al.*, 1981).

Infants must master several complex tasks. They must either stay in or be able to locate the nest or home area; keep warm; establish and maintain contact with the dam; and locate, suckle, and withdraw milk from available teats. The nature of these demands changes during the nesting period. With increased motor development, for example, the infant can more readily leave its base and fall victim to predators. Maternal attentiveness and accessibility also changes. The dam is less fastidious in maintaining an insulating nest, visits the nest and young less frequently, and initiates nursing behavior with greater reluctance (Rosenblatt and Lehrman, 1963). The means by which infant altricial mammals meet these obligations is now discussed, with special reference to olfactory considerations.

II. Orientation

Orientation consists of (1) approaching an odor source and (2) reducing activity. The two components have been distinguished empirically. Schapiro and Salas (1970) demonstrated that nest odors significantly reduced activity in day-6 rats. According to Gregory and Pfaff (1971) and Nyakas and Endroczi (1970), preferential approach to nest odors does not appear in albino rats until about 10–12 days of age. It is now clear, however, that albino rats as young as 1 day of age can orient to an odor. Orientation to a nonpreferred odor can be induced by pairing it with the delivery of liquid diet to the mouth in warm ambient temperature (Johanson and Hall, 1982; Johanson and Teicher, 1980). One- and 3-day-old rats can also discriminate between two odors emanating from a punctuate source. Johanson and Hall (1979) demonstrated that pups would probe a lever whose scent was associated with food delivery. This contrasts with earlier reports in which approach to home nest material was obtained in infant albino rats starting at days 10–12. Testing conditions in the earlier studies differed importantly from those of Johanson, however, as infants were tested at room temperature and were nondeprived (Gregory and Pfaff, 1971; Nyakas and Endroczi, 1970). As will be discussed below, an important boundary condition that must be met for both acquisition of a relationship and performance of an instrumental task is that infants must be in a state of high activation.* Thus, the home orientation studies should be replicated using either moderately deprived animals or those tested in the heat. The issue might be moot regarding biological adaptation because altricial young enjoy very close contact with the dam, who spends anywhere from 70–90% of her time with them in the nest in laboratory cages (Ader and Grota, 1970; Plaut and Davis, 1972) and even in a seminaturalistic environment in which the nest occupies only 10% of the total area and the dam has easy egress to an open area (B. A. MacFarlane and E. M. Blass, unpublished observation, 1983).

Olfactory orientation to the nest or to the dam has been reported in diverse altricial infants, although tactile (Tobach *et al.*, 1967), thermal (Fowler and Kellogg, 1975; Leonard, 1974; Okon, 1971; Rosenblatt, 1971), and auditory (Arnold *et al.*, 1975) stimuli are also utilized. Both the settling and approach components of orientation are seen during the first 8–10 days postpartum in kittens (Rosenblatt, 1971; Rosenblatt *et al.*, 1969), puppies (Scott *et al.*, 1974), and sheep (Poindron, 1976). Two experiments (MacFarlane, 1975; Schaal *et al.*, 1980) have reported that 8-day-old human infants preferentially orient to the soiled breast pad of their own mother over that of mothers in a similar lactational status. The olfactory relationship is reciprocal as these mothers, on the basis of olfactory cues alone, can recognize their own infants (Schaal *et al.* 1980).

Olfaction appears to be the only modality utilized in establishing reciprocal orientation in certain species. Goats accept their own young and reject others on the basis of olfactory cues deposited via licking during or shortly after birth (Gubernick, 1981). Anosmia markedly reduces discrimination between own and alien young. Similar mechanisms seem to underlie acceptance and rejection in sheep. Only 20–30 min of contact is sufficient to allow ewes to reject alien young (Poindron, 1976; Smith, 1965). Again, the phenomenon appears to have an olfactory basis, as anosmia attenuates or eliminates

*We still lack an adequate measure of activation. Behavioral activation, whether caused by infusions of food into the mouth (Hall, 1979b), electrical stimulation of the medial forebrain bundle (Moran *et al.*, 1983a,b), or whole-body stroking, is unmistakable and striking in rats younger than 9–10 days of age. Yet amphetamines and food deprivation, which elevate adult activity, are not obviously activating behaviorally in neonates. These manipulations are treated as activating because of their known effects on adults and because they are sometimes interchangeable with stimulation that is blatantly activating (see below). The inadequacy of this justification highlights the pressing need for reliable and sensitive indices of arousal.

maternal discrimination between own and alien young. The same holds for domestic pigs. According to Meese and Baldwin (1975), olfactory bulbectomy prevents the sow from rejecting alien young.

Squirrel monkey infants raised on terry cloth surrogate mothers of different color and odor distinguish among "mothers" on the basis of odor and not vision (Kaplan and Russell, 1974). It is also likely that squirrel monkey dams, which carry their young on their backs and therefore have minimal eye contact, can identify their young on the basis of olfactory cues, although to my knowledge this has not been studied.

Precocial rodents such as spiny mice (*Acomys cahirinus*) and guinea pigs are extraordinary in their ability to establish olfactory bonds quickly and to make very subtle distinctions on olfactory bases. Thus spiny mice, after being exposed to an artificial odor during the first hour after birth, will orient to that odor even after 3 days of separation (Porter and Etscorn, 1974). This orientation appears to be time-linked in its development. It can be obtained by exposing spiny mice to a novel odor during the first 3 days after birth but not on days 4 and 5 (Porter and Etscorn, 1974). This is the first example of a recurrent theme in this chapter, namely, that many olfactory cues can gain control over the expression of a behavior during the earliest manifestation of that act but not after the act has been "labeled" by recurrent exposure to a particular olfactory complex.

In short, evidence for early orientation, discrimination, and recognition by a number of mammals points to reciprocal olfactory mechanisms between mother and young. Mothers appear to attend more selectively their own infants. Likewise, the infants either orient to their own mother (nest) and are quieted by maternal (nest) odors and textures. This evidence must be tempered, however, by the limited number of species studied and by the fact that primary reciprocal recognition in many species is achieved later in ontogeny by vision and audition.

Orientation during the Weaning Period in Altricial Rodents

The demands placed upon altricial mammals during the weaning period differ markedly from those experienced earlier. These slightly older animals occupy a precarious position. Although they start to thermoregulate physiologically, they cannot survive protracted exposure in even moderate temperatures. Weanling rats, for example, start to eat spontaneously in the nest area (Babicky *et al.*, 1973), yet not enough to sustain themselves. They start to engage actively in play (Bolles and Woods, 1964), yet this must be confined to the nest and its surrounds lest it attract predators. Thus, on the one hand, the beginning of weaning is the start of liberation from mother and surrounds. On the other hand, weanlings are not free to exploit fully this newly found independence. Although much remains to be learned about how this balance is struck, the work of Leon (1974) and colleagues has demonstrated how young rats are attracted to and stay in the nest and its immediate area during the period of weaning (~ days 16–28 postpartum).

Leon and Moltz (1971, 1972) demonstrated that rats 16–28 days of age were selectively attracted to an airstream that passed over dams of same-age pups. Leon (1974) reported that the soiled bedding of such mothers was sufficient to elicit differential orientation and approach. The material that soiled these nests was a bacteria-rich gelatinous substance, cecotrophe, released by the cecum. According to Leon (1974) day–16–28 pups stimulate increased prolactin release in their dams, which, in turn, causes hyperphagia with secondary hyperdipsia. Much of the additional food is lodged in the cecum, where it is broken down to release a highly volatile odor that, among other things, serves to attract 16- through 28-day-old pups. Either blocking prolactin release by ergocornine or preventing bacterial degradation of food by neomycin was sufficient to block preference for the dam's odor. It is worth noting, however, that cecotrophe is not the only biological attractant, as, according to Galef and Kaner (1980), maternal feces are also a very potent attractant in some rat strains, even more so than cecotrophe.

It seems that cecotrophe may serve a digestive as well as social function. Moltz and Lee (1981) suggest that the excess food eaten by the dam, which yields more cecotrophe than she herself ingests, provides cecotrophe eaten by the developing young. Moltz and Lee (1981) speculate that bile acids contained in the ingested cecotrophe protect against necrotizing enterocolitis and might actually help in development of the central nervous system.

Leon *et al.* (1972) investigated the mechanism underlying attraction to the home odor. They isolated 14-day-old pups for 3 h and exposed them to an airstream laced with peppermint. Rats so treated chose peppermint-scented air over clean air. Thus, simple exposure, even during the stress of isolation, reduced neophobia and promoted a preference, at least, over clear air and contrasts with the elevated levels of activity needed in younger rats (~ days 1–5) for a behavior to come under olfactory guidance.

The proximal mechanism for inducing orientation in day 10–16 rats is not known. Also, we do not know how enduring the peppermint preference was, nor do we know if there are more optimal means of establishing a preference. Furthermore, there is the possibility that certain substances are differentially preferred (i.e., more readily conditioned) over others (Galef and Kaner, 1980). Finally, no studies, to my knowledge, have assessed the ease by which these preferences for novel odors might be established during the 12-day period of attraction to maternal cecotrophe starting at day 16. This is a particularly important issue for, as shown above, in regards to spiny mice, it is increasingly difficult with age to establish control by a new odor over a given behavior. Given that the preference for maternal odor disappears by the end of weaning in albino rats as well as spiny mice and that the preference for the odor may serve to "tether" the rat to the nest area, that is, a time-limited function (Alberts and Leimbach, 1980), it would be surprising indeed if a strong preference for novel odors would be easily obtained by limited exposure later in the weaning period.*

III. Contact Behavior (Huddling)

Newborn altricial mammals face the same challenges as reptiles with regard to thermoregulation. They cannot maintain normothermia physiologically and metabolically when placed outside of a very restricted thermoneutral zone. They are ectotherms. Reptiles solve the problem of ectothermia behaviorally by occupying nonstressing microclimates during periods of extreme high or low ambient temperatures (Schmidt-Nielsen, 1964). Indeed, lizards can learn operant tasks that allow them transport to a thermoneutral ambient temperature (Hammel *et al.*, 1967). Altricial infant mammals also utilize behavior to achieve thermoneutrality. A drop in body temperature elicits ultrasonic calling in mice (Okon, 1971, 1972) that retrieves the dam to the nest where she warms her young by contact conduction and through her milk. Maternal nest departure elicits thermal self-protection by the pups. This is achieved by huddling.

Alberts (1978a,b) and colleagues have revealed the dynamics of huddling in albino rats and the role of olfaction in this behavior. Less huddle surface is exposed at low ambient temperatures (TA) and more at elevated TA. This is especially of benefit to younger pups. Five-day-old rats maintained in isolation at 24°C suffer a 6°C drop in rectal temperature relative to huddled littermates. The huddle is actively maintained to minimize exposure of individual rats at low TA and to maximize individual exposure in the heat. By including an anesthetized sibling in the huddle and utilizing time-lapse video recording, Alberts (1978b) showed that the anesthetized pup "sank" to the bottom of the huddle at high TA, that is, pups were able to maximize heat loss by actively maintaining maximal contact with the air. In the cold the anesthetized pup "floated" to the top of the huddle; that is, nonanesthetized pups minimized heat loss by actively seeking the huddle's

*This must be tempered, however, by the recent finding of Rodriguez Echandia *et al.* (1982), who showed that rats exposed to lemon nest odors from birth till weaning (day 21) showed an enduring preference for lemon odor (over clean shavings) till 121 days of age.

center. Olfaction is important in the expression of huddling in rats younger than day 10. Both peripherally (Singh and Tobach, 1975) and centrally (Tobach *et al.*, 1967) induced anosmia severely disrupted huddling in the nest in rats 3–10 days of age. To the extent that huddling choice is under thermal control, we may conclude that olfactory stimuli act permissively, possibly by allowing changes in thermotactile stimulation to gain control over the behavior. Alternatively, olfactory stimulation may tonically activate normal rats to a level necessary for the sustained expression of this organized behavior.

Although huddling is under the control of thermal conditions throughout the nursing period, there is a major transition in the objects with which infant rats will huddle. Given a choice between huddling with a warm, fur-covered tube versus an inanimate conspecific, 5- and 10-day-old rats chose the warm tube, whereas 15- and 20-day-old rats preferred to maintain contact with the conspecific. This preference was based on the olfactory characteristics of the conspecific. It was eliminated in older pups by anosmia caused by the intranasal infusion of zinc sulfate. Furthermore, rats reared in a nest laced with artificial odors preferred to huddle with rats bearing these odors, providing strong evidence for experience conferring control upon an odor. As in our discussion of orienting behavior, however, the ease with which the "preferred" huddling stimulus can be altered during the course of ontogeny remains an open and important question.

In one series of experiments, Alberts (1981) demonstrated that exposing day-15 pups for 4 h to a foster dam or to siblings bearing an artificial odor induced a huddling preference for that odor versus a novel one. It is of interest that natural nest odors continued to be preferred over the artificial ones, possibly because of the rat's greater exposure to them.

In order to assess the potency of various classes of contacts with the dam, Alberts (1981) exposed rats on alternate days from birth till day 14 to two odors for 4 h daily. On one day rats were exposed to a lactating foster dam that had the odor painted on her ventrum. On the alternate day the pups were exposed as a litter to the second odor. When tested on day 15 pups demonstrated a strong preference for the mother-associated odor over that of the pup-associated odor. It seems that the critical maternal dimension for the formation of odor preference under the conditions just described is that of the dam's temperature. Specifically, pups did not prefer to huddle with test stimuli bearing the odor of a scented lactating dam over that of the scented nonlactating dam with whom they had lived on the alternating days. Moreover, as shown in Fig. 1, when alternate days featured exposure to a warm versus a cool plastic tube, pups preferred the odor associated with the warm tube. The idea that warmth was the key feature was derived from failure of pups to prefer the scent of an odorized lactating dam over that of a warm tube to which they had been exposed on alternate days (Fig. 1C). These

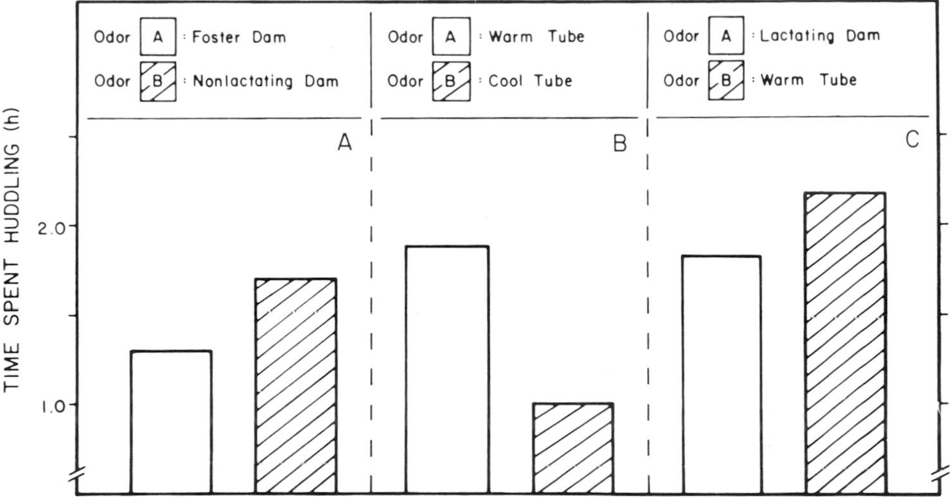

Fig. 1. Time spent huddling by day-15 rats with various scented objects. See text for experimental conditions. From Alberts (1981).

findings suggest that normal filial preferences that form developmentally may reflect the union of an odor with the positive affect of warmth that is derived by the pup by actively maintaining contact with its mother and siblings.

IV. Suckling

Suckling is the most highly specialized organized behavior of infant altricial mammals. It consists of nipple location followed by a stereotyped grasping response of licking the nipple, erecting it, and then taking it into the mouth. Once the nipple is seated in the mouth, the pup relaxes until milk is let down, at which time the rat exhibits a dramatic hyperextension response and withdraws milk from the engorged teat (Drewett *et al.*, 1974). The factors that determine milk intake from the nipple have been reviewed (Blass and Cramer, 1982) as have those concerned with nipple location and apprehension (Blass and Teicher, 1980; Pedersen and Blass, 1981). Because of the vital role of olfaction in rodent suckling, I will briefly summarize olfactory control and then discuss new findings concerned with the role of experience in such control.

Evidence for a role of olfaction in nipple attachment was first brought to light by Tobach and colleagues (Tobach *et al.*, 1967; Singh and Tobach, 1975), who demonstrated that olfactory bulbectomy was fatal to infant rats. Such rats gradually lost weight postoperatively and died of starvation within 2–4 days for not having suckled. The lesions were effective because of sensory rather than central damage, because destroying the peripheral mucosa with bilateral intranasal irrigation of zinc sulfate also proved lethal (Singh and Tobach, 1975).

Washing the nipples of dams prevented their 4-day-old pups from suckling even when the pups were hand-held in contact with the nipple. Such pups would suckle, though, if saliva derived from the mouth or salivary glands of same-age pups was painted on the washed teats (Teicher and Blass, 1976). This phenomenon held for rats throughout the nursing-suckling period (i.e., 1–30 days of age) even after eye opening at day 15 (Blass *et al.*, 1977; Bruno *et al.*, 1980; Hofer *et al.*, 1976).

The first suckling bout is elicited by nipples coated during parturition with saliva* or amniotic fluid. The effectiveness of amniotic fluid seems to be specific as a number of attractive (to us at least) scents such as vanilla and amyl acetate (a banana odor) did not elicit suckling. The fact that neither parturient female urine, which is rich in the hormones surrounding birth, nor saliva of virgin females elicited nipple attachment spoke to the specificity of the effect (Teicher and Blass, 1977).

The potency of amniotic fluid raised the possibility that experience determines the efficacy of an odor in the first suckling experience. Amniotic fluid almost continuously undergoes significant modifications as a result of the fetus's swallowing and excreting it (Lev and Orlic, 1972). The fluid also changes near delivery, reflecting the dam's altered hormonal profiles. Also, to the extent that amniotic fluid is a plasma filtrate, then, like other plasma filtrates, it should change in accordance with the varied diet of the omnivorous rats.

Blass and Pedersen (1980) devised a technique to alter the characteristics of amniotic fluid *in utero*. On the twentieth day of a timed pregnancy 0.2 cc of citral, a tasteless, lemon-scented fluid, was injected through the distended uterus into the visible individual amniotic sacs. Control pups underwent the same surgical procedure but received isotonic saline vehicle. The litters were delivered 2 days later by Cesarean section and vigorously stroked by a soft artist's brush in the presence of either citral or isotonic saline (Pedersen and Blass, 1982). The results of these pre- and postnatal citral and isotonic saline treatments on the first nipple attachment are presented in Fig. 2. Only rats that had experienced citral *both* pre- and postnatally suckled washed nipples coated with citral (Fig. 2A). Moreover, only rats so treated failed to suckle amniotic fluid–coated nipples that normally elicited suckling. Thus, the predominant odor experienced by a rat

*Parturient dam saliva has not been tested without its amniotic contaminant. This should be done by placing a salivary gland-derived extract on washed nipples and determining nipple attachment in newborn pups.

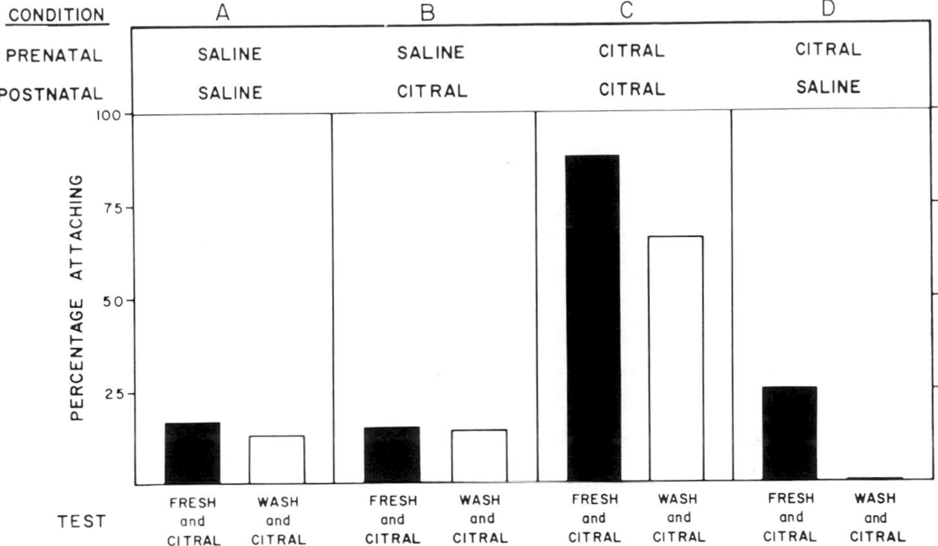

Fig. 2. The first nipple attachment of rats normally reared (A) or reared with citral treatment *in utero* and *ex utero* (C) (see text for details) and tested on normal or washed nipples in (C) citral-scented air. From Pedersen and Blass (1982). Copyright (1982) John Wiley & Sons, Inc.

during the last 2 days of gestation and during the period immediately after birth (while the animal is being stimulated) elicits the first nipple attachment. Specifically, citral permitted suckling and amniotic fluid did not in rats exposed to citral as above. Amniotic fluid may normally be effective by virtue of its having been experienced pre- and postnatally.*

Our finding that olfactory control of initial suckling could develop experientially led Pedersen *et al.* (1983) to determine if suckling continues to reflect changes in infantile olfactory stimulation. They exploited the natural chain of events that precede a suckling bout to obtain conditioning to a novel odor. Under normal circumstances the mother, upon returning to her nest to nurse, activates her nested young by stepping on them and, especially, by licking their anogenital areas (Bolles and Woods, 1964; Rosenblatt and Lehrman, 1963). She then settles over the litter to nurse. Pedersen *et al.* (1983) mimicked the sequence by activating pups via stroking with a soft artist's brush. They also injected *d*-amphetamine (0.5–1.0 mg/kg body weight) in

*A number of studies in birds (Impekoven 1970) and rats (Smotherman, 1982; Stickrod *et al.*, 1982) have presented good evidence for prenatal conditioning when stimuli are presented during the natural stream of behavioral events prior to birth (or hatching). These studies are the opening paragraph in what promises to be an interesting and important chapter in developmental psychology.

the presence of either citral or benzaldehyde, an almond scent. Pups were allowed to then suckle washed nipples in the presence of either citral or benzaldehyde. Pedersen *et al.* (1983) found that animals treated in the presence of an odor suckled washed nipples scented with that odor but not washed nipples bearing the other scent. The percentage of treated rats that suckled reflected the interaction of ambient temperature, odor, drug dosage, and stroking. This is seen in Fig. 3, where the efficacy of the combination of the stroking and amphetamine treatments was dependent upon the background in which these treatments had occurred. Thus, combining stroking with amphetamine in a relatively dense (.05 cc citral: 300 cc clean shavings), warm (33°C) ambience actually eliminated nipple attachment, whereas the same treatment in a cooler (23°C), less dense citral ambience was extraordinarily effective, leading to "normal" levels of nipple attachment to citral-scented, washed nipples.

I suggest, therefore, that the rule utilized early in development for nipple attachment is the following: Arousal in an odor predicts the presence of a nipple scented with that odor. Our preliminary findings suggest that this form of conditioning is probably limited to the first week postnatally. The phenomenon is very difficult to obtain in rats 7 days of age, and we have not been able to get it in 10-day-old rats (P. E. Pedersen, C. L. Williams, and E. M. Blass, unpublished observations). As in the case of orientation in spiny mice, the window for odor substitution closes with extended experience. Thus, older albino rats seem to work on absolute terms, that is, a specific nonsubstitutable odor elicits suckling.

Two sources of data offer support for the speculation that stimulation by odor, temperature, stroking, and amphetamine share a common catecholaminergic pathway in neonatal rats. First, caffeine, which seems to arouse day 3 rats, is ineffective throughout a wide range of doses in leading to olfactory conditioning, whereas amphetamine, a catecholaminergic agonist, is effective. Second, electrical stimulation of the medial forebrain bundle, a catecholaminergic pathway, causes a pattern of motor behavior bearing a strong resemblance to that seen when the rat is stroked anogenitally (Moran *et al.*, 1983a,b). Because the efficacy of amphetamine, a known catecholamine agonist, depends upon the intensity of olfactory, thermal, and mechanical stimulation in establishing conditioning (Pedersen *et al.*, 1983), one might speculate that the latter sources of stimulation are mediated by common neurochemical and neuroanatomical pathways. These findings and speculations may provide fertile territory for the collaborative efforts of psychobiologists and neurochemists who are interested in the central bases of affect, conditioning, and performance during ontogeny.

Preference for novel odors presented after nipple attachment has also been

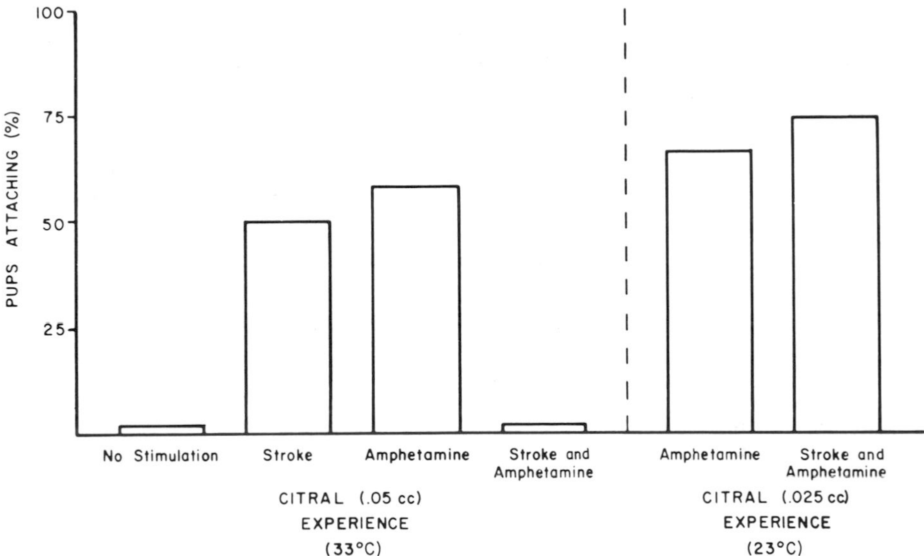

Fig. 3. Percentage nipple attachment to washed nipples scented with citral in day-3 rats treated with *d*-amphetamine or stroked at various temperatures in different citral concentrations (see text for details). From Pedersen *et al.* (1982).Copyright 1982 by the American Psychological Association. Reprinted by permission of the authors.

established. Brake (1981) demonstrated that 11- through 14-day-old, 24-h deprived rats preferred an odor paired with milk delivery during suckling. This finding is especially important in the present context because it demonstrates (1) that conditioning can occur during suckling and (2) that extremely high levels of activation are not necessary for conditioning to occur during suckling in slightly older rats. It is of considerable interest, therefore, to determine if this form of conditioning could be obtained in younger rats that remain quiescent while receiving milk infusions while suckling. From Johanson and Teicher (1980) we know that oral milk injections paired with an odor in very young rats cause a preference for that odor. The preference does not develop in a cool ambient temperature where the injections do not seem to be arousing.

In summary, suckling behavior is clearly under olfactory control, but this control shifts during ontogeny. At first, a number of stimuli can elicit nipple attachment. This narrows markedly with age. As in other cases of instrumental and classical conditioning in neonatal rats (see below), activation, which can come from a variety of sources, seems critical for such conditioning to form. At this point, we may assume that activation occurs normally in the warm nest when the mother licks the pups to introduce a suckling bout. The neurochemical mechanism mediating activation may be of catecholaminergic origin.

Nipple odors, in addition to eliciting probing and contributing to the olfactory configuration of the nest, influence suckling permissively. Specifically, pups placed in actual contact with washed but otherwise apparently normal nipples do not lick, erect, or suckle them. These behaviors appear to be under tactile control that is mediated via the infraorbital branch of the trigeminal nerve. Severing or anesthetizing this branch prevents nipple attachment (Hofer *et al.*, 1981). Yet such rats locate, orient to, and probe into the nipple region. Taken together, studies on suckling by normal rats on washed dams and of nerve V–damaged rats on normal dams suggest that olfaction facilitates nipple location, elicits probing, and permits the tactile properties of the nipple to elicit, via the infraorbital nerve, nipple licking, erection, and attachment.

Finally, there is no established relationship between odors that elicit nipple attachment and those that foster orientation to the nest. The latter odors appear considerably more volatile than the former, and pups orient to them at a distance. Nipple odors become effective only when the infant is within 1–2 cm of the nipple. Also, the mechanisms of conditioning seem to differ. For preference to be expressed in older rats, simple exposure suffices. More is needed for control over suckling. The mechanism also seems to differ for establishment of huddling preference as described above (Alberts, 1981). The important issue remains open, of course, as to whether the earlier conditioning (of suckling or huddling) experiences, when the rats appear to learn about the relationship between odors, states, and behavioral outcome, are a prerequisite for the development of social or olfactory preferences at the start of the weaning period.

V. Olfactory Conditioning outside the Nest

Olfactory conditioning has been obtained away from dam and nest area in altricial rodents under a variety of circumstances during the first week of age. Animals allowed to perform different tasks under conditions of activation brought upon by high ambient temperatures, strong olfactory stimulation, maternal and food deprivation, or various combinations of the above, learn tasks that are instrumental to the delivery of milk to the mouth (Johanson and Hall, 1979), delivery of current to the medial forebrain bundle (Moran, *et al.*, 1981), and suckling nonlactating nipples (Amsel *et al.*, 1976; Kenny and Blass, 1977). The discriminatory capabilities of these animals are likewise impressive. Based on olfactory cues, day-1 rats distinguish between paddles that cause milk delivery to the mouth from those that do not. By utilizing spatial information, 3-day-old rats more often push paddles that cause the delivery of current to the MFB than those that do not. Also, basing their tests on the rats' differential running speeds, Amsel *et al.* (1976) showed that by day 10 rats can assess the quality of the dam at the end of a runway, that is, they move fastest when rewarded with milk from the nipple, less quickly to nonlactating nipples, and slower yet when allowed only to contact the dam without suckling. Olfactory preference, as judged by the amount of time spent over an odor, has been obtained by the delivery of milk while suckling (Brake, 1981) as well as delivery of milk when not suckling (Johanson and Teicher, 1980; Johanson and Hall, 1982).

When properly stimulated, young pups can also learn about negative facets of an odor. Rudy and Cheatle (1977, 1979), for example, caused the scent of lemon to be aversive to day-2 rats by pairing it with lithium chloride-induced illness. The aversion was retained for at least 6 days and could also be extinguished by exposing pups to the lemon odor without any aversive consequence. The ability of day-2 rats to manifest important properties of classical conditioning is impressive. In fact, Rudy and Cheatle (1979) were not able to detect any differences between rats 2 and 14 days of age in their simpler classical conditioning paradigms. Rudy and Cheatle (1979), however, did find age-related differences in the ability of animals to maintain information about the conditioning stimulus prior to the onset of the unconditioned stimulus (UCS). Simply placing day-2 rats with their dam for 25 sec between conditioned stimulus (CS) and UCS prevented formation of a learned aversion. Day-8 rats, however, integrated a CS–UCS delay period of even 90 min. This finding of disruption is remarkable. It is not clear, however, if the disruption is time-related or if the information was disrupted by the pups interacting with their dam during the 25-sec interval.

In keeping with the notion that 8-day-old rats can retain information over long periods of time, Rudy and Cheatle (1979) found that preexposing them to an odor reduced its conditioning efficacy. In fact, such rats retained information concerning odors over long periods of time and utilized it to form second-order aversions. That is, when the original CS was paired with a second odor, orange in this case, the orange took on aversive qualities. Others have demonstrated that odors and tastes can take on aversive qualities in very young animals exposed to very brief, intensely cold (10°C) conditions (Alberts, 1981) or made sick by virtue of feeding (Gemberling *et al.*, 1980). In fact, pairing an odor with LiCl toxicity *in utero* confers a negative affect upon the odor, at least until 10 days of age, as judged by slower running speeds in the presence of the "toxic" odor (Smotherman, 1982). Although much remains to be learned concerning the particulars of olfactory conditioning, the literature suggests that under conditions of appropriate stimulation, classical conditioning in neonatal rats appears to have many important adult characteristics.

Nest or conspecific odors can either facilitate or hinder conditioning, depending upon the task requirements (Smith and Spear, 1978, 1980). Nest odors seem to provide a biological context that modulates the availability of certain patterns of responses. The task now is to identify more fully the nest events that give rise to odor familiarity and to the classes of behavior that become more or less available to the animal as a result of the nesting experience. In this regard, the complexity of the olfactory milieu affecting activity on the mother has recently been demonstrated by MacFarlane *et al.* (1983). Day-7 rats were extremely active on female rats that presented incongruous olfactory fields, such as ovariectomized virgin females bearing nipple-wash extract or normal mothers with nipples that had been washed free of scent. The former rats presented a familiar "figure" scent, on an unfamiliar "background." The washed dams presented a familiar background but no familiar figure. Day-7 rats were inactive on nonscented, ovariectomized females and equally inactive on unwashed (i.e., normal) dams. These two classes of rats present very different but congruous—from the point of the infant's experience—stimulus fields. These finding in conjunction with the experiments of Smith and Spear (1978, 1980) emphasize the perceptual richness of the infant's olfactory field and stress that conditioning successes and failures must be evaluated within this context.

The above sections have reviewed demonstrations of the unexpected ability of altricial animals to form associations under circumstances that bear important properties of the nest. As discussed in the introductory section, there are multiple sources of comfort or "reward" to the infant. Importantly, specific sources appear to differentially affect specific physiological (e.g., the Ackerman *et al.* (1978) demonstration of milk's reducing ulceration in deprived rats) or behavioral processes (see Alberts, 1978a, on the sufficiency of warmth for establishing olfactory-determined huddling preferences). The dam or caretaker is composed of many dimensions and of varied facets within a dimension. Different aspects of the mother seem to be preferentially utilized during the developmental period to calm the infant. I tentatively suggest that specific classes of behaviors can become differentially conditioned to certain stimuli but not others arising from the dam. Arousal provided by the mother or experimenter seems necessary for learning to occur in the newborn. Whether this reflects the reinforcing qualities of activation or its allowing the animal to detect and utilize stimulus information remains an important question that can be answered empirically.

VI. Neural Development of the Olfactory System

The behaviors described above demand an underlying olfactory system that at birth has a number of functional connections with the more caudal areas that organize suckling, huddling, and orienting behavioral patterns (Pieper, 1963). Yet the neural system must remain sufficiently unspecified to allow olfactory experiences to shape future behavior. Central olfactory development meets these two needs admirably. The following summarizes evidence that points to anatomical maturity in certain portions of the olfactory system before birth but a profound lack of synaptic development at that time and a very slow development of inhibitory interneurons. This conceptualization may provide a neural basis for selectivity of behavioral control for a progressive narrowing of the range of olfactory stimuli that can manage a particular behavior and for establishing new olfactory–behavioral relationships.

Hinds (1972a,b) and Cuschieri and Bannister (1975a,b) used light and electron microscopic techniques to trace the development of olfactory mucosa in mice. Both sets of investigators found differentiation of epithelial cells in the olfactory mucosa by embryologic day 10 (E10). By E17 the mucosal layer had assumed its mature form. A full complement of cilia appear on the mucosal surface. There are mitochondria and Golgi apparati, glial cells secrete mucosal substances, receptor cells are fully differentiated, and receptor cell axons orient and travel to the bulb in glia-encapsulated sheaths.

Hinds (1968a,b), utilizing autoradiographic techniques for dating birth dates, proliferation, and migration, demonstrated that the major centripetal units of mitral and tufted cells are born between E10 and E16 in the main and accessory olfactory bulbs. Figure 4, which is Albert's (1976) summary of Hinds (1968a,b) data, demonstrates that the very large mitral cells have all been born by E15, the tufted cells by E18. It is of considerable interest that the birth dates of granule cells, the principal interneurons of the olfactory system, are delayed until completion at birth of the long-axoned mitral and tufted neurons. Indeed, a substantial portion of granule cell birth occurs well after parturition, which may allow these cells to be most influenced by postnatal events (Doving and Pinching, 1973; Meisami, 1978).

There is anatomical evidence for functional organization within the bulb during suckling. Teicher *et al.* (1980), utilizing the 2-deoxyglucose autoradiographic technique that detects areas of differential central glucose utilization, demonstrated high levels of ^{14}C activity confined to a modified glomerular portion of the olfactory bulb during suckling in day-9 rats. This modified glomerular complex is displaced relative to the bulb's glomerular

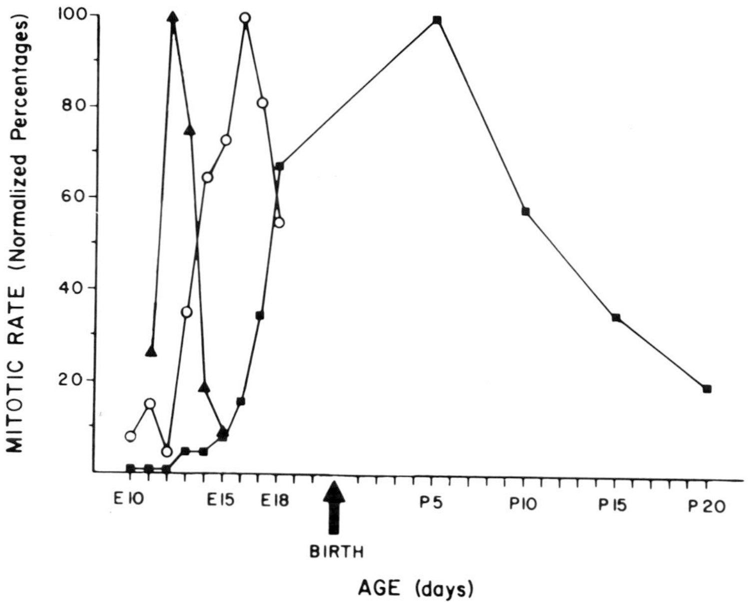

Fig. 4. "Birthdays" of three types of olfactory bulb neurons. To equate for differences in the absolute size of each neural population, mitotic rate is shown as a percentage of the peak rate for each cell type. (▲——▲) Mitral cells; (○——○) tufted cells; (□——□) granule cells. From Alberts (1981), based on data presented by Hinds (1968a).

layer and is adjacent to projections from the vomeronasal (VMN) organ. As of now, however, no evidence has been found for VMN projections to this modified glomerular area (G. M. Shepherd, personal communication, 1982).

Pedersen *et al.* (1982) have extended these findings to reveal specific activity in the modified glomerular complex during the rat's first suckling episode on a normal nipple. It is of considerable interest that rats treated in the manner of Pedersen and Blass (1982), suckling a citral-coated nipple, do not show this activity but show activity in the glomerular areas that normally are activated by citral (Pedersen *et al.*, 1982).

Hypotheses concerning functional capability of the olfactory system at different developmental epochs must be presented very tentatively. Differing conclusions have been reached utilizing different techniques. There is, however, agreement on the finding of mature synapses in the pyriform (olfactory) cortex in rodents at birth, as reported by Hinds (1972a) and Westrum (1975). The synapses arise in pyriform cortex from the bulb via the anterior olfactory nucleus and the lateral olfactory tract. Confirming evidence has been presented by Schwob and Price (1978), using tritiated leucine autoradiography in day-1 rats. They also found labeled fibers in the amygdala. These fibers were both of main and accessory olfactory bulb origin. By postnatal day 9 the pattern of the transported label and its extent were adultlike.

Additional evidence for functional direct projections from the olfactory bulb have recently been provided by Astic and Saucier (1982) who exposed rats 1, 9, and 21 days of age at room temperature to air streams containing "no odor," ethyl acetoacetate, or home shavings. In all three ages there was increased autoradiographic (2DG) activity within the bulb, anterior olfactory nuclei, lateral olfactory tract nucleus, the anterior portion of the olfactory tuburcle, and the pyriform cortex. By day 9 secondary and tertiary areas also showed considerable activity and started to approach adult complexity by day 21.

Future studies utilizing autoradiographic procedures will have to be conducted with animals tested during the 45-min odor exposure periods under conditions that more closely approximate the nest, thereby providing animals with high ambient temperatures and with the type of stimulation afforded by the mother. When Horwitz *et al.* (1982) took the temperature factor into consideration in their assessment of the ontogeny of nigrostriatal dopaminergic systems, they found qualitative similarities in the biochemical responses between 1-, 4-, and 10-day-old rats, raising the possibility of some form of striatal function at birth. They revised their earlier conclusions, based on studies conducted in room temperature, that dopaminergic ontogeny occurs in a stepwise manner. It would not be surprising, therefore, to find evidence for function in the secondary and tertiary olfactory projection systems earlier than day 9, given the Horwitz *et al.* (1982) findings and the very complex behavior that infant rats have exhibited.

VII. A Model for Olfactory Modification of Behavior during Ontogeny

This review has made it clear that olfactory stimulation can gain control over certain classes of effector mechanisms during ontogeny. Olfactory stimulation does not appear to organize these behaviors but does seem to direct, invigorate, or permit their expression (cf. sections on huddling and suckling). The opportunities for new odors to establish control over particular behaviors are restricted during infancy, especially by time (experience). Moreover, an association between a motor pattern and an odor can be established when the reinforcing event *linked to that pattern* during that developmental epoch is available. Events that permit establishing one class of social bond do not necessarily support formation of other social bonds. Thus, huddling preferences for objects bearing a particular odor develop in infants that had previously huddled with warm objects (even plastic tubes) that bore the odor. Other sources of stimulation that may be either sufficient or necessary for the development of different olfactory–behavior associations may not be adequate for this synthesis. Accordingly, the mother, although conceptually unitary, presents to the infant a montage of behaviorally and physiologically comforting stimuli consisting of smells, touch, warmth, nipples, milk, and the timing and patterning in which these stimuli are made available to the infant (Hofer and Shair, 1982). A major goal for developmental psychobiologists and neurobiologists, therefore, is to identify opportunities of conditioning availability for various behaviors, the circumstances that support such conditioning, how the circumstances change during ontogeny, and the longer term functions, if any, of a particular association.

These considerations give rise to a model for the olfactory conditioning of social attraction, defined in terms of the behaviors that have been discussed previously.

The square in Fig. 5 designated olfaction incorporates a polygon that progressively narrows over time. This narrowing represents an ontogenetic restriction of stimuli that can channel certain behaviors, as these behaviors become increasingly directed by familiar, naturally occurring stimuli. The model suggests that, at the beginning of expression of a motor pattern, control by naturally occurring stimuli occurs through their availability in the nest setting. Again, time and experience constrain the domain of odors to which behavior patterns can become associated (see Rescorla, 1978, on prediction in classical conditioning for a fuller development of this idea).

The nest provides the setting for the development of olfactory control over behavior. The mechanism for establishing control is maternal contact, especially the dam's arousing her pups. Moderate levels of activity per se, whether through anogenital licking, moderately strong odors, or high ambient temperatures, appears to be reinforcing to very young animals who perform acts that are instrumental to obtaining stimulation. One might speculate that at first all of these stimuli are behaviorally effective through a common catecholaminergic substrate but become progressively differentiated and nonsubstitutable during ontogeny.

Contact is broadly defined functionally. Some of its components are provided in the diamond portion of Fig. 5. The (+) symbols represent contact (and activity) support of olfactory conditioning through the dam's ability to satisfy or normalize infant internal or peripheral demands. When such needs are *not* met, infants act in ways to reestablish contact with the mother.

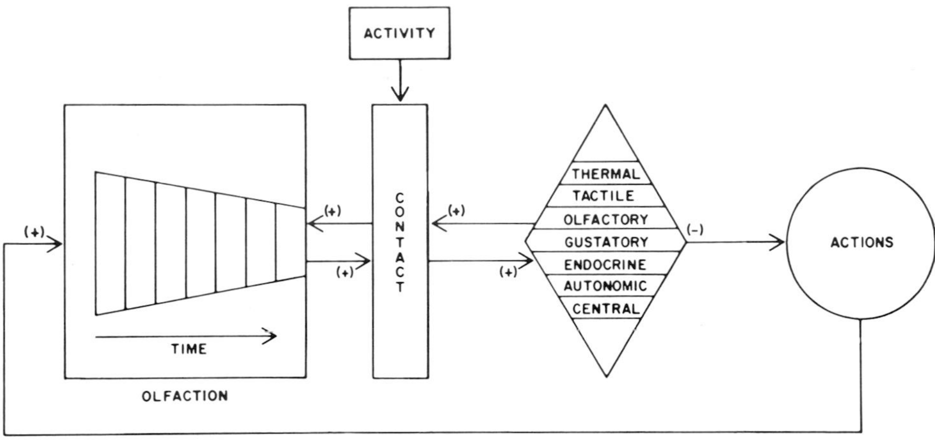

Fig. 5. Model for behavioral conditioning.

Younger animals cry or emit ultrasonic calls. More mobile animals act by locomoting to the mother either in the large nest area or by approaching her outside of the nest (Alberts and Leimbach, 1980). Once contact has been reestablished, indeed, through its reestablishment, the dynamics of conditioning start again.

This model calls attention to the progressive closing of the "olfactory window" and invites an assessment of the mechanisms that lead to this closing. It also makes explicit the different pathways through which reinforcing stimulation becomes available to the infant at different epochs of development. The field of developmental psychobiology has progressed to the point where we know the deceptiveness of the term "atricial." If the model is of use, it will help us to determine when and how infants utilize specific facets of the rich stimulus array provided by the mother, siblings, and nest. It can also help focus on developmental bases of adult adaptive behaviors. To reparaphrase Bekoff (1981), we can much better understand mechanisms of familiality once we understand those of familiarity in such behaviors as kin recognition and mate selection.

VIII. Conclusions and Future Directions

Research during the past decade has provided the neurophysiologist both functional and experimental behavioral phenomena against which the functional development of the nervous system may be evaluated. As indicated, these phenomena include an identification of nest circumstances that permit behavioral expression, examples of conditioning, and an appreciation of precocious development of certain behavioral systems, feeding, for example (Hall, 1979a,b). The advances are reciprocal, as the types of questions asked by the behaviorist must be guided by the rapidly growing information provided by studies of the developing mammalian brain.

I have not focused on ultimate mechanisms of causation during development because, for the olfactory system at least, the road toward adult specification of recognition, selection, and so on, has not been sufficiently specified. As psychobiologists become more invested in function in the strong sense of the term (Hinde, 1975) and, reciprocally, as sociobiologists seek an understanding of the proximal mechanisms through which developmental events shape phenotypic expression (see Marler, 1975, and Nottebohm, 1981, for reviews) and how different behavioral strategies may be adaptive during particular developmental epochs, the type of progress that one anticipates through the collaborative efforts of neurophysiologists and psychobiologists can also be realized through joint approaches to even more molar behavioral problems.

The task of the developmental psychobiologist has really just been formulated. As indicated in Fig. 5, we must now start to focus on the classes of stimuli that can gain control over behavior and the mechanisms through which such control is achieved. Implicit in this approach is a need to determine what limits the spontaneous appearance of organized behaviors such as feeding that can be seen only under special circumstances. At its best this approach will reveal the successive steps in the utilization of capabilities during ontogeny and the induction, through these experiences, of the adult expression of these capabilities (Gottlieb, 1973).

Acknowledgments

Research from the author's laboratory was supported by grant in aid of research AM 18560 from the National Institute of Arthritis, Metabolism, and Digestive Diseases.

References

Ackerman, S. H., Hofer, M. A., and Weiner, H. (1978). *Science* **201**, 373–376.

Ader, R., and Grota, L. J. (1970). *Anim. Behav.* **18**, 144–150.

Alberts, J. R. (1976). *In* "Mammalian Olfaction, Reproductive Processes, and Behavior" (R. L. Doty, ed.), pp. 67–94. Academic Press, New York.

Alberts, J. R. (1978a). *J. Comp. Physiol. Psychol.* **92**, 220–230.

Alberts, J. R. (1978b). *J. Comp. Physiol. Psychol.* **92**, 231–240.

Alberts, J. R. (1981). *In* "The Development of Perception: Psychobiological Perspectives" (R. N. Aslin, J. R. Alberts,and M. R. Petersen, eds.), pp. 321–357. Academic Press, New York.

Alberts, J. R., and Gubernick, D. J. (1983). *In* "Mother-Infant Symbiosis" (H. Moltz and L. Rosenblum, eds.), pp. 7–44. Plenum, New York.

Alberts, J. R., and Leimbach, M. (1980). *Dev. Psychobiol.* **13**, 417–429.

Amsel, A., Burdette, D. R., and Letz, R. (1976). *Nature (London)* **262**, 816–818.

Arnold, G. W., Boundy, C. A. P., Morgan, P. D., and Bartle, G. (1975). *Appl. Anim. Ethol.* **1**, 167–176.

Astic, L., and Saucier, D. (1982). *Dev. Brain Res.* **2**, 141–156.

Babicky, A., Ostadalova, J., Parizek, J., Kolar, R., and Bibr, B. (1973). *Physiol. Bohemoslov.* **22**, 449–456.

Bekoff, M. (1981). *In* "Parental Care in Mammals" (D. J. Gubernick and P. H. Klopfer, eds.), pp. 307–346. Plenum, New York.

Blass, E. M. (1980). *In* "Neurophysiological Mechanisms of Goal Directed Behavior and Learning" (R. Thompson, L. Hicks, and V. Shvyrkov, eds.), pp. 461–469. Academic Press, New York.

Blass, E. M., and Cramer, C. P. (1982). *Proc. Inst. Neurol. Sci. Symp. Neurobiol. 1st,* 503–523.

Blass, E. M., and Pedersen, P. E. (1980). *Physiol. Behav.* **25**, 993–995.

Blass, E. M., and Teicher, M. H. (1980). *Science* **210**, 15–22.

Blass, E. M., Teicher, M. H., Cramer, C. P., Bruno, J. P., and Hall, W. G. (1977). *J. Comp. Physiol. Psychol.* **91**, 1248–1260.

Bolles, R. C., and Woods, P. J. (1964). *Anim. Behav.* **12**, 427–441.

Brake, S. C. (1981). *Science* **211**, 506–508.

Bruno, J. P., Teicher, M. H., and Blass, E. M. (1980). *J. Comp. Physiol. Psychol.* **94**, 115–127.

Cuschieri, A., and Bannister, L. H. (1975a). *J. Anat.* **119**, 277–286.

Cuschieri, A., and Bannister, L. H. (1975b). *J. Anat.* **119**, 471–498.

Doving, D. R., and Pinching, A. J. (1973). *Brain Res.* **52**, 115–129.

Drewett, R. F., Statham, C., and Wakerley, J. B. (1974). *Anim. Behav.* **22**, 907–913.

Fowler, S. J., and Kellogg, C. (1975). *J. Comp. Physiol. Psychol.* **89**, 738–746.

Freeman, N. C. G., and Rosenblatt, J. S. (1978a). *Dev. Psychobiol.* **11**, 459–468.

Freeman, N. C. G., and Rosenblatt, J. S. (1978b). *Dev. Psychobiol.* **11**, 437–457.

Friedman, M. I., Bruno, J. P., and Alberts, J. R. (1981). *J. Comp. Physiol. Psychol.* **95**, 26–36.

Galef, B. G. (1981). *In* "Parental Care in Mammals" (D. J. Gubernick and P. H. Klopfer, eds.), pp. 211–241. Plenum, New York.

Galef, B. G., and Kaner, H. C. (1980). *J. Comp. Physiol. Psychol.* **94**, 588–596.

Gemberling, G. A., Domjan, M., and Amsel, A. (1980). *J. Comp. Physiol. Psychol.* **94**, 734–743.

Gottlieb, G. (1973). *In* "Studies on the Development of Behavior and the Nervous System: Behavioral Embryology" (G. Gottlieb, ed.), pp. 3–45. Academic Press, New York.

Gottlieb, G. (1981). *In* "The Development of Perception: Psychobiological Perspectives" (R. N. Aslin, J. R. Alberts, and M. R. Peterson, eds.), pp. 5–44. Academic Press, New York.

Gregory, E., and Pfaff, D. (1971). *Physiol. Behav.* **6**, 573–576.

Gubernick, D. J. (1981). *In* "Parental Care in Mammals" (D. J. Gubernick and P. H. Klopfer, eds.), pp. 243–305. Plenum, New York.

Hall, W. G. (1979a). *Science* **205**, 206–208.

Hall, W. G. (1979b). *J. Comp. Physiol. Psychol.* **93**, 977–1000.

Hammel, H. T., Caldwell, F. T., Jr., and Abrams, R. M. (1967). *Science* **156**, 1260–1262.

Hinde, R. A. (1975). *In* "Function and Evolution in Behaviour: Essays in Honour of Professor Niko Tinbergen" (G. Baerends, C. Beer, and A. Manning, eds.), pp. 3–15. Oxford Univ. Press (Clarendon), London and New York.

Hinds, J. W. (1968a). *J. Comp. Neurol.* **134**, 287-304.
Hinds, J. W. (1968b). *J. Comp. Neurol.* **134**, 305-322.
Hinds, J. W. (1972a). *J. Comp. Neurol.* **146**, 233-252.
Hinds, J. W. (1972b). *J. Comp. Neurol.* **146**, 253-276.
Hofer, M. A., and Shair, H. (1982). *Dev. Psychobiol.* **15**, 229-243.
Hofer, M. A., Shair, H., and Singh, P. (1976). *Physiol. Behav.* **17**, 131-136.
Hofer, M. A., Fisher, A., and Shair, H. (1981). *J. Comp. Physiol. Psychol.* **95**, 123-133.
Horwitz, J., Heller, A., and Hoffman, P. C. (1982). *Brain Res.* **235**, 245-252.
Impekoven, M. (1970). *Anim. Behav.* **19**, 475-480.
Jacobson, M. (1974). *In* "Aspects of Neurogenesis: Studies of Behavior and the Nervous System" (G. Gottlieb, ed.), pp. 151-166. Academic Press, New York.
Johanson, I. B., and Hall, W. G. (1979). *Science* **205**, 419-421.
Johanson, I. B., and Hall, W. G. (1982). *Dev. Psychobiol.* **15**, 379-397.
Johanson, I. B., and Teicher, M. H. (1980). *Behav. Neural Biol.* **29**, 132-136.
Kaplan, J., and Russell, M. (1974). *Dev. Psychobiol.* **7(1)**, 15-19.
Kenny, J. T., and Blass, E. M. (1977). *Science* **196**, 898-899.
Leon, M. (1974). *Physiol. Behav.* **13**, 441-443.
Leon, M. (1978). *Adv. Study Behav.* **8**, 117-153.
Leon, M., and Moltz, H. (1971). *Physiol. Behav.* **7**, 265-267.
Leon, M., and Moltz, H. (1972). *Physiol. Behav.* **8**, 683-686.
Leon, M., Galef, B. G., and Behse, J. H. (1972). *Physiol. Behav.* **18**, 387-391.
Leonard, C. M. (1974). *J. Comp. Physiol. Psychol.* **86**, 458-469.
Leonard, C. M. (1981). *In* "The Development of Perception: Psychobiological Perspectives" (R. N. Aslin, J. R. Alberts, and M. R. Peterson, eds.), pp. 363-410. Academic Press, New York.
Lev, E., and Orlic, O. (1972). *Science* **177**, 522-524.
MacFarlane, A. (1975). *In* "Parent-Infant Interaction" (M. A. Hofer, ed.), pp. 103-113. Elsevier, Amsterdam.
MacFarlane, B. A., Pedersen, P. E., Cornell, C. E., and Blass, E. M. (1983). *Anim. Behav.* **31**, 462-471.
Marler, P. (1975). *In* "Function and Evolution in Behaviour: Essays in Honour of Professor Niko Tinbergen" (G. Baerends, C. Beer, and A. Manning, eds.), pp. 254-275. Oxford Univ. Press (Clarendon), London and New York.
Meese, G. B., and Baldwin, B. A. (1975). *Appl. Anim. Ethol.* **1**, 379-386.
Meisami, E. (1978). *Prog. Brain Res.* **48**, 211-230.
Moltz, H., and Lee, T. M. (1981). *Physiol. Behav.* **26**, 301-306.
Moore, C. L. (1981). *Anim. Behav.* **29**, 383-386.
Moore, C. L., and Morelli, G. A. (1979). *J. Comp. Physiol. Psychol.* **93**, 677-684.
Moran, T. H., Lew, M. F., and Blass, E. M. (1981). *Science* **214**, 1366-1368.
Moran, T. H., Schwartz, G. J., and Blass, E. M. (1983a). *J. Neurosci.* **3(1)**, 10-19.
Moran, T. H., Schwartz, G. J., and Blass, E. M. (1983b). *Dev. Brain Res.* **7**, 197-204.
Nottebohm, F. (1981). *Prog. Psychobiol. Physiol. Psychol.* **9**, 85-124.
Nyakas, C., and Endroczi, E. (1970). *Acta Physiol. Acad. Sci. Hung.* **38**, 59-65.
Okon, E. E. (1971). *J. Zool.* **164**, 227-237.
Okon, E. E. (1972). *J. Zool.* **168**, 139-148.
Oppenheim, R. W. (1981). *In* "Maturation and Development: Biological and Psychological Perspectives" (K. J. Connolly and H. F. R. Prechtl, eds.), pp. 73-109. Spastics International Medical Publications. Lavenham Press, Suffolk, U. K.
Pedersen, P. E., and Blass, E. M. (1981). *In* "The Development of Perception: Psychobiological Perspectives" (R. N. Aslin, J. R. Alberts, and M. R. Peterson, eds.), pp. 359-381. Academic Press, New York.

Pedersen, P. E., and Blass, E. M. (1982). *Dev. Psychobiol.* **15**, 349–355.
Pedersen, P. E., Greer, C. A., Stewart, W. B., and Shepherd, G. M. (1982). *Assoc. Chemorecept. Sci. Annu. Meet. Fourth* (Abstr.).
Pedersen, P. E., Williams, C. L., and Blass, E. M. (1983). *J. Exp. Psychol., Anim. Behav. Processes* **8(4)**, 329–341.
Pieper, A. (1963). "Cerebral Function in Infancy and Childhood." Consultants Bureau Enterprises, New York.
Plaut, S. M., and Davis, J. M. (1972). *Physiol. Behav.* **8**, 43–51.
Poindron, P. (1976). *Biol. Behav.* **2**, 600–602.
Porter, R. H., and Etscorn, F. (1974). *Nature (London)* **250**, 732–733.
Rescorla, R. A. (1978). *In* "Cognitive Processes in Animal Behavior" (S. H. Hulse, H. Fowler, and W. K. Honig, eds.), pp. 15–50. Erlbaum, Hillsdale, New Jersey.
Rodriguez Echandia, E. L., Foscolo, M., and Broitman, S. T. (1982). *Physiol. Behav.* **29**, 47–49.
Rosenblatt, J. S. (1970). *In* "Development and Evolution of Behavior: Essays in Memory of T. C. Schneirla" (L. R. Aronson, E. Tobach, D. S. Lehrman, and J. S. Rosenblatt, eds.), pp. 489–515. Freeman, San Francisco.
Rosenblatt, J. S. (1971). *In* "The Biopsychology of Development" (E. Tobach, L. R. Aronson, and E. Shaw, eds.), pp. 345–410. Academic Press, New York.
Rosenblatt, J. S. (1976). *In* "Growing Points in Ethology" (P. P. G. Bateson and R. A. Hinde, eds.), pp. 345–383. Cambridge Univ. Press, London and New York.
Rosenblatt, J. S., and Lehrman, D. S. (1963). *In* "Maternal Behavior in Mammals" (H. L. Rheingold, ed.), pp. 8–57. Wiley, New York.
Rosenblatt, J. S., Turkewitz, G., and Schneirla, T. C. (1969). *Trans. N. Y. Acad. Sci.* **31**, 231–250.
Rudy, J. W., and Cheatle, M. D. (1977). *Science* **198**, 845–846.
Rudy, J. W., and Cheatle, M. D. (1979). *In* "Ontogeny of Learning and Memory" (N. E. Spear and B. A. Campbell, eds.), pp. 157–188. Erlbaum, Hillsdale, New Jersey.
Schaal, B., Montagner, H., Hertling, E., Bolzoni, D., Moyse, A., and Quichon, R. (1980). *Reprod. Nutr. Dev.* **20**, 843–858.
Schapiro, S., and Salas, M. (1970). *Physiol. Behav.* **5**, 815–817.
Schmidt-Nielsen, K. (1964). "Desert Animals." Oxford Univ. Press, London and New York.
Schwob, J. E., and Price, J. L. (1978). *Brain Res.* **151**, 369–374.
Scott, J. P., Stewart, J. M., and DeGhett, V. J. (1974). *Dev. Psychobiol.* **7**, 489–513.
Singh, P., and Tobach, E. (1975). *Dev. Psychobiol.* **8**, 151–164.
Smith, F. V. (1965). *Anim. Behav.* **13**, 84–86.
Smith, G. J., and Spear, N. E. (1978). *Science* **202**, 327–329.
Smith, G. J., and Spear, N. E. (1980). *Behav. Neural Biol.* **28**, 491–495.
Smotherman, W. P. (1982). *Physiol. Behav.* **29**, 769–771.
Sperry, R. W. (1971). *In* "The Biopsychology of Development" (E. Tobach, L. R. Aronson, and E. Shaw, eds.), pp. 27–44. Academic Press, New York.
Stickrod, G., Kimble, D. P., and Smotherman, W. P. (1982). *Physiol. Behav.* **28**, 5–7.
Teicher, M. H., and Blass, E. M. (1976). *Science* **193**, 422–425.
Teicher, M. H., and Blass, E. M. (1977). *Science* **198**, 635–636.
Teicher, M. H., Stewart, W. B., Kauer, J. S., and Shepherd, G. M. (1980). *Brain Res.* **194**, 530–535.
Tobach, E., Rouger, Y., and Schneirla, T. C. (1967). *Am. Zool.* **7**, 792–793.
Westrum, L. E. (1975). *Exp. Neurol.* **47**, 442–447.

16

New Views of Parent–Offspring Relationships

JEFFREY R. ALBERTS
Psychology Department
Indiana University
Bloomington, Indiana

I. Introduction

Mammalian reproduction is based on production of live-born offspring and on a postnatal period during which there are usually intense and specialized interactions between parents and offspring. The fundamental nature of parent–offspring relationships, the stimuli that regulate interactions, the mechanisms that mediate them, and the functions that are subserved are central topics psychobiology. The parent–offspring aggregate is a "system" that can be viewed and characterized from several perspectives. The main topic of the present chapter is to survey alternative views of parent–offspring relationships, data that support each view, and some implications for further analyses offered by each perspective.

Maternal behavior of the Norway rat (*Rattus norvegicus*) will be used to trace the formulation of basic concepts that have been important in the empirical analysis of parental behavior. Similarly, postnatal ontogenesis of the rat pup will be used to review alternative perspectives on infancy and to discuss recent trends in characterizing postnatal development. With these overviews as foundation, the parent–offspring relationships will be considered and discussed. Each model offers a distinct perspective and emphasizes different features of the relationship. Discussion of these models is also used as a vehicle for presenting a sample of diverse approaches and contemporary data.

II. Some Seminal Concepts: Antecedents to New Views

A. Parental Behavior

Parental behavior can be defined functionally, in terms of outcome, or it can be defined operationally, in terms of specific activities. A broad, functional definition of parental behavior is ". . . behavior that increases the survival probability of fertilized eggs or offspring that have left the female's body" (Lott, 1973, p. 240). Although such functional definitions are important in the establishment of an overall analytic framework, operational definitions have had greater impact on empirical analysis of parental behavior. Nevertheless, operational definitions do not exist *in vacuo*; they fit within larger frameworks. In the case of parent–offspring relationships, the larger framework is usually a broad, functional type, based on outcome.

The study of parental behavior in the Norway rat has focused almost entirely on the adult female because the reproductive strategy of the species, like that of most mammals, does not typically involve extensive paternal behavior; males mate with more than one female (and vice versa) and do not attend exclusively to any single female's offspring. The concepts derived from this work can be applied to most forms of parental behavior. The data may be specific to rats, but the approach is not.

B. Maternal Behavior

Four distinct, stereotyped behavioral activities, namely, nursing, transport, nest-building and licking of the young, together constitute a generalized profile of maternal behavior in many mammals (Rheingold, 1963a; Weisner and Sheard, 1933). Figure 1 depicts these four activities in the Norway rat. Each activity is relatively stereotyped, although the rat can adapt them to numerous kinds of environments ranging from subterranean burrows to plastic boxes in a laboratory. Frequency, duration, and intensity of each of these activities can be quantified. The idea that operational definitions of behavioral acts could be assembled into a unified profile that characterized maternal behavior was an important concept that helped in the early effort to analyze systematically maternal behavior in animals and to establish a common paradigm for cross-species study.

C. Maternal Behavior Cycle

It was soon recognized that maternal behavior is not a fixed entity. Instead, it is a profile of activities and tendencies that undergoes orderly changes over time. This temporal pattern constitutes a cycle, termed the maternal behavior cycle. Figure 2 depicts the cycle of maternal behavior in the laboratory rat but it is generally applicable to nearly all mammals, with adjustments for time scale and ecological specializations. Maternal behavior in the rat can be divided into at least three phases: initiation, maintenance, and decline. Rosenblatt and associates have studied extensively the sensory and hormonal characteristics of each stage and have also explored the role of environmental and experiential factors in the onset and expression of maternal behavior (Rosenblatt, 1965, 1969; Rosenblatt and Lehrman, 1963; Rosenblatt and Siegel, 1981).

The concept of a cycle of of maternal behavior was significant because it led to formulation of *developmental* perspectives. Infant development was no longer seen as occurring within a stable context of parental care. Parent and offspring are both dynamic. Thus the developmental sequence involves simultaneous ontogenies: the development of the offspring and the development of the cycle of maternal behavior. The seemingly simple concept of a maternal cycle had profound implications that were reflected in further refinements, described below.

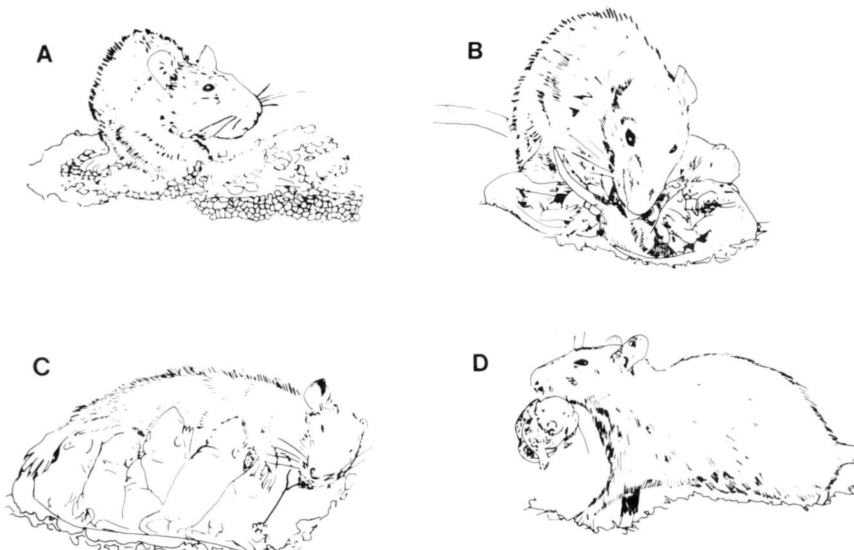

Fig. 1. Maternal behavior in the Norway rat. Nest-buiilding (A), licking (B), nursing (C), and transport (D) are the four major maternal activities in this species as in many mammals. They can be defined operationally and quantified. Adapted from Alberts and Gubernick (1983).

D. Maternal Responsiveness

Changes in a mother's behavior, as highlighted by the concept of a maternal cycle, are not readily interpretable because the mother's behavior can reflect either (1) new responses to the changing characteristics of her offspring, or (2) motivational changes endogenous to the mother. This is a basic interpretive dilemma, common to developmental investigations of interactive, dyadic systems such as that of parents and offspring. To solve the dilemma, there was introduced a set of standardized tests of maternal activities, performed in a test environment, with "constant-age" infants (e.g., Rosenblatt, 1965; Rosenblatt and Lehrman, 1963).

The use of standard stimulus infants controlled for the maturational changes in the offspring that normally confounded interpretation of changes in the female. Maternal responsiveness is a motivational concept, designed to characterize a state of internal readiness to display maternal behavior. The concept, however, is defined and quantified according to standard, objective measures. In a balanced experimental design, additional studies are conducted in which the condition of the mother is held constant to reveal independent stage-related processes in the offspring. It was thus established that there

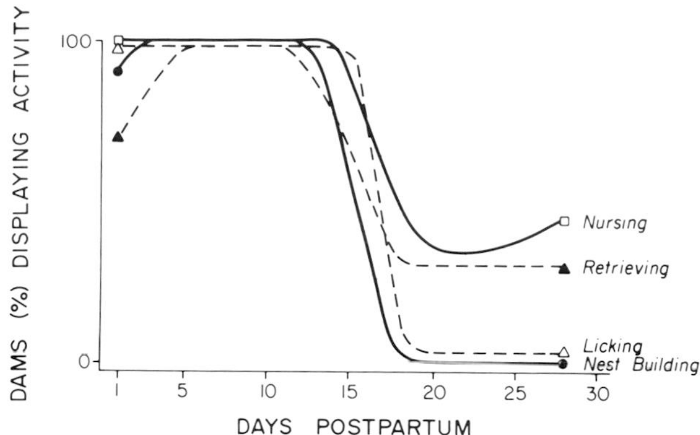

Fig. 2. The cycle of maternal behavior in the rat consists of regular changes in the frequency and intensity of the various maternal activities. Adapted from Alberts and Gubernick (1983).

exists a temporal cycle of maternal responsiveness, evidenced by changes in the adult's motivation to display parental responses that are independent of the maturational changes in offspring (Rosenblatt, 1969), that are usually correlated with changes in maternal responsiveness (see cycle in Fig. 1).

E. Synchrony

Synchrony is another temporal concept. It is based on a synthesis of data on maternal behavior and pup development (e.g., Rosenblatt, 1965). Simplified, the concept of synchrony represents a descriptive account of the "goodness of fit" between expression of maternal behavior at each point in the cycle, and the developmental stages of the offspring. The concept of synchrony between maternal behavior and offspring development was important because it allowed the experimenter/interpreter to appreciate the ways in which maternal activities of the dam are adapted to the maturational, stage-related needs of the offspring. Thus, for example, when the infants are most "helpless," maternal responsiveness is at its highest levels and the dam is maximally attendant. As the offspring's sensory and motor competence improves, making it more independent, maternal responsiveness diminishes appropriately.

F. Reciprocity

What is the basis of developmental synchrony between mother and offspring? There are at least four possible answers: (1) Synchrony exists between maternal behavior and offspring development because each is "triggered" by a common event (e.g., birth) and then is timed to proceed at similar rates. (2) Synchrony exists because the mother actively matches her activities to those of the offspring. As the pups change, the dam alters the quantity and quality of her maternal activities. (3) The pups control the activities of the dam and her maternal characteristics. Synchrony is based on control and manipulation of the mother by the offspring. (4) Synchrony is achieved through mutual controls exerted by parents and offspring.

In the analysis of parent–offspring relationships there are data that can be used to support each of the four possible kinds of control. Nevertheless, the rule that seems to characterize the overall orchestration is that of mutual control and reciprocal relations (Rosenblatt, 1965, 1969). Reciprocity is a general pattern of vital significance to appreciating the nature and basis of many aspects of parent–offspring relationships.

III. Concepts of Infancy and Development

Many of the issues that pervade studies of infancy and early postnatal maturation are pertinent to the study of parent–offspring relationships. Parental behavior occurs in relation to the developing offspring; our perspectives on the nature of the offspring and their development strongly influence our views of parent–offspring relations. The mammalian infant differs from its adult counterpart in anatomical form, physiological function, and behavioral repertoire. Traditionally, the newborn has been regarded as a relatively helpless being, dependent on its parents for warmth, sustenance, and protection. In these terms, the infant is often seen as a "primitive" precursor to the adult, dominated by simple reflexes and lacking many of the subtleties of complex and regulated integration of activities.

The traditional views have been seriously challenged. In numerous instances there is startling counterevidence to the data that suggest that the infant is primitive, reflexive, dependent, and helpless. I will also review some concepts that can be used to synthesize these apparently contradictory views.

A. The Infant as a Primitive, Incomplete Analog of the Adult: Thermoregulation

The thermally fragile newborn can be characterized as an *ectotherm*, dependent on heat from its environment (including the parent) to maintain vital physiological function. In contrast, the adult mammal is classified as an *endotherm* because it generates its own body heat and, by balancing heat production with heat loss, is able to maintain a fairly constant body

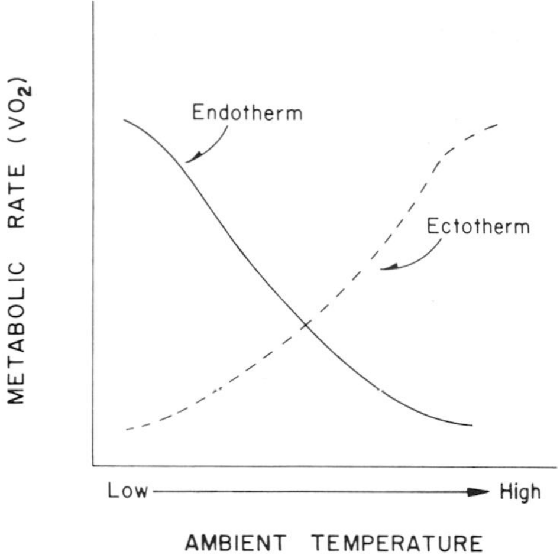

Fig. 3. Contrasting patterns of metabolic rate or heat production (oxygen consumption) as a function of environmental temperature in ectothermic and endothermic animals.

temperature. Postnatal development, then, involves a transition from ectothermy to endothermy. Figure 3 illustrates these two contrasting strategies in terms of metabolic rate (oxygen consumption) as a function of temperature. By definition, ectotherms are organisms in which metabolic rate is a direct function of environmental temperature; $\dot{V}O_2$ is very low at low temperatures and higher at warmer temperatures. Endotherms show the opposite pattern, that is, high metabolic heat production in cold temperatures and correspondingly lower metabolic rates in warmer environments. By analogy, this developmental picture runs parallel with the *phylogenetic* representation of homeothermy. Reptiles and other phylogenetically old and conservative lifeforms are ectothermic. The more recently evolved vertebrates, that is, the birds and mammals, are dedicated to endothermy throughout their adult lives.

It has been conventional to think and speak of the infant in terms used for adults. There is often an implicit comparison with the adult, and this leads to the explicit bias embodied in statements such as "The infant lacks the ability to maintain its body temperature." Rat pups inhabit a different niche than the adults; perhaps they should be investigated with different procedures. Pups live in a nest, among a clump or huddle of siblings. Adults tend to spend more time outside the nest and engage in many activities that do not keep them in physical contact with other rats. To study metabolic responses to cold in single, isolated pups it to treat the pup as an adult. When the metabolism of the developing infant rat is analyzed in a more typical context, namely, in the context of a huddle of littermates, a different picture emerges. In a huddle, oxygen consumption by rat pups is inversely related to temperature, that is, the pups display the mammalian, adultlike, endothermic response (Alberts, 1978). Alone, the pups behave like ectotherms; together they use an endothermic strategy (see Fig. 4). In this case, the whole (the litter) differs from the sum of its parts (the individual parts).

The ability of a rat pup to increase metabolic heat production has long been overlooked by many workers (see Hull, 1973). Bignall and his associates (Bignall *et al.* 1975, 1977), however, have argued that the infant's decreased metabolic rate is a regulated response to cold challenge. When descending neural input is eliminated by spinal transection, 10-day-old infants showed a sustained elevation of heat production in the cold (Bignall *et al.*, 1975). This implies that the decreased metabolic rate in the intact infant can be an active, regulated inhibition of heat production, rather than a passive reflection of ambient temperature.

Rat pups in huddles both receive and provide insulative protection from the other members of the huddle, so they can more efficiently use their metabolic heat for body temperature. I have shown that huddles of infant rats interact behaviorally to create and regulate the thermal characteristics in the huddle (Alberts, 1978). This has been termed *group regulatory behavior*. Rats regulate the exposed surface area of the huddle and conserve metabolic energy. Huddles are small and tight in a cool environment, and become larger, with greater surface area, as the environmental temperature becomes warmer. This regulatory ability cannot be seen in the singleton.

Within the huddle, pups constantly change positions. These movements establish a continuous flow of bodies, termed *pup flow*, which can be observed and quantified through time-lapse video techniques (Alberts, 1978). The rate and direction of pup flow can be influenced by nest temperature. Flow is in the downward, central direction in cool nests, and is upward and to the periphery in warm nests. Figure 4 depicts pup flow in the downward direction. Through these two behavioral mechanisms, pup flow and group behavioral regulation, the infant rat is able to demonstrate previously unknown regulatory competence.

B. Reflexes, Regulation, and the Development of Ingestion

Infant mammals obtain all the nutrients, water, and electrolytes by suckling. Weaning involves a maturational transition from sole reliance on ingestion of teat milk to a diet based on the separate ingestion of a variety of nutritive substances and drinking water. The development of ingestion

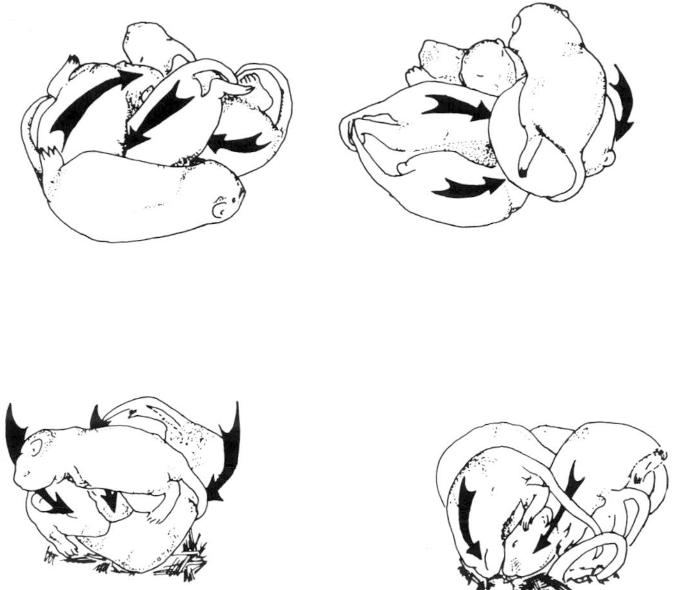

Fig. 4. Pup flow within a litter or rats. The movements of pups within the huddle establish convection currents of bodies. The rate and direction of pup flow are determined by the temperature of the nest environment. Adapted from Alberts (1978).

can be viewed as a kind of model of development in mammals, representing the process by which independence is achieved.

The earliest work focused on the invariant, reflexive aspects of early ingestions. Suckling was studied as an innately organized behavior, based on reflexive responses such as rooting and stereotyped responses to nipple cues. It was posited that infants do "regulate" their intake of mother's milk. Satinoff and Stanley (1963) preloaded puppy dogs with milk and then gave them access to a bitch from which they could nurse. The puppies continued to gorge themselves. Hall and Rosenblatt (1977) installed a cannula in the tongue of suckling pups, so that milk could be delivered to the pup's mouth while it suckled. They describe the behavior of bloated pups that struggled repeatedly to reattach to nipples and suckle more, gasping for air after having ingested vast quantities of milk. Friedman (1975) suggested that under normal conditions, the major limitation on pup ingestion of milk is the amount of milk available from the mother. Together, these data were regarded as evidence that ingestion by the infant rat is largely an automatic, reflexive affair. The development of ingestion, then, appeared to be a problem of how reflexive suckling expands and becomes more finely regulated until it encompasses the simultaneous balancing of numerous nutrients and water.

More recently, however, a stunning observation was reported. Newborn rat pups (3 days old) were deprived of mother's milk for 24 h and then placed in a warm, moist environment that contained puddles of milk on the floor. The nutrient-deprived newborns showed a previously unknown capability under these conditions: they independently ingested the milk from the cage floor (Hall and Bryan, 1980)! Such independent ingestion by a suckling rat pup has never before been observed. In contrast to intake by the suckling infant, this novel form of ingestion appeared to be regulated by postingestional cues. After consuming about 10% of their body weight, ingestions ceased. Preloading with milk inhibited this form of ingestion.

These dramatic phenomena can be interpreted as evidence of discontinuity in the development of ingestion by rats. It is argued that suckling and feeding are two distinct, separable behaviors. The development of ingestion is discontinuous because it is based on the dimunition of the suckling–ingestion system and its replacement by the emergent feeding-ingestion system. This vague-sounding assertion is made poignant and meaningful by a sizable body of empirical data that distinguishes suckling and feeding in terms of the sensory controls of each behavior, the internal stimuli that affect each behavior, and central anatomical systems (Hall and Williams, 1983). Pharmacological manipulations such as amphetamine injections that decrease food intake can have different effects on suckling (Raskin and Campbell, 1981).

It has now been suggested that the complex process of weaning is based on the maturation of a central serotonergic mechanism. Blockade of serotonin receptors by methysergide injection reinstates neonatal suckling behavior (Williams *et al.*, 1979), suggesting the emergence of a serotonergic inhibitory mechanism around the time of weaning. Serotonergic blockade has no effect on suckling and does not affect solid food intake until after day 15.

C. The Dual Infant

The infant possesses, simultaneously, at least two modes of operation. It can operate in its "infantile" mode or it can, under special circumstances, display a distinctly "adult" mode. Research on early ingestion and thermoregulation suggests that this is a valid distinction. Under normal developmental circumstances, only the suckling mode is displayed during early life. Laboratory manipulations have stripped away infantile mechanisms, and have revealed the presence of separate, adultlike functions in the infant. It is in this sense that the concept of the dual infant has reality. Postnatal development does not necessarily entail direct, irreversible transitions from one mode to another or the transmutation of one system into another. Two separate modes may exist simultaneously, but one is actively inhibited.

D. Ontogenetic Adaptations

Ontogenetic adaptation is a more sophisticated concept than is the simpler idea of duality, and it offers a functional rather than a descriptive account of developmental change. Ontogenetic adaptation is easily appreciated by considering an organism's progress through developmental stages to be a sequence of adaptations, shaped by natural selection, by which the myriad proximate adjustments are made to changing habitats so that the organism can survive and reproduce. The emphasis here is on the sequence of requirements imposed by the developmental milieu. In mammals, for instance, a sudden adaptive shift is required to change from life in the aqueous environment of the uterus to life in the extrauterine world. The ability to suckle must be prepared before it is needed. There follows a series of environmental changes in the nutritive, thermal, social, and behavioral surround. The use of this adaptive framework has helped to provide useful perspectives for the study of ontogenetic change (Alberts and Gubernick, 1984; Galef, 1981; Oppenheim, 1981).

IV. Models of Parent–Offspring Relationships

Progress in the analysis of parental behavior and greater sophistication in discerning the capacities of the developing infant have contributed to a proliferation of data and concepts on the nature of parent–offspring relationships. To survey some of this research and thinking, I have constructed a set of five models of parent–offspring relationships. Each model provides a distinct perspective. These perspectives are not mutually exclusive, but they offer strikingly different emphases on the organization of the parent–offspring relationship and the factors that establish, maintain, and regulate it. The immediate goal of the discussion of these models is not to discern which is "true" or "best." These models do not necessarily exist beyond the pages of this chapter. In most cases they did not explicitly guide the work of the investigators whose research is cited. Instead, I am using these models as a vehicle for presenting a diversity of data and ideas that pertain to a common theme.

A. Parental Provision Model

The parental provision model gives primacy to the vital resources that pass from parents of offspring. This can be an exquisitely detailed perspective. Its persuasive power is enhanced by formal awareness of the needs of offspring and the manner in which the various resources from parents are well suited to meet the offspring's requirements.

The parental provision model probably represents the most prevalent and traditional view of parent–offspring relationships. It recognizes synchrony between parent and offspring and the dynamic adjustments required to achieve appropriate interactions within the dyad. Although interaction is not denied or excluded, the outstanding feature of this model is that parent–offspring relations involve a fundamentally unidirectional flow of resources from parent to offspring.

The parental provision model is based on an awareness of the ways in which the infant is dependent upon the parents and, in turn, the ways in which the parents meet these needs. As a starting point for appreciating the myriad ways in which adults provide for infants, we can examine some of the supporting data.

1. Nutrition

The hallmark of parent–infant relations in mammals is that of the nursing relationship. In most altricial forms the infant appears totally dependent on mother's milk for nutritive energy used in growth and development. Babicky *et al.* (1972) fed a radioactive label to lactating rats, determined the concentration of the label in their milk, and thereby estimated the amount of milk transferred from mother to offspring through the nursing period. Total quantity of milk transferred to pups increased during the first 2 weeks, corresponding to the increased size and metabolic requirements of the pups. These requirements are formidable, considering that by day 15, the litter, as a whole, may weigh as much as the mother.

Mother's milk is an energy-rich nutritive substrate. Estimates for rat milk range from 1.6-2.3 kcal/g (Brody, 1945; Stolc *et al.*, 1966; Hahn and Koldovsky, 1966). In rat, milk contains a high proportion of fats and proteins and is relatively low in carbohydrate content (Luckey *et al.*, 1954). In addition, the milk contains the pups' complete early postnatal ration of water, minerals, and electrolytes needed for growth and development.

2. Immune Competence

Infants cannot produce antibodies and they therefore can be highly susceptible to pathogens. Mothers often provide the offspring with natural "inoculations" of antibodies. In rat, most of the transfer of immunity occurs postnatally (Brambell, 1970; Culbertson, 1938). Maternal immunoglobulins are provided via colustrum and later in milk. Suckling rats continue to absorb antibodies selectively from immune milk for up to 18 days postnatally, after which the absorptive quality is lost though maternal antibodies remain present in the milk. This decline in absorption correlates with the ability of the pups to synthesize their own antibodies (Brambell, 1970).

3. *Thermal Protection*

Due to their size and immaturity, infants are more vulnerable to temperature challenges than are adults (Hull, 1973). A small body has a large surface area relatively to its mass and therefore heat is lost more rapidly. In addition, the thermogenic capacity of infants in limited and they usually lack the insulation provided by subcutaneous fat, fur, or feathers.

Parental provisions relevant to the offspring's thermal needs include brooding, nest building, and maintenance. Physical contact between parent and infant almost always leads to conductive transfer of heat from the parent to offspring. Such contact also reduces the exposed surface area of the infant's body, and hence reduces heat loss. The infant needs this heat for maintenance of body temperature and for growth and development. Physical contact with larger, warmer adult bodies provides valuable thermal energy to the offspring, thus reducing extrinsic demands on its metabolic resources. Heat conservation by nest insulation is a significant factor in the energetics of development for many species. Construction and maintenance of the natal nest is a prominent aspect of parental behavior in many species. The insulative value of the nest is regulated by the amount of materials used and the configuration of the construction. Nest building in many species is appropriately regulated by ambient temperature.

4. *Mechanical Stimulation*

Adults provide a variety of forms of mechanical stimulation to the young. One of the most dramatic forms is licking, an almost ubiquitous feature of parental behavior in mammals (e.g., Ewer, 1968). Licking can also provide thermal stimulation, through the warmth of the tongue and the effects of evaporative cooling afterward (Barnett and Walker, 1974). The infant's suckling can be activated by being licked; it has been suggested that licking may be used by parents to compensate for initial differences in vigor among offspring (Klopfer and Klopfer, 1977; Leuthold, 1977; Rheingold, 1963b).

Maternal licking begins at birth. The dam licks each offspring as it passes through the birth canal, assisting its exit and removing fetal membranes and birth fluids. Thereafter, most licking of juveniles is focused on their anogenital area. In Fig. 1A there is depicted a mother rat amid her offspring, holding one pup in her paws while she licks its anogenital region. The mechanical stimulation from anogenital licking triggers reflexive urination (micturition) and may facilitate urine formation by young rats (Capek and Jelinek, 1956). The graph in Fig. 2 shows the cycle of anogenital licking (Gubernick and Alberts, 1983; see also Fleming and Rosenblatt, 1974). The developmental diminution in maternal licking corresponds to the increase in spontaneous, independent urination by 15- to 20-day-old pups.

5. Transport

Another resource that parents provide is that of the mechanical energy that transports the offspring from one place to another. Most altricial infants need such assistance because they usually have limited locomotory abilities during early life. Many mammalian species display transport, sometimes called *retrieval,* as a regular feature of parental behavior. Infants are grasped, most often by the skin of dorsal neck or shoulders, and carried (see Fig. 1D).

There are numerous forms and examples of transport in the world of parental care. The female scorpion carries as many as 20 young on her back. Male sea horses use a specialized brood pouch to carry their young (Cloudsley-Thompson, 1960). Some bats regularly carry their young in flight (Bradbury, 1977). A mother sea otter carries her infant on her chest, holding it with her forepaws and swimming on her back (Fisher, 1940). Primate parents from numerous genera display transport of their young (Hediger, 1955; Rheingold and Keene, 1963).

B. Conflict Model

The conflict model is based on a prediction arising from a sociobiological perspective. To appreciate the logic and derivation of this view, one must review some of the essential tenets of the sociobiological perspective. The crucial component is *parental investment theory* (Trivers, 1972). Parental investment is defined by Trivers as "any investment by the parent in an individual offspring that increases that offspring's chance of surviving (and hence reproductive success) at the cost of the parent's ability to invest in other offspring" (p. 139). Thus, parents "invest" in their own reproductive success by producing offspring (and hence replicating part of their own genetic material). The more offspring that survive to reproduce, the more "successful" is the parent. Parental investment is part of an animal's total reproductive effort, which includes all other facets of the reproductive cycle, including the time and energy involved in finding and procuring mates. Within this framework it is assumed that animals operate on a limited budget of time, energy, and resources for reproduction. Natural selection, therefore, has favored strategies that maximize the cost (effort): benefit (number of offspring) ratio.

Trivers views each offspring as an investment independent of other offspring, so that increased investment in one implies a decreased investment in others. The size of parental investment is "measured" in terms of its detrimental effect on the parent's ability to invest in other offspring. From this perspective, there arises an interesting, novel view of parent–offspring relationships. I have labeled this view as the conflict model, because it predicts a fundamental conflict involving the interests of the parent versus those of the offspring. We can see this most clearly by assuming the perspectives of the parents and the offspring.

From the parent's point of view, the path to reproductive success is paved with large numbers of offspring, each of which, in turn, reproduces. An efficient parental investment is one that achieves survival of an offspring, without provision of excess resources. In this way, the remaining reproductive potential can be used for additional offspring. Once survival is maximally assured, further parental resources would be better invested in another offspring. From the parent's point of view, the investment should be withdrawn from the one offspring as soon as it is viable, so the remaining reproductive energy can be transferred to another.

In contrast, from the offspring's point of view, the withdrawal of parental resources does not look good. The parents' reinvestment in another individual will not benefit the initial offspring as much as direct provision. To maximize its own reproductive success, the offspring should protest and attempt to retain the parental resources, whereas the parents' tendency is to shift them. This represents a fundamental conflict. The conflict model, then, predicts a period of weaning conflict around the time when parents should want the young to achieve independence and when the young should want to continue enjoying the free provision of parental resources. There are numerous observations that are consistent with predictions from the conflict model.

C. Autoregulation Model

Most of the models of parent–offspring relationships that I discuss in this chapter treat the dyad [i.e., the parent(s) and offspring] as a unified system regulated by common controls. In contrast, the view illuminated by the autoregulation model is that parent and offspring operate as two autonomous systems. This is a matter of emphasis. No model treats parents and offspring as an entirely unified system or as completely autonomous ones. The autoregulation view makes its contribution by highlighting parsimony within complexity.

Consider the pattern of nest attendance exhibited by the rat dam during the cycle of maternal behavior (see Fig. 2). One of the striking regularities of the cycle is the gradual decrease in the daily frequency and duration of nest bouts. When the pups are very young and most thermally fragile, the dam is maximally attendant; she may spend 70–90% of each day in contact with them during the first week postpartum. As the pups grow and mature, there follows a steady decline in the number and average duration of next bouts. There is a compelling correlation between reduced nest time by the dam and improved thermoregulation by the litter. Such synchrony between modification of parental behavior and changing needs of the offspring is appealing, because it seems to have adaptive value.

If nest bout duration is determined by the temperature requirements of the litter, then dams should spend more time with cooled pups and less time with warm pups. They do not. Leon and his colleagues (Leon *et al.*, 1978; Woodside and Leon, 1980) scrutinized the control of maternal nest bout duration in rats. They examined empirically the idea that rat dams regulation the duration of the nest bouts according to the specific and immediate needs of the offspring. Manipulations such as raising the environmental temperature, or reducing loss from the dam's tail, both of which increase her heat load, resulted in shorter bouts. In these situations, the dam's ventral skin temperature was a better correlate of nest bout termination than was the nest temperature or the pups' temperature.

Further experimentation confirmed that the best predictor of nest bout termination is the mother's temperature, not the temperature of the litter. To test more rigorously the notion that nest bout duration is determined by regulation of maternal temperature, Leon *et al.* (1978) experimentally manipulated the heat load to rat dams during nesting. Factors that decreased the dam's heat load, such as shaving her back and lowering environmental temperature, increased the duration of nest bouts. Conversely, factors that increased her heat load, such as excising the heat-dissipating tail or raising the ambient temperature, reduced nest duration. It appears that the dam merely regulates her own temperature, and the pups' temperature is maintained as a correlate, not as a proximate cause of her behavior. Elucidation of the cues that regulate an individuals's behavior can reveal simple mechanisms for complex activities.

Consider next the group regulatory behavior of rat pups, mentioned above (Section IIIA). The surface area of the huddle appears regulated because it covaries with ambient temperature in a manner that conserves metabolic energy (Alberts, 1978). In addition, individual pups exchange positions in the huddle, so that each participant both gives and receives insulation. What are the cues that guide the mechanisms for this behavior?

I have found that group behavior is based on autonomous adjustments made by each individual. The sum total of the individual adjustments is the so-called group phenomenon. Group size varies during (is "regulated" by) continuous movements of pups within the huddle (Fig. 4). I quantified pup flow in the huddle, and showed that it is activated by individual thermoregulatory adjustments (Alberts, 1978). More recently, we observed such movements among pups beneath the mother's body (unpublished). Each pup, in effect, regulates itself and, by so doing, contributes to a functional coherence of the group. The important distinction is that group coherence is not directly regulated.

Autoregulation, as used in the present context, emphasizes the idea that complex, patterned interactions between two or more players can be mediated by simple mechanisms inherent in only one of the players. In autoregulation models, emphasis is on the proximate controls, particularly controls that are endogenous to one player or to one side of a dyadic relationship. Sometimes we can become too enamored with the beauty of interindividual coordination and overlook instances of simple, self-regulation that are not necessarily part of a larger, integrated system.

D. Parasitism Model

The sight of 8 or 10 infant rat pups simultaneously attached to the teats of an adult female is both a picture of nurturance and an image of wholesale consumption of the mother by the offspring. It is clear that the mother is functionally an environment for the offspring, and they they act on their environment to extract from it resources such as nutritive energy and water.

According to traditional definition, parasitism refers to an association between members of two different species in which one member, usually the smaller one, is metabolically dependent upon the other. The relationship is often, but not necessarily, detrimental to the host. There are in nature many types of parasitic relationships. Table I gives a sample of some common categories of parasitism.

Although the most familiar and frequent examples of parasitism involve heterospecific associations, it is not necessary for the individuals involved to be members of different species. One type of same-species parasitism cited by Galef (1981) is displayed by male angler fish. In the groups displaying this form of sexual parasitism (Villwock, 1973), males are much smaller than females. The parasitic male spends much of its life attached to the body of a female angler fish. These males derive their food directly from the bloodstream of their female host(ess) and they display a profile of specializations typical of other parasites that occupy similar niches, that is, reduced dentition and simplified gastrointestinal apparatus.

Galef (1981) suggested a broad use of parasitism, proposing that the altricial mammal is a reproductive parasite on its mother. In this stimulating essay,

Table I

Categories of Parasitism

Category	Description
Endoparasite	Parasite internal to its host
Ectoparasite	Parasite external to its host
Obligate parasite	Parasite is completely metabolically dependent on its host
Facultative parasite	Parasite is partially dependent on its host
Constant parasite	Parasite is in continuous contact with its host
Periodic parasite	Parasite is in intermittent contact with its host

Galef explored mother–litter interaction in the Norway rat as an analog to the traditional host–parasite relationship. The mother rat was treated as a host that provides an environment to her offspring. The maternal environment offers milk, heat, and mechanical energy (the resources discussed earlier in the context of the parental provision model). As we have seen, there are profound changes during the developmental cycle in the dam's willingness to provide her resources. Thus, from the point of view of the parasite, the environment changes considerably over time.

The life strategy of the rat pup can be seen, by analogy, as a sequence of parasitic adaptations. During early life, the embryo and fetus live as endoparasites, attached to the wall of the uterus and then via the placenta to the dam's bloodstream. Oxygen and nutrition are taken directly from the dam. There is complete metabolic dependence.

Birth requires a transition to a new phase of parasitic adaptation. The newborn is an ectoparasite, attached to the dam's nipple for most of each day and deriving thermal energy by conductance. Recall the range of commodities the infant obtains from the dam. In the case of the parental provision model, we emphasized the parent's role in delivering these resources. Within the present framework, however, emphasis is on the role of the offspring and the manner in which they adapt to their environment and make the sequential transitions from one type of maternal milieu to another.

The transition from intrauterine life as an ectoparasite to the early postnatal environment places the pup in the niche of a passive, obligate parasite. The host (mother) approaches the offspring and establishes contact. She is a willing host.

Between 2 and 3 weeks postpartum, the developing offspring face a distinctly different niche. The dam in no longer as willing to approach the pups; indeed, she evades their attempts to suckle. Heat and mechanical stimulation from the mother are no longer as available. The maternal milieu has changed considerably. The offspring can ingest solid food on their own, but this requires considerably different and greater kinds of expenditures on the part of the offspring. To feed, they must leave the nest and forage for a feeding resource. Independent feeding means a new and broader range of gastrointestinal adaptations for nutrient utilization as well as defense against the environmental challenges of the world beyond the confines of the nest. During this phase, which lasts from about day 18–28, the rat pup lives as an active, facultative parasite. The rat pup is prepared for its new lifestyle. Its growth and maturation provide the ability to ambulate independently; its sensory systems provide the receptors to find food and avoid predation; and fur and insulation provide the pup with the thermal resiliance to allow forays from the nest. Nevertheless, at this stage in its development, the pup continues to exploit the resources of the dam as well as to derive food and other resources from new portions of its environment.

At about 4 weeks postpartum, the dam's milk production has ceased. The 28-day-old offspring can ingest and utilize a full range of nutrients. Thermogenic capacity, increased size, and insulation make the juvenile fairly competent at defending body temperature against cold challenge. Similarly, the juvenile displays great speed and coordination in its movements and behavioral repertoire, including bouts of vigorous social play that contain elements of fighting and sexual movements. Galef characterizes the juvenile rat as a commensal parasite. Recall that commensalism refers to a relationship in which participants feed together and share other resources.

Commensalism begins around day 18 in the rat, as the offspring begin to ingest nutrients other than mother's milk. The earliest bites of solid food are almost always taken in the presence of the mother, sometimes literally out of her mouth. At his early stage the pups both suckle and feed. Other commensal habits at this stage include the use of the burrow itself, and the protection it provides from the surface environment and predators.

Sometime after the fourth week, pups emerge from the burrow. When studied under seminatural conditions in the laboratory the pups emerge and egress from the nest at an earlier time but the sequence of events is similar. At this stage they are about to embark on independent foraging, seeking, and ingesting food at some distance from the maternal burrow. If conditions permit, the pups may also take up residence in other burrows of the colony. Galef interprets this stage of life strategy as one in which ". . . the juvenile pup has extended its commensal relationships, with respect to food and shelter, from its dam to others of its species" (p. 232). The commensal quality aspects of the pup's strategy, even at 40 days of age, are apparent in the details of its existence. Modes of ingestion and intake regulation appear adultlike. Nevertheless, site selection for feeding and the choice of diet are influenced by adults. It has been shown, for instance, that separable mechanisms act on the pup to guide it to areas where adults have fed, to follow adults around a territory known to the senior members of the colony, and to ingest foods known to the adults to be safe (Galef, 1977a,b). These mechanisms have been interpreted as adaptations favoring survival by avoidance of ingestion of toxic substances.

Embedded in the metaphor of the offspring as a reproductive parasite is the concept of ontogenetic adaptation, discussed earlier in this paper. The stagewise sequence of adaptation to local conditions is represented in the metaphor of the reproductive parasite, moving through a sequence of feeding niches to which it is adapted first as an endoparasite, and then as an ectoparasite: obligate, faculative, and commensal. In the course of creating this clear view of the ontogenetic adaptation on the part of the offspring, one effect of the parasitic model is to relegate the parents to a relatively passive role in the developmental cycle. Although the parasite model recognizes sequential changes in the probability of willingness of the parent to play "host," the parental role is that of environment. The offspring are the active extractors.

E. Symbiosis Model

There is in the biological world another set of phenomena which can be organized into yet another metaphor and applied to the parent–offspring relationship. I have saved this one for last, because it is the most inclusive of those to be discussed here and because it is one that has captivated my attention in the past few years.

The concept of symbiosis has a century-old legacy in biology and a derived meaning in some areas of psychological thinking. I shall discuss its relatively recent application to psychobiology of parent–offspring relations in the rat.

The term symbiosis was coined by Heinrich Anton de Bary (1879), a German botanist. He used the term to describe "the living together of unlike organisms." de Bary applied the term in its most general sense—to encompass a wide range of types of associations and degrees of relationship, including parasitism, inquilism, mutualism, and commensalism (see Hertig *et al.*, 1937, and Starr, 1975, for analytical discussions of these terms). Historically, it became common for biologists to restrict the term symbiosis to associations of mutual benefit. Table II lists a set of criteria that we use to represent the meaning of symbiosis in common biological usage.

In my laboratory, we have been applying this metaphor of symbiosis to mother–litter relations in the Norway rat. The metaphor of symbiosis has heuristic value; it can attune us to levels of interaction that might otherwise escape our awareness.

Table II
Common Characteristics of Symbiotic Associations

Proximity
Interaction and close integration
Interdependence (sharing of resources)
Some degree of mutual benefit

I have discussed (Alberts and Gubernick, 1983) the act of maternal licking, as displayed by rat mothers as part of the profile of maternal behavior (Fig. 1B). The parental provision model helped elucidate the numerous functional contributions made by the simple act of anogenital licking: hygienic needs of the litter are served, the pups are activated, and the micturition reflex is stimulated, and the offspring void. How nicely met are the needs of the offspring! To follow this traditional framework, we conceive of the relevant mother–litter events as consisting of maternal provision of milk, heat, and stimulation that ensure that the pups ingest the milk and metabolize it, and provision of mechanical stimulation to the pups' anogenital region to void the waste, all on behalf of the pups. The conflict/parental investment perspectives would train our attention on the benefits to the parent of most of these functions, because they are directly relevant to the survival and reproductive success of the offspring, making the immediate investment worthwhile.

Our preconceptions were jostled, however, by reports that adult dingoes and rats ingest the urine of their offspring. Could the flow of resources be more complex than we had envisioned? We have examined the phenomenon, using the metaphor of symbiosis as a stimulus for new probes into the nature of the mother–litter relationship.

The basic paradigm for many of the experiments is shown in Fig. 5. Radioactively labeled (tritiated) water in injected into some of the pups in a litter. The next day the tritium label is detected in the blood plasma of the dam. If urination by injected pups is blocked by ligation of the urethra, the tritium label remains and does not transfer. By analyzing the concentration of tritium in the bladder urine of injected pups of various ages and by determining volume of total body water in the mother, we were able to determine the amount of labeled urine the dam had consumed by the diluted values obtained from her blood plasma (Gubernick and Alberts, 1983). Figure 6 shows the results of this quantitative study, which indicates that a mother rat consumes as much as 40 ml of water in a day by licking her pups and consuming their urine. The peak level of water transfer from pups to mother is reached on day 15, which is also the stage of peak milk production. We

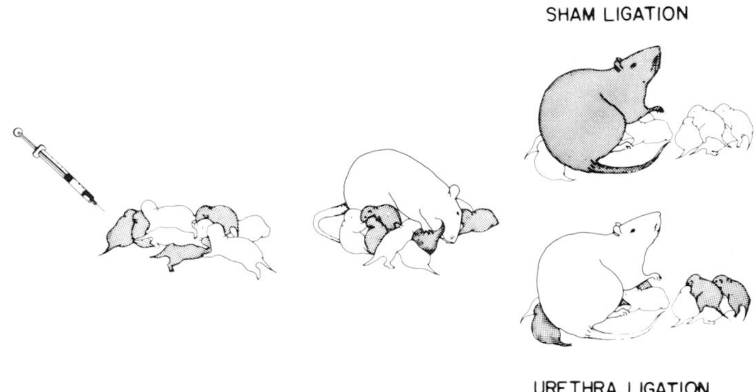

Fig. 5. Basic pattern of results obtained in tritium transfer studies of pup licking by mothers. Some of the pups in a litter are injected with the radioactively labeled water. After interacting with the mother and noninjected littermates, the radioactive label is found in the dam. Urethra ligation prevents the transfer of tritium, indicating that it is carried in the pups' urine. See Gubernick and Alberts (1983) for a complete methodological description and quantification of transfers.

have estimated that the dam reclaims about two-thirds of the water transferred to her litter via mother's milk by licking the pups and ingesting their urine.

In a related series of experiments, we examined the behavioral and physiological significance of urine ingestion by dams (Friedman *et al.*, 1981). Mother rats deprived of water for 24 h lose body water from the extracellular compartment and become hypovolemic. Mother rats deprived of both water and pup urine, become twice as hypovolemic. Thus, to the dam, pup urine represents a physiologically significant quantity of water.

We also showed that the dam's licking behavior is controlled by hydrational cues, and that she uses the litter as a source of repletion. Thirsty mothers licked their babies more than did water-sated dams. In water-replete mothers, licking was determined by urine availability. Dry pups received less licking stimulation than pups that offered urine. Dams were deprived of pup urine and given simultaneous access to both water and .15 *M* saline. Urine-deprived dams increased their consumption of .15 *M* saline, but not water, suggesting that the mother uses pup urine as a source of both water and electrolytes (Friedman *et al.*, 1981). Similarly, when we gave dams access to .15 *M* NaCl, they licked pups less than did the watered controls, even when fluid intake was equated. This effect of salt intake on licking appears to be sodium-specific; in a parallel experiment with KCl instead of NaCl there was no effect on maternal licking (Gubernick and Alberts, 1983).

The metaphor of symbiosis can be applied comprehensively to numerous

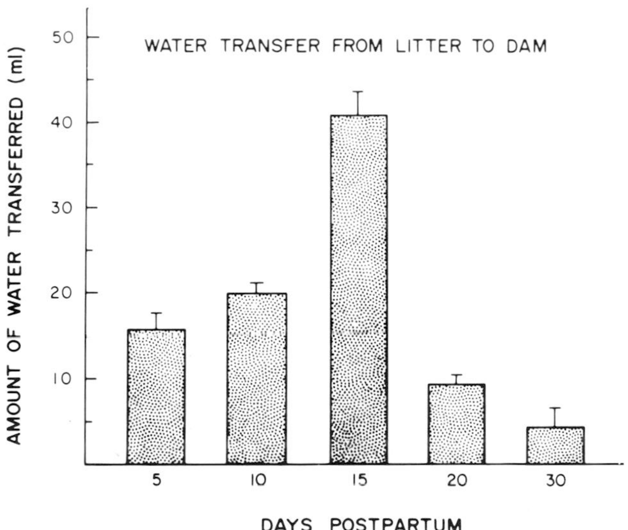

Fig. 6. Average amount of water transferred from a litter of eight pups to their mother, at different points during the lactational cycle. Water transfer was estimated by injecting some pups with tritium and measuring the amount of radioactivity recovered 24 h later in the mother. From Gubernick and Alberts (1983).

types and levels of interactions between a rat dam and her litter (Alberts and Gubernick, 1983). In addition to the bidirectional exchanges of water and electrolytes, described above, the model of symbiotic association sheds light on potentially significant exchanges of other physiological commodities. Thermal energy can be viewed as a commodity that is exchanged between dam and her litter. In this case, heat flow is predominantly unidirectional for the first 14 days of the postnatal relationship. At about this point the directionality of heat flow can reverse and the larger, more thermogenic, and better insulated litter conducts heat to the mother (Leon *et al.*, 1978).

There is also a set of interactions whereby various gastrointestinal commodities are exchanged that lead to mutual regulation of acid–base levels in the mother and offspring. Briefly, the emission of an olfactory attractant by lactating dams (Leon and Moltz, 1971) is regulated by the pups that are attracted to it. When pups are 14 days old, their mother excretes a soft, odorous feces that can attract offspring. If, by litter substitution, the dam is kept with pups of constant age, she does not emit the attractant after 2 weeks of lactation (Leon and Moltz, 1972). The attractiveness of the mother's feces has been related to the onset of feeding in the offspring. Exclusive consumption of mother's milk supports bacterial populations in pup intestines that render the gut acidic. Gut acidity is reduced when solid food is added to the diet. Rat dams ingest their pups' feces. When offspring excreta is acidic, the dam's gut becomes acidic. The acid pH suppresses cecal bacteria that metabolize cholic acid in the dam's intestine to produce the attractant. When pups begin to eat some solid food, they provide the dam with more alkaline excreta and "disinhibit" the mother's cecal bacteria. Appropriate manipulations, such as providing dams with feces from young or old pups, regulated her emission of the attractant in the predicted manner (Moltz and Lee, 1983).

Pup feces give the dam an intestinal milieu that promotes her attractiveness to offspring. In return, Moltz and Lee (1983) suggest, the dams give the pups enteric immunocompetence and the bile acids necessary for conversion of triglycerides into brain myelin. They have demonstrated that deprivation of maternal feces and the deoxycholic acid therein renders pups more susceptible to enteric diseases and stress-related death.

Moltz and Lee also note that peak rates of myelin deposition occur in rat brain from days 15–30. The long-chain fatty acids that predominate in the lipid fraction of myelin cannot be absorbed into the pups' bloodstream unless bile acids are present to prepare the triglycerides for hydrolysis and eventual absorption. Bile acids are not well-developed in pups until after day 30. Thus, the bile in the pup's gut is maternal bile, provided by her attractive feces. Pups denied access to maternal feces and, presumably, to the deoxycholic acid have smaller brains and are reported to show a profile of neurobehavioral deficits, compared to control animals that had normal access to maternal feces.

F. Which Model Is Best?

The five models of parent–offspring relationships discussed in this chapter were not presented with the goal of discerning which is closest to the truth. None of the models to date have been put forth in a manner designed for rigorous testing. Instead, these models have served as guides to researchers, helping them perceive aspects of relations and to formulate rigorous, testable hypotheses about the nature of parent–offspring interactions.

It is clear that some models are more inclusive than others and may be "better" in terms of providing a unified framework that can incorporate more disparate data points on parent–offspring interactions. The broadest views of the set presented above were probably those of the parental provision and the symbiosis models.

On the other hand, one virtue of models such as those reviewed here is that they reveal or clarify aspects of a phenomenon that would otherwise be hidden or opaque. Notable instances of specialized focus were offered by the autoregulation model and by the emphasis on bidirectionality in the symbiosis model. The conflict model offers a special connection to broad levels of evolutionary thinking and an evolutionary-genetic level of causal analysis that is not relevant to most of the other models. At this stage in the development and articulation of alternative views, the quality of a model is best judged in terms of the number and quality of hypotheses that it generates.

V. Form and Functions of Parent–Offspring Relationships

The five models discussed above provide a variety of alternative views of the forms that the relationship between parent and offspring can take and of its changes through developmental time. Recent research has also provided many new insights into the function of parent–offspring relations. I will not discuss the obvious function of ensuring the survival of the offspring. This can be considered its "ultimate" function. In addition, there are numerous other types of function that are important to recognize. Some of these have direct bearing on further refinement that are important to recognize. Some of there have direct bearing on further refinement of the way we can discern and describe the form of the interactions.

A. Multiple, Simultaneous Functions

Faced with a complex, multifaceted phenomenon such as parent–offspring relationships, we are justly tempted to isolate discrete units for study. Historically, it was important and rewarding to organize maternal behavior in mammals into a profile of activities or functional units, such as nursing, nest building, licking, and transport. We are wrong, however, in treating these units as functional entities to which we ascribed a singular function or role in the relationship. The rule may be that there are multiple functions of many, if not most, aspects of parent–offspring relations. Maternal licking, a discrete and punctuate component of parent–offspring interactions, is an excellent example of the multiple function principle.

When rat dams lick their pups, about 80% of their licking is focused on the anogenital region of the pup. This licking stimulates the pup's micturition reflex and it voids. Infants do not void spontaneously, so licking facilitates this vital function. It has also been suggested that licking stimulates various physiological processes such as peripheral blood flow and urine formation. In addition, there are hygienic functions served by such licking. I have described in some detail the functional significance of maternal licking with respect to the reclamation of lactational water and electrolytes. Gubernick (1980, 1981) has shown that maternal licking by goats of their kids provides an important chemical ''label'' on the kids that is the necessary cue for identification and the establishment of a profound and long-lasting bond between mother and kid.

Moore and colleagues (Moore, 1981, 1984; Moore and Morelli, 1979) have proposed that anogenital licking by rat dams exerts a masculinizing influence on the offspring. This provocative suggestion stems from the initial observation that rat dams spend more time licking their male offspring than they do their female offspring. The gender difference is attributed to an androgen-dependent substance present in the urine of the male pups. The dams presumably find it attractive and lick it more. Androgenized female pups receive malelike levels of maternal licking. Castrated male pups are licked less than intact males and testosterone replacement reinstates the gender-typical amount of licking. An olfactory mask, such as perfume on the pups, obviates the sex difference. Moore has gone on to report that pups that receive higher levels of anogenital stimulation are more masculinized. Behaviorally, this is most evident in copulatory performance. Males that received abnormally low levels of maternal licking as pups, due to olfactory impairment of the dam, subsequently displayed longer latencies to ejaculation and prolonged refractory periods after copulation. These are behavioral correlates of diminished testicular function. One implication is that testosterone may masculinize the developing pup not only by its direct action on the developing nervous system, but also by altering the maternal environment and thus the quantity and quality of the environmental stimuli it receives.

B. Nongenetic, Transgenerational Effects

There is a class of psychobiological analyses demonstrating the existence of mechanisms whereby phenotypic characteristics of the parental generation are transmitted to the offspring via nongenetic avenues. If the members of a colony of wild rats learn that one of the foods in their environment is toxic, their offspring are observed to avoid completely the ingestion of the potentially toxic food without ever tasting the food of experiencing its toxic effects (Galef and Clark, 1971). Genetic transmission from the parental generation does not mediate this feeding phenotype, because it is not based on a response to the specific food cues. The pups' behavior is dependent on the presence of adults that have learned to avoid the food. Galef and co-workers have unraveled this and other instances of specific feeding phenotypes and have identified a set of parent–offspring attributes that mediate social learning among rats that determine their feeding habits—in terms of site selection (Galef and Heiber, 1976), food preferences, and aversions (Galef and Clark, 1971; Galef and Henderson, 1972), and even the modes of foraging that are incorporated (Galef, 1982).

The voluminous literature on the effects of early handling, separation, and stress contains numerous examples of the pervasive effects that can be exerted on an offspring generation by the nature of their early environment (e.g., Denenberg, 1964; Harlow and Harlow, 1962; Levine *et al.*, 1967; Newton and Levine, 1966). Even more dramatic and startling is the demonstration that the second generation offspring can also reflect the early experiences of their grandparents as a result of their interactions with their parents. This earlier work, in my opinion, suffered from manipulations and measures that were too general to be well understood. Nevertheless, important contributions were made. There were clear demonstrations that experiential effects endured by members of one generation could be represented in subsequent generations without genetic transmission and without repetition of the initial, inductive experiences.

One contemporary synthesis of the early experience literature with the transgenerational paradigms is the analysis of the transmission of susceptibility to stress ulceration. It has been found in rats that premature separation results in adults that are especially susceptible to ulcers (Ackerman *et al.*, 1975). The early experience/separation effect is interesting, but the spectacular finding is that the offspring of these rats also show heightened susceptibility (Skolnick *et al.*, 1980)! This effect appears to based on prenatal influences (an important period of mother–offspring interaction). Offspring of prematurely separated dams reflect the mother's experiences, even if they are cross-fostered and weaned normally. Final identification of the causal mechanisms remains, as does clarification of the remarkable specificity of these effects. It is less surprising that prenatal conditions can affect later behavior than to find the same, specific effect echoed so faithfully.

Ongoing autonomic function and hormonal states in infants are linked to relatively specific features of parental interaction. Episodes of separation cause massive changes in heart rate, sympathetic tone, and respiration (Hofer, 1973). Such effects have been shown to be related to nutritive aspects of parental provision. The ability to reverse these changes with controlled administration of nutrients has been valuable in revealing specific interoceptive loci and mechanisms that control vital aspects of cardiovascular function and development (Hofer and Weiner, 1971, 1975). Such research may clarify major questions relevant to early experiential effects, psychosomatics, and general forms of disease susceptibilities.

Nevertheless, some effects of separation are not linked to simple nutritive mechanisms. These effects resemble in many ways the "failure to thrive" syndromes seen in clinical separations of human infants. In rats, even brief episodes of separation lead to rapid decreases in growth hormone and ornithine decarboxylase (ODC), an enzyme essential for synthesis of brain proteins (Butler *et al.*, 1978). Decreased ODC levels were not rectified by controlling for nutritive or thermal deficits. Provision of a passive (anesthetized) mother did not ameliorate the effect, even if the pups were fed in her presence. The exciting finding was that an active foster mother could reverse the ODC effect—even if this foster mother could not deliver milk due to nipple ligation. Evidently, there are crucial aspects of parental interaction that regulate important developmental features of autonomic and neural function.

Infants can have similar kinds of effects on parents. Recent studies have identified infant stimuli that regulate autonomic stress and which may "buffer" the dam from stress (Smotherman *et al.*, 1976). Nipple stimulation is an important channel of such effects. Similar mechanisms have also been implicated in the hormonal induction and maintenance of maternal states that control nest defense and related motivational states. These isolated points of data represent additional evidence that we are just beginning to tap the range of functional connections between parents and offspring.

Acknowledgments

Preparation of this chapter and research from the author's laboratory reported herein were supported, in part, by grant MH 28355 and Research Scientist Development Award MH 00222, both from the National Institute of Mental Health to J. R. Alberts. I am grateful for this support.

References

Ackerman, S. H., Hofer, M. A., and Weiner, H. (1975). *Psychosom. Med.*, **37**, 180–184.

Alberts, J. R. (1978). *J. Comp. Physiol. Psychol.* **92**, 231–240.

Alberts, J. R., and Gubernick, D. J. (1983). *In* "Symbiosis in Parent–Offspring Interactions" (L. A. Rosenblum and H. Moltz, eds.), pp. 7–44. Plenum, New York.

Alberts, J. R., and Gubernick, D. J. (1984). *Learn. Motivation* **15**, 334–359.

Babicky, A., Pavlik, L., Parizek, J., Ostadolova, I., and Kolar, J. (1972). *Physiol. Bohemoslov.* **21**, 467–471.

Barnett, S. A., and Walker, K. Z. (1974). *Dev. Psychobiol.* **7**, 563–577.

Bignall, K. E., Heggeness, F. W., and Palmer, J. E. (1975). *Exp. Neurol.* **49**, 174–188.

Bignall, K. E., Heggeness, F. W., and Palmer, J. E. (1972). *Am. J. Physiol.* **2**, R23–R29.

Bradbury, J. W. (1977). *In* "Biology of Bats" (W. A. Wimsatt, ed.), pp. 1–72. Academic Press, New York.

Brambell, F. W. R. (1970). *Front. Biol.* **18**.

Brody, S. (1945). "Bioenergetics and Growth." Van Nostrand-Reinhold, Princeton, New Jersey.
Butler, S. R., Suskind, M. R., and Schanberg, S. M. (1978). *Science* **199**, 445–447.
Capek, K., and Jelinek, J. (1956). *Physiol. Bohemoslov.* **5**, 91–96.
Cloudsley-Thompson, J. L. (1960) "Animal Behaviour." Oliver & Boyd, Edinburgh.
Culbertson, J. T. (1938). *J. Parasitol.* **24**, 65–82.
de Bary, A. (1879) "Die Erscheinung der Symbiose." Trubner, Strassberg.
Denenberg, V. H. (1964). *Psychol. Rev.* **71**, 335–354.
Ewer, R. F. (1968) "Ethology of Mammals." Elek Science, London.
Fisher, E. M. (1940). *J. Mammal.* **21**, 132–137.
Fleming, A. S., and Rosenblatt, J. S. (1974). *J. Comp. Physiol. Psychol.* **86**, 957–972.
Friedman, M. I. (1975). *J. Comp. Physiol. Psychol.* **89**, 636–647.
Friedman, M. I., Bruno, J. P., and Alberts, J. R. (1981). *J. Comp. Physiol. Psychol.* **95**, 26–35.
Galef, B. G., Jr. (1977a). *J. Comp. Physiol. Psychol.* **91**, 1136–1140.
Galef, B. G., Jr. (1977b). *In* "Learning Mechanisms in Food Selection" (L. M. Barker, M. Best, and M. Domjan, eds.), pp. 123–148. Baylor Univ. Press, Waco, Texas.
Galef, B. G., Jr. (1981). *In* "Parental Care in Mammals" (D. J. Gubernick and P. H. Klopfer, eds.), pp. 211–241. Plenum, New York.
Galef, B. G., Jr.(1982). *Dev. Psychobiol.* **15**, 279–295.
Galef, B. G., Jr., and Heiber, L. (1976). *J. Comp. Physiol. Psychol.* **90**, 727–739.
Galef, B. G., Jr., and Henderson, P. W. (1972). *J. Comp. Physiol. Psychol.* **78**, 213–219.
Gubernick, D. J. (1980). *Anim. Behav.* **28**, 124–129.
Gubernick, D. J. (1981). *Anim. Behav.* **29**, 1.
Gubernick, D. J., and Alberts, J. R. (1983). *Physiol. Behav.* **31**, 593–601.
Hahn, P., and Koldovsky, O. (1966). *In* "Utilization of Nutrients during Postnatal Development" Pergamon, Oxford.
Hall, W. G., and Bryan, T. E. (1980). *J. Comp. Physiol. Psychol.* **94**, 746–756.
Hall, W. G., and Rosenblatt, J. S. (1977). *J. Comp. Physiol. Psychol.* **91**, 412–427.
Hall, W. G., and Williams, C. L. (1983). *Adv. Study Behav.* **13**, 219–254.
Harlow, H. F., and Harlow, M. K. (1962). *Sci. Am.* **207**, 137–146.
Hediger, H. (1955). "Studies of the Psychology and Behavior of Captive Animals in Zoos and Circuses" (G. Sircom, transl.), Criterion, New York.
Hertig, M., Taliferro, W. H., and Schwartz, B. (1937). *J. Parasitol.* **23**, 326–329.
Hofer, M. A. (1973). *Physiol. Behav.* **10**, 423–427.
Hofer, M. A., and Weiner, H. (1971). *Psychosom. Med.* **33**, 353–362.
Hofer, M. A., and Weiner, H. (1975). *Psychosom. Med.* **37**, 8–24.
Hull, D.J. (1973). *In* "Comparative Physiology of Thermoregulation" (C. G. Whittow, ed.), Vol. 3. Academic Press, New York.
Klopfer, P., and Klopfer, M. (1977). *Anim. Behav.* **25**, 286–291.
Leon, M., and Moltz, H. (1971). *Physiol. Behav.* **7**, 265–267.
Leon, M., and Moltz, H. (1972). *Physiol. Behav.* **8**, 683–686.
Leon, M., Croskerry, P. G., and Smith, G. K. (1978). *Physiol. Behav.* **21**, 793–811.
Leuthold, W. (1977) "African Ungulates." Springer-Verlag, Berlin and New York.
Levine, S., Haltmeyer, G. C., Karas, G. G., and Denenberg, V. H. (1967). *Physiol. Behav.* **2**, 55–66.
Lott, D. F. (1973). *In* "Perspectives in Animal Behavior" (G. Bermant, ed.), pp. 239–279. Scott, Glenview, Illinois.
Luckey, T. S., Mende, T. J., and Pleasants, J. (1954). *J. Nutr.* **54**, 345–359.
Moltz, H., and Lee, T. M. (1983). *In* "Symbiosis in Parent–Offspring Interactions" (L. A. Rosenblum and H. Moltz, eds.), pp. 45–60. Plenum, New York.
Moore, C. L. (1981). *Anim. Behav.* **29**, 383–386.

Moore, C. L. (1984). *Dev. Psychobiol.* **17**, 347–356.
Moore, C. L., and Morelli, G. A. (1979). *J. Comp. Physiol. Psychol.* **93**, 677–684.
Newton, G., and Levine, S., eds. (1968) "Early Experience and Behavior." Thomas, Springfield, Illinois.
Oppenheim, R. W. (1981). *In* "Maturation and Behavior Development" (K. Connolly and H. Prechtl, eds.). Spastics Soc. Publ., London.
Raskin, L. A., and Campbell, B. A. (1981). *J. Comp. Physiol. Psychol.* **95**, 425–435.
Rheingold, H. L., ed. (1963a) "Maternal Behavior in Mammals." Wiley, New York.
Rheingold, H. L. (1963b). *In* "Maternal Behavior in Mammals" (H. L. Rheingold, ed.), pp. 169–202. Wiley, New York.
Rheingold, H. L., and Keene, G. C. (1965). *In* "Determinants of Infant Behaviour" (B. M. Foss, eds.), Vol. 3, pp. 87–110. Methuen, London.
Rosenblatt, J. S. (1965). *In* "Determinants of Infant Behaviour" (B. M. Foss, ed.), pp. 3–41. Wiley, New York.
Rosenblatt, J. S. (1969). *Am. J. Orthopsychiatry* **39**, 36–56.
Rosenblatt, J. S., and Lehrman, D. S. (1963). *In* "Maternal Behavior in Mammals" (H. L. Rheingold, ed.), pp. 8–57. Wiley, New York.
Rosenblatt, J. S., and Siegel, H. I. (1981). *In* "Parental Care in Mammals" (D. J. Gubernick and P. H. Klopfer, eds.), pp. 14–76. Plenum, New York.
Satinoff, E., and Stanley, W. C. (1963). *J. Comp. Physiol. Psychol.* **56**, 66–68.
Skolnick, N. J., Ackerman, S. H., Hofer, M. A., and Weiner, H.(1980). *Science* **208**, 1161–1162.
Smotherman, W. P., Weiner, S. G., Mendoza, S. P., and Levine, S. (1976). *Ciba Found. Symp.* **45**, 5–22.
Starr, M. P. (1975). *Symp. Soc. Exp. Biol.* **29**, 1–20.
Stolc, V., Knopp, J., and Stolcova, E. (1966). *Physiol. Bohemoslov.* **15**, 219–225.
Trivers, R. L. (1972). *In* "Sexual Selection and the Descent of Man" (B. Campbell, ed.), pp. 136–178. Aldine, Chicago.
Villwock, W. (1973). *In* "Grzimek's Animal Life Encyclopedia" (B. Grzimek, ed.), Vol. 4, pp. 398–404. Van Nostrand-Reinhold, Princeton, New Jersey.
Weisner, B. P., and Sheard, N. M. (1933). *In* "Maternal Behavior in the Rat." Oliver & Boyd, London.
Williams, C. L., Rosenblatt, J. S., and Hall, W. G. (1979). *J. Comp. Physiol. Psychol.* **93**, 414–429.
Woodside, B., and Leon, M. (1980). *J. Comp. Physiol. Psychol.* **94**, 41–60.

Index

G

H

I

L

M

P

R

S

T

V

W

X

Y

Z